www.brookscole.com

www.brookscole.com is the World Wide Web site for Brooks/Cole and is your direct source to dozens of online resources.

At *www.brookscole.com* you can find out about supplements, demonstration software, and student resources. You can also send email to many of our authors and preview new publications and exciting new technologies.

www.brookscole.com
Changing the way the world learns®

About the Authors

DR. BERNARD W. PIPKIN is Professor Emeritus at the University of Southern California. He received his doctorate from the University of Arizona and is a licensed geologist and certified engineering geologist in the state of California. After graduation he worked for the U.S. Army Corps of Engineers on a variety of military and civil projects, from large dams to pioneer roads in rough terrain for microwave sites. He has been a consulting engineering geologist as well as a university teacher since 1965.

Dr. Pipkin is past president of the National Association of Geology Teachers and is a Fellow in the Geological Society of America. He hosted the PBS 30-part program *Oceanus* that won a local Emmy for Best Educational Television Series. He recently shared the Clare Holdredge Award (1995) with Richard Proctor from the Association of Engineering Geologists for their book *Engineering Geology Practice in Southern California*.

Dr. Pipkin is a private pilot with a flight instructor's rating; he took many of the aerial photos in the book. He and his wife Faye have three grown children and live south of Los Angeles in Palos Verdes.

DR. D. D. "DEE" TRENT has been working at, or teaching, geology since 1955. After graduating from college, he was employed by an oil company where his geological skills were sharpened with projects in Utah, Arizona, California, and Alaska. When the company decided to send him to Libya, he decided it was time to become a college geology teacher.

He has taught 28 years at Citrus Community College in Glendora, California, and along the way has worked with the National Park Service, picked up a Ph.D. at the University of Arizona, done field research on glaciers in Alaska and California, visited numerous mines in the United States and Germany, mapped deserts in California and Arizona, and served as an adjunct faculty member at the University of Southern California, where he taught Field Geology. When not involved with geology, he's either skiing or playing the banjo in a dixieland band.

Dr. Trent and his wife Pat have two grown children and they live in Claremont, California.

DR. RICHARD W. HAZLETT is an Associate Professor in the Geology Department and Coordinator of the Environmental Analysis Programs at Pomona College in Claremont, California. Prior to entering academia, Dr. Hazlett worked as a ranger naturalist and exhibit planner for the U.S. National Park Service in Hawaii and as a research technician at the Hawaiian Volcano Observatory. His teaching experience includes tours of duty at Occidental College and the University of Hawaii at Hilo. He holds graduate degrees from Dartmouth College and the University of Southern California. His main field of expertise is volcanology, with field projects conducted in Nicaragua, Iceland, Italy, Hawaii, the Aleutian Islands, the Pacific Northwest, and the Mojave Desert. In recent years Dr. Hazlett has also become increasingly involved with public education, assisting in the scripting and development of the PBS television series "Earth Revealed," and writing two books of regional interest, "Roadside Geology of Hawaii" (co-authored with Donald H. Hyndman), and "Joshua Tree National Park Geology" (co-authored with D. D. Trent). His environmental interests include patterns of energy consumption and land-use issues in the American West. Outside interests include spelunking, swimming, and landscape sketching. Dr. Hazlett lives in Claremont.

Fourth Edition

Geology and the Environment

BERNARD W. PIPKIN
University of Southern California

D. D. TRENT
Citrus College

RICHARD HAZLETT
Pomona College

THOMSON

BROOKS/COLE

Australia • Canada • Mexico • Singapore • Spain
United Kingdom • United States

Earth Sciences Editor: Keith Dodson
Development Editor: Marie Carigma-Sambilay
Assistant Editor: Carol Ann Benedict
Editorial Assistant: Melissa Newt
Technology Project Manager: Ericka Yeoman-Saler
Marketing Manager: Kelley McAllister
Marketing Assistant: Leyla Jowza
Advertising Project Manager: Nathaniel Bergson-Michelson
Project Manager, Editorial Production: Hal Humphrey
Senior Art Director: Vernon T. Boes
Print/Media Buyer: Judy Inouye

Permissions Editor: Sarah Harkrader
Production Service and Copy Editor: Helen Walden
Text Designer: Lisa Buckley
Photo Researcher: Helen Walden
Illustrator: Precision Graphics
Cover Designer: Denise Davidson
Cover Image: © Daryl Benson/Masterfile www.masterfile.com;
Waterton Lakes National Park in autumn, Alberta, Canada
Compositor: Parkwood Composition
Printer: Quebecor World–Dubuque

Library of Congress Control Number: 2003114789

Student Edition: ISBN 0-534-49051-4
Instructor's Edition: ISBN 0-534-49058-1

Brooks/Cole—Thomson Learning
10 Davis Drive
Belmont, CA 94002
USA

Asia
Thomson Learning
5 Shenton Way #01-01
UIC Building
Singapore 068808

Australia/New Zealand
Thomson Learning
102 Dodds Street
Southbank, Victoria 3006
Australia

Canada
Nelson
1120 Birchmount Road
Toronto, Ontario M1K 5G4
Canada

Europe/Middle East/Africa
Thomson Learning
High Holborn House
50/51 Bedford Row
London WC1R 4LR
United Kingdom

Latin America
Thomson Learning
Seneca, 53
Colonia Polanco
11560 Mexico D.F.
Mexico

Spain/Portugal
Paraninfo
Calle Magallanes, 25
28015 Madrid, Spain

Brief Contents

Contents

Chapter 4
Earthquakes and Human Activities 72

Chapter 5
Volcanoes 111

Chapter 6
Soils, Weathering, and Erosion   148

Chapter 7
Mass Wasting and Subsidence   175

Chapter 8
Fresh-Water Resources   205

Chapter 9
Hydrologic Hazards at the Earth's Surface 237

Chapter 10
Coastal Environments and Humans 261

Chapter 11
Glaciation and Long-Term Climate Change 289

Chapter 12
Arid Lands and Desertification   328

Chapter 13
Mineral Resources and Society   343

Preface

Our Changing Planet

Environmental geology is the study of the relationship between humans and their geologic environment. An underlying assumption is that this relationship is interactive. Not only do naturally occurring geologic phenomena affect the lives of people each day, but also human activities affect geological processes, sometimes with tragic consequences. Renowned historian Will Durant noted that our physical environment "exists by geological consent, subject to change without notice." We may regard this as the historian's assurance that during our lifetimes, we will probably be subject to an earthquake, a flood, a landslide, volcanic activity, or some other significant geologic hazard. No one is truly educated until she or he knows about our geological environment and these hazards.

This book deals with the geology of now and not some distant past. The reader will find that the seismic chapter deals with earthquakes as recent as 2003 and emphasizes that earthquakes don't kill people; the collapse of human-built structures is the danger that seismic hazards present to our existence. The devastating magnitude-7.7 earthquake near Gujarat, India, in 2001 that killed close to 20,000 people is the example presented of the danger that poorly constructed buildings present to humans. An earthquake in the central United States in 2003 reminds us also that the most widely felt and perhaps the strongest earthquake in U.S. history was centered near New Madrid, Missouri, which has a geologic setting similar to Gujarat, India. Gujarat is worthy of study for this reason, and the U.S. Geological Survey believes the Gujarat earthquake to be an important event that can expand our understanding of mid-continent seismicity. The human misery and economic impact of natural disasters is difficult to evaluate, but the book emphasizes that practically no area is free from geologic hazards.

Fresh water is now thought to be the limiting factor to man's existence on earth, and this and the hydrologic hazards presented by too much water are emphasized. In late 2003 a group of young people were camped at the foot of the mountains in southern California that had been subject to an intense forest fire several months earlier. A debris flow caught them by surprise and took many lives. It is the sincere hope of the authors that the students who take this course and read this book will be more aware of the hazard that denuded slopes present during the rainy season. The same might be said of rip currents (erroneously called rip tides) in the surf zone discussed in the chapter on coastal environments. To be aware of your geologic surroundings will lead to a safer exis-

tence in an increasingly complex and dangerous world, and has the added benefit of being satisfying intellectually.

In a society where science and technology are interwoven with economics and political action, an understanding of the sciences is increasingly important. The National Science Foundation, the National Center for Earth Science Education, and several other prestigious earth-science organizations are promoting earth-science literacy in the United States for students from elementary school through college. The hope is that through education, today's students will become better stewards of our planet than their predecessors have been. How long can civilization sustain itself if it continues to harvest the earth's natural resources at the current rate? How much responsibility should we as individuals take for maintaining the earth that we share, and how much can we accomplish as individuals? Earth-science literacy opens our minds to such questions and allows us to give them better answers. It helps us to appreciate the earth's beauty, but also to recognize its limitations.

One thing is certain: the earth can sustain just so much life, and it is now being pushed to the limit. The fragility of earth was described very eloquently by *Apollo 14* astronaut Edgar Mitchell:

> It is so incredibly impressive when you look back at our planet from out there in space and you realize so forcibly that it's a closed system—that we don't have unlimited resources, that there's only so much air and so much water. You get out there in space and you say to yourself "That's home. That's the only home we have, and the only one we're going to have for a long time." We had better take care of it. We don't get a second chance.

About This Book

In this new edition, we are pleased to welcome our longtime friend and professional colleague, Rick Hazlett, to the author team. All of us have drawn upon our many field studies in the writing of this text. One of us (Pipkin) has specialized in the study of geologic hazards and their mitigation—more specifically, in the safe siting of engineering works ranging from dams and tunnels to single-family dwellings. Dee Trent has focused on geologic fieldwork, exploration for natural resources, and glacier studies. Rick Hazlett, formerly a ranger at Hawaii Volcanoes National Park, has maintained his interest and his research on volcanoes and is active in environmental studies. As colleagues in the teaching profession, it seemed natural for us to combine our past experiences and mutual interests in teaching and environmental protection in a textbook.

Geology and the Environment, Fourth Edition, is intended to fulfill the needs of a one-semester college course for students with little or no science background. It examines geologic principles, processes, and phenomena and relates them to human activities. What you learn in this course will be useful to you throughout your life. It will serve you as you form opinions about environmental issues, select a home site, or evaluate real property in a business venture. At the very least, when you understand how the earth "works," any fears you have about such things as earthquakes, hurricanes, and volcanic eruptions should be lessened.

This book presents geology that can be applied to improving the human endeavor. Chapter 1 presents an overview of the underlying cause of many of our present environmental problems. The stress of overpopulation on the environment has resulted in water and air pollution, land degradation, occupation of lands subject to geological hazards, and uncontrolled extraction of resources. The impact of overpopulation and the relatively new awareness of the environment are reasons that colleges are adding "environmental" courses such as environmental geology to their curriculums.

Because this book is intended for students who are not (yet!) geology majors, Chapters 2 and 3 offer some basic information useful for "getting around in geology." Chapter 2 presents information that is needed for the book's subsequent examinations of geological hazards and processes, mineral resources, and so on. The emphasis is that the earth is an interacting "system." Students will see that one cannot divorce atmospheric and oceanic processes from processes that occur on the solid earth surface, nor moving tectonic plates (Chapter 3) from volcanic and seismic activity.

Following these groundwork chapters, the focus shifts to various types of geologic processes and hazards. Earthquakes and volcanoes are two of the most spectacular expressions of the energy contained *within* the earth, and Chapters 4 and 5 examine how this energy is transferred to the earth's surface—where we live—and how we can learn to recognize the signals that "something's up" and work to minimize the damage when the inevitable occurs.

The next two chapters focus on geologic processes that occur on or very near the surface of the earth. Much of the soil that we build our homes on and grow our food in today is bedrock of the far distant past that has been chemically and physically altered through weathering processes (Chapter 6). Present-day rocks and soils continue to undergo these processes, in some cases assisted by human activities such as construction, mining, and extraction of water from the earth. The results—mass wasting (landslides) and ground subsidence and collapse—cause deaths and millions of dollars of damage each year. These ground-failure problems are examined in detail in Chapter 7.

Earth is sometimes described as the water planet, and the presence or absence of water is essential in defining the geological and biological environments in which we live. Chapter 8 deals with fresh-water resources, with the emphasis on underground water and how we can protect and conserve this most essential resource. Streams, rivers, and lakes are also vital water and esthetic resources. They sustain life along their banks, and some of them are important means of transportation. They also pose hazards, most notably floods, to people who live along them. These hazards are the subject of Chapter 9.

In Chapter 10 we look at the effects of the largest bodies of water on earth, the oceans, on life. Coastal geologic processes, beaches, and hurricanes—the most life-threatening of all natural hazards—are examined. Because oceans play an important role in global climate, the phenomenon known as El Niño is discussed.

Glaciers and glaciation and their impacts on humans are the subject of Chapter 11. This chapter logically includes a discussion of long-term climate change, which leads naturally to the consideration of the potential for global warming that is so much on people's minds today. The theme of climate extremes continues in Chapter 12, which deals with arid environments and the consequences of human-induced desertification.

Geology and the Environment, Fourth Edition, closes with a look at mineral resources—their origins, limits, extraction, and processing—and at the potential health hazards to humans from mining and smelting activities (Chapter 13). All sources of energy that can be utilized by humans, their geology and limitations, are considered in Chapter 14. The geology of disposing of waste materials generated by affluent, energy-rich societies is discussed in the final chapter.

Distinctive Features of the Book

At most schools, the course for which this book is used is a component of the general education curriculum. General education broadens and enriches students' lives and minds beyond the specialization of their major interest. To this end, **Galleries** of photos at the end of each chapter illustrate many geologic wonders and some oddities. For example, at the end of Chapter 5 (Volcanoes), there are photographs of often-visited scenic volcanic features with explanations of how they formed. In addition, we also show some jarring images highlighting the often catastrophic collision of humans with the volcanic environment. The Galleries are intended to stimulate students' curiosity and appreciation for natural geologic wonders.

Case Studies in each chapter highlight the relevance of the text discussion. These cover a broad spectrum of subjects and geographic areas, but many of them focus on the causes and aftereffects of bad environmental and geological planning. Also within each chapter are several provocative

Consider This questions. They require students to apply the information just presented in the text and, thus, reinforce the learning. The questions will stimulate classroom discussion as well.

At the end of each chapter is a list of **Key Terms** introduced in the chapter, a **Summary** of the chapter in outline form, **Study Questions** geared to test understanding of the chapter's key concepts, and a list of related books and articles (**For Further Information**).

The Fourth Edition

- In this fourth edition of *Geology and the Environment,* we have attempted to incorporate as many of the reviewers' suggestions as possible. For example, the Case Studies were moved to the end of each chapter to make it easier for the student to follow the flow of text. It has also enabled the locating of figures closer to their discussions in the text and generally made the text more user friendly.

- The new edition has been significantly enhanced with the introduction of a powerful new interactive media program called Environmental GeologyNow, which has been seamlessly integrated with the text, enhancing students' understanding of important geological processes. It brings geology alive with animated figures based directly on figures in the text (Active Figures), media-enhanced activities, tutorials, and personalized learning plans. And like other features in our new edition, it encourages students to be curious, to think about geology in new ways, and to connect their new-found knowledge of the world around them to their own lives.

- Chapter 4 (Earthquakes and Human Activities) now provides a comparison of four extremely strong earthquakes (Gujarat, India, 2001; Alaska, 2002; Colima, Mexico, 2003; and Northridge, California, 1994) and the different impacts they had on nearby inhabitants due to population density near the source. A discussion of impulsively generated waves (tsunami) now takes place in Chapter 10 (Coastal Environments and Humans) and has been expanded to include the danger, to all coastal dwellers, of submarine landslide-generated tsunami.

- Chapter 11 (Glaciation and Long-Term Climate Change) contains significant revisions to discussions of climate change and evidence of global warming as well as completely new material on the shrinking Arctic ice cap. Chapter 12 (Arid Lands and Desertification) contains new sections on arid regions, winds, and human health, and an update on the drought in Africa's Sahel.

- The chapters dealing with water are now presented as Fresh-Water Resources (Chapter 8) with the addition of lakes. Flooding by both streams and hurricanes as hazards at the earth's surface are to be found in Chapter 9 (Hydrologic Hazards at the Earth's Surface).

- The text continues to emphasize remediation and prevention, an outgrowth of the authors' professional geological experiences. All of the chapters on geological hazards have dedicated sections on mitigation options, and resource and pollution issues are considered both in terms of the problems we face and of the potential ways to help forestall or lessen the impacts of these problems.

- We also continue to employ the systems approach in this edition—the idea that all of the earth's reservoirs (atmosphere, hydrosphere, solid earth, biosphere, and extraterrestrial) and the processes acting within them and interconnected.

- Environmental legal issues are discussed in the text where they are applicable, rather than placed in a separate chapter near the end of the text.

- All chapters have been updated in terms of data (where available), art, and photos. New Case Studies have been included and existing ones have been revised.

Supplements

Environmental GeologyNow

As described above, we now have a powerful new interactive media program called Environmental GeologyNow, which has been seamlessly integrated with the text, enhancing students' understanding of important geological processes. It brings geology alive with animated figures, media-enhanced activities, tutorials, and personalized learning plans. This student tutorial system is available through the Book Companion Web site and is free to students with each purchase of a new textbook. In addition to Environmental GeologyNow, students using the book's Web site will have access to Web links, Internet exercises, learning objectives, and other enhancement materials.

Thomson Brooks/Cole Earth Science Resource Center

This full-service Web site offers an array of teaching and learning resources. It contains online testing, tutorials, discussion forums, maps, and many more features to make your course an interactive experience.

InfoTrac® College Edition

Packaged free with each copy of the text, InfoTrac is an online library offering the full text of articles from hundreds of scholarly and popular publications. It is updated daily, and adopters and their students receive unlimited access for one academic term (four months).

Study Guide

The study guide is designed to help students master the course, with key concepts, terms, and review questions.

Acknowledgments

We gratefully acknowledge the thoughtful and helpful reviews by the following individuals, some of whom have also reviewed for earlier editions:

Richard M. Allen, *University of Wisconsin-Madison*
Elizabeth Catlos, *Oklahoma State University*
Kevin Cornwell, *California State University-Sacramento*
James Cotter, *University of Minnesota, Morris*
Larry Fegel, *Grand Valley State University*
George H. Myer, *Temple University*
Jennifer Shosa, *Colby College*
Edward Shuster, *Rensselaer Polytechnical Institute*
Neptune Srimal, *Florida International University*
Peter J. Thompson, *University of New Hampshire*

We also acknowledge the contributions of other reviewers and those who have contributed published and unpublished materials and photographs as this book has evolved to its Fourth Edition. These people have truly helped us build this book. The list is long, and our thanks are many to:

Lloyd Olson, Earl Francis, Peter Kresan, and Kevin Lamb, who all made their photos available and enriched the fourth edition.

Thomas B. Anderson, *Sonoma State University*
Robert Boutilier, *Bridgewater State College*
Kathleen M. Bower, *Eastern Illinois University*
David Bowers, *Montana Dept. of Environmental Quality*
Christopher Cirmo, *State University of New York, Cortland*
Rachael Craig, *Kent State University*
Charles DeMets, *University of Wisconsin, Madison*
John Field, *Western Washington University*
Josef Garvin, *Eden Foundation, Falkenberg, Sweden*
Tark S. Hamilton, *Consultant*
Gilbert Hanson, *State University of New York, Stony Brook*
Douglas W. Haywick, *University of South Alabama*
David D. Jackson, *UCLA*
Hobart King, *Mansfield University*
Robert Kuhlman, *Montgomery County Community College*
Kenneth A. LaSota, *Robert Morris College*
Berry Lyons, *University of Alabama*
Harmon Maher, *University of Nebraska, Omaha*
William Mode, *University of Wisconsin, Oshkosh*
Robert Sanford, *University of Southern Maine*
Steven Schafersman, *Miami University*
Hongbing Sun, *Rider University*

Terry Swanson, *University of Washington*
Glenn D. Thackray, *Idaho State University*
James L. Baer, *Brigham Young University*
William B. N. Berry, *University of California, Berkeley*
Lynn A. Brant, *University of Northern Iowa*
Don W. Byerly, *University of Tennessee, Knoxville*
Susan M. Cashman, *Humboldt State University*
Robert D. Cody, *Iowa State University*
Mark W. Evans, *Emory University*
Robert B. Furlong, *Wayne State University*
Bryan Gregor, *Wright State University*
Roger D. Hoggan, *Ricks College*
Alan C. Hurt, *San Bernardino Valley College*
Michael Lyle, *Tidewater Community College*
Lawrence Lundgren, *University of Rochester*
Garry McKenzie, *Ohio State University*
Robert Meade, *California State University, Los Angeles*
Marie Morisawa, *State University of New York, Binghamton*
William J. Neal, *Grand Valley State University*
Michael J. Nelson, *University of Alabama, Birmingham*
June A. Oberdorfer, *San Jose State University*
Charles Rovey, *Southwest Missouri State University*
Susan C. Slaymaker, *California State University, Sacramento*
Frederick M. Soster, *DePauw University*
Douglas J. Lathwell, *Cornell University*
Jack A. Muncy, *Tennessee Valley Authority*
T. K. Buntzen, *Alaska Department of Natural Resources*
Chuck Meyers, *U.S. Dept. of the Interior, Surface Mining Reclamation and Enforcement*
Ed Nuhfer, *University of Colorado, Denver*
Peter W. Weigand, *California State University, Northridge*
Joe Snowden, *University of Southeastern Missouri*
James L. Schrack, *Arco, Anaconda, Montana*
Marcie Kerner, *Arco, Anaconda, Montana*
Herbert G. Adams, *California State University, Northridge*
Lisa DuBois, *San Diego State University*
Edward B. Evenson, *Lehigh University*
Raymond W. Grant, *Mesa Community College*
Roger D. Hoggan, *Ricks College*
Steve Kenaga, *Grand Valley State University*
Barbara Hill, *Onondaga Community College*
Rita Leafgren, *University of Northern Colorado*
Joan Licari, *Cerritos College*
Larry Mayer, *Miami University*
David McConnell, *University of Akron*
James Neiheisel, *George Mason University*
Geoffrey Seltzer, *Syracuse University*
Albert J. Robb, III, *Mobil Exploration and Producing U.S., Inc., Liberal, Kansas*
Joan Van Velsor, *California Department of Transportation*
Lydia K. Fox, *University of the Pacific*
Bill Kane, *University of the Pacific*
Randall Jibson, *U.S.G.S.*
John E. Gray, *U.S.G.S.*
Edwin Harp, *U.S.G.S.*

Lynn Highland, *U.S.G.S.*

Robert Schuster, *U.S.G.S.*

Pam Irvine, *California Division of Mines and Geology*

Siang Tan, *California Division of Mines and Geology*

John Gamble and Ed Belcher, *Wellington, New Zealand*

Rick Giardino, *Texas A&M University*

Kenneth Ashton, *West Virginia Geological Survey*

Peter Martini, *University of Guelph*

Ward Chesworth, *University of Guelph*

Marion M. Gallant, *Colorado Department of Public Health and Environment*

Simon Young, *Montserrat Volcano Observatory*

Special thanks are due Helen Walden who copyedited and managed the production of this edition, and Carol Benedict who rode herd over three authors who were not always synchronized. Finally, we thank Keith Dodson, Science Editor at Brooks/Cole, for encouraging Richard Hazlett to become one of our authors and for his guidance during the preparation of the manuscript.

BERNARD W. PIPKIN
D. D. TRENT
RICHARD HAZLETT

Humans, Geology, and the Environment

EARTH SCIENCES, AND GEOLOGY IN PARTICULAR, pervade all aspects of our lives. The water we drink, the soil we depend upon to grow our food, and the natural hazards that impact many of us fall under the realm of geology. Many public buildings, including our nation's beautiful Capitol building, are largely composed of geologic materials.

Building the Capitol began in 1793 when President George Washington placed the cornerstone at a site chosen by the city's designer, Pierre L'Enfant. The first structure was destroyed by the British during the War of 1812, and it is suspected that terrorists planned to destroy the Capitol building on September 11, 2001. The stunning white dome of the Capitol and the wings that house the Senate and House of Representatives were built of native rocks from many states: marble from Maryland, Virginia, Georgia, Tennessee, Colorado, and Massachusetts; sandstone from Maryland and Virginia; and granite from Minnesota and North Carolina. In fact, most of the buildings along the Mall, from the Washington Monument to the Capitol building, are composed of sandstone, granite, limestone, and marble, all quarried from within the lower 48 states.

"The greatest shortcoming of the human race is our inability to understand the exponential function."

Professor Albert A. Bartlett
Department of Physics,
University of Colorado

GETTY

ENVIRONMENTAL
Geology ⇌ Now™

This icon, appearing throughout the book, indicates an opportunity to explore interactive tutorials, animations, or practice problems available on the GeologyNow Web site at http://earthscience.brookscole.com/pipkingeo4e

With the passing of the year 2000 (Y2K), we are now into the new "millennium." This second millennium is based upon a widely used calendar that reckons the birth date of Christ as the year "zero." However, the year 2000 of the Christian calendar is the year 1377 for Muslims; Jews view it as the calendar year 5759; and Hindus say it is the year 5101. Keep in mind that at the first millennium, the year 1000 of the Christian calendar, there were no printing presses, calendars were rare, almost all information was disseminated by word of mouth, and few people had much to celebrate.

How things have changed in a millennium! It is estimated that a quarter-billion people lived on earth 2,000 years ago. It took 15 centuries from year A.D. 1 to double the population to a half-billion people. The next doubling took only two to three centuries, as the population passed one billion in about 1830. That population doubled in only one century, to two billion in 1930, and the next doubling, to four billion people, took only 44 years (1977). The doubling to eight billion is expected to occur by about the year 2020 (◆Figure 1.1). If we could shrink the earth's population to 100 people with all the population ratios remaining the same, it might look something like the following: 57 Asians, 21 Europeans, 14 from the Western Hemisphere both North and South, and 8 Africans. Seventy people would be non-white, 30 would be Christian, 50 would suffer from malnutrition, 70 would be unable to read, 1 (only 1) would have a college education, and 1 (yes, 1) would own a computer.

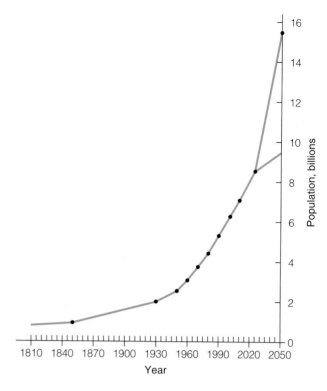

◆ FIGURE 1.1 World population projections to the year 2050 based on minimum and maximum birthrate estimates. Each tick on the time scale is 5 years. (Blue Planet Group)

What Is Environmental Geology?

Geology (Greek *geo*, "earth"; *logos*, "knowledge of") is the study of the earth, and **environmental geology** is the application of geological principles to the effects of humans on their physical environment. Environmental geology is the geology of "now," as opposed to what happened five million years ago, and seeks ways for the earth's inhabitants to interact safely with their geological surroundings. There is much evidence that the environmental problems of today have been developing for a long time, not the least of which is the explosive world-population growth since the early 1900s. In 2002 the

U.S. population reached 289 million people, many of whom had experienced the effects of seven of the country's ten most costly natural disasters in history in the preceding decade—the 1994 Northridge, California, earthquake (Chapter 4) and Florida's Hurricane Andrew in 1992 (Chapter 10).

The financial cost of natural disasters in the United States rose to about a billion dollars per week during the 1990s, a cost that many experts say will continue to rise (◆Figure 1.2). Phenomena such as El Niño and global warming amplify the impact of flooding and hurricanes, and inflation diminishes the value of the dollar, but those things do not adequately explain the huge increase in U.S. costs for

◆ FIGURE 1.2 The cost of natural disasters in the United States in five-year increments, 1952–1996. The graph would appear slightly different if different five-year periods were used, but the sharp increase in the 1990s is obvious. (From "Why the U.S. Is Becoming More Vulnerable to Natural Disasters" by G. van der Vink and others, 1998, *EOS* 70, no. 44)

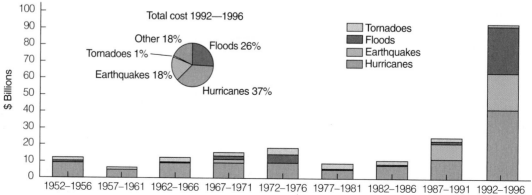

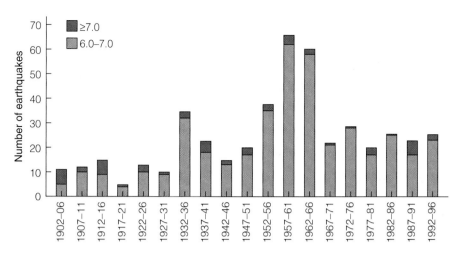

◆ FIGURE 1.3 Earthquakes in the United States with Richter magnitudes of 6.0 and above, 1902–1996. Earthquakes of greater than Richter-magnitude 7.0 are indicated. Magnitudes of 6.0 and above are designated as "damaging" earthquakes; those of 7.0 or more, "major" ones. (From "Why the U.S. Is Becoming More Vulnerable to Natural Disasters" by G. van der Vink and others, 1998, *EOS* 70, no. 44)

natural disasters in the 1990s. As ◆Figure 1.3 shows, there has been no obvious increase in the number or severity of strong earthquakes that can account for the increased costs. It appears that nature is not becoming more violent; rather, we are becoming more vulnerable because of population and land-occupation trends in our society. That is, we are occupying more lands that are in harm's way. Most natural disasters are not random acts as far as impacts on humans are concerned. Rather, they are the result of using land that is inappropriate, whether of necessity or lack of scientific knowledge of the geological setting. This could be rephrased, "Earthquakes don't kill people; buildings do." As populations grow, more buildings can be expected to topple on more people during earthquakes.

Some of life's choices have geological ramifications. For example, where one decides to live may depend upon the stability of a hillside, a neighborhood's low potential for flooding, or a community's distance from an active earthquake fault. A person's ability to make a sound decision in such a case would be enhanced by knowledge of geology.

Environmental geology is largely concerned with surface geologic processes, processes that not only make human life possible but that also may threaten life. Some of these processes are soil formation and erosion, water supply and floods, landslides, and natural and human-made pollution. It applies the science of geology to problems arising from the complex interactions of water, air, soil, solid earth, and life. As such, it encompasses all the traditional branches of geological study such as mineralogy, petrology (rocks), geomorphology (landforms), hydrology (water), geophysics, and geochemistry. ◆Figure 1.4 illustrates environmental geology's relationship to the traditional branches of geologic study and practice.

The Environmental Geologist

Environmental geology is a science, and like other scientists, the environmental geologist formulates hypotheses and then tests them. He or she uses observation, mapping and sampling of geologic materials, measurements, and sometimes experiments to collect data. In all sciences the goal is to obtain results that can be reproduced by other scientists. The next step of the process is data interpretation, a totally different activity that occurs in the brain of the investigator. From this interpretation, we may arrive at conclusions. This is much the same process that a medical doctor follows with a patient, and the geologist, just as the physician, does not know for sure that a conclusion is correct until "the patient is opened up." Similarly, conclusions are not cast in concrete; they may have to be changed when additional data become available. These data may come from new excavations, drill holes, roadcuts, or geophysical investigations.

Oil and mining companies were the biggest employers of geologists from about 1900 until the late 1950s. As cities grew,

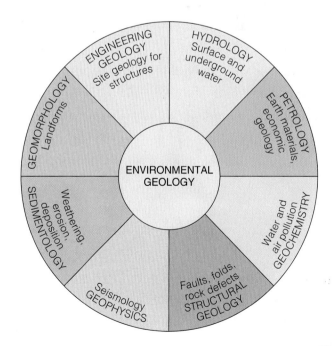

◆ FIGURE 1.4 The relationship between environmental geology and other geological sciences. Environmental geology incorporates many skills and specialties of the broad discipline of geology.

however, the need to protect public health and welfare created a large demand for **engineering geologists,** specialists who work with engineers on development and construction projects. Many older, uncontrolled projects were becoming public-health hazards, but they lay undiscovered. Then, with the discoveries of buried toxic waste at Love Canal in 1978, of leaking oil and gasoline tanks all over the country, and of agricultural chemicals in rivers and lakes, the specialty we now call *environmental* geology evolved. Its work is protecting existing environments and correcting past transgressions. Many environmental geologists work in areas of high geologic-hazard potential, such as sites subject to earthquakes or landslides (◆Figure 1.5). Their efforts have significantly reduced injuries and economic losses. In the past much of this work was strictly within the realm of the engineering geologist, but today the distinction between "engineering" and "environmental" geology is blurred if it exists at all. (In fact, the old *Bulletin of the Association of Engineering Geologists* bears a new name, *The Bulletin of Environmental and Engineering Geosciences.*)

The work of many environmental geologists today is in identifying adverse geologic conditions and potential problems in areas being considered for development. Others work in resource planning. The demand for resources increases as cities grow. (See ◎ Case Study 1.1 on page 12.) This growth requires advance planning for local sources of building materials, adequate good-quality drinking water, and determining optimal uses for given parcels of land. Many cities' growth problems are reducible to engineering or geological ones.

Tools of the Trade

Multipurpose maps of areas being considered for development are essential to effective planning, and they must be available long before the bulldozers arrive. Maps (◆Figure 1.6) can be generated that show areas of unstable ground, locations of active earthquake faults, areas with high flood potential, the depth to underground water, and possible mineral resources. Such maps can also indicate areas that are unsuitable for residential or commercial development but which could be used for parks and recreation. The generation of a multipurpose map usually begins with a geologic map that shows the distribution and spatial relationships of rocks below and at the surface of the earth and the area's drainage patterns. Geologic mapping is a fundamental part of most earth science studies. It is used in such fields as applied-geology investigations; water-resource planning; and land-use planning for waste-disposal sites, housing developments, and transportation. Geologic maps can be adapted to the special requirements of a given project or long-term plan.

Geographic Information System (GIS)

Derivative maps—that is, maps derived for a special use—can be assembled from known data using the Geographic Information System (GIS). In a strict sense, the GIS is a computer system capable of assembling, storing, manipulating, and displaying geographically referenced information—that is, data identified according to their locations. In practice, we recognize the total GIS as the personnel that operate the system and the data that go into it, as well as the role of the computer. GIS technology has refined and enhanced the power of traditional mapping by producing images and animations, as well as maps. The images thus produced allow researchers and nonscientists to view the subject in ways they have never seen before. By putting satellite data into GIS, for example, we can evaluate how plant health changes as rainfall increases or decreases throughout the year. Although vegetation changes are perhaps an obvious use, we can generate something as simple as slope maps that may indicate which slopes in a given map region are very steep and should be avoided for development. This information can be reinforced if used in conjunction with a geologic map that shows the type of geologic material underlying the slope. GIS is becoming an essential tool in environmental research and also in our effort to understand the magnitude and impact of global change.

◆ FIGURE 1.5 Geologists use tilt-up drilling rigs on hillsides and in close quarters such as people's backyards. At this residence a landslide is encroaching upon the swimming pool.

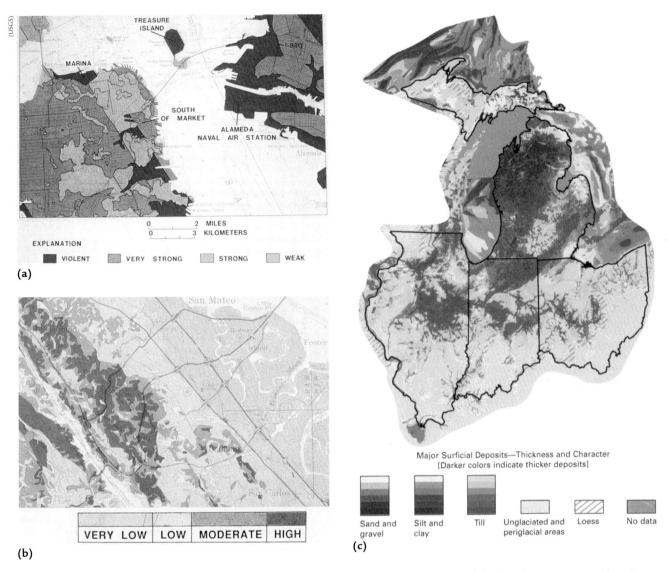

(a)

(b)

(c)

Major Surficial Deposits—Thickness and Character
[Darker colors indicate thicker deposits]

Sand and gravel   Silt and clay   Till   Unglaciated and periglacial areas   Loess   No data

◆ FIGURE 1.6 Hazard maps of the San Francisco Bay region. (a) Maximum predicted ground shaking due to strong earthquake motion. Ground shaking is a function of the geologic foundation at the site, such as soft bay muds in the Marina district (violent shaking) or hard bedrock on Nob Hill (weak shaking). (b) Risk of earthquake-induced landslides along the San Andreas fault west of San Francisco Bay. (c) Map of the surface deposits of Illinois, Indiana, Ohio, and Michigan. Lake Michigan and part of Lake Erie also appear on the map. Glacial deposits comprise most of the surficial geology shown. Such maps are indispensable to planners and indicate also where further detail and mapping are needed. (U.S. Geological Survey)

Case Histories

Nature is a good teacher, and many significant advances in environmental geology have resulted from studying "case histories," serious attempts to describe and analyze geologic events that have occurred. Several strong earthquakes that caused extensive damage initiated reevaluations of what had been considered state-of-the-art earthquake-resistant design of freeways, high-rise buildings, parking structures, apartments, and even single-family dwellings. They were the earthquakes in San Francisco (1989); Northridge, California (1994); Kobe, Japan (1995); and Izmit, Turkey (1999). Although none of these was a "great" one (>8.0 on the Richter magnitude scale), the Kobe earthquake cost $100 billion, the world record for monetary losses from a natural disaster as of the year 2002. Hurricane Andrew cost an estimated $25–30 billion, and the Northridge earthquake an estimated $25 billion, huge sums but much less than Kobe's. It follows that research and design improvements that cut damage losses are well worth the expenditure. The eruption of Mount Pinatubo in the Philippines is another example. It was considered a dormant volcano when the U.S. Air Force built the immense Clark Air Force Base nearby. Its 1991 eruption forced abandonment of the air base (see Chapter 5).

Other advances in environmental geology are developed in response to problems that become apparent less catastrophically. U.S. laws now require underground gasoline-storage tanks to be examined for leaks. If leaks are found, the contaminated soil must be "cleaned" and new leak-proof tanks must be installed. New technologies for sanitizing hydrocarbon-

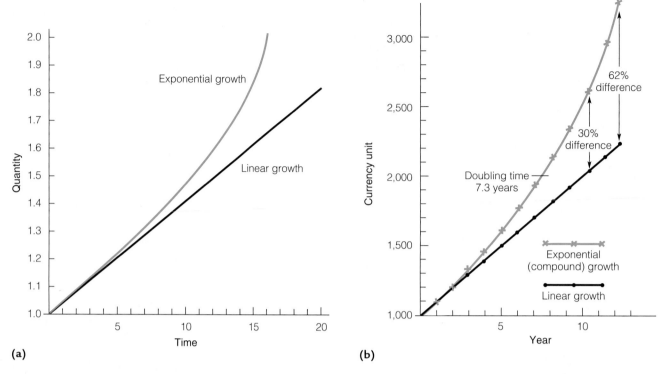

◆ FIGURE 1.7 Linear versus exponential growth. (a) The principle. (b) The growth of a savings account at 10 percent interest per year. Although the text example uses the dollar as the currency unit, the unit could be pounds, yen, or rupees—and the principle is the same for population growth. Note that the doubling time for exponential growth at a 10 percent annual rate is about 7 years. (USGS)

contaminated underground water and soils have been developed in response to the laws (see Chapter 8). The deadline for replacing identified leaking tanks—which became known as "yank-a-tank" in the environmental field—was the end of 1998. Service stations that were not compliant by then were no longer permitted to store or dispense gasoline.

Solid-waste disposal is an ongoing major problem. It requires identifying geologically acceptable disposal sites, including sites judged to be safe for sequestering hazardous substances (see Chapter 15).

Extracting water from beneath the land surface for human use may seem harmless enough, but in Venice, Italy, this is causing the land beneath the city to subside. This sinking causes periodic flooding, with serious consequences for the residents and their largest source of revenue, tourism. Similarly, subsidence due to extraction of underground water has altered the flow of surface water and storm drains over large areas of the U.S. Gulf Coast and the Great Valley of California.

At the core of these and many other environmental problems is the burgeoning world population. This growing population exists on a planet with limited air, land, water, and mineral resources.

Too Many People

The concept of **carrying capacity**—the idea that a given amount of land or other critical resource can support a limited number of people or animals—has intrigued scientists, economists, and philosophers for centuries. Thomas Malthus postulated in 1812 that populations grow geometrically, whereas food supplies increase only linearly, which leads to a gap between demand (population) and supply (food) that eventually results in famine. Human life is inextricably linked to resource utilization, environmental quality, and economic well-being. Thus, a good starting point for any environmental study is to learn how populations grow, a study referred to as **population dynamics.**

World population increased dramatically in the twentieth century. In 1900, the population of the earth was estimated at 1.6 billion people. By 2000 this figure had quadrupled, to 6 billion. Stated another way, it took about 10,000 generations for humankind to reach the first two billion in population, but only one generation to add the most recent two billion. Pogo, a popular comic strip character of the 1960s, once said, "Yep, son, we have met the enemy, and he is us," and this applies to the global population dilemma. Much of the recent dramatic increase in population can be explained by a reduction in the death rate due to overall better health globally.

Exponential Growth and the Population Explosion

Growth and increases are familiar processes to all of us; we see them in our national debt, taxes, rents, food costs, and so on. Increases that occur in nearly equal increments are described as *linear.* Examples include your electric bill variation with usage of electricity and growth of the grass in your lawn between weekly cuttings. Some other things grow by increments that become larger with time, and this kind of growth is described as *geometric* or *exponential*. Examples include the expansion of weeds in a neglected garden and historic increases in the national debt. The growth of a savings account also is exponential, assuming no withdrawals. A thousand dollars invested at 10 percent annual interest rate earns $100 the first year and becomes $1,100. During the second year $1,100 at 10 percent earns $110, and the value of the account grows to $1,210, and so on in each succeeding year. Thus, interest is earned on interest (compounded) so that at the end of ten years, for instance, the account balance is not just $1,000 *plus* 10 percent interest *times* 10 years, or $2,000, but $2,593.74.

It may seem incongruous that to our well-being and way of life the most important increases are not interest rates or the national debt, but the explosive growth of world population and natural-resource consumption. The graphs of ◆Figure 1.7 contrast linear (straight) and exponential (**J**-shaped) growth curves. Although a population growth rate of 2 percent per year or a similar increase in the consumption of a particular natural resource may seem trivial, when we realize that this growth is exponential, our assessment changes. Valid concerns for the future of the human condition arise. As world population increases exponentially, natural resources are rapidly being depleted; even if geological exploration can double or triple known oil or mineral reserves, it will make little difference in the useful lifetime of those resources (see ○ Case Study 1.2 on page 13).

Population growth rate is determined by subtracting a population's death rate from its birthrate, with death and birth rates expressed as deaths and live births per 1,000 people per year, respectively. For example, a birthrate of 20 *minus* a death rate of 10 would yield a growth rate of 10 per 1,000 persons, or 1.0 percent. World growth rates were low in 1900 (about 0.8%), and they have fluctuated with time. They reached a high of about 2.06 percent during the 1960s and then have slowly declined to the year 2000. At first glance you might think that a growth rate of 1 percent would yield a population **doubling time** of 100 years (since 1% per year × 100 years = 100%). This is not the case, however. Populations grow like savings accounts with compounded interest. Just as you can earn interest on interest, the people that are added produce more people, resulting in a shorter doubling time. A very close approximation of doubling time can be obtained by the "rule of 70"; that is,

Population doubling time = 70 ÷ growth rate (%)

Thus, the doubling time for a 1 percent growth rate would be 70 years; that for a 2 percent growth rate would be 35 years, and so forth. ✳ Table 1.1 provides some examples. The global population growth and the average annual growth rate for 1950 to 2000 are shown in ✳ Table 1.2. Although some progress has been made in slowing the growth rate, the number of people added each year continues to grow larger. The world population grew by almost a billion people in the decade 1990–2000, the largest increment ever for a single decade.

✳TABLE 1.1	How Populations Grow
Growth Rate, %	**Doubling Time, Years***
1.0	70.0
2.0	35.0
3.0	23.3
4.0	17.5
5.0	14.0
6.0	11.7
7.0	10.0

*Calculated by using the formula 70 ÷ growth rate (%), which yields a close approximation up to a growth rate of 10%.

U.S. Population Growth and Immigration

In 2002 an estimated 289 million people were living in the United States. They generated 15.9 live births per thousand people and a death rate of 9.0 per 1,000, for a growth rate of 6.9 per 1,000, or 0.69 percent. At first glance these figures appear to be real progress toward **zero population growth,** a condition that exists when, on average, as many people die as are born each year. However, even at this low rate the population would double in 100 years, without accounting for immigration—which adds significantly to the U.S. population, particularly in Texas, Florida, and California. From the 1920s through the 1960s about 200,000 immigrants came to the United States each year. Since then this number has grown dramatically for a variety of political and economic reasons. By 1993 there were 810,000 legal immigrants *plus* an estimated

✳TABLE 1.2	World Population and Growth Rate, 1950–2002	
Year	**Population, Billions**	**Growth Rate, %***
1950	2.52	
1955	2.75	1.77
1960	3.03	1.95
1965	3.34	1.99
1970	3.77	1.90
1975	4.08	1.84
1980	4.45	1.81
1985	4.85	1.75
1990	5.30	1.70
1995	5.76	1.68
2002	6.26	1.56

*Average annual rate for the previous five-year period.
Sources: Estimates in United Nations, *Demographics Yearbooks* for 1985 and 1990, and in *World Resources, 1994–95* (New York: Oxford University Press).

200,000 illegal ones, for a total of about a million new persons—five times the annual immigration of 30 years earlier. From the year 2000 through the end of the twenty-first century, it is estimated that 90 percent of the population growth in the United States will be immigrants and their progeny.

World Population Growth, Resources, and the Environment

In 1900 the world population growth rate was 1 percent. It climbed to a historical high of 2.2 percent in 1964, and then gradually dropped to 1.4 percent in 1997—a good sign. If the 1964 growth rate had remained at that level instead of decreasing, the earth's population would now be over 8 billion people. Some experts expect this growth rate to remain fairly steady for at least two decades and then to drop. Other experts expect education and birth-control measures to lower the growth rate steadily. The world continues to grow by 75 million people each year, the equivalent of almost five New York cities, with all the increases in the least developed countries. Forty eight of these underdeveloped nations are expected to triple in population by 2050, and in many more the population could easily double. The most significant world trend is that death rates are falling in poor and rich countries alike, while birthrates remain high in most poor countries and low in most rich ones. Exceptions are the generally higher death rates of Africa and the high birthrates of the rich oil-producing countries.

Due to varying estimates of *growth rate momentum,* world population projections for 2050 range widely. The United Nations Population Division in 2002 projected a 2050 world population of 8.9 billion people, 2.6 billion higher than today's population but 400 million lower than its previous projection. About half of the decrease is due to deaths from the HIV/AIDS pandemic and the lower than expected birthrates. The U.N. report says that fertility levels in most developing countries have fallen from six children per woman in 1950 to three children per woman today. China provides a good example of population momentum. Utilizing education and incentives, the country cut its growth rate dramatically in a few decades to 1.1 percent from more than 2.0 percent. With a population in excess of 1.3 billion, however, 11–13 million people are added every year even with a low growth rate! Keep in mind that the population growth curve is the typical **J-curve** for exponential growth of anything—slow growth at first followed by compounding and very rapid growth (see Figure 1.1).

The tripling of the population in the twentieth century has placed extreme demands on soil, water, forest, energy, and mineral resources. As a population increases, marginal lands are tilled because of the need for more food, and thus, poor agricultural practices become exaggerated. In addition, between 1940 and 1990, atmospheric carbon dioxide (CO_2) levels increased 13 percent, which may have implications for global warming, and the protective stratospheric ozone layer decreased 2 percent worldwide (more than that over Antarctica; see Chapter 11). None of these problems is insurmountable, but human ingenuity will certainly be taxed to overcome the challenges they present.

Overpopulation Scenario One: The Sky Is Falling

This scenario is fairly simple. The earth has too many people and finite resources for sustaining them. The carrying capacity of the earth has either been reached or is fast approaching, and the basic resources for life—soil, water, energy, and clean air—are being taxed to their limit. Students of population dynamics, economics, and environmental science generally agree that the earth will be hard pressed to sustain the 9–10 billion people projected for 2050, based upon the *lowest* of the projected world growth rates. If nothing is done to stabilize population, future population reduction will occur by increased fatalities from natural disasters, mass starvation, epidemics, and ecological degradation.

Overpopulation Scenario Two: The Gaia Hypothesis

The name of Gaia, the Greek goddess of the earth, has been applied by James Lovelock and U.S. biologist Lynn Margulis to their hypothesis that the earth is a superorganism whose environment is controlled by the plants and animals that inhabit it, rather than vice-versa. According to this hypothesis, the planet is "self-adjusting," so to speak. The superorganism regulates its environment with a complex system of mechanisms and buffers, just as an animal adjusts to varying temperatures, chemistry, and other environmental factors. For example, carbon dioxide, CO_2, a gas that is required for photosynthesis and therefore all of life, also acts as an insulating blanket in that it holds the sun's heat close to the earth. Terrestrial and marine plants act as regulators that "pump out" excess CO_2 from the atmosphere when great amounts are added by volcanic activity, which, if it were to accumulate, would smother humans and other animals.

According to the Gaia hypothesis, the last ice age ended abruptly when atmospheric CO_2 doubled due to a sudden failure of the "pumps," not by the slow processes of geochemistry as invoked by conventional scientific theory.

The Gaia hypothesis is appealing to some people because it seems to offer an easy solution for our environmental problems. However, toward the end of the twenty-first century the amount of CO_2 in the atmosphere will be doubled, and even Lovelock admits that "it is near certain that the new state will be less favorable for humans than the one we enjoy now."

Toward a Sustainable Society

A **sustainable society** (see Case Study 1.2) is one that satisfies its needs without jeopardizing the needs of future generations. The current generation *should* strive to pass along its legacy of food, fuel, clean air and water, and mineral resources for material needs to the generations of the twenty-first century. We know what needs to be done to accomplish this, but knowing what to do and implementing a long-range plan for carrying it out are entirely different things. A promising development in

*TABLE 1.3	Examples of Glaciers Wasting
Name	**Described Loss**
Greenland	Thinning at a rate of about a meter a year at its southern and eastern edges.
Glacier National Park	Since 1850, the number of glaciers has dropped from 150 to less than 50.
Alps	Glacial volume has shrunk by more than 50 percent since 1850.
Kilamanjaro (Tanzania)	Glacial ice has shrunk almost one-third since 1989.
Alaska (Glacier Bay)	Ice was at the entrance to the bay in 1794. Today, the Grand Pacific Glacier has retreated 105 km (65 mi) up the bay (see Chapter 11, Figure 11.4).

Various sources.

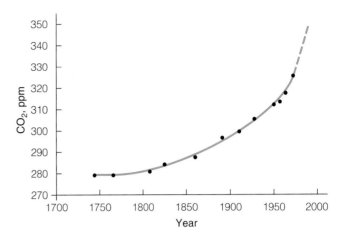

◆ FIGURE 1.8 Estimated atmospheric carbon dioxide changes from 1700 to 1980 based on data from air bubbles trapped in Antarctic ice cores and on atmospheric measurements. (NOAA data)

industrialized nations is the growing respect for basing an economy on steady-state conditions, rather than on continual growth. The conceptual ideal is sometimes referred to as the *spaceship economy,* since all resources must be conserved, shared, and recycled in space travel. This, together with population control, could lead in time to a sustainable society.

Too Many People and Signs of a Stressed Earth

Global Warming and Energy

In August 2000, the *New York Times* reported that a cruise ship sailed to the North Pole in open water. The Arctic Ocean at the pole is usually frozen over, year round. It was the rule in the past that captains of adventure cruises allowed the passengers to disembark in order to be photographed on the ice near the pole—but not during the summer of 2000. Early explorers hiking over the ice that summer would have had to swim the last few miles to the North Pole. It is noteworthy that the 14 warmest years since 1866 occurred between 1980 and 2001.

Melting ice is one of the most visible results of warmer temperatures (see ✿ Case Study 1.3) on page 13. In late 1991 an intact 5,000-year-old man was discovered protruding from a glacier on the Austrian–Italian border. He became known as the "Ice Man" and was quite famous. In general, glaciers and sea ice worldwide are wasting and are a sensitive indicator of global warming. Selected examples are shown in ✳ Table 1.3, and the subject is covered in more detail in Chapter 11. Besides providing scenic beauty, glaciers store water in winter that is released in spring and summer to run power stations and provide fresh water. Switzerland, Scandinavia, and to some extent the United States depend on glaciers as a source of water.

Global warming is potentially the most serious long-term environmental problem facing humankind. It is closely associated with the increase in atmospheric carbon dioxide produced by the burning of coal, oil, and natural gas (◆Figure 1.8). Other greenhouse gases, such as methane, nitrous oxide, ozone, and chlorofluorocarbons (CFCs), are also increasing in the atmosphere, and this, too, is closely associated with population growth. It is estimated that these gases may contribute to about half of the global warming projected for the twenty-first century.

Carbon-dioxide concentrations in the atmosphere today (as measured atop Hawaii's Mauna Loa volcano; see Chapter 5) are at their highest levels of the past 160,000 years, as indicated also by CO_2 in air bubbles in Antarctic deep ice cores (Chapter 11). If the goal is to stabilize climate, calculations indicate that the CO_2 emissions must be lowered from their 1990 level of 22 billion tons per year to 8 billion tons per year. In developed countries this would require an 85 percent per capita reduction in fossil-fuel consumption, which is unrealistic in the short term. As a start, a 5.2 percent average reduction has been agreed to by 38 industrialized nations, and there is some doubt that even this can be accomplished. Industrial nations consume energy at a prodigious rate—2–4 times the global per capita mean and 10–20 times that of Africa and Australasia (◆Figure 1.9). In the early 1990s half a billion motor vehicles were registered in the world, and this figure is expected to double by 2025—each consuming an average of two gallons of gasoline per day. Thus, other means of powering those vehicles must be developed—perhaps fuel cells; liquid or gaseous fuels derived from coal, natural gas, or biomass (alcohol); or even solar energy (see Chapter 14).

As developing nations raise their standards of living, their energy consumption will also increase, and competition between the developed and the developing nations for energy supplies will increase. This competition will surely increase as the world's coal and oil reserves decrease. The developing nations were using 30 percent of the world's energy in the early 1990s, a usage that was predicted to double by 2015.

The only options for large-scale power production sufficient to drive economic growth and human requirements in the near future are solar-energy conversion and nuclear energy, including both fission and fusion. Solar-derived electricity at present costs one-third less than electricity from a nuclear power plant and it lacks the hazards associated with radioactive substances. On the negative side, with present technology, solar-energy conversion requires vast areas of land and an effective storage system (see Chapter 14).

Rising Sea Level

A sensitive indicator of global warming is the rising sea level. It is affected by melt water from land-based glaciers, thermal expansion of seawater, and to a lesser extent the volume of the container. During the twentieth century sea level rose 10–20 centimeters (4–8 in) and it could rise as much as 1 meter during the twenty-first century. The consequences are loss of land, salt-water contamination of coastal underground fresh water (see Chapter 8), and beach erosion as waves break further inland (see Chapter 10). We must add to this the loss of agricultural productivity. For instance, Bangladesh would lose half its rice production by a 1-meter rise of sea level. It is not hard to visualize the economic and social consequences of millions of climate refugees.

The Land

The amount of cultivatable soil limits the number of humans the earth can support, because soil determines grain production and therefore food supplies. Arable soils worldwide suffered slight to severe degradation between 1945 and 1990 due to poor agricultural practices, natural erosion, and erosion accelerated by deforestation (✳ Table 1.4). The world's farmers

Region	Degraded Land Area as a Percentage of Vegetated Land		
	Total	Light Erosion*	Moderate to Extreme Erosion**
World	17	7	10
Asia	20	7	13
South America	14	6	8
Europe	23	6	17
Africa	22	8	14
North America (U.S., Canada)	5	1	4
Central America, Mexico	25	1	24

✳TABLE 1.4   Estimated World Soil Degradation, 1945–1990

*Light: crop yields reduced less than 10 percent.
**Moderate: crop yields reduced 10–50 percent. Severe: crop yields reduced more than 50 percent. Extreme: no crop growth possible. About 9 million hectares worldwide exhibit extreme erosion, less than 0.5 percent of all degraded lands.
Source: Various sources compiled by World Resources Institute and the United Nations, World Resources 1992–93 (New York: Oxford University Press, 1992).

lost an estimated 500 million tons of topsoil in the 1970s and 80s, an amount equal to the tillable area of India and France combined. Waterlogging and salt-contamination of soils are reducing the productivity of at least a fourth of the world's cropland, and human-generated smog (ozone) and acid rain also are taking their toll on crops. It is the lack of soil, sometimes combined with drought, that is the cause of famine, not soil infertility. (Soils are discussed in detail in Chapter 6.)

CONSIDER THIS   *A limiting resource is one that controls a population's growth and stability; for instance, the abundance of nitrates and phosphates in the oceans limits plant growth. Many earth scientists and biologists believe water is the limiting resource for humans. Can you think of any other resource or substance that might be limiting to human populations? Why might it be?*

Water—A Limiting Resource

Mark Twain's famous quote, "Whiskey is for drinking; water is for fighting over," was prophetic. Little water from the Nile River makes it to the Mediterranean, the Ganges River is a

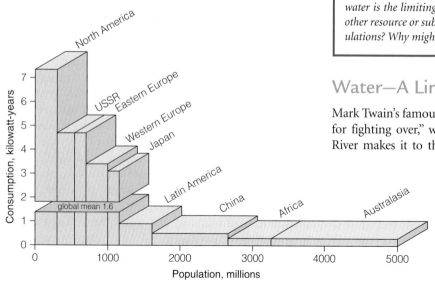

◆FIGURE 1.9 Per capita energy consumption in kilowatt-years (kwy). Note that the industrialized nations consumed 2–4 times the global per capita mean. The data are for 1987 but had not changed much by 2003 except for China, which has greatly increased industrial production. (Blue Planet Group)

trickle in the dry season, and the Colorado River is a joke when it crosses the international border with Mexico. With the populations of Egypt and her neighbors projected to double between 2000 and 2050, and as the inevitable growth of India and Bangladesh occurs, there is bound to be competition for the use of water on the Nile and the Ganges. Two-thirds of the world's irrigated lands lie in Asia, where populations and water needs also are greatest. The Yellow River ran dry in 1972 for the first time in China's 3,000-year history, and water tables fell on every continent in the early 2000s. Water withdrawals exceed water recharge in the southern Great Plains and the entire southwestern United States (see Chapter 8).

Seventy percent of water that is diverted from rivers or pumped from underground is used for irrigating crops upon which entire cultures depend. It is said that India is on a temporary "free ride," expanding irrigation at the expense of stored underground water. Underground water needs to be managed, as it is not a limitless resource. When this fact strikes home, India's grain production could fall as much as 25 percent. For a country that adds 16 million people every year, the consequences could be disastrous.

CONSIDER THIS *Few people doubt the existence of ozone holes in our atmosphere and the increased flux of ultraviolet radiation to the earth's surface due to ozone depletion by CFCs in the stratosphere. Global warming and its cause, however, are subjects of heated (no joke intended) debate. The earth is getting warmer, and the big question is whether the cause is anthropogenic (humans' production of greenhouse gases), or just part of a natural climatic rhythm of warming and cooling. Think hard about the long-term impact of global warming on the hydrosphere, atmosphere, biosphere, and on humans, and then answer the question, Is it wise to wait, perhaps several decades, to know for certain whether the warming is natural or anthropogenic before reducing greenhouse-gas emissions?*

Forests

Deforestation continues to be one of the environmental issues of the day. Forests cover 30 percent of the earth's land surface, or 3.9 billion hectares (abbreviated ha; each ha is about 2.5 acres) and each year forests are diminished by man or by fires. According to the Food and Agricultural Organization (FAO) of the United Nations, between 1990 and 2000 there was a loss of 94 million hectares (235 million acres). The emerging nations lost 130 million hectares (325 million acres), and the industrial countries gained 36 million hectares (90 million acres), mostly by conversion of agricultural land. Simply put, every 3 years the developing countries lose 2 percent of their forestland. Deforestation leads to loss of habitat and decreased biodiversity, contributes to climate change by adding CO_2 to the atmosphere, and often results in soil degradation.

One reason for clearing forests is to convert land to agricultural use. More than half of the tropical-forest lands destroyed each year are in Brazil. It should be noted, however, that Brazil plans to set aside 1.6 million square kilometers (618,000 mi^2), about 20 percent of its land area, as public reserves and parks. The Indonesian government, in a full-page advertisement in a U.S. magazine in 1989, explained that the aspirations of its 170 million people were the same as those of citizens in developed nations, and that in order to better their lives, 20 percent of Indonesian forests were to be converted to plantations of teak, rubber, rice, and coffee. A poor African farmer, on the other hand, has little choice but to clear land to grow crops just for subsistence. How a country uses its land is a complex political and sociological issue, and there are no simple ways to reconcile the conflicting human needs that the issue involves.

Logging is another cause of deforestation, but it has been demonstrated that selective logging in tropical forests is sustainable if the logged lands are allowed to rejuvenate. In 1988 Thailand was inundated with 40 inches of rain in five days. Thousands of logs left on hillsides to dry floated off in mud and water that engulfed entire villages, killing 350 people. The Thai government banned logging, and was caught between the companies that wanted compensation for lost business and the landless poor who were given farms in the previously logged areas.

A third cause of deforestation is the demand for wood fuel and forest products. Many of the forests in India and the Sahel of West Africa have been decimated for use as domestic fuel and in light industry.

Deforestation is second only to the burning of fossil fuels as a source of atmospheric CO_2. Forests store 450 billion metric tons of carbon. When cleared, they can no longer sequester the carbon, and it is released rapidly to the atmosphere when trees are burned, or slowly if they are left to decay.

Other Resources

Our throw-away society, though certainly a convenient one, has become a recycling society as natural resources dwindle and waste-disposal sites become scarce and expensive to operate. Most of what we used to discard after one use—paper, plastic, glass, aluminum, and even steel—is now recycled. Trash disposal has become so expensive, at least in urban settings where most trash is generated, that recycling is close to being cost-effective. Unless alternative energy sources are maximally utilized, oil reserves will be exhausted in your lifetime or in your children's. United States coal reserves are estimated at a 400-year reserve, but such pollution considerations as smog, acid rain, and global warming will continue to limit coal's use. World reserves of metallic and nonmetallic resources, many of which are in abundant supply at present, can certainly be extended by recycling. The geological and environmental consequences of mining and energy consumption and the problems associated with waste disposal are considered in Chapters 13 through 15.

CONSIDER THIS *Name some of the things that earth's inhabitants could do now to develop a sustainable society for future generations.*

The Road to Seven Billion

Sometime during October 1999, the United Nations announced that the world population had reached 6 billion, only 12 years after the population reached the 5 billion mark. According to the U.N.'s Population Division projections, the 7 billionth person could be born as early as 2011 or as late as 2015, depending upon birth-rate trends in India or China. What is more disturbing, to us at least, is the trend toward urbanization. In the 1950s there were only 16 cities with more than a million people. London was the leader with 7 million, and only 7 percent of the population could be described as urban. The earth's population has doubled in the past 40 years, but city populations have increased five-fold. According to the U.N., in the next five years over half the population of the globe will be living in cities, with inadequate infrastructures and too few jobs (see the figure). Megacities are heat sinks that promote smog, trigger thunderstorms, and reduce the productivity of the land. At the same time, rise in sea levels due to global warming will threaten coastal areas and river deltas where most large cities are located. How can megacities exist with the need to protect the environment? This is one of the big questions debated at the Johannesburg Summit Meeting (see ✪ Case Study 1.2). However, with good planning and an eye for aesthetics, city dwelling of the future need not be bleak. The good news is that there is a trend for women to have one or two children in those nations with over 80 percent of the population.

Top 10 Countries in Population Growth
Net annual additions, in millions, 1995–2000

Country	Millions
India	16.0
China	11.4
Pakistan	4.0
Indonesia	2.9
Nigeria	2.5
United States	2.3
Brazil	2.2
Bangladesh	2.1
Mexico	1.5
Philippines	1.5

Largest Urban Areas, 2002
Population in millions

City	Millions
Tokyo	28.8
Mexico City	17.8
São Paulo, Brazil	17.5
Bombay, India	17.4
New York	16.5
Shanghai	14.0
Los Angeles	13.0
Lagos, Nigeria	12.8
Calcutta, India	12.7
Buenos Aires	12.3

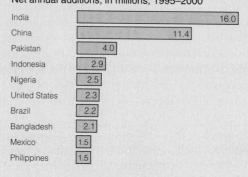

CASE STUDY 1.2

RIO + 10

The World Summit on Sustainable Development (also known as Rio + 10) met on September 2–11, 2002 in Johannesburg, South Africa. Ten years before, the Earth Summit on the Environment and Development met in Rio de Janeiro, Brazil, and 30 years before that the U.N. Stockholm Conference (1972) had focused on environmental issues and international pollution. The Rio meeting produced a 231-page document titled *Agenda 21,* which has not been altered and remains the call to save the environment. As a result of all these meetings, nations agreed and funded projects and research in the earth sciences that involved global warming and greenhouse gases, desertification, fisheries, the transport and disposal of hazardous waste, the ozone layer, management of the ocean, and freshwater resources. The main concerns of *Agenda 21* developed in 1992 reads like the table of contents for an environmental science textbook. Other concerns taken up at these meetings involved biological diversity, population, human health, women's rights, and poverty.

The entire meaning of the summit meetings has been captured in the definition of "sustainable development." It is the *"development that meets the needs of the present without compromising the ability of future generations to meet their own needs."* At about the same time was the announcement by President Bush's representatives that he had decided not to endorse the Kyoto Treaty (1997), which requires reduction of greenhouse-gas production. The United States produces about 25 percent of the world's greenhouse emissions, and even though Canada, Russia, Japan, and China have signed the agreement, there is some doubt that they can meet the required reductions.

CASE STUDY 1.3

Twenty-Five Years of "Thinking Green"

During the 1960s a relatively small group of environmentalists, conservationists, and scientists, all widely regarded at the time as silly do-gooders and "tree huggers," began raising the public awareness that there is such a thing as "the environment" and that it needs protection. One of them was an outspoken biologist named Barry Commoner. In his 1971 book, *The Closing Circle,* Commoner advanced four laws of ecology that have had great impact on our thinking:

1. Everything is connected to everything else.
2. Everything must go somewhere.
3. Nature knows best.
4. There is no such thing as a free lunch.

Today, most of us appreciate Commoner's first law, the concept that all elements of an ecosystem are related and that disrupting any of them may adversely impact the entire system. The second law stresses that humans do not destroy matter but simply alter it into forms that may be detrimental to health and the environment. Consider toxic waste and greenhouse gases, for example. Number three, "Nature knows best," seems simple enough, but it presents an enormous educational problem because it conflicts with the normal human tendency to attempt to control nature. The last law puts the teeth in the first three; it assures us that we will pay for any ill-gotten benefits. The cost of polluting streams, oceans, or soil or of overpopulating unfit lands is eventual remediation expense, resource depletion, health problems, natural disasters, global warming, or the like.

Before 1970 little serious thought was given to environmental education. Today, however, one would be hard pressed to find a college, university, law school, or graduate program that does not offer courses dealing with environmental issues. "Green" thinking is now ingrained in policies, practices, and activities of industry, government, and the public. Some successes of the past 25 years are:

- Hydrocarbon emissions from automobiles have decreased by almost half.
- The number of large cities violating clean-air standards has dropped from 40 to 9.
- The use of unleaded gasoline has reduced lead emissions by 98 percent.
- Industrial toxic-waste spills have declined 43 percent.
- The California gray whale, the bald eagle, and the American alligator have been removed from the endangered species list.
- Mandatory recycling programs have been implemented in 6,600 U.S. cities. The amount of waste that is recycled increased from 17 percent in 1990 to 22 percent in 1995.
- 100 million acres have been set aside as wilderness areas.
- Since 1979 the country has invoked energy savings valued at almost five times all the new sources of energy combined. Some people call these savings "negawatts."
- The number of environmental groups on college and university campuses grew from 50 in 1989 to more than 2,000 in 1995.

Underlying these achievements is a large body of federal and state laws enacted since the 1960s to clean up and protect the environment.

Overpopulation Problems

As emphasized in this chapter, overpopulation is responsible for environmental damage, societal problems, and human suffering. In some countries population growth is utterly out of control. In other countries—China, for instance—measures and incentives are in place for curbing the growth and achieving a sustainable population. This photo gallery highlights the dilemma.

◆ FIGURE 1 Coney Island beach in the 1930s, when the U.S. population was a fraction of what it is today. Imagine this beach on a summer weekend today.

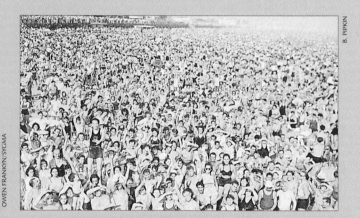

(a)

(b)

◆ FIGURE 2 Deforestation near the city of Manaus in Amazonia, Brazil's largest state. (a) False-color satellite photo of the confluence of the Rio Negro (*black*) and the sediment-laden Amazon River (*light green*). Healthy rain forest is shown in red, and deforested areas appear in light colors. Manaus is on the point of land that extends into the Rio Negro. The waters of the Rio Negro are recognizable for more than 80 kilometers (48 mi) downstream in the Amazon. (b) Adjacent rain forests have been removed as the population of the area has grown. Manaus's population is now more than a million people, and the city is rapidly expanding into areas that were once sparsely populated.

◆ FIGURE 3 A model Chinese family in front of a billboard encouraging the one-child family; Beijing. Although the People's Republic of China is pushing population control, its population of more than a billion people makes continued growth inevitable.

Summary

Environmental Geology

Defined
The interaction of humans and the geological environment.

Utilizes
Most of the traditional geologic specialties such as petrology, structural geology, sedimentology, economic geology, engineering geology, geophysics, and geochemistry.

Areas of Interest
Geologic hazards such as volcanism, earthquakes, and flooding; mineral and water resources; and land-use planning—should I build my house on this hillside?

Goals
To predict or anticipate geologic problems by collecting data in the field and analyzing it in the laboratory. Data may be portrayed on special-use maps and utilized by planners and public officials.

Case Histories
Nature is a good teacher, and geologists learn much by studying case histories of geologic events.

Population Dynamics

Carrying Capacity
Defined for our purposes as the number of people the earth can support. This number is uncertain but lies somewhere between the 1995 population of 5.76 billion and the 10–15 billion projected for the year 2050.

Population Growth Rate
Defined as live births per 1,000 persons (that is, the birthrate) minus the death rate per 1,000. A birthrate of 20 *minus* a death rate of ten *equals* a growth rate of 10 per 1,000 persons, or 1%.

Doubling Time
The number of years in which a population is doubled. It can be projected closely by dividing 70 by the percentage birthrate. The 1995–96 world growth rate of 1.68 percent yields a doubling time of 40 years.

Overpopulation Scenarios

Uncontrolled Growth, Worst-Case Scenario
The earth becomes uninhabitable as resources dwindle, sea-level rises due to global warming and inundates coastal cities, and hazardous ultraviolet radiation strikes the earth's surface due to depletion of the ozone layer.

Gaia Hypothesis
Life is an important regulator of the earth environment. If left alone, natural systems are self-regulating; for example, as carbon dioxide builds up in the atmosphere, plant growth and the oceans will remove it, thereby reducing global warming. Human intervention into natural systems (by deforestation, for example) has rendered a basic premise of this hypothesis untenable.

Sustainable Society

Defined
Society's current needs are satisfied without jeopardizing future generations. To accomplish this, population control is needed.

Water Implications
Water is a finite resource and needs to be managed. Simply put: no water, no food.

Energy Implications
Increasing energy demands dictate that we seek alternatives to burning the earth's reserves of limited fossil fuels. Alternatives include solar, wind, geothermal, and nuclear energy.

Land Implications
Soil erosion and the occupation of marginal lands subject to geologic hazards are problems that must be addressed.

Forests Implications
Deforestation is the second-greatest contributor to the problem of human-induced carbon dioxide accumulating in the atmosphere. It results in loss of habitat for humans and animals and contributes to desertification.

Other Resources Implications
We must invest our efforts in recycling and conserving all resources.

Key Terms

carrying capacity	environmental geology	population dynamics	sustainable society
doubling time	J-curve	population growth rate	zero population growth
engineering geologist			

Study Questions

1. Physical geology is the study of earth materials, the processes that act upon them, and the resulting products. How does environmental geology differ from physical geology? In order to answer this question you might compare textbooks in the two subject areas.

2. Distinguish between engineering geology and environmental geology.

3. Which natural resources are most important in determining the carrying capacity of the earth?

4. The United States has a current growth rate of 0.7 percent. At this rate how much time is required for a population to double? Discuss the impact of a doubling of your community's population in that length of time.

5. Of what value are case histories in solving environmental geological problems?

6. What is the impact of overpopulation on energy use and global warming? How does this bode for the well-being of future generations? In what parts of the world will the largest percentage increases in energy consumption probably occur in the next decade, and why?

7. What are some of the adverse consequences of clearing tropical rain forests?

For Further Information

Books and Periodicals

Brown, L., and others. 1999. *The state of the world,* 1999. New York: W. W. Norton and Co.

Brown, Lester. 2001. *Eco-economy—building an economy for the earth.* New York: W. W. Norton and Co.

Cohen, J. E. 1999. A global garden, *The ZPG reporter* 31, no. 1 (February).

Ehrlich, Paul R., and Anne Ehrlich. 1970. *Population, resources, and environment: Issues in human ecology.* New York: W. H. Freeman and Co.

————————. 1987. *Earth.* New York: Franklin Watts.

Flavin, Christopher, Hillary French, Gary Gardner, and others. 2002. *The state of the world, 2002.* New York: W. W. Norton and Co.

Joseph, Lawrence E. 1990. *Gaia: The growth of an idea.* New York: St. Martin's Press.

Klesius, Michael. 2002. *The state of the planet,* Washington, D.C.: *National geographic,* September.

Knickerbocker, Brad. 1995. Earth Day at 25: A U.S. environmental report card. *Christian Science monitor,* April 18.

Lovelock, James. 1988. *The ages of Gaia: A biography of the living earth.* New York: W. W. Norton and Co.

Parfit, Michael. 1998. *Living with natural hazards,* Washington, D.C.: *National geographic,* July.

Silver, Cheryl S., and Ruth S. DeFries. 1990. *One earth one future: Our changing global environment.* Washington, D.C.: National Academy of Sciences, National Academy Press.

U.S. Geological Survey, and others. 1999. *Sustainable growth in America's heartland, 3-dimensional geologic maps as the foundation.* Circular 1190, Washington, D.C.: U.S. Government Printing Office.

Van der Vink, G., and others. 1998. Why the United States is becoming more vulnerable to natural disasters. *EOS* 79, no. 44 (November 3).

World Resources Institute and the United Nations Development and Environment Programs. 1992–93. *Toward sustainable development: A guide to the global environment.* New York: Oxford University Press.

World Resources Institute and the United Nations Development and Environment Programs. 1994–95. *People and the environment: Resource consumption, population growth, and women.* New York: Oxford University Press.

Zwingle, Erla. 2002. *Cities,* Washington, D.C.: *National geographic,* November.

ENVIRONMENTAL
Geology ⇌ Now™

Assess your understanding of this chapter's topics with additional quizzing and comprehensive interactivities at http://earthscience.brookscole.com/pipkingeo4e as well as current and up-to-date Web links, additional readings, and InfoTrac College Edition exercises.

Getting Around in Geology

*Barnacle Bill, Yogi Bear, and Scooby Doo—
Planetary Rock Stars*

ABOUT FIVE BILLION YEARS AGO AN INTERSTELLAR CLOUD OF GAS and dust began to collapse, probably under the force of its own gravity. As it shrank it started spinning, which caused the cloud to flatten into a disc shape. Eventually, the compression accompanying the collapse created temperatures and pressures in the center of the cloud sufficient to initiate nuclear reactions. Hydrogen atoms fused, forming helium and releasing enormous energy. Our primordial sun began to shine. Around the sun a cooler disc, the solar nebula, contained the elemental leftovers of the solar implosion that eventually would form the planets. Low-boiling-point compounds such as methane, ammonia, and various ices condensed in the cold far reaches of the nebula to form the outer planets, the very large *Jovian planets.* Closer to the sun, metallic and rocky substances with high melting points condensed and collided to form the four dense inner planets, the *terrestrial planets.* Of that group, the third and fourth planets from the sun are most interesting to geologists.

◆ FIGURE 1 *A view in the vicinity of the* Mars Pathfinder *landing site. The "marscape" seen is a portion of Ares Vallis (Mars Valley), with the rover Sojourner backed up to the rock named Yogi. Sojourner has placed an alpha proton X-ray spectrometer instrument against Yogi in order to determine its elemental composition. Yogi is about 6.5 meters from the* Mars Pathfinder *lander and stands about 1 meter (3 ft) high. Sojourner is 30 centimeters (1ft) tall. On the lower left is the ramp that Sojourner used to exit the lander. On the horizon, about 150 meters (500 ft) distant, is a rock known as the Couch.*

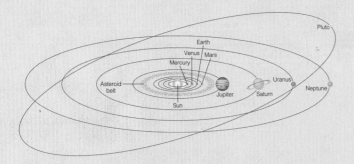

◆ FIGURE 2 *Diagram of the solar system showing the relative sizes and positions of the planets. Note Pluto's discordant orbit. It causes astronomers to debate whether Pluto should be considered a planet.*

On the fourth of July, 1997, transmissions from NASA's *Mars Pathfinder* space vehicle that had just landed on the surface of Mars showed an alien landscape once ravaged by floods and now swept by wind-blown sand. Aha! Mars did indeed have an atmosphere and liquid water at one time. *Mars Pathfinder* carried the Jet Propulsion Laboratory's (JPL) remote rover, Sojourner, a cross between a dune buggy and a scientist's laboratory. The funny-looking little ten-inch-high rover carried an X-ray spectrometer and other instruments for analyzing specimens of Martian rocks and soil on the spot. *Mars Pathfinder* was the media event of the summer. Its Web site took 265 million hits, and toy stores' stocks of Mattel's lander and rover models sold out quickly.

During Sojourner's first week on Mars the little rover collected data from soil and a football-size rock affectionately dubbed "Barnacle Bill." Other rocks named Yogi and Scooby Doo made headlines. The naming of rocks goes back to the 1970s *Viking* missions. Investigators found that naming items in the transmitted photographs made it easier for them to remember the geography of the sample field.

Most rocks on Earth, the Moon, and Mars are silicates, composed of silica (SiO_2) and aluminum. The amount of SiO_2 in a rock that was once molten is a key indicator of how "distilled" the melt that cooled to form the rock was. Low-silica rocks ($<50\%$), such as the black lava basalt found around the Hawaiian Islands, come from "primitive" melts, melts that have not lost any of their original high-temperature, low-silica minerals. This is what researchers expected to find on Mars, because Mars has huge volcanoes resembling those making up Hawaii. Barnacle Bill, however, contained moderate amounts of SiO_2, similar to rocks found in many continental volcanic regions on Earth, like the Andes (see Chapter 3). The mystery is that Mars evidently has no Earth-like continents or continent-forming processes. Sojouner gave us some answers, and also some new questions for exploring the Red Planet.

A Martian day, known as a "sol," is 24 hours and 37 minutes long. *Mars Pathfinder* was designed to transmit for about 30 sols before going dead. It blinked out on September 27, 1997, almost two months longer-lived than expected. NASA and the JPL were tickled Martian pink.

Geologic materials such as rocks, valuable mineral deposits, and soils are an integral part of the human geologic environment. Rocks form the geologic foundations and provide building material for many of our structures. In addition, rocks are involved in many surface earth processes that are hazardous to humans, such as landslides and rockfalls. In this chapter we will discuss briefly how the planet Earth formed and how its rocky crust and atmosphere arrived at their present condition. Environmental geologists are often asked about future geologic events, and in order to better understand the future a geologist must learn as much as possible about the past. An important concept in geology known as the Law of Uniformitarianism is that all geologic processes active today have been active in the past, only at differing rates. The law can be paraphrased as "the present is the key to the past." For example, by studying the geologically recent past, we have learned that North America was subject to multiple glacial advances in the past 1.6 million years, a period of time known as the Ice Ages. As a result, scientists monitor and computer-model weather and climate in order to anticipate long-term trends and perhaps even the next ice age (see Chapter 11). In this regard, we might also say "the present is the key to the future."

The Earth System

Planet Earth functions like a finely tuned machine, a natural spaceship for its inhabitants that is powered by nuclear energy from the sun and deep interior. Just as the elements of human physiological systems are connected and interdependent, so on earth there are linkages and interactions between the rocky surface and the atmosphere and oceans, and between these systems and all plant and animal life. Our interest extends from the ecosphere, where life evolves, to the reactions between the crust and its fluid and gaseous envelopes, and even to the earth's core, where a magnetic field that protects most organisms from lethal radiation is generated. An example of a little-known connection is the one that exists between dust storms in the Gobi and Takla Makan deserts in China and the North Pacific Ocean. Airborne silt from those deserts is carried all the way to the North Pacific, where it increases the nutrient content and fertility of this otherwise barren oceanic realm. The periodic weather phenomenon known as *El Niño*, which heavily impacts the west coasts of South and North America, has its origins in ocean–atmosphere interactions called the *Southern Oscillation* in the western Pacific thousands of miles away. We now know that

◆ FIGURE 2.9 Some common ferromagnesian silicates: (a) olivine; (b) augite, a pyroxene-group mineral; (c) hornblende, an amphibole-group mineral; and (d) biotite mica. Courtesy Sue Monroe

can scratch glass ($H = 5\frac{1}{2}–6$) and calcite ($H = 3$), and your fingernail can scratch gypsum ($H = 2$).

Cleavage refers to the characteristic way particular minerals split along definite planes as determined by their crystal structure. Mica has perfect cleavage in one direction; this is called *basal* cleavage, because it is parallel to the basal plane of the crystal structure. Feldspars split in two directions; halite, which has *cubic* structure, in three directions; and so

on, as shown in ◆Figure 2.10. Minerals that have perfect cleavage can be split readily by a tap with a rock hammer or even peeled apart, as in the case of mica. Some minerals do not cleave but have distinctive **fracture** patterns that help one to identify them. The common crystal forms, shown in ◆Figure 2.11, can be useful in identifying a particular mineral. Color tends to vary within a given mineral species, so it is not a reliable identifying property. **Luster**—how a mineral reflects light—is useful, however. We recognize metallic and nonmetallic lusters, the latter being divided into types such as glassy, oily, greasy, and earthy. (✪ Case Study 2.2 on page 40 discusses some gem minerals.)

Rocks

Rocks are defined as consolidated or poorly consolidated aggregates of one or more minerals, glass, or solidified organic matter (such as coal) that cover a significant part of the earth's crust. There are three classes of rocks, based upon their origin: igneous, sedimentary, and metamorphic. **Igneous** (Latin *ignis*, "fire") rock is crystallized from molten or partly molten material. **Sedimentary rocks** include both *lithified* (that is, turned to stone) fragments of preexisting rock and rocks that were formed from chemical or biological action. **Metamorphic** rocks are those that have been changed, essentially in the solid state, by heat, fluids, and/or pressure within the earth.

The **rock cycle** is one of many natural cycles on earth. The illustration of it in ◆Figure 2.12 shows the interactions of energy, earth materials, and geologic processes that form and destroy rocks and minerals. The rock cycle is essentially a closed system; "what goes around comes around," so to speak. In a simple cycle there might be a sequence of formation, destruction, and alteration of rocks by earth processes. For example, an

✱TABLE 2.2 Some Common Rock-Forming Minerals

Mineral	Abundance in Crust, %	Rock in Which Found
Plagioclase*	39	Igneous rocks mostly
Quartz	12	Detrital sedimentary rocks, granites
Orthoclase**	12	Granites, detrital sedimentary rocks
Pyroxenes	11	Dark-colored igneous rocks
Micas	5	All rock types as accessory minerals
Amphiboles	5	Granites and other igneous rocks
Clay minerals	5	Shales, slates, decomposed granites
Olivine	3	Iron-rich igneous rocks, basalt
Others	11	Rock salt, gypsum, limestone, etc.

*A series of six minerals within the plagioclase group from albite ($NaAlSi_3O_8$) to anorthite ($CaAl_2Si_2O_8$).
**Feldspar group of minerals.

✱TABLE 2.3 Mohs Hardness Scale

Hardness	Mineral	Common Example
1	Talc	Pencil lead 1–2
2	Gypsum	Fingernail 2½
3	Calcite	Copper penny 3 Brass
4	Fluorite	Iron
5	Apatite	Tooth enamel Knife blade Glass 5½–6
6	Orthoclase	Steel file 6½
7	Quartz	
8	Topaz	
9	Corundum	Sapphire, ruby
10	Diamond	Synthetic diamond

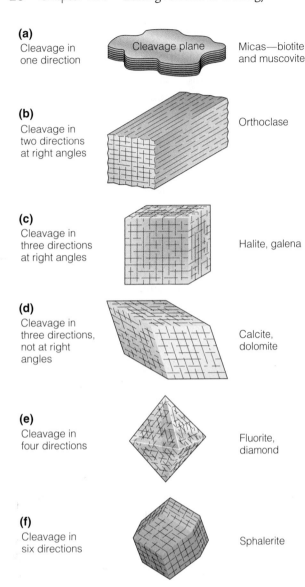

(a)
Cleavage in
one direction

Cleavage plane

Micas—biotite
and muscovite

(b)
Cleavage in
two directions
at right angles

Orthoclase

(c)
Cleavage in
three directions
at right angles

Halite, galena

(d)
Cleavage in
three directions,
not at right
angles

Calcite,
dolomite

(e)
Cleavage in
four directions

Fluorite,
diamond

(f)
Cleavage in
six directions

Sphalerite

◆ FIGURE 2.10 Types of cleavage and typical minerals in which they occur.

igneous rock may be eroded to form sediment that subsequently becomes a sedimentary rock, which may then be metamorphosed by heat and pressure to become a metamorphic rock, and then melted to become an igneous rock.

Throughout this book we emphasize those rocks that, because of their composition or structure, are involved in geologic events that endanger human life or well-being or that are important as resources.

CONSIDER THIS *Your school district is told it must remove all asbestos insulation from its older buildings at considerable cost to the taxpayers, and at the expense of teachers' salaries and educational materials. What questions should be asked of the government agency requiring this action, and what position will you take if you learn that the asbestos is the long-chain, fibrous type, rather than the sheet-structure, mica type (Case Study 2.1)?*

Igneous Rocks

Igneous rocks are classified according to their texture and their mineral composition. A rock's texture is a function of the size and shape of its mineral grains, and for igneous rocks this is determined by how fast or slowly a melted mass cools. If **magma**—molten rock within the earth—cools slowly, large crystals develop, and a rock with coarse-grained, **phaneritic texture** such as granite is formed (see Figure 2.8). The resulting large rock mass formed within the earth is known as a **batholith.** Greater than 100 square kilometers (39 mi²) in area by definition, batholiths are mostly granitic in composition and show evidence of having invaded and pushed aside the *country rock* into which they intruded. Also, many batholiths have formed in the cores of mountain ranges during mountain-building episodes. Batholiths are a type of igneous mass referred to as *plutons*, and rocks of batholiths are described as **plutonic** (after Pluto, the Greek god of the underworld), because they were formed at great depth. The Sierra Nevada of California is an example of an uplifted and eroded mountain range with an exposed core composed of many plutons (◆Figure 2.13).

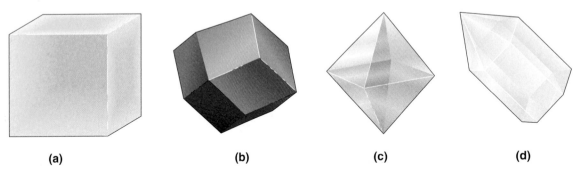

(a)   **(b)**   **(c)**   **(d)**

◆ FIGURE 2.11 Mineral crystal shapes: (a) cube (halite), (b) 12-sided dodecahedron (garnet), (c) 8-sided octahedron (diamond, fluorite), and (d) 6-sided hexagonal prism (quartz).

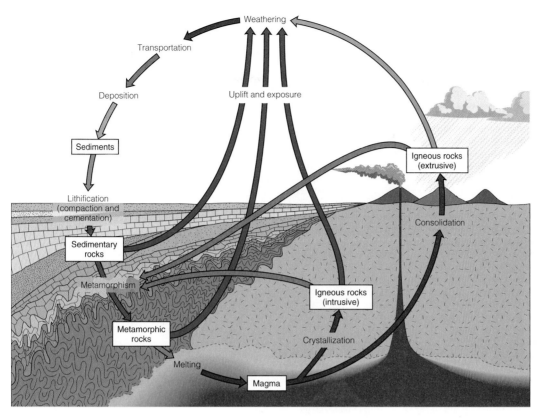

◆ FIGURE 2.12 The rock cycle. The three rock types are interrelated by internal and external processes involving the atmosphere, ocean, biosphere, crust, and upper mantle.

Most igneous rocks—whether on Earth, Moon, or Mars—are composed of silicate-mineral-containing silicon, oxygen, and aluminum. Silicon combined with oxygen is the compound *silica*, SiO_2. The percentage of silica in an

◆ FIGURE 2.13 The Sierra Nevada batholith and Mount Whitney, looking west from Owens Valley; east-central California.

igneous rock is a measure of how "distilled," or fractionated, its original magma was. The higher the percentage of silica in an igneous rock, the more steps of fractionation it has gone through. Minerals that are rich in magnesium and iron and low in silica (minerals such as pyroxene, amphibole, and olivine) will crystallize and settle out of the magma first. The resulting rocks, relatively rich in magnesium and iron and poor in silica, are thus designated **mafic** (*ma-* for magnesium, *-fic* for iron). The remaining magma has less magnesium and iron and a greater percentage of silica. When the next batch of minerals settles out, perhaps with quartz crystals, rocks of *intermediate* composition are formed. During the fractionation process the magma becomes richer and richer in silica and more impoverished in magnesium and iron. This late-stage magma is largely silica, alumina, and other elements that combine at lower temperatures. The resulting rocks are rich in feldspar and silica, and are thus described as **felsic** (*fel-* for feldspar, *-sic* for silica). "Primitive" magmas (<50% SiO_2) produce mafic gabbro and its volcanic equivalent, basalt, the rock common to Hawaii volcanoes. More "evolved" magmas (>50% SiO_2) produce rocks of intermediate composition, such as diorite and its volcanic equivalent, andesite. Highly evolved magmas produce felsic rocks such as granite and its volcanic equivalent, rhyolite. Mafic rocks are black or dark-colored, and felsic rocks are whitish and pinkish. Intermediate

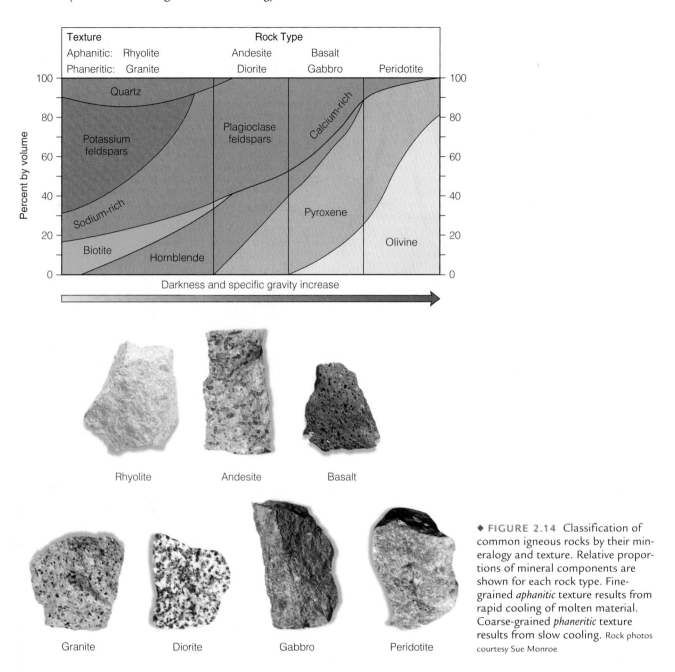

Rhyolite Andesite Basalt

Granite Diorite Gabbro Peridotite

◆ **FIGURE 2.14** Classification of common igneous rocks by their mineralogy and texture. Relative proportions of mineral components are shown for each rock type. Fine-grained *aphanitic* texture results from rapid cooling of molten material. Coarse-grained *phaneritic* texture results from slow cooling. Rock photos courtesy Sue Monroe

igneous rocks vary in color between these extremes according to their percentage of silica.

Lava is molten material at the earth's surface produced by volcanic activity. Lava cools rapidly, resulting in restricted crystal growth and a fine-grained, or **aphanitic,** texture. An example of a rock with aphanitic texture is rhyolite, which has about the same composition as granite but a much finer texture. The grain size of an igneous rock can tell us whether it is an **intrusive** (cooled within the earth) or an **extrusive** (cooled at the surface of the earth) rock. The classification by texture and composition of igneous rocks is shown in ◆Figure 2.14. Note that for each composition there are pairs

of rocks that are distinguished by their texture. Granite and rhyolite, diorite and andesite, and gabbro and basalt are the most common pairs found in nature. The end member shown in Figure 2.14, peridotite, is rarely found because it originates deep within the earth. Three important and relatively common glassy (composed of amorphous or finely crystalline SiO_2) volcanic rocks are obsidian, which looks like black glass; pumice, which has the composition of glass but is not really glassy-looking; and tuff, consolidated volcanic ash or cinders (◆Figure 2.15). Tuff has a **pyroclastic** (literally "fire-broken") texture, resulting from fragmentation during violent volcanic eruptions.

(a)

(b)

(c)

◆ **FIGURE 2.15** Extrusive igneous rocks: (a) glassy obsidian; (b) pumice, which is formed from gas-charged magmas; and (c) tuff. The lavas that form obsidian and pumice cool so rapidly that crystals do not grow to any size. The reference scale in each photo is 5 centimeters long; 5 centimeters is about 2 inches.

Sedimentary Rocks

Sediment is particulate matter derived from the physical or chemical weathering of materials of the earth's crust and by certain organic processes. It may be transported and redeposited by streams, glaciers, wind, or waves. Sedimentary rock is sediment that has become **lithified**—turned to stone—by pressure from deep burial, by cementation, or by both of these processes. **Clastic sedimentary rocks** are those that are composed of *clasts,* fragments of preexisting rocks and minerals. Clastic sedimentary rocks are classified according to their grain size. Thus sand-size sediment lithifies to sandstone, clay to shale, and gravel to conglomerate (◆Figures 2.16 and 2.17 a, b, and c). Other sediments result from chemical and biological activity. **Chemical sedimentary rocks,** which may be clastic or nonclastic, include chemically precipitated limestone, rock salt, and gypsum. **Biogenic sedimentary rocks** are produced directly by biological activity, such as coal (lithified plant debris), some limestone and chalk ($CaCO_3$, shell material), and chert (siliceous shells) (◆Figure 2.17 d and e). ✳ Table 2.4 presents the classification of sedimentary rocks.

A distinguishing characteristic of most sedimentary rock is bedding, or **stratification** into layers. Because shale is very thinly stratified, or *laminated,* it splits into thin sheets. Some sedimentary rocks, such as sandstone and limestone, may occur in beds that are several feet thick. Other structures in these rocks give hints as to how the rock formed. *Cross-bedding*—stratification that is inclined at an angle to the main stratification—indicates the influence of wind or water currents. Thick cross-beds and frosted quartz grains in sandstone indicate an ancient desert sand-dune environment (◆Figure 2.18). Some shale and claystone show polygonal *mud cracks* on bedding planes, similar to those found on the surface of modern dry lakes; these indicate desiccation in a subaerial environment (◆Figure 2.19).

CONSIDER THIS *Forensic geologists are specialists that are sometimes called upon to help solve crimes. They can, for example, compare the mineralogy of soil on a suspect's shoes with the soil at the scene of the crime. Also, they use geologic clues to reconstruct events that happened millions of years ago. For instance, huge round scars (craters) on the surface of the earth can indicate ancient meteorite impacts. What are some of the "clues" in sedimentary rocks that tell us about the ancient environments in which they formed?*

B. PIPKIN

◆ FIGURE 2.16 How sediments are transformed into clastic sedimentary rocks. The sequence of lithification is *deposition* to *compaction* to *cementation* to hard rock.

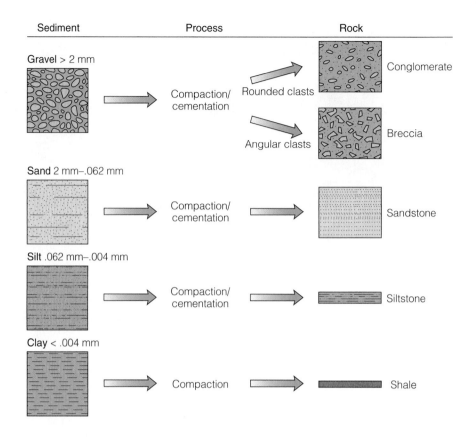

Sediment	Process	Rock
Gravel > 2 mm	Compaction/cementation	Rounded clasts → Conglomerate
		Angular clasts → Breccia
Sand 2 mm–.062 mm	Compaction/cementation	Sandstone
Silt .062 mm–.004 mm	Compaction/cementation	Siltstone
Clay < .004 mm	Compaction	Shale

Ripple marks, low, parallel ridges in deposits of fine sand and silt, may be asymmetrical or symmetrical (◆Figure 2.20). Asymmetrical ripples indicate a unidirectional current; symmetrical ones, the back-and-forth motion produced by waves in shallow water. Only in sedimentary rock do we find abundant fossils, the remains or traces of life.

(a)

(b)

5 CM.

(c)

5 CM.

(d)

(e)

◆ FIGURE 2.17 Common *clastic* sedimentary rocks: (a) shale, (b) sandstone, (c) conglomerate. *Biogenic* sedimentary rocks: (d) limestone composed entirely of shell materials (called *coquina*), (e) coal.

B. PIPKIN

*TABLE 2.4 Classification of Sedimentary Rocks		
Detrital Sedimentary Rocks (clastic texture)		
Sediment	**Description**	**Rock Name**
Gravel (>2.0 mm)	Rounded rock fragments	Conglomerate
	Angular rock fragments	Breccia
Sand (0.062–2.0 mm)	Quartz predominant	Quartz sandstone
	>25% feldspars	Arkose
Silt (0.004–0.062 mm)	Quartz predominant, gritty feel	Siltstone
Clay, mud (<0.004 mm)	Laminated, splits into thin sheets	Shale
	Thick beds, blocky	Mudstone
Chemical Sedimentary Rocks		
Texture	**Composition**	**Rock Name**
Clastic	Calcite ($CaCO_3$)	Limestone
	Dolomite [$CaMg(CO_3)_2$]	Dolostone
Crystalline	Halite (NaCl)	Rock salt
	Gypsum ($CaSO_4 \cdot 2H_2O$)	Rock gypsum
Biogenic Sedimentary Rocks		
Texture	**Composition**	**Rock Name**
Clastic	Shell calcite, skeletons, broken shells	Limestone, coquina
	Microscopic shells ($CaCO_3$)	Chalk
Nonclastic (altered)	Microscopic shells (SiO_2), recrystallized silica	Chert
	Consolidated plant remains (largely carbon)	Coal

Metamorphic Rocks

Rocks that have been changed from preexisting rocks by heat, pressure, or chemical processes are classified as metamorphic rocks. The process of metamorphism results in new structures, textures, and minerals. **Foliation** (Latin *folium,* "leaf") is the flattening and layering of minerals by nonuniform stresses. Foliated metamorphic rocks are classified by the development of this structure. Slate, schist, and gneiss are foliated, for example,

◆ FIGURE 2.18 Wind-blown sandstone exhibiting large cross-beds; Zion National Park, Utah.

◆ FIGURE 2.19 Mud cracks in an old clay mine; Ione, California.

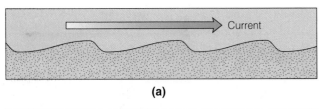

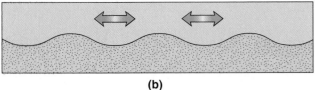

◆ FIGURE 2.20 Ripple marks found in sedimentary rocks. (a) Asymmetrical ripple marks are common in streambeds and are also found on bedding planes in sedimentary rocks. (b) Symmetrical ripple marks are due to oscillating water motion; that is, wave action.

(a)

(b)

(c)

◆ FIGURE 2.21 Metamorphic rocks: (a) coarsely foliated gneiss, (b) mica schist showing wavy or crinkly foliation, (c) finely foliated slate showing slaty cleavage.

and they are identified by the thickness or crudeness of their foliation (◆Figure 2.21). Metamorphic rocks may also form by **recrystallization.** This occurs when a rock is heated and strained by uniform stresses so that larger, more perfect grains result or new minerals form. In this manner a limestone may recrystallize to marble or a quartz sandstone to quartzite, one of the most resistant of all rocks (◆Figure 2.22). ✳ Table 2.5 summarizes the characteristics of common metamorphic rocks.

Rock Defects

Some rocks have structures that geologists view as "defects"; that is, surfaces along which landslides or rockfalls may occur. Almost any planar structure, such as a bedding plane in sedimentary rock or a foliation plane in metamorphic rock, holds potential for rock slides or falls. The orientation of a plane in space, such as a stratification plane, a fault, or a joint, may be defined by its **dip** and **strike** (Appendix 3). **Joints** are rock fractures without displacement and they occur in all rock types. They are commonly found in parallel sets spaced several feet apart (◆Figure 2.23). **Faults** are also fractures in crustal rocks, but they differ from joints in that some movement or displacement has occurred along the fault surface. Faults also are found in all rock types and are potential surfaces of "failure." We will examine the relationships of rock defects to geologic hazards when we discuss landslides, subsidence, and earthquakes.

CONSIDER THIS *Rock climbing is a popular sport in many parts of the world. A favorite spot in the U.S. West is Yosemite Valley, a picturesque area carved by glaciers into a uniform granite body known as the Cathedral Peak Granite (see Gallery Figure 3). Why do you suppose climbers would prefer to climb near-vertical cliffs of granite, as opposed to similar cliffs composed of schist, gneiss, or shale?*

Geologic Time

Interest in extremely long periods of time sets geology and astronomy apart from other sciences. Geologists think in terms of billions of years for the age of the earth and its oldest rocks—numbers that, like the national debt, are not easily comprehended. Nevertheless, the time scales of geologic activity are important for environmental geologists because they provide a way to measure human impacts on the natural world. For example, we would like to know the rate of natural soil from solid rock to determine whether topsoil erosion from agriculture is too great or not. Likewise, understanding how climate has changed over millions of years is vital to properly assess

(a)

Metamorphism

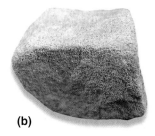
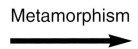

(b)

Metamorphism

◆ FIGURE 2.22
(a) Limestone recrystallizes to form white marble. (b) Quartzite, the hardest and most durable common rock, is metamorphosed sandstone. Rocks courtesy Sue Monroe

current global warming trends. Clues to past environmental change are well-preserved in many different kinds of rocks.

Geologists evaluate the age of rocks and geologic events using two different approaches. **Relative age dating** is the technique of determining *a sequence of geological events,* based upon the structural relations of rocks. **Absolute age dating** provides the *actual ages* for rocks in years before the present. Relative age is determined by applying geologic laws based upon the structural relations of rocks. For instance, the Law of Superposition tells us that in a stack of undeformed sedimentary rocks, the stratum (layer) at the top is the youngest. The Law of Cross-cutting Relationships tells us that a fault is younger than the youngest rocks it displaces or cuts. Similarly, we know that a pluton is younger than the rocks it intrudes (◆Figure 2.24).

Using these laws, geologists arranged a great thickness of sedimentary rocks and their contained fossils representing an immense span of geologic time. The geologic age of a particular sequence of rock was then determined by applying the Law of Fossil Succession, the observed chronologic sequence of life-forms through geologic time. This allows fossiliferous rocks from two widely separated areas to be correlated by matching key fossils or groups of fossils found in the rocks of the two areas (Figure 2.24). Using such indicator fossils and radioactive dating methods, geologists have developed the geologic time scale to chronicle the documented events of earth history (◆Figure 2.25). Note that the scale is divided into units of time during which rocks were deposited, life evolved, and significant geologic events such as mountain building occurred. Eons are the longest time intervals, followed, respectively, by eras, periods, and epochs. The Phanerozoic ("revealed life") Eon began 570 million years ago with the Cambrian Period, the rocks of which contain the first extensive fossils of organisms with hard skeletons. Because of the significance of the Cambrian Period, the informal term *Precambrian* is widely used to denote the time before it, which extends back to the formation of the earth 4.6 billion years ago. Note that the Precambrian is

✳TABLE 2.5	Common Metamorphic Rocks	
Rock	**Parent Rock**	**Characteristics**
Foliated or Layered		
Slate	Shale and mudstone	Splits into thin sheets
Schist	Fine-grained rocks, siltstone, shale, tuff	Mica minerals often crinkled
Gneiss	Coarse-grained rocks	Dark and light layers of aligned minerals
Nonfoliated or Recrystallized		
Marble	Limestone	Interlocking crystals
Quartzite	Sandstone	Interlocking, almost fused quartz grains

◆ FIGURE 2.23 Joints in basalt flows near Chico, California.

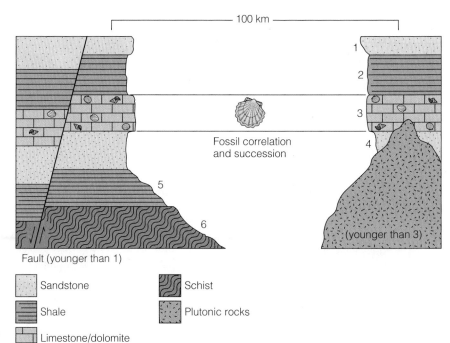

◆ FIGURE 2.24 Geologic cross sections illustrating the Laws of Superposition, Cross-cutting Relationships (intrusion and faulting), and Fossil Succession. The limestone beds are correlative because they contain the same fossils. The numerals indicate relative ages, with 1 being the youngest.

divided into the Archean and Proterozoic Eons, with the Archean Eon extending back to the formation of the oldest known in-place rocks about 3.9 billion years ago. Precambrian time accounts for 88 percent of geologic time, and the Phanerozoic for a mere 12 percent. The eras of geologic time correspond to the relative complexity of life forms: Paleozoic (oldest life), Mesozoic (middle life), and Cenozoic (most recent life). Environmental geologists are most interested in the events of the past few million years, a mere heartbeat in the history of the earth.

Absolute age-dating requires some kind of natural clock. The ticks of the clock may be the annual growth rings of trees or established rates of disintegration of radioactive elements to form other elements. At the turn of the twentieth century, American chemist and physicist Bertram Borden Boltwood (1870–1927) discovered that the ratio of lead to uranium in uranium-bearing rocks increases as the rocks' ages increase. He developed a process for determining the age of ancient geologic events that is unaffected by heat or pressure— **radiometric dating.** The "ticks" of the radioactive clocks are radioactive decay processes—spontaneous disintegrations of the nuclei of heavier elements such as uranium and thorium to lead. A radioactive element may decay to another element, or to an isotope of the same element. This decay occurs at a precise rate that can be determined experimentally. The most common emissions are alpha particles ($^4_2\alpha$), which are helium atoms, and beta particles (ß^-), which are nuclear electrons. New *radiogenic* "daughter" elements or isotopes result from this **alpha decay** and **beta decay** (◆Figure 2.26).

For example, of the three isotopes of carbon, ^{12}C, ^{13}C, and ^{14}C, only ^{14}C is radioactive, and this radioactivity can be used to date events between a few hundred and a few tens-of-thousands of years ago. Carbon-14 is formed continually in the upper atmosphere by neutron bombardment of nitrogen, and it exists in a fixed ratio to the common isotope, ^{12}C. All plants and animals contain radioactive ^{14}C in equilibrium with the atmospheric abundance until they die, at which time ^{14}C begins to decrease in abundance and, along with it, the object's radioactivity. Thus by measuring the radioactivity of an ancient parchment, log, or piece of charcoal and comparing the measurement with the activity of a modern standard, the age of archaeological materials and geological events can be determined (◆Figure 2.27). Carbon-14 is formed by the collision of cosmic neutrons with ^{14}N in the atmosphere, and then it decays back to ^{14}N by emitting a nuclear electron (ß^-).

$$^{14}_7N + \text{neutron} \longrightarrow \; ^{14}_6C + \text{proton}$$
$$^{14}_6C \longrightarrow \; ^{14}_7N + \text{ß}^-$$

Both carbon—the radioactive "parent" element—and nitrogen—the radiogenic "daughter"—have 14 atomic mass units. However, one of ^{14}C's neutrons is converted to a proton by the emission of a beta particle, and the carbon changes (or *transmutes*) to nitrogen. This process proceeds at a set rate that can be expressed as a **half-life,** the time required for half of a population of radioactive atoms to decay. For ^{14}C this is about 5,730 years.

Radioactive elements decay exponentially; that is, in two half-lives one-fourth of the original number of atoms remain, in three half-lives one-eighth remain, and so on (◆Figure 2.28). So few parent atoms remain after seven or eight half-lives (less than 1 percent) that experimental uncertainty creates limits for the various radiometric dating methods.

Whereas the practical age limit for dating carbon-bearing materials such as wood, paper, and cloth is about 40,000 to 50,000 years, ^{238}U disintegrates to ^{206}Pb and has a

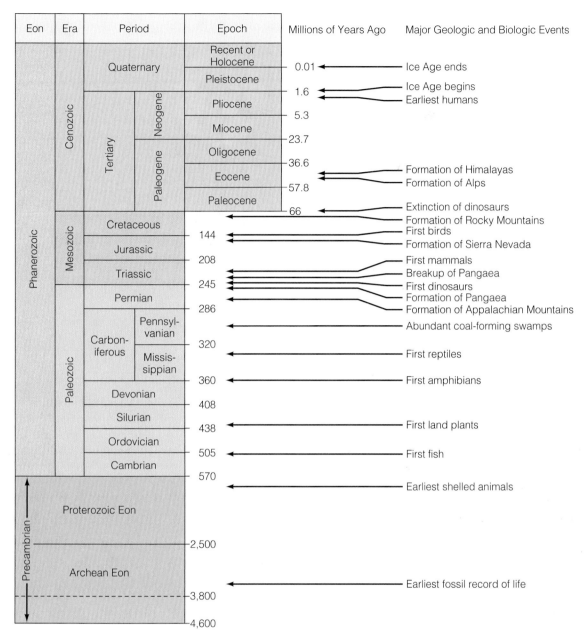

◆ FIGURE 2.25 Geologic time scale. Adapted from A. R. Palmer, Geology, 1983, 504.

half-life of 4.5×10^9 years. Uranium-238 emits alpha particles (helium atoms). Since both alpha and beta disintegrations can be measured with a Geiger counter, we have a means of determining the half-lives of a geological or archaeological sample. ✳ Table 2.6 shows radioactive parents, daughters, and half-lives commonly used in age-dating.

Age of the Earth

Before the advent of radiometric dating, determining the age of the earth was a source of controversy between established religious interpretations and early scientists. Archbishop Ussher (1585–1656), the Archbishop of Armagh and a professor at Trinity College in Dublin, declared that the earth was formed in the year 4004 B.C. Ussher provided not only the year but also the day, October 23, and the time, 9:00 A.M. Ussher has many detractors, but Stephen J. Gould of Harvard University, though not proposing acceptance of Ussher's date, was not one of them. Gould pointed out that Ussher's work was good scholarship for its time, because other religious scholars had extrapolated from Greek and Hebrew scriptures that the earth was formed in 5500 B.C. and 3761 B.C., respectively. Ussher based his age determination on the verse in the Bible that says "one day is with the Lord as a thousand years" (2 Peter 3:8). Because the Bible also says that God created heaven and earth in 6 days, Ussher arrived at 4004 years

◆ FIGURE 2.26 (a) Alpha emission, whereby a heavy nucleus spontaneously emits a helium atom and is reduced 4 atomic mass units and 2 atomic numbers. (b) Emission of a nuclear electron (ß⁻ particle), which changes a neutron to a proton and thereby forms a new element without a change of mass.

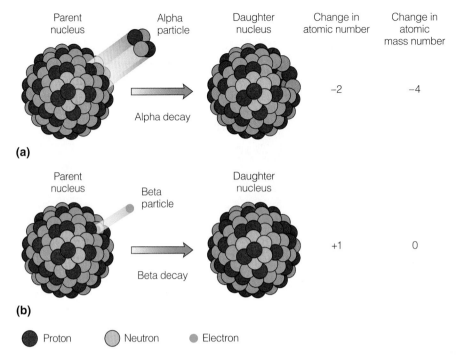

B.C.—the extra four years because he believed Christ's birth year was wrong by that amount of time. Ussher's age for the earth was accepted by many as "gospel" for almost 200 years.

By the late 1800s geologists believed that the earth was on the order of 100 million years old. They reached their estimates by dividing the total thickness of sedimentary rocks (tens of kilometers) by an assumed annual rate of deposition (mm/year). Evolutionists such as Charles Darwin thought that geologic time must be almost limitless in order that minute changes in organisms could eventually produce the present diversity of species. Both geologists and evolutionists were embarrassed when the British physicist William Thomson (later Lord Kelvin) demonstrated with elegant mathematics how the earth could be no older than 400 million years, and maybe as young as 20 million years. Thomson based this on the rate of cooling of an initially molten earth and the assumption that the material composing the earth was incapable of creating new heat through time. He did not know about radioactivity, which adds heat to rocks in the crust and mantle.

✱TABLE 2.6 Isotopes Used in Age-Dating

Isotopes		Parent's Half-Life, Years	Effective Dating Range, Years	Material That Can Be Dated	
Parent	*Daughter*				
Uranium-238	Lead-206	4.5 billion	10 million to 4.6 billion	Zircon	
				Uraninite	
Uranium-235	Lead-207	704 million			
Thorium-232	Lead-208	14 billion			
Rubidium-87	Strontium-87	48.8 billion	10 million to 4.6 billion	Muscovite	
				Biotite	
				Orthoclase	
				Whole metamorphic or igneous rock	
Potassium-40	Argon-40	1.3 billion	100,000 to 4.6 billion	Glauconite	Hornblende
				Muscovite	Whole volcanic rock
				Biotite	
Carbon-14	Nitrogen-14	5,730 years	less than 100,000	Shell, bones, charcoal	

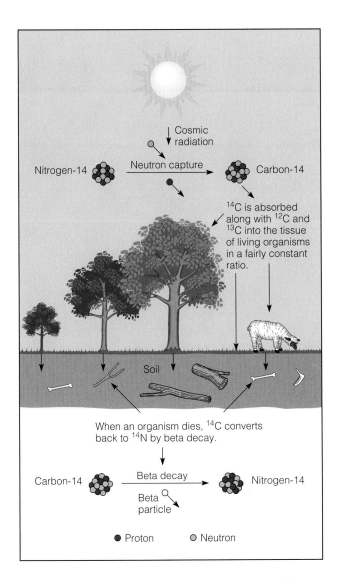

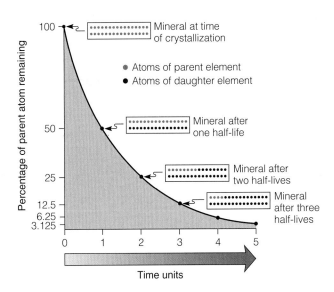

◆ FIGURE 2.28 Decay of a radioactive parent element with time. Each time unit is one half-life. Note that after two half-lives one-fourth of the parent element remains, and that after three half-lives one-eighth remains.

◆ FIGURE 2.27 Carbon-14 is formed from nitrogen-14 by neutron capture and subsequent proton emission. Carbon-14 decays back to nitrogen-14 by emission of a nuclear electron (ß⁻).

Bertram Boltwood postulated that older uranium-bearing minerals should carry a higher proportion of lead than younger samples. He analyzed a number of specimens of known relative age, and the absolute ages he came up with ranged from 410 million to 2.2 billion years old. These ages put Lord Kelvin's dates based on cooling rates to rest and ushered in the new radiometric dating technique. By extrapolating backward to the time when no radiogenic lead had been produced on earth, we arrive at an age of 4.6 billion years for the earth. This corresponds to the dates obtained from meteorites and lunar rocks, which are part of our solar system. The oldest known intact terrestrial rocks are found in the Acasta Gneiss of the Slave geological province of Canada's Northwest Territories. Analyses of lead to uranium ratios on the gneiss's zircon minerals indicate that the rocks are 3.96 billion years old. However, older detrital zircons on the order of 4.0–4.3 billion years old have been found in

western Australia, indicating that some stable continental crust was present as early as 4.3 billion years ago (✳ Table 2.7). Suffice it to say that the earth is very old and that there has been abundant time to produce the features we see today (◆Figure 2.29).

Environmental geology deals mostly with the present and the recent past—the time interval known as the *Holocene Epoch* (10,000 years ago to the present) of the *Quaternary Period*. However, because the Great Ice Age advances of the Pleistocene Epoch of the Quaternary Period (which preceded our own Holocene Epoch) had such a tremendous impact on the landscape of today, we will investigate the probable causes of the "ice ages" in order to speculate a bit about what the future might bring. In addition, Pleistocene glacial deposits are valuable sources of underground water, and they have economic value as sources of building materials. We will see that "ice ages" have occurred several times throughout geologic history (see Chapter 11).

What is most impressive about geologic time is how short the period of human life on earth has been. If we could

✳TABLE 2.7	Earth's Oldest Known Materials	
Material	**Location**	**Age, Billions of Years**
Crust	Zircon minerals in rocks of western Australia	4.0–4.3
Rock	Zircon minerals in the Acasta Gneiss, N.W. Terr., Canada	3.96
Sedimentary rock	Isua Greenstone Belt, Greenland	3.7–3.75
Fossils	Algae and bacteria	3.5

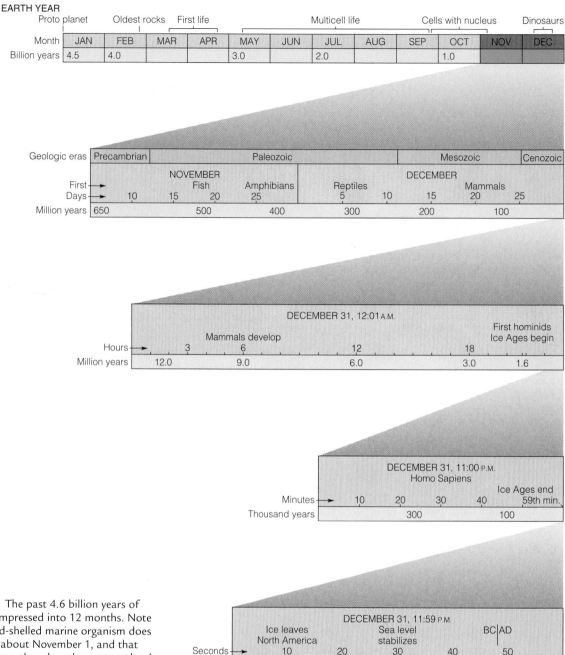

◆ FIGURE 2.29 The past 4.6 billion years of geologic time compressed into 12 months. Note that the first hard-shelled marine organism does not appear until about November 1, and that humans as we know them have been around only since the last hour before the New Year.

compress the 4.6 billion years since the earth formed into one calendar year, *Homo sapiens* would appear about 30 minutes before midnight on December 31 (see Figure 2.29). The last ice-age glaciers would begin wasting away a bit more than 2 minutes before midnight, and written history would exist for only the last 30 seconds of the year. Perhaps we should keep this calendar in mind when we hear that the dinosaurs were an unsuccessful group of reptiles. After all, they endured 50 times longer than hominids have existed on the planet to date. At the present rate of population growth, there is some doubt that humankind as we know it will be able to survive anywhere near that long.

The earth we see today is the product of earth processes acting on earth materials over the immensity of geologic time. The response of the materials to the processes has been the formation of mountain ranges, plateaus, and the multitude of other physical features found on the planet. Perhaps the greatest of these processes has been the formation and movement of platelike segments of the earth's surface, which influence the distribution of earthquakes and volcanic eruptions, global geography, mineral deposits, mountain ranges, and even climate. These *plate tectonic* processes are the subject of the next chapter.

Minerals, Cancer, and OSHA—Fact and Fiction

When an international health organization designated quartz as "probably" carcinogenic (cancer-causing) to humans, the U.S. Occupational Safety and Health Agency (OSHA) immediately went into action. It issued a mandate requiring products containing more than one-tenth of one percent (0.1%) of quartz or "free silica" (SiO_2) to display warning signs. Because quartz is chemically inert and because it is the most common mineral species, questions arose as to whether this requirement extended to sandpaper, beach sand, and unpaved roads that may produce clouds of fine-grained quartz dust. Questions—and more—arose when some Delaware truck drivers were "ticketed" for not displaying the required "quartz on board" signs during transport of crushed rock to construction sites. Such well-meaning OSHA actions have been perceived as ridiculous. They illustrate what happens when a government authority establishes health regulations without understanding the substances involved. In this case OSHA did not understand the where, how, and chemistry of the mineral—its mineralogy.

Unless one has had a long-term craving to inhale beach sand or crushed rock, there are no health hazards in the examples cited. Nonetheless, because exposure to very large quantities of any substance, even table salt or vitamins, can be harmful to health, it is certain that long-term workplace exposure to dusts from some minerals can pose health problems.

The World Health Organization's Group 1 classification of "known" carcinogenic minerals includes asbestos minerals (there are six), erionite (a zeolite), and minerals containing arsenic, chromium, and nickel. Its Group 2 of "probable" carcinogenic minerals includes all radioactive minerals and minerals containing lead, beryllium, and silica. The classification is based upon suspected relationships between specific diseases and workplace exposure to the identified substances and on laboratory experiments with animals.

A major human health concern is the relationship between lung cancer and inhaling the fine fibers of asbestos minerals. The long, fibrous crystal forms of asbestos belonging to the amphibole group (see hornblende structure, Figure 2.6) are considered the most carcinogenic. Of these, blue asbestos, crocidolite (pronounced krō-sid′-əl-īt; ◆Figure 1), is thought to be most dangerous and is believed to cause mesothelioma, a relatively rare cancer of the lining of the heart and lungs. Although crocidolite constitutes only 5 percent of all industrial asbestos, its morbidity statistics are grim indeed. For example, a 1990 survey of 33 men who, in 1953, worked in a factory where crocidolite was utilized in manufacturing cigarette filters found that 19 of them had died of asbestos-related diseases.

Common white asbestos, chrysotile (pronounced kris′-ə-tīl; ◆Figure 2), is a sheet-structure mineral similar to mica. It appears under high magnification as long fibers, which are in reality rolled sheets, much like long rolls of gift-wrapping paper. Chrysotile constitutes 95 percent of all industrial asbestos and has never been demonstrated to cause cancer at the levels found in most schools and public buildings. A study conducted in Thetford, Quebec, Canada, evaluated persons who had been exposed to high concentrations

CONTINUED

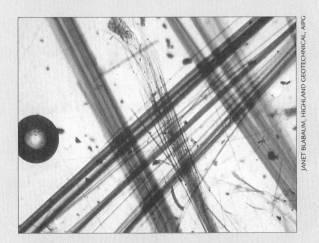

◆ FIGURE 1 Deadly crocidolite, "blue asbestos," magnified about 100 times. The mineral appears in its characteristic blue and green colors when viewed through a petrographic microscope. It was used in only a small percentage of commercial products.

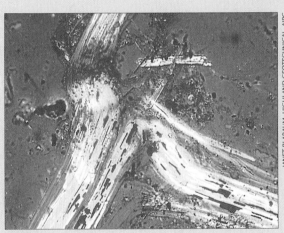

◆ FIGURE 2 Chrysotile asbestos (×100) has many uses in homes, offices, and public buildings. Although it is the target of removal in thousands of structures, at low levels it poses no documented risk. Note the difference between chrysotile and crocidolite's crystal structures.

JANET BLABAUM, HIGHLAND GEOTECHNICAL, AIPG

of white asbestos in their workplace and found no deaths attributable to it. Many of the diagnosed chrysotile-related lung problems have been traced back to improper handling of the material and inadequate respiratory protection during asbestos removal procedures. Experimental studies at Virginia Tech have shown that respirable-sized 1-nannometer-diameter chrysotile fibers dissolve in about nine months under conditions expected in lung tissue. This very short lifetime is difficult to reconcile with the general observations that asbestos-related diseases appear in asbestos workers only after many years on the job.

Inhaling finely divided airborne asbestos fibers in large quantities is definitely a health hazard. Under the Toxic Substance Control Act of 1972, the EPA mandated in 1986 that *all* asbestos minerals be treated as identical health hazards, regardless of their mineralogy, and that they be removed. By the mid 1990s the cost of removal had amounted to hundreds of millions of dollars, and the total cumulative cost for remediation of rental and commercial buildings, litigation, and enforcement was estimated at $100 billion. The EPA requirements were sweeping. They applied to buildings containing the relatively benign white asbestos and to those where airborne concentrations were so low that they could not be measured. The fear of asbestos and the EPA regulations are based on the idea that there is no safe "threshold" concentration—that even one fiber can kill. No doctor or public health official would agree with this. A New York school district's administration was besieged by irate parents when it admitted that it could not *guarantee* that *all* asbestos had been removed from its buildings. The district had spent millions of dollars to remove it.

Industry and government can ill afford to squander capital and tax dollars on poorly conceived hazard-mitigation requirements based upon the "no threshold" theory of health risk; that is, that one molecule of pesticide or one asbestos fiber can cause health problems. Many government regulations regarding "toxic substances" address risks to humans that are no greater than those incurred by routinely drinking several cups of coffee each day or eating a peanut butter sandwich for lunch. The "no threshold" criterion forces "carcinogen" classification upon otherwise benign minerals and many other useful substances and needlessly increases public anxiety. (Remember when synthetic sweeteners and cellular telephones were labeled carcinogenic?) The costs of removal and liability protection in such cases are horrendous, and a fiscal crisis in the environmental field is acknowledged. Liability extends to building owners, real estate agents who sell the buildings, banks that lend money for property purchases, and purchasers of land upon which any material classified as hazardous is found.

A person's chances of being struck by a lightning bolt are about 35 per 1 million lifetimes, and the risk of a nonsmoker's dying from asbestos exposure is about 1 in 100,000—about a third of the chance of being struck by lightning. Incidentally, the risk of dying from cigarette-related diseases is about 1 in 5 for smokers. Certainly, some substances in the environment pose dangers, and the key to risk reduction and longevity is awareness. Just as a reasonable person would not play golf during a violent thunderstorm because of the risk of being struck by lightning on the course, one should not handle hazardous substances without wearing protective gear.

CASE STUDY 2.2

A Girl's Best Friend and Other Gems

Rarity and beauty are what make precious and semiprecious gems so desirable. Those shown in ◆ Figure 1 are but a sample of several hundred valued gem minerals and their varieties. The largest uncut diamond in the world, the Cullinan diamond (◆ Figure 1a), was recovered in 1905 from the Premier Mine at Kimberley, Transvaal, South Africa; the huge, colorless stone weighed 3,106 carats (1.3 pounds). The smooth face of the Cullinan replica shown in the photograph replicates the stone's natural cleavage surface, which indicates that it was not the entire stone. Unfortunately, the remainder of the Cullinan lies undiscovered in a deep vol-canic pipe—which is where all diamonds originate. The Transvaal government bought the diamond for $1.6 million and gave it uncut to King Edward VII of England. After months of study, an Amsterdam diamond cutter's decisive blow yielded nine large stones and about 100 smaller ones, all of them flawless. All are now in the British crown jewels collection.

Diamonds are the high-pressure form of carbon and are the hardest natural substance known. Curiously, graphite, the high-temperature, low-pressure form of carbon, is among the softest minerals known.

Diamonds are found mostly in Africa, Australia, Russia, India, and South America (◆ Table 1). Near Yellowknife, Northwest Territories in Canada, the Ekati diamond mine opened in the late 1990s with great expectations. Diamond finds in the United States are shown on the map of ◆ Figure 2. Most are solitary finds, and their sources are unknown. At Murfreesboro, Arkansas, however, thousands of diamonds have been recovered from an eroded diamond "pipe," one weighing 40 carats. For the price of admission, one can dig for diamonds here in the "blue ground" of the eroded pipe. One lucky hunter uncovered a 15-carat stone.

Precious opal (Sanskrit *upâlâ*, "precious stone"; ◆ Figure 1b) occurs in shal-

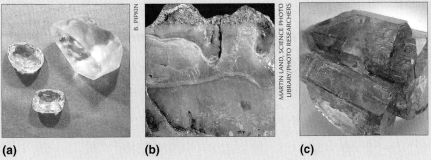

(a)   (b)   (c)

◆ FIGURE 1 (a) Exact replica of the Cullinan diamond with the two largest stones cut from it, the Great Star of Africa (530.2 carats) and the Second Star of Africa (317 carats), which now reside in the British crown jewels collection. (b) Uncut precious opal. (c) Uncut emerald.

* TABLE 1	Major World Diamond Producers, 1997*
Country	**Value, $ millions**
Botswana	$1,461
Russia	1,210
Angola	1,035
South Africa	971
Congo	715
Namibia	410
Australia	288

*Just seven countries account for 90% of all rough-diamond production.
Source: Larry Gordon, *Los Angeles Times*, November 4, 1998.

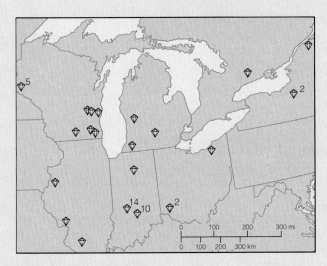

◆ FIGURE 2 Reported diamond finds in glacial drift in the Great Lakes region. Numerals indicate the number of diamonds found at the location; single finds have been reported at the other locations.
After C.B. Gunn, *Gems and Gemology* 12, nos. 10 and 11 (1968), pp. 297–303, 333–334, © Gemological Inst. of America

low surface diggings, and because it lacks crystalline form, it actually is not a mineral. It is amorphous hydrated silica and gets its "fire" from an orderly arrangement of silica spheres interspersed with water molecules. Black and fire opals are the most desirable, and each stone is priced individually.

Emerald, the most valuable mineral carat-for-carat, is the precious variety of the silicate mineral beryl (◆ Figure 1c). It is found in coarse-grained granites called *pegmatites* (see Chapter 13) and in mica schist.

Quartz, plain old SiO_2, one of the most common minerals, has a host of semiprecious varieties such as purple

amethyst, banded agate, yellow citrine, and smoky and rose quartz (◆ Figure 3). Searching for polishing-quality quartz varieties is the passion of many "rock hounds," and good crystal specimens also are cherished.

CONTINUED

◆ FIGURE 3 Some varieties of quartz: (a) colorless crystals, (b) smokey quartz, (c) amethyst, (d) agate, and (e) rose quartz.
(a) Smithsonian Institute; (b)–(e) Sue Monroe

(a)

(b)

(c)

(d)

(e)

Minerals, Cats, and the Litter Belt

Many of Lovell, Wyoming's 2,100 inhabitants once depended upon the oil industry for their livelihood. The people here mine bentonite, which is used in oil drilling. Mixed with water, bentonite forms a "mud" that is circulated through the drill pipe and back to the ground surface outside the pipe. It forms a filter cake on the walls of the hole that prevents entry of water and loss of the drilling fluid into porous formations. It also cools the drill bit as the drill churns through hard rock and carries the drilled rock debris up to the surface. Oil prices dropped in the late 1980s, and Lovell all but dried up (Chapter 14).

Bentonite is formed by geochemical alteration of volcanic ash to form clay minerals, the most important of which is *smectite*. One of smectite's unusual properties is that it can absorb water equivalent to many times its original volume. The clay may swell as it absorbs water, forming a gooey mess, or it may crumble into fluffy pellets, depending on the amount of water and the initial size of the clay aggregates. Land areas underlain by

bentonite may absorb so much water in the rainy season that they become impassable. Also, bentonitic soils can swell and cause severe damage to structures founded in it or on it (see Chapter 6). It has many uses. Native Americans once used it to remove oil and dirt from buffalo hides. It is used today in the landscaping of golf greens, to filter and clarify wine, to line landfills and ponds to prevent water loss, and in the manufacture of crayons.

◆ FIGURE 1 © 1994 MARGARET MILLER/ THE NATIONAL AUDUBON SOCIETY COLLECTION/PHOTO RESEARCHERS

Lovell was a potential ghost town without oil drilling. Bentonite to the rescue! It began when kitty litter replaced the sand box in 1987. The new product utilized a variety of clay-type minerals in the litter mix to absorb moisture. An enterprising chemist recognized the potential of bentonite, with its great moisture-absorbing properties, as cat litter. He reasoned that, with the proper grain size, the bentonite would clump when the cat did his duty, so that the soiled clumps could easily be removed from the litter pan. He was right, and the rest, as they say, is history (◆ Figure 1).

Today, almost all cat litter contains bentonite, and the Lovell mine is back in full-swing operation. The plant employs 60 people, and new houses are being built in the community to accommodate them. Bentonites in Wyoming are the result of volcanic eruptions in Cretaceous time, when dinosaurs roamed the land. The swath of mineable bentonites from Lovell to Casper has been dubbed the "Litter Belt" in Wyoming.

A Rock Collection

Specific rocks have had historical, aesthetic, and practical significance to humans; Plymouth Rock, the Rock of Gibraltar, the biblical Rock of Ages, and the Rosetta Stone are examples. Various geologic agents of erosion have shaped the rocks of the Grand Canyon, Bryce Canyon, Yosemite, and other national parks into areas that we admire for their great beauty. Early humans lived in rock caves and threw rocks to defend themselves against their enemies and to kill animals for food. Over time humans came to build crude rock shelters, and then rock fences, and eventually beautiful castles and abbeys. Today, rock and rock products are important mineral resources, used mostly as building materials. The beauty and usefulness of rocks are celebrated in these photographs.

(a)

◆ FIGURE 1 Ayers Rock, the world's largest sandstone monolith, is one of Australia's leading tourist attractions. An erosional remnant of a previous highland, it rises abruptly above the surrounding flat land. In geological terminology it is called an *inselberg* (German "island mountain") because of its resemblance to an island rising from the sea. Aborigines (native Australians) call Ayers Rock *Uluru*. In 1958 Ayers Rock and the rocks of the Olgas nearby were incorporated into Uluru National Park. (a) Ayers Rock rises 348 meters (1,141 ft) above the surrounding plain, and is scaled by 80 percent of the visitors to the park. (b) The Olgas, a circular grouping of more than 30 red domes, is a subdued but still spectacular sandstone inselberg in Uluru National Park. Its highest point, Mt. Olga, rises 460 meters (1,500 ft) above the plain. The Olgas' Aboriginal name is *katajuta,* "many heads." Ayers Rock is in the background.

(b)

FROM J. MONROE AND R. WICANDER, *THE CHANGING EARTH*, BROOKS/COLE, 1997.

◆ FIGURE 2 Sixty-foot-high representations of the heads of four presidents were carved into Mount Rushmore's Harney Peak Granite in southwestern South Dakota between 1927 and 1941. The great work was possible because the Precambrian granite is relatively homogeneous (uniform) and not riddled with defects. Why would such a project not have been considered if Mount Rushmore were composed of slate, shale, or schist? Can you identify the presidents honored here?

THEODORE ROOSEVELT COLLECTION, HARVARD COLL LIBRARY

◆ FIGURE 3 President Theodore Roosevelt (*left*) and naturalist John Muir at Yosemite Valley in 1903. The rock on which they were photographed is Cathedral Peak Granite, one of the many granite intrusions that make up the Sierra Nevada batholith. Note the rounded exfoliation domes in the background (see Chapter 6).

B. PIPKIN

◆ FIGURE 4 A skillfully assembled sandstone wall in Yorkshire, England. Yorkshire is the largest county (shire) in England, and literally thousands of kilometers of these walls are found there. This commonly occurring sandstone breaks readily into platy and blocky pieces, which makes it perfect for wall building.

Summary

Solar System

Description

Formed 5 billion years ago from an interstellar cloud (nebula).

Probable Cause

The nebula collapsed as a result of a gravitational attraction. The nebula started to spin and condense. The inner part of the cloud became the hot protosun and it was surrounded by a cooler, disc-shaped cloud, the solar nebula, which formed the large outer, Jovian planets. Rocky material closest to the sun formed the inner, terrestrial planets. Mars rocks are similar to Earth rocks.

Earth System

The earth system has five interconnecting subsystems, or reservoirs, sometimes called "spheres." They are the solid earth (includes lithosphere and soil), the atmosphere, the hydrosphere, the biosphere, and the extraterrestrial. Most are closed systems, such as the rock cycle and the hydrologic cycle, but energy is the exception. Continual interaction and recycling by these reservoirs is vital to the dynamic earth system.

Earth Materials

Crust

Outermost rocky layer of earth composed of rocks that are aggregates of minerals.

Minerals

Naturally occurring inorganic substances with a definite set of physical properties and a narrow range of chemical compositions.

Classification By

Chemistry of anions (negative ions or radicals), such as oxides, sulfides, carbonates, etc. Silicates are the most common and important mineral group and are composed of silica tetrahedra $(SiO_4)^{-4}$ units that are joined to form chains, sheets, and networks.

Identification

Physical properties such as hardness, cleavage, crystal structure, fracture pattern, and luster.

Rocks

Consolidated or poorly consolidated aggregates of one or more minerals or organic matter.

Classes

Igneous—rock formed by crystallization of molten or partially molten material.
Sedimentary—layered rock resulting from consolidation and lithification of sediment.
Metamorphic—preexisting rock that has been changed by heat, pressure, or chemically active fluids.

Classification by

Texture and composition.

Structures

Many planar rock structures may be viewed as defects along which landslides, rockfalls, or other potentially hazardous events may occur. Stratification in sedimentary rocks and foliation in metamorphic rocks are common planar features. Faults and joints occur in all classes of rock.

Rock Cycle

A sequence of events by which rocks are formed, altered, destroyed, and reformed as a result of internal and external earth processes.

Geologic Time

Relative Time

The sequential order of geologic events established by using basic geologic principles or laws.

Geologic Time Scale

The division of geologic history into eons, eras, periods, and epochs. The earliest 88 percent of geologic time is known informally as Precambrian time. *Precambrian* rocks are generally not fossiliferous. Rocks of the Cambrian Period and younger contain good fossil records of shelled and skeletonized organisms. Environmental geology is concerned mainly with geologic events of the present epoch, the Holocene, and the one that preceded it, the Pleistocene.

Absolute Age

Absolute dating methods use some natural "clock." For long periods of time, such clocks are known rates of radioactive decay. A parent radioactive isotope decays to a daughter isotope at a rate that can be determined experimentally. This established rate, expressed as a half-life, enables us to determine the age of a sample of material.

Age of the Earth

Comparing present U/Pb ratios with those in iron meteorites, we can extrapolate backward and obtain an earth age of 4.6 billion years old.

Rocks and Crust

The oldest known in-place rocks are 3.96 billion years old. Dating of minerals derived from older crust yields an age of 4.0–4.3 billion years for stable continental crust.

Key Terms

absolute age dating	chemical sedimentary	foliation	luster	radiometric dating
alpha decay	rocks	fracture	mafic	recrystallization
amorphous	clastic	half-life	magma	relative age dating
aphanitic texture	clastic sedimentary	igneous rocks	metamorphic rocks	rock cycle
atom	rocks	intrusive	mineral	rocks
atomic mass	cleavage	ion	Mohs hardness scale	sedimentary rock
atomic number	dip	isotope	nucleus	stratification
batholith	element	joint	phaneritic texture	strike
beta decay	extrusive	lava	plutonic	
biogenic sedimentary	fault	lithified	pyroclastic	
rocks	felsic			

Study Questions

1. What physical and chemical factors are the bases of the rock-classification system? What are the three classes of rocks, and how does each form? Identify one characteristic of each class that usually makes it readily distinguishable from the others.

2. How and where do batholiths form? What type of rock most commonly forms in batholiths?

3. What are some of the planar surfaces in rocks that may be weak and thus lead to various types of slope failure?

4. How may structures in sedimentary rocks be used to reconstruct past environments?

5. How may absolute age-dating techniques be used to the betterment of human existence?

6. The earth is 4.6 billion years old, and humans have been on earth for only a few hundred thousand years. What changes of a global nature have humans invoked on the earth's natural systems (water, air, ice, the solid earth, and biology) in this short length of time? Which impacts are reversible, and which ones cannot be reversed or mitigated?

7. What is the most common mineral species? Name several rocks in which it is a prominent constituent.

8. What is cleavage? How can it serve as an aid in identifying minerals?

9. The most common intrusive igneous rock is composed mostly of (1) the most common mineral and (2) a mineral of the most common mineral group. Name the rock and its constituent minerals.

10. What mineral is found in both limestone and marble?

For Further Information

Books and Periodicals

Albritton, C. C. 1984. Geologic time. *Journal of geological education* 32, no. 1: 29–47.

Bowring, S. A., I. S. Williams, and W. Compston. 1989. 3.96 Ga gneisses from the Slave province, Northwest Territories, Canada. *Geology* 17: 971–975.

Brown, V. M., and J. A. Harrell. 1991. Megascopic classification of rocks. *Journal of geological education* 39: 379.

Dietrich, R. V., and Brian Skinner. 1979. *Rocks and rock minerals.* New York: John Wiley and Sons.

Edgett, Ken. 1997. Mars Pathfinder, Sojourner, and a cast of rock stars captivate earthlings. *Earth in space* 10, no. 3: 1–16.

Eicher, D. L. 1976. *Geologic time,* 2d ed. Englewood Cliffs, N.J.: Prentice-Hall, Inc.

Gunter, Mickey Eugene. 1994. Asbestos as a metaphor for teaching risk perception. *Journal of geological education* 42: 17.

Harvey, Carolyn, and Mark Rollinson. 1987. *Asbestos in the schools.* New York: Praeger.

Hurley, Patrick. 1959. *How old is the earth?* New York: Anchor Books.

Libby, W. F. 1955. *Radiocarbon dating.* Chicago: University of Chicago Press.

Mackenzie, Fred T. G., and Judith A. Mackenzie. 1995. *Our changing planet: An introduction to earth system science and global environmental change.* Englewood Cliffs, N.J.: Prentice-Hall, Inc.

National Research Council, National Academy of Sciences. 1993. *Solid-earth sciences and society.* Washington, D.C.: National Research Council Commission on Geosciences, Environment, and Resources, National Academy of Sciences Press.

Newcott, William R. 1998. Return to Mars. *National geographic* 194, no. 2: 2–29.

Nuhfer, E. B., R. J. Proctor, and Paul H. Moser. 1993. *The citizen's guide to geologic hazards.* Arvada, Colo.: American Institute of Professional Geologists.

Parker-Pope, T. 1997. Cat litter breathes new life into region of bentonite mines. *The Wall Street journal,* April 1.

Skinner, H. Catherine W., and Malcom Ross. 1994. Geology and health. *Geotimes* (Washington, D.C.: American Geological Institute), January: 11–12.

———. 1994. Minerals and cancer. *Geotimes,* January: 13–15.

Snow, T. P. 1993. *Essentials of the dynamic universe.* St. Paul, Minn.: West Publishing Company.

Wicander, Reed, and James Monroe. 1999. *Essentials of Geology,* Pacific Grove, California: Brooks/Cole.

ENVIRONMENTAL
Geology⇌Now™

Assess your understanding of this chapter's topics with additional quizzing and comprehensive interactivities at http://earthscience.brookscole.com/pipkingeo4e as well as current and up-to-date Web links, additional readings, and InfoTrac College Edition exercises.

Plate Tectonics

"And the Walls Came Tumbling Down"

THE PROMINENT LINEATION DOWN THE CENTER of this satellite photo is the Dead Sea rift zone, an active tectonic plate margin (◆Figure 1). It is the Holy Land of biblical times and an area of political and religious conflict. Included in the view are the lands of Jordan, Israel, Syria, and part of Saudi Arabia and the Jordan River, the Mediterranean Sea, the Gulf of Aqaba, and the Dead Sea. The Dead Sea's shoreline is the lowest point on the earth's surface, averaging 396 meters (1,300 ft) below sea level. This valley was part of the sea 65 million years ago, and its desiccation may be the source of salt deposits in the region. The Dead Sea is fed by the Jordan River, but its level has dropped 19 meters (63 ft) in the past hundred years due to development along the river.

A well-known Old Testament story tells us that the Lord appointed Joshua to lead the Israelites to Canaan, the land of milk and honey. That "Promised Land" was in the vicinity of present-day Jericho. Jericho is at least 8,000 years old, the oldest almost continuously occupied city yet discovered by archaeologists. Karnak, Giza, and the tomb of Tutankamen are all young in comparison. The scriptures also tell us (in Joshua 6) that Joshua's orders were to march his army around the walled city for six days, blowing seven ram's-horn trumpets. On the seventh day, as the trumpets were blown, the walls of the city would fall flat, and the city could then be destroyed.

Jericho (see map) is in a rift valley formed by lateral slip on faults between the African plate to the west and the Arabian plate to the east (◆Figure 2). Although both plates are moving northward, the African plate is moving more slowly, so the Arabian plate's *relative* motion is to the north, and the African plate's is to the south (◆Figure 3). Motion of this type on faults results in a "pull-apart" basin, or **graben,** as shown in the figure. This explains how the floor of the valley came to be 400 meters (1,300 ft) below sea level (◆Figure 4).

The ruins of Jericho included warehouses that contained grain. It was not the Israeli *modus operandi* to attack and

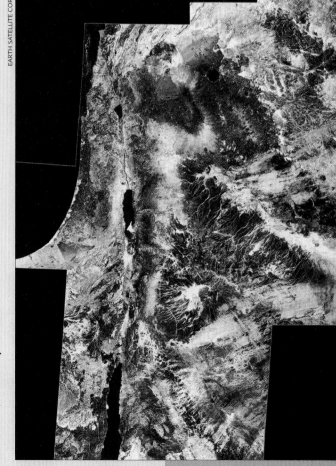

EARTH SATELLITE CORP.

◆ FIGURE 1 *Dead Sea rift zone*

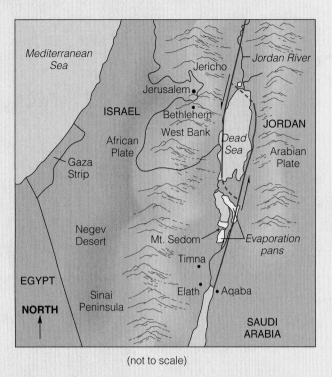

(not to scale)

◆ FIGURE 2 *The Dead Sea, Jordan River, and biblical villages and lands surrounding the sea.*

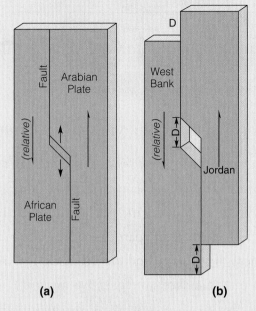

◆ FIGURE 3 *(a) The Dead Sea transform fault, a special variety of strike-slip fault, separates two northward-moving plates, the slower-moving African plate (left) and the faster-moving Arabian plate. (b) Overlapping faults between the West Bank and Jordan outline a pull-apart basin in which the Dead Sea resides. D indicates displacement on the fault.*

destroy after a grain harvest. They usually attacked just before grain was harvested so that a siege period would be effective and they could later harvest the grain for themselves.

Similar plate boundaries at other places in the world have produced destructive earthquakes.

Examples include California and Turkey. An alternative explanation for the Israelites' success at Jericho, and perhaps a more plausible one, is that the walls were felled by an earthquake caused by the Dead Sea faults so that the earth shook and the mud-brick "walls came tumbling down."

(a)

(b)

◆ FIGURE 4 (a) The Dead Sea viewed from the western shore in Israel. This is the lowest dry land in the world. (b) The Gulf of Aqaba from Aqaba, Jordan—an example of a seaway, or arm of the ocean that floods a young rift valley. Someday the gulf of Aqaba may grow so wide that one can no longer see all the way across it at this location.

Continental Drift

From the time that the first complete maps of the Atlantic Ocean were published in the seventeenth century, thoughtful minds (including Francis Bacon) were struck by the similarity of the coastlines of Africa and South America. Various explanations for this included a popular one that the Atlantic is a gigantic river bed cut by Noah's flood! The banks of rivers, after all, tend to be parallel to one another. By 1910, however, enough scientific evidence had collected to begin pondering this similarity more seriously. Alfred Wegener (1880–1930), a German meteorologist, presented a paper to the Frankfurt Geological Society in which he expressed his opinion that Africa and South America, together with other continents, had once been joined together in a single gigantic landmass—a supercontinent—that somehow split apart. The fragments then separated thousands of kilometers from one another over a period of about 100 million years to produce the world map we see today. He called this hypothesis *die Verschiebung der Kontinente,* meaning "continental displacement," which English-speaking scientists translated to **continental drift,** the name that stuck. Wegener had no well-grounded explanation for how continental drift took place. He suggested speculatively that the spin of the Earth might have something to do with it. In any event, he thought that as the Americas drifted westward *(Westwanderung)* from the Old World, their edges crumpled into the mountains of the Andes and Rockies. This deformation resembles that seen in a rug slid against a wall. Likewise, as Australia drifted apart from southern Africa the Great Dividing Range grew along the Australian east coast.

Wegener and many Southern Hemisphere geologists including Victor Linz cited as field evidence for continental drift the distribution of late Paleozoic (Permian Period) glacial deposits and plant fossils known collectively as the **Gondwana Succession** (named after the Gond tribe of central India, whose region contains many of these fossils). These distinctive ancient land plants (◆Figure 3.1a) and their associated glacial deposits (◆Figure 3.2) are found only in South America, Africa, India, Australia, and Antarctica, which comprised the southern part of the original supercontinent. Wegener named this giant ancient landmass **Pangaea** from the Greek word meaning "All Earth" (◆Figure 3.3).

One famous fossil associated with the break-up of Pangaea is the freshwater reptile *Mesosaurus,* which inhabited Permian rivers, lakes, and swamplands in what are now Brazil and South Africa (Figure 3.1b). It is inconceivable that these freshwater creatures swam or island-hopped across the salty Atlantic Ocean to an identical freshwater habitat on the opposite side. It seems much more reasonable that *Mesosaurus* inhabited a single large homeland that geographically broke apart.

The evidence isn't restricted to fossils and glacial deposits alone. Belts of mineral deposits, and groups of similar rocks of the same age—even whole pre-Mesozoic mountain ranges—can be fit back together hand-in-glove by reassembling the continents as Wegener imagined.

Had most of the world's geologists and academic institutions been located in the Southern Hemisphere, Wegener and his advocates might have been taken far more seriously. But in the Northern Hemisphere evidence for continental drift wasn't so clear-cut because the shorelines of continents there did not fit together as well. Enormously thick ice-age glaciers had pressed the land down in northern latitudes, causing the ocean to flood continental margins in a highly irregular way (just glance at a map of western Europe!). This made it difficult to discern the critical transoceanic linkages in fossils and other features that had reportedly been seen in the warmer latitudes south of the equator. Since geologists did not travel as widely to do research then as they do now, the development of knowledge was sluggish by today's standards. To make matters worse, Wegener was not even a certified geologist—the

PATRICIA GENSEL, UNIV. OF N. CAROLINA

(a)

(b)

◆ FIGURE 3.1 (a) Fossil leaves of the Gondwana Succession plant *Glossopteris,* from the Upper Permian Dunedoo Formation, Australia; (b) *Mesosaurus,* a freshwater reptile that inhabited Gondwana ecosystems.

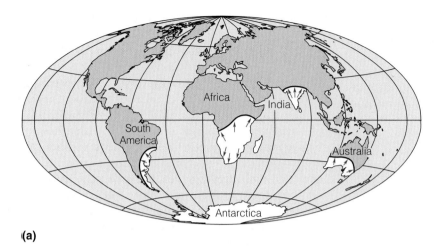

(a)

◆ FIGURE 3.2 (a) The directions of glacial grooves preserved in bedrock on continents as they are now positioned show the direction of glacier movement during Permian (Gondwana Succession) time. For a large glacier or a number of glaciers to produce the directions of the observed grooves would otherwise require a source area in the Indian and South Atlantic Oceans. (b) With Gondwana continents reunited and an ice sheet over a south pole in present-day South Africa, the directions of glacier motion are resolved. (c) Glacial grooves at Hallet's Cove, Australia. Formed in Permian time, these grooves are more than 200 million years old. The width of the area shown in the photo is approximately 2 meters.

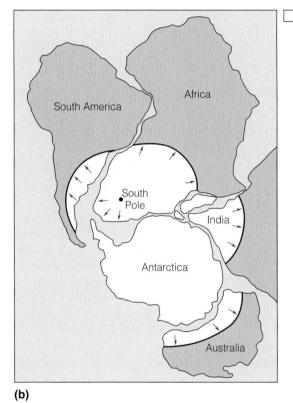

☐ Glaciated area
Arrows indicate the direction of glacial movement based on striations preserved in bedrock.

(b)

(c)

courageous German spent the last years of his life defending his hypothesis in scientific papers and five editions of his classic book *The Origin of Continents and Oceans.* He disappeared in 1930 on an expedition to Greenland, where, it is said, his theory was born decades earlier when he saw a giant iceberg float away from a calving glacier.

Wegener Vindicated

For various cultural reasons, American resistance to Wegener's hypothesis was much greater than it was in Europe. Despite this discouragement, Professor Arthur Holmes, a British geologist and possibly the finest geology teacher of his day, developed an explanation for continental drift in 1926–1927 that ultimately proved to be spectacularly correct! Holmes suggested that the continents don't drift by plowing through sea-floor crust, like ships moving through sea ice. Rather, the continents move apart as new ocean floor grows *in between* them. The Atlantic and Indian oceans began as weak seams fracturing Pangaea. These ocean basins grew wider due to the eruption of basaltic lava along the crests of **mid-ocean ridges,** the belt of broad volcanic high-

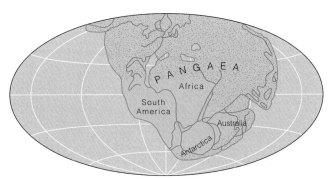

Late Carboniferous Period (285–300 million years ago)

ENVIRONMENTAL
◆ **Geology ⇌ Now**™ ACTIVE FIGURE 3.3 The earth of 285–300 million years ago according to Alfred Wegener. The supercontinent Pangaea began to break up about 100 million years ago, and the individual continents, roughly outlined as we know them today, have since then "drifted" to their present positions. The stippled areas represent widespread shallow seas over the continents in which sediments and fossils accumulated.

lands that bisect the sea floor worldwide like the seams on a giant baseball. He further speculated that very old oceanic crust must be destroyed by sinking back into the deep earth through **marine trenches,** the steep-walled abyssal valleys fringing some ocean floors. Internal convection, a churning of hot rock within the earth, could account for this activity. The evidence to test Holmes' bold explanation lay on the sea floor, a region largely out of reach in the 1920s, and not likely to get any closer without strong international collaboration and new technology.

As often happens in history, military conflict led to new inventions with unforeseen applications. The ship-borne magnetometer proved valuable in the detection of mines and Axis submarines threatening sea lanes during the Second World War. For the first time, too, a detailed map of the ocean floor could be made by means of echo sounding—a way of measuring the depth of the ocean by sending sound pulses to the bottom and retrieving the echoes on return. These techniques allowed oceanographers to map the seabed as never before, both bathymetrically and magnetically. The Cold War that followed provided many opportunities to apply this new technology cooperatively as Americans and Western Europeans sought military advantage and favor with Third World countries through prestigious scientific research.

The study of lava flows using the magnetometer in the 1950s led to several important breakthroughs. Each involved **magnetic polarity,** a property created by the alignment of tiny iron-bearing mineral grains within earth's magnetic field as molten lava cools into solid rock. The minerals act like natural compass needles, recording the direction toward the north magnetic pole existing at the time that they harden

into place. They are "fossil magnetic field indicators." Because the earth as a whole acts like a giant bar magnet, with a north (positive) magnetic pole presently situated in Northern Canada and a south (negative) pole near Antarctica, any change in the past orientation of the planetary magnetic field will be recorded in the polarity of ancient rocks formed just before and after a field change.

Geophysicists correctly assumed that the positions of earth's magnetic poles almost continuously drift within just a few degrees latitude of the planet's geographic (rotational) poles. But researchers discovered that the magnetic polarity of different equivalent-aged ancient rocks on opposite sides of the Atlantic gave *different locations* for each of the magnetic poles. In other words, there appeared to be *multiple* north poles and south magnetic poles each paired to a separate continent, and each with its own history of wandering through many latitudes across the face of the globe (◆Figure 3.4). Try as people might, this simply did not make any sense—there can only be two magnetic poles at any given time for a planetary magnetic field. However, if the Atlantic was made smaller and the continents rotated back closer to one another at specific increments of time, then the multiple-wandering-pole problem disappeared. This was indirect evidence of continental drift—but very compelling indeed!

Stimulated by the magnetic findings, U.S. geologist Harry Hess revived and advanced Holmes' speculations in 1960, to suggest a mechanism for continental drift that he called **sea-floor spreading**—the growth of new oceanic crust along the crests of mid-ocean ridges. He also supported the notion that oceanic crust must be destroyed as it grows old, cold, and heavy, nosediving back into the plant through marine trenches. Perhaps wary of Wegener's professional fate, Hess referred to this model as "geopoetry," but there was growing evidence that he was correct.

In addition to the magnetic data, seismology delivered some clues. In 1927, Japanese seismologist Kiyoo Wadati discovered that a zone of frequently occurring earthquakes, including some very destructive and powerful tremors, slants under Japan from the trench along the archipelago's eastern coast. Hugo Benioff later mapped similar zones worldwide, which are now called **Benioff–Wadati zones.** They mark the paths of sinking oceanic material extending hundreds of kilometers into the planet, mostly around the edge of the Pacific Basin (which is growing smaller as the Atlantic and Indian oceans grow wider). Fragments of broken slabs may actually settle several thousand kilometers, all the way to Earth's core! Geologists now call this process **subduction,** meaning "under-moving" (◆Figure 3.5). Further evidence for the correctness of the Holmes model comes from a simple consideration of the age of the sea floor. Whereas we've discovered continental rocks as old as 4 billion years, the oldest intact sea floor has an age of less than 200 million years. Where did all of the older material go, if it wasn't swallowed back up by the planet?

POLAR WANDERING VERSUS...

CONTINENTAL DRIFT

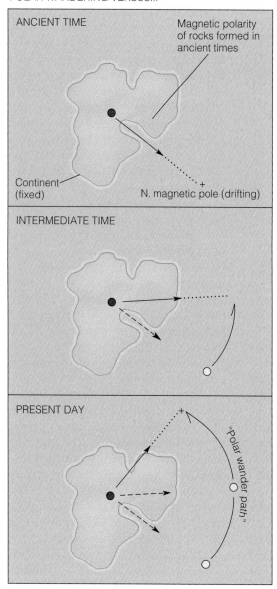

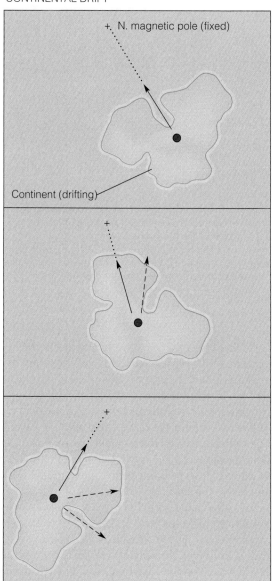

◆ FIGURE 3.4 Two interpretations from the 1950s of magnetic polarities found in ancient rocks. The column to the left presumes that the continents remained fixed as the magnetic poles drifted to account for the variation seen in rock magnetization over time. The column to the right presumes just the opposite—that the continents drift through a magnetic field with fixed poles. Both models lead to the same present-day magnetic pattern for any given continent.

Researchers were close to verifying continental drift, but they still needed more evidence. They found it again as they continued studying patterns of natural rock magnetism. Geophysicists discovered that every few hundred thousand to few million years something truly spectacular happens; Earth's magnetic poles swap positions entirely—in other words, the entire magnetic field suddenly flips over, and the poles become *reversed*. As a reversal approaches, the strength of the field simply weakens to the point of temporarily disappearing. When the field returns, each pole reappears in the opposite hemisphere from its former position. We regard the present time as having a "normal" magnetic field simply because of our place in time, with the field position "reversed" in other epochs (◆Figure 3.6). (One can only speculate about what happens during a reversal to birds and marine creatures that depend upon the magnetic field to assist their migrations!)

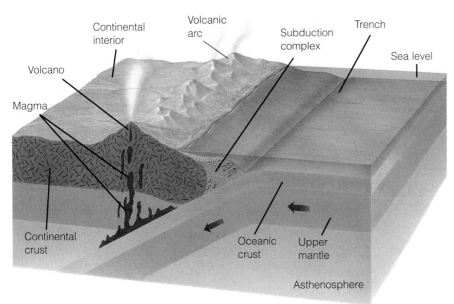

Continental interior

Volcanic arc

Subduction complex

Trench

Sea level

Volcano

Magma

Continental crust

Oceanic crust

Upper mantle

Asthenosphere

ENVIRONMENTAL

◆ Geology⇌Now™ ACTIVE FIGURE
3.5 The process of subduction is the sinking of the sea floor and a thick layer of underlying rock back into the planet. This process can generate huge earthquakes, and can create magma to feed a chain of volcanoes near the edge of the overthrusting landmass. Whereas seafloor spreading at mid-ocean ridges create oceanic crust, subduction destroys it.

From B. Murphy and D. Nance, *Earth Science Today*, 2nd ed., Brooks/Cole.

In 1963, British researcher Fred Vine and Drummond Matthews and Canadian Lawrence Morley proved that rocks that formed when earth's magnetic field was the same as today's yield a strong magnetic signal—or **positive anomaly**—whereas those that formed when the field was the opposite of today's field yield a weak signal—a **negative anomaly.** Vine, Matthews, and Morley reasoned that a striped anomaly pattern of sea-floor magnetization would develop if sea-floor spreading really happened. The pattern should be geometrically symmetrical about the mid-ocean ridges, assuming that sea-floor spreading rates are the same on both sides of a ridge crest. Oceanographers leapt to the challenge, and in 1966 confirmed the presence of this pattern in the northern Atlantic (◆Figure 3.7). Because over 170 reversals are known to have occurred during the past 76 million years, the striped magnetic record of sea-floor spreading is strikingly detailed, and geologists the world over gasped in unison.

All at once, the human race had its first holistic understanding of the physical world. Coupled with the first pictures of Earth from space, scientists began to look at our planet in terms of a *single system of integrated processes*, rather than an assortment of independent phenomena. This had important consequences politically as well as scientifically, including a surge in popular environmental interest. It is no coincidence that the first Earth Day took place in 1970.

The story of scientific breakthroughs leading to this revolution differs notably from that of Darwinian biology or Newtonian physics. Scientists from many diverse backgrounds independently provided vital clues to complete the picture. No one person can claim overwhelming credit for revolutionizing our understanding of the earth.

ENVIRONMENTAL
Geology⇌Now™

Click **Geology Interactive** to work through an activity on plate boundaries through Plate Tectonics.

Plate Tectonics

In 1926, Vening Meinesz, a Dutch surveyor, introduced the term **plate** to describe broad regions of structurally distinctive crust, referring specifically to the Caribbean plate in a study that he had done for the U.S. Navy. In the late 1960s, Canadian geophysicist J. Tuzo Wilson revived Meinesz's work and helped popularize the term **plate tectonics** to give people a quick way of describing our new view of the physical world. "Tectonics" derives from the Greek word, *tektonikos,* meaning "builder." The term "plate" refers to a rigid fragment of earth's outermost skin. Plate tectonics radicalized notions about the forces responsible for shaping earth's myriad landforms. Prior to the 1960s, most geologists thought primarily in terms of vertical force to account for the construction of mountain and valleys and other topographic features. But plate tectonics shows that vertical motions in earth's crust are mostly a product of far greater *horizontal* movements (◆Figure 3.8). One reason the new theory challenged many older geologists was that their thinking had to be turned literally 90°.

There are seven major plates making up the surface of the earth, and numerous smaller ones ranging in width from several hundred kilometers to only a few tens of kilometers

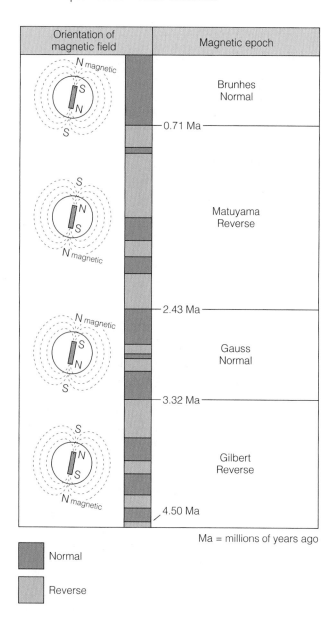

Orientation of magnetic field	Magnetic epoch
	Brunhes Normal
	— 0.71 Ma —
	Matuyama Reverse
	— 2.43 Ma —
	Gauss Normal
	— 3.32 Ma —
	Gilbert Reverse
	— 4.50 Ma —

Ma = millions of years ago

■ Normal

■ Reverse

◆ FIGURE 3.6 The earth's magnetic field is much like the field that would be generated by a bar magnet inclined at 11° from the earth's rotational axis. Reversals of the field have occurred periodically, leaving "fossil" magnetism in rocks that can be dated. Each epoch is named after an important student of earth magnetism.

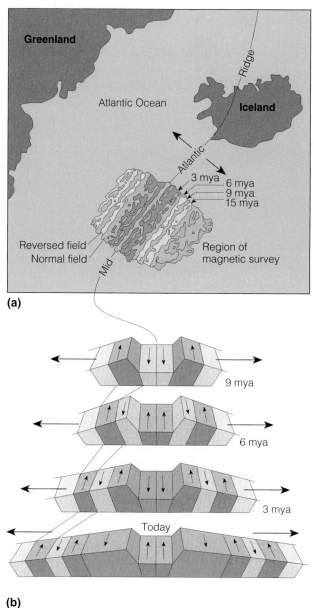

(a)

(b)

↑ = Normal magnetic polarity

↓ = Reversed magnetic polarity

← → = Direction of plate movement

mya = Millions of years ago

ENVIRONMENTAL
◆ Geology⇌Now™ ACTIVE FIGURE 3.7 (a) Symmetric pattern of magnetic variation across part of the Mid-Atlantic Ridge southwest of Iceland. Colors indicate sea-floor rocks with matching polarity. (b) The origin of the magnetic anomaly pattern across mid-ocean ridges. Lava erupting along the crest of the ridge records the earth's prevailing magnetic polarity as it cools and solidifies. Sea-floor spreading then carries the rock away from the ridge in both directions as new lava erupts on the ridge axis.

(◆Figure 3.9). The plates interact with one another along their margins. As you will learn, all of human civilization and its environmental conditions are vitally linked to what happens along these margins. Essentially, three types of plate interactions take place (◆Figure 3.10):

■ At **divergent boundaries** there is *tension;* stresses from the earth's interior act to move plates apart along these

boundaries. Divergence within the continental crust creates features such as the East African Rift Valley, while in the oceans, sea-floor spreading along mid-ocean ridges

(a)

(b)

(c)

ENVIRONMENTAL
◆ **Geology⇌Now™** ACTIVE FIGURE 3.8 Paleogeographic reconstruction of continents' movements over the past 180 million years (a) during the early Triassic Period, (b) late in the Cretaceous Period, and (c) as it is today. Note that the Panthalassa Ocean was the "one" ocean dominating the globe, as Pangaea was the "one" continent. The maps show present-day coastlines; ancient coastlines differed from these. Compare these maps with Figure 3.3, which was originally published in 1915 and revised by Wegener in 1929 just before his death.

takes place. *Seaways* represent an intermediate stage of development between continental and oceanic divergent boundaries (Figure 4 in the chapter opener and ◆Figure 3.11). Frequent small earthquakes and mild volcanic eruptions characterize divergent margins. Divergent boundaries are also called *spreading centers*.

■ At **convergent boundaries** there is *compression;* stresses converge to drive plates together. Where a plate capped with oceanic crust collides with a plate made up of continental material, as around much of the Pacific Rim, subduction takes place (Figure 3.5). This greatly disturbs the underlying mantle. The world's deadliest earthquakes and very explosive volcanic eruptions result from subduction. Volcanic mountain ranges such as the Andes mark the tectonic disturbance at the edge of the overriding plate along this convergent boundary.

■ Where *two* plates made up of continental material collide, a great range of mountains rises. The Himalaya and the Alps mountains formed in this way. Very powerful earthquakes often occur as the intracontinental mountains grow, but volcanic activity is essentially nonexistent, because the underlying hot mantle is little disturbed by this kind of ("Himalayan-style") convergence.

■ At **transform boundaries** *shear* forces cause plates to slide horizontally past one another (◆Figure 3.12). Long, linear faults mark transform boundaries in many places. Outstanding examples include the Alpine fault in New Zealand and the San Andreas fault in California. The strong earthquakes along both faults ease the world's biggest plate, the Pacific plate, past two other giant plates, the India–Australian and North American plates. Whereas destructive earthquakes are a strong risk from these faults, virtually no volcanic activity takes places along transform boundaries.

Plate boundaries may also be described by what happens to the crust at that plate margin. Where plates diverge, new ocean crust is created, and we have a *constructive* plate boundary. Where an oceanic plate sinks and disappears at a subduction zone, the plate boundary is described as *destructive*. Finally, at transform boundaries crust is neither created nor destroyed, and we have a *conservative* boundary. Because these designations describe how the margin is modified, they also serve as a memory aid for what processes take place there (◆Figure 3.13).

CONSIDER THIS *It is said that all knowledge of the natural world is potentially beneficial to humanity, even knowledge about past conditions that have subsequently changed. How might our knowledge of past and present plate positions and subsequent motions be applied to benefit the human race?*

ENVIRONMENTAL
Geology⇌Now™

Click **Geology Interactive** to work through an activity on
■ Plate boundaries and locations through Plate Tectonics
■ Distribution of Volcanism
■ Earthquakes and Tsunamis
■ Seismic Risk

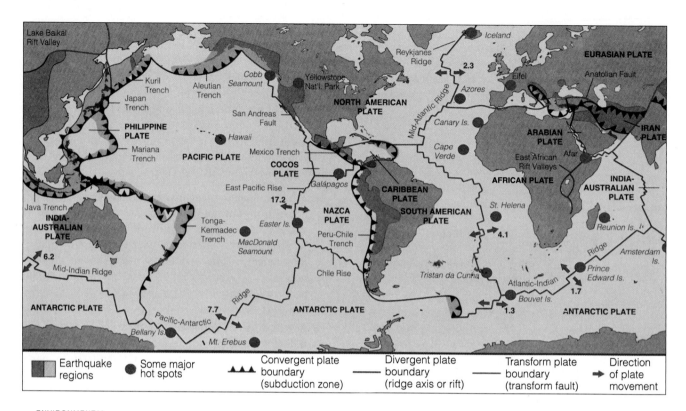

ENVIRONMENTAL
◆ Geology⇌Now™ ACTIVE FIGURE 3.9 The earth's major tectonic plates and types of plate boundaries. Regardless of the type of plate boundary, each one is delineated by earthquake epicenters (not shown). The arrows indicate the present direction of plate motion; the numbers, the rate of spreading in centimeters per year. The rate of spreading is determined by dividing the distance to a magnetic anomaly (or rock) of known age in kilometers ($\times 10^5$ = cm) by the age of that anomaly in years (= cm/y).

How Fast Do the Plates Move?

The rate at which the sea floor is spreading away from a mid-ocean ridge may be calculated by measuring the distance from the ridge to lava of known age that presumably erupted from that ridge. The magnetization of the volcanic rock can help verify its age and origin. Simply divide the distance traveled from the ridge by the age of the rock. That gives you the drift rate of the rock.

Plates typically move on the order of centimeters per year (about the same rate as human fingernail growth). For example, Honolulu, Hawaii, now lies about 6 meters (20 ft) farther west-northwest than it did a hundred years ago. The East Pacific Rise is spreading at the fastest rate, up to 18 centimeters (7 in) per year, whereas the northern Mid-Atlantic Ridge is slowest, at about 2 centimeters (3/4 in) per year (see Figure 3.9). It may be that ocean basins fringed with subduction zones, like the Pacific, tend to have faster spreading centers within them, as the sinking plates around the edges strongly tug at the centers of the basins. Techniques using very-long-baseline radio interferometry (VLBI), satellite laser ranging (SLR), and the global positioning system (GPS) are now providing direct measurements that confirm the longer-term magnetic measurements. As these techniques'

use expands, it will be possible to monitor surface deformation directly. This will contribute to our understanding of strain buildup that leads to earthquakes, volcanic eruptions, and plate movements.

> **CONSIDER THIS** *Remembering that the "earth system" is composed of many different cycles, such as the rock cycle and the supercontinent cycle, predict the general arrangement of oceanic and continental plates in, say, 75 million years. In particular, will the continents be more or less consolidated than they are now?*

What's Causing All This Restlessness?

Wegener had difficulty defending his hypothesis because he was haunted by the questions of why and how continental drift takes place. In contrast, plate tectonicists have had abundant geophysical evidence to explain the plate motions that we observe. Important clues lie in the earth's deep interior. Seismologists have constructed an overview of the deep earth based upon the way that earthquake shock waves travel through it. When waves hit a boundary between different layers, they reflect or refract back to the surface in mathe-

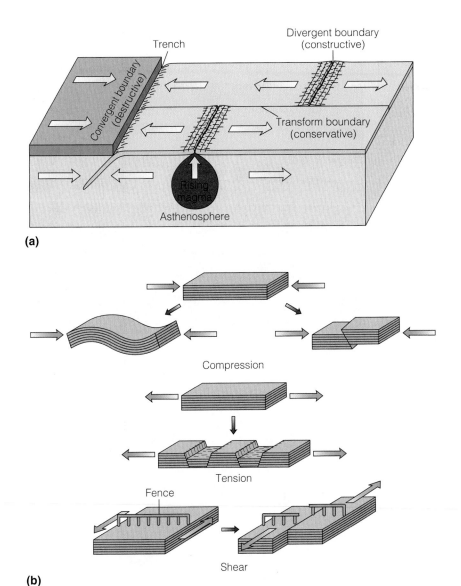

(a)

Trench

Divergent boundary
(constructive)

Convergent boundary
(destructive)

Transform boundary
(conservative)

Rising
magma

Asthenosphere

Compression

Tension

Fence

Shear

(b)

◆ FIGURE 3.10 (a) Divergent, convergent, and transform plate boundaries. Each type of boundary is characterized by a particular type of plate motion and stress. (b) Stresses found at plate boundaries: (1) *Compression.* Two forces acting in opposite directions in the same plane results in shortening of crust by folding and faulting. (2) *Tension.* Extension, or pulling the crust apart, causes faulting. (Rifting is shown.) (3) *Shear.* Two forces acting in opposite directions on different planes causes faulting and displacement between those planes.

matically predictable ways. The speed and type of seismic waves depend upon several factors, including the density of a rock layer and whether or not it is partly or wholly molten.

When the earth first formed, the densest material accumulated at the center of the planet, forming a metallic iron–nickel-rich **core.** Around this ball a thick shell of magnesium-silicate matter developed that we call the **mantle.** This is by far the thickest layer making up the planet (◆ Figure 3.14). A thin "skin" of relatively lightweight aluminum and alkali (sodium and potassium)-rich silicate rocks, termed the **crust,** formed atop the mantle. The crust is still evolving through processes of igneous activity, weathering, sedimentation, and metamorphism. There are two basic kinds of crust—*oceanic* and *continental.* Oceanic crust consists largely of basaltic rocks resulting from sea-

floor spreading, and is typically 5–10 kilometers (3–6 mi) thick. The continental crust, made up largely of silica-rich igneous and sedimentary rocks (like granite and sandstone) and related metamorphic rocks ranges from 20–90 kilometers (12–56 mi) thick, with the greatest thicknesses found beneath mountain ranges. Oceanic rocks are typically darker and somewhat heavier (3.0 gm.cm^3) than continental rocks (2.7 gm.cm^3), because of a higher content of magnesium and iron. This difference in relative density helps explain why oceanic crust subducts during plate convergence. Because continental crust is thicker as well as lighter weight, it also stands topographically higher, mostly rising above the sea.

In addition to the compositional layers above, seismologists have also discovered other kinds of boundaries inside the

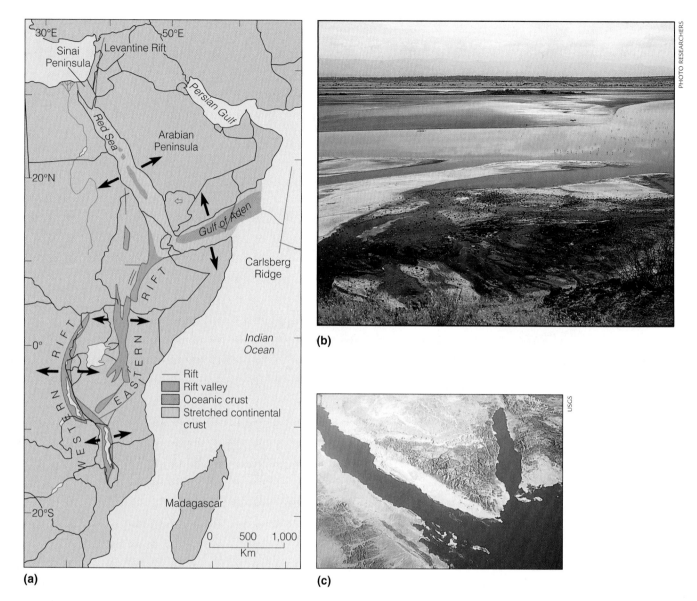

◆ FIGURE 3.11 (a) Two large rift valleys are forming in East Africa as continental crust stretches along a divergent plate margin. The large lake between the rifts is Lake Victoria. The Red Sea and the Gulf of Aden exemplify more advanced ("seaway") stages of rifting and ocean development. (b) Looking down the Great Rift Valley of East Africa with Lake Magadi, Kenya, in the foreground. Because of high evaporation rates and lack of exterior drainage Lake Magadi is a soda lake rich in sodium carbonate, a salt that whitens mud flats on the lake's shore. (c) Satellite photo of the Sinai Peninsula, looking north. The Red Sea is at the bottom, the Gulf of Suez on the left, and the Gulf of Aqaba on the right. Very likely, the Red Sea resembles the South Atlantic Ocean of 270 million years ago.
(a) From R. Wicander and J. Monroe, *Essentials of Geology,* 2nd ed., Brooks/Cole, 1999.

earth. Within the upper mantle, typically between 100 and 350 kilometers (60–210 mi) down, is a seismic zone known as the **asthenosphere** (Greek, "weak sphere"). In the upper part of the asthenosphere is a zone of low seismic wave speeds, indicating that the mantle here is close to its melting point and may actually contain melted portions. It is a zone of mobility that detaches the overlying rigid outer shell of the planet from the hotter, more plastic interior. The rigid outer shell is called the

lithosphere (Greek, "rocky sphere"), and it consists not only of the crust but also of **mantle lid,** a narrow layer of rigid upper mantle rock (◆Figure 3.15). Tectonic plates are actually fragments of lithosphere floating atop the mushy asthenosphere.

The asthenosphere owes its existence to the release of heat from the earth's interior. Most of this heat originates from the slow decay of radioactive elements, such as uranium, lying in the core and lower mantle. The pressure in the

RICHARD HAZLETT

◆ FIGURE 3.12 Dartmouth College geology students standing next to a fresh rupture of the Motagua fault, the transform boundary between the North American and Caribbean plates in Guatemala. The land on which the students are standing is shifting to the left through time relative to the land in the foreground.

like the churning in a pot of rolled oats cooking on a stove (◆Figure 3.16). Hot rock wells up from the depths of the mantle, becomes partially molten and weakens in the asthenosphere, loses heat to the overlying lithosphere, and being denser because it is cooler, sinks back down into the depths. The heat imparted to the lithosphere causes magma to form, contributes to earthquakes and volcanic eruptions, and slowly leaks from the ground into the air. (The percentage of this "earth heat" is trivial in most places, however, compared to the warmth of ordinary sunlight).

What drives plate motion is still being debated, but it is most certainly a combination of gravity acting on the lithosphere (subduction) and thermal convection currents in the asthenosphere that drag the overlying plates along with them as though on a conveyor belt.

ENVIRONMENTAL
Geology⇌Now™

Click **Geology Interactive** to work through an activity on the earth's layers

lower mantle is so great that the rock there remains solid, despite temperatures in the range of 3500–4000°C—hot enough to melt steel at the earth's surface! In the upper mantle, however, pressure is reduced just enough that the hot rock can become partly molten, even though its temperature is a cooler 1500°C at that depth.

The most efficient way for the earth's heat to escape is through the slow convective churning of mantle rock, much

The Dance of the Continents

The process of subduction means that the rocks of the ocean basins have been recycled back into the mantle many times. Therefore, no rocks or sediments older than the current cycle of sea-floor spreading would be expected. In fact, if we multiply the average spreading rate by the length of the mid-ocean ridge system, we find that new oceanic crust is forming

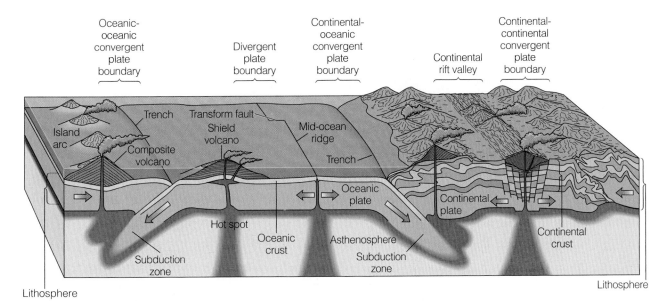

◆ **Geology⇌Now**™ ACTIVE FIGURE 3.13 The plate tectonic model. Divergent and convergent boundaries and their relationships to sea-floor and continental volcanoes are shown. Transform faults on the sea floor are transform boundaries linking different segments of mid-ocean ridge. Rift-valley formation as a result of divergence under the continent is also shown.

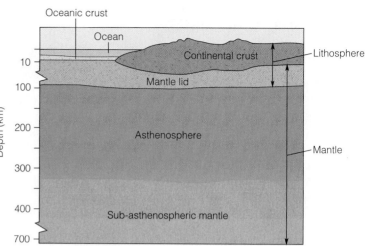

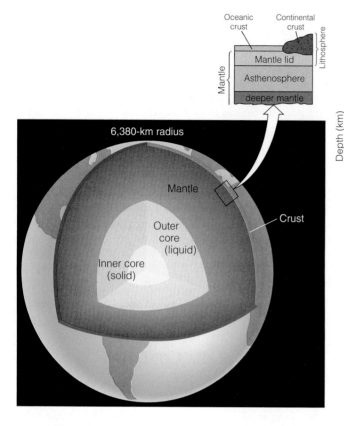

6,380-km radius

Mantle

Outer core (liquid)

Inner core (solid)

Crust

ACTIVE FIGURE 3.14 The internal structure of the earth. The lower part of the figure shows the layers that are defined by contrasts in density, as deduced from observations of seismic waves. The upper part of the figure illustrates the relative thickness of the crust in continental areas as compared with the sea floor.
The lithosphere is comprised of the rigid tectonic plates moving on a plastic asthenosphere.

◆ FIGURE 3.15 The earth's outer layers are divided into *crust* and *mantle* based on rock type, and into *lithosphere* and *asthenosphere* based on rigidity.

at a rate of 2.8 square kilometers (1.1 mi²) per year. Because our oceans cover an area of 310 million square kilometers (120 million mi²), their beds could have formed in as little as 110 million years, and over the past two billion years, as many as 20 global ocean floors may have been created and destroyed! Land geology also records at least two and as many as five complete openings and closings of all the ocean basins.

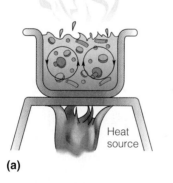

Heat source

(a)

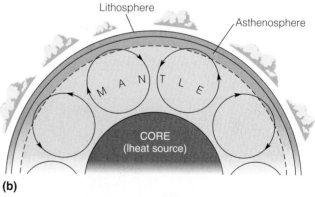

Lithosphere

Asthenosphere

MANTLE

CORE (heat source)

(b)

◆ FIGURE 3.16 Convection (a) in a pot of soup or rolled oats on a stove; (b) in the mantle, involving "whole mantle" overturn; and (c) an alternate model of mantle convection involving multiple layers, which is probably closer to reality than (b).

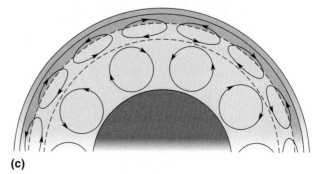

(c)

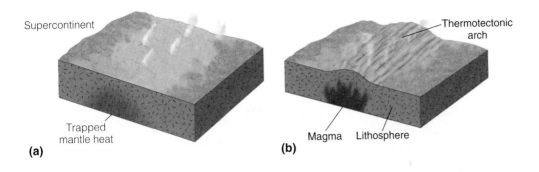

Supercontinent

Trapped mantle heat

(a)

Thermotectonic arch

Magma Lithosphere

(b)

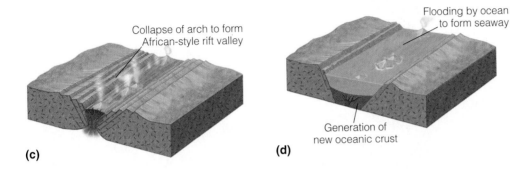

Collapse of arch to form African-style rift valley

(c)

Flooding by ocean to form seaway

Generation of new oceanic crust

(d)

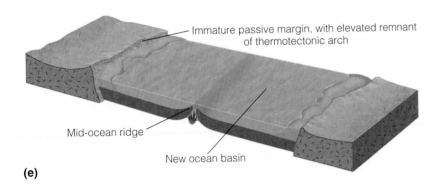

Immature passive margin, with elevated remnant of thermotectonic arch

Mid-ocean ridge

New ocean basin

(e)

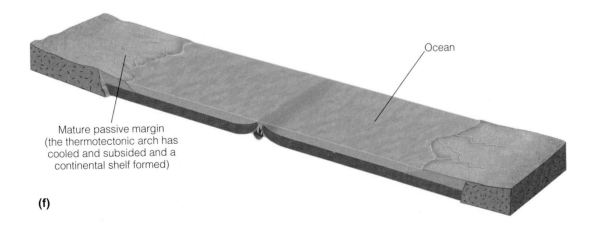

Ocean

Mature passive margin (the thermotectonic arch has cooled and subsided and a continental shelf formed)

(f)

◆ FIGURE 3.17

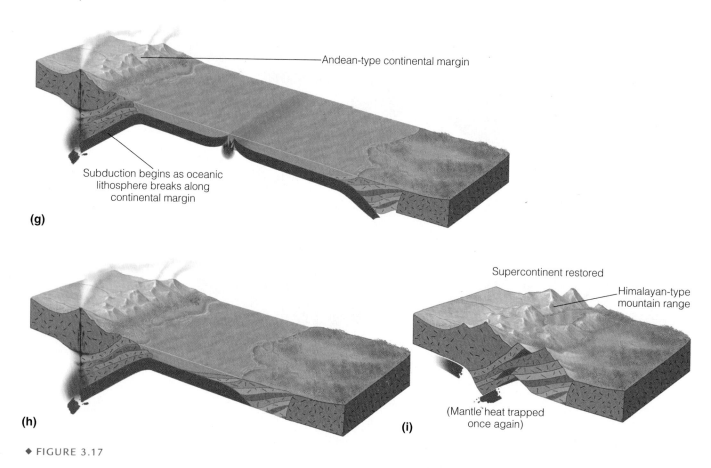

Andean-type continental margin

Subduction begins as oceanic lithosphere breaks along continental margin

(g)

(h)

Supercontinent restored

Himalayan-type mountain range

(Mantle heat trapped once again)

(i)

◆ FIGURE 3.17

The *Wilson cycle* is an evolutionary sequence for oceans named for its developer, J. Tuzo Wilson. He identified five stages in the tectonic birth and death of oceans (◆Figure 3.17):

Evolutionary stage	Example
Embryonic (Fig. 3.17 a–c)	Rift valleys of East Africa
Youthful (Fig. 3.17d)	Gulf of California, Gulf of Aqaba, Red Sea
Mature (Fig. 3.17e–f)	Atlantic Ocean (growing larger)
Declining (Fig. 3.17g)	Pacific Ocean (becoming smaller)
Terminal (Fig. 3.17 h–i)	Mediterranean Sea (closing, almost extinct)

One explanation for the cyclical nature of supercontinent formation is that there is an ongoing tug-of-war between the forces of gravity and heat flow in the upper mantle. A supercontinent like Pangaea acts like a giant lid to trap heat trying to escape earth's interior. (It is generally much easier for heat to escape through oceanic lithosphere than through thicker continental lithosphere). As the trapped heat accumulates, the supercontinent wells up into a volcanically active *thermotectonic arch*, weakens, and rifts apart, creating a new divergent boundary. The sundered continental fragments drift away as a vigorous new Atlantic-type ocean grows, eventually becoming thousands of kilometers wide. The rifted continental margins fringing the ocean initially stand high as coastal mountain ranges, but as the lithosphere cools these ranges subside, and broad coastal plains replace them, an example being the Eastern Seaboard of the United States. These coastlines are termed **passive margins,** in contrast to active continental margins such as the west coast of South America, where vigorous mountain building and volcanism are taking place. An ocean can only grow so large, however, before the sea floor grows too old and heavy for the mantle to support it. Heavy loads of sediment dumping onto an aged sea floor from nearby passive margins help initiate subduction; and once subduction begins the ocean starts to close up. Subduction ultimately draws the sundered continental fragments back together to form a new supercontinent transected by a huge mountain range marking the line of collision. Traces of unsubducted oceanic crust typically suture the range axis, and the cycle starts all over again. Europe and Africa are examples of former pieces of Pangaea in the process of rejoining as the Mediterranean Basin disappears between them, thanks to subduction beneath Italy and Greece. Elsewhere, Pangaea's fragments continue to drift apart.

The formation of a supercontinent (and associated mountain building) occurs about every 400–500 million years in earth history. Geologists have discovered only scraps of evidence for the supercontinent preceding Pangaea, a

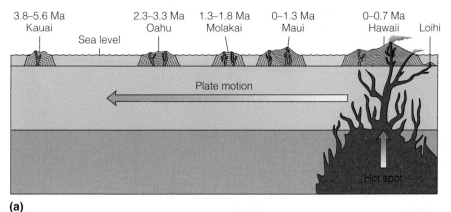

3.8–5.6 Ma Kauai   Sea level   2.3–3.3 Ma Oahu   1.3–1.8 Ma Molakai   0–1.3 Ma Maui   0–0.7 Ma Hawaii   Loihi

Plate motion

Hot spot

(a)

ENVIRONMENTAL
◆ **Geology ⇌ Now™** ACTIVE
FIGURE 3.18 (a) Cross section of six islands formed and forming over the Hawaiian "hot spot." The Hawaiian Islands are sequentially older in a northwest direction. The new volcano Loihi is forming underwater off the southeast end of the Hawaiian chain. (b) Kinked hotspot trajectories in the Pacific. The kink in each island chain occurred about 40 million years ago due to a change in the Pacific plate's direction of motion. (b) B. Murphy and D. Nance, *Earth Science Today*, 2nd ed., Brooks/Cole

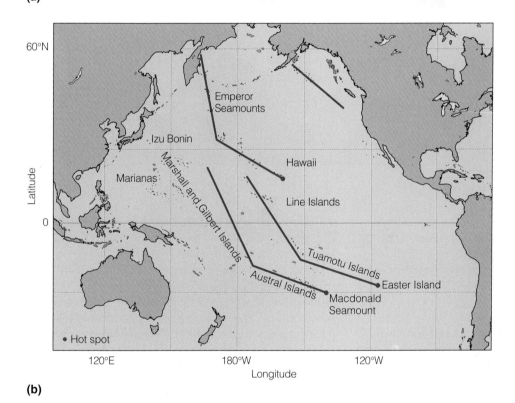

(b)

landmass of about a billion years age that researchers have dubbed *Rodinia*. At that time, Antarctica, Australia, North America, and Europe were joined together. The Pacific Ocean did not yet exist.

CONSIDER THIS *A stockbroker might speculate on what the market will do the next day, a business person might try to forecast six-month trends, but only an earth scientist will speculate on what the face of our planet will look like a million years from now. Here's your chance! What will be the fate of (1) the Mediterranean Sea, (2) the Atlantic and Pacific Ocean basins, and (3) the west coast of the United States exclusive of Alaska if the present plate motion continues for the next 20 million years? Hazard a guess.*

Plate Tectonics and the Rock Cycle

The theory of plate tectonics also provides an excellent model for explaining the rock cycle, described in Chapter 2. Igneous rocks form from cooling magma bodies at divergent and convergent plate boundaries. Uplift at convergent boundaries exposes this rock to the harsh surface agents of weathering and erosion, which yields up the loose sediment needed to make sedimentary rocks. Further plate convergence metamorphoses the sedimentary rock, which under intense conditions of heat and pressure causes it to partially melt. This renews the material transfer of the rock cycle.

◆ FIGURE 3.19 Desiccation of the eastern and western Mediterranean basins about ten million years ago (Miocene time), when its connection with the Atlantic Ocean through the Strait of Gibraltar was blocked. (a) Total evaporation took as little as 1,000 years, during which salt was deposited and dry lakes or salt lakes existed in the basins. Some spillover from the Red Sea occurred in the eastern Mediterranean. (b) Two-kilometer-thick deposits of sediments and evaporites (especially gypsum and halite) beneath the floors of these basins indicate that the cycle of filling, evaporation, and refilling was repeated many times (2 km ≈ 6,500 ft or 1¼ mi).

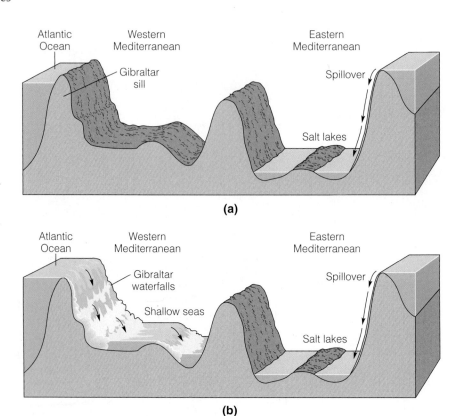

Hot Spots

Within some ocean basins there are chains of aligned volcanic islands that become progressively older in one direction. The Hawaiian Islands, Line Islands, and Tuamotus are such chains in the Pacific Ocean. Although the Galápagos Islands off the coast of Ecuador—made famous by the writings of Charles Darwin—are only roughly aligned, they are of similar origin. These lines-of-islands are not associated with plate boundaries, but have formed over areas of unusually hot rock, or **hot spots,** fixed in the asthenosphere. Many (though not all) geologists believe that hot spots are fed by plumes of material welling up from the deep mantle. These rising intraplate plumes penetrate the general convective circulation that drives plate motion and are independent of spreading centers. As a tectonic plate moves over a hot spot, the plate carries with it the volcanoes that form there, with new volcanoes growing in place of the older ones that are carried away (◆Figure 3.18). For example, Maui and Hawaii,

One tectonic consequence of the rock cycle is that a submerged continental shelf of soft muds and sands in one era can become a towering granitic mountain range in the next, with a net addition of new, hard crust to a continent. As a result, the total land area of the continents has enlarged over the past few billion years, with new growth concentrated along convergent plate boundaries. Earth is gradually becoming more of a land planet and less of a water world, although the growth in continents may be compensated by an overall deepening trend in the seas as well.

the islands closest to the Hawaiian hot spot, are the only ones with active volcanism today. A line drawn from the Island of Hawaii to Kure Atoll, the oldest of the chain, would represent an arrow marking the direction and rate of movement of the Pacific plate in this region. The track of the Hawaiian Islands forms a "lazy *L*" with the Emperor Seamount chain, a line of submerged volcanoes that extends northward from Midway, tracing a different direction of plate movement prior to 40 million years ago. Of greater environmental concern because of population densities are hot spots that occur beneath the continents. Yellowstone National Park, for example, is situated above a midplate hot spot with a potential for destructive volcanic activity. About a third of the more than 120 identified hot spots that have been active within the last ten million years penetrate continental areas (See Figure 3.9).

ENVIRONMENTAL
Geology⇌Now™

Click **Geology Interactive** to work through an activity on the distribution of volcanism.

Plate Tectonics and the Environment

Plate tectonic movements have shaped and changed past natural environments throughout the world. Some tectonically forced environmental changes have impacted the entire world, while others have focused only upon particular regions. An excellent example comes from research done by the Deep Sea Drilling Project of the U.S. National Science Foundation. During a series of cruises in the Mediterranean Sea, the researchers discovered thick salt deposits beneath the sea floor, indicating that about ten million years ago (Miocene time) the sea was isolated from the Atlantic Ocean and that the sea totally evaporated (◆Figure 3.19). Access of Atlantic Ocean water to the Mediterranean through the Strait of Gibraltar was cut off by uplift as the African plate moved northward, and also by a global lowering of sea level at the time. Further drilling yielded sediments indicating that the floors of the two Mediterranean basins once resembled the great dry lakes of present-day deserts in the western United States. Occasionally they were occupied by water when the connection with the Atlantic was periodically reestablished. About five million years ago the Strait of Gibraltar reopened, and the Atlantic Ocean flowed into the Mediterranean Sea as a waterfall or cascade of great magnitude (◆Figure 3.19b). This inflow filled the Mediterranean in a few thousand years. The sediments on top of the five-million-year-old salt-flat deposits contain fossils of marine organisms, indicating rapid filling with normal seawater, and no abnormal salinity, which would have occurred if the filling were slow.

Plate Tectonics and the Human Environment

Plate tectonic activity not only shapes the natural world, but also explains the global geography that has strongly influenced human history and the present distribution of the world's population. The origins of many natural resources are tied to plate movements. For example, valuable mineral deposits such as zinc and copper ore form around sea-floor hot springs and volcanic vents at divergent boundaries. Upheaval of ancient oceanic lava flows rich in copper formed the Island of Cyprus in the eastern Mediterranean, which was a vital source of metal during the Bronze Age. One of the oldest known smelter sites in the world, at Timna in southern Israel (Figure 2, p. 48), extracted copper from the Dead Sea rift valley over 5,000 years ago. Metal from this site helped make Egypt a powerful kingdom during the time of Moses.

Along convergent boundaries, volcanoes erupt billowing clouds of volcanic ash, which fertilizes surrounding lands for agriculture. Largely because of the rich farms, about one out of every ten people in the world lives in a convergent plate setting today. On the Island of Java, which is notorious for its volcanic activity, there are as many as three harvests a year.

On neighboring Borneo, however, a harvest may be possible only once every one or two years, because that island lies outside of the nearby volcanic zone. Gold, silver, tin, and mercury deposits form along convergent boundaries, too, helping support wild mining booms. The conquistadores braved treacherous conditions in Central and South America to overwhelm indigenous cultures in the quest for tectonic treasures. The unique character of modern Latin America begins essentially with its geology.

Passive margins have been even more vital to human affairs. Mature passive margins allow large rivers to form large floodplains and deltas as they flow to the sea, providing the most fertile environments for agriculture. Except for occasional bad weather and flooding, people living along passive margins have little to fear from natural disasters, and generally have plenty to eat—to begin with at least. (Unfortunately, coastal floodplains are also prone to become the most crowded places in the world!). The list of early civilizations below shows the influence of passive margins at the dawn of human history.

Civilization	Size of origin	Tectonic setting
Greco–Minoan	Small coastal floodplains	Convergent boundary
Roman	Near a small floodplain	Convergent boundary
Aztecs and Incas	Mountain valleys, lakeshore	Convergent boundary
Chinese	Initially fertile highlands and long, broad floodplains	Passive margin near convergent boundary
Sumerian	Long, broad floodplain	Passive margin near convergent boundary
Egyptian	Long, narrow floodplain	Passive margin
Harappan (Pakistan)	Long, broad floodplain	Passive margin

Mature passive margins have become even more important in recent times, because they also provide ideal locations for petroleum and natural gas reservoirs, upon which the modern world economy and political stability depend. Geologists use plate tectonic concepts in exploring for oil and gas, focusing especially on the sedimentary rocks of passive margins. Our technical civilization, standing on the two mutually dependent legs of metals and fuel, could never have developed without the phenomenon of plate tectonics.

Exotic Terranes—A Continental Mosaic

Because continental plates divide, collide, and slip along faults, it is not surprising that bits and pieces of continental crust existing within the oceans ultimately collide with continents at subduction zones and become stuck there. Because of their low density, they stand high above oceanic crust (due to isostatic equilibrium), and these microplates or microcontinents become plastered, *accreted,* to larger continental plates when they collide with them. The accreted plates are known as **terranes,** defined as fault-bounded blocks of rock with histories quite different from those of adjacent rocks or terranes.* In size they may be several thousand square kilometers or just a few tens of square kilometers, and they may become part of a continent composed of many terranes. Oceanic crust and the sediment resting on it may be scraped up onto the continents. Terranes may consist of almost any type of rock, but each one is fault-bounded, has a paleomagnetic signature indicating a distant origin, and has little geology in common with adjacent terranes or with the continental **craton**—the continent's stable core.

The terrane concept developed in Alaska when geologic mapping for land-use planning revealed that the usually predictable pattern of rocks and structures was not valid for any distance. In fact, the rocks the geologists found a few kilometers away were almost always of a "wrong" composition and age. Further studies revealed that Alaska is a collection of microplates (terranes)—tectonic flotsam and jetsam, if you wish—that have mashed together over the past 160 million years and that are still arriving from the south (◆Figure 1). One block, the Wrangellia terrane, was an island during Triassic time and it has a paleomagnetic signature of rocks that formed 16° from the equator. It is not known whether Wrangellia formed north or south of the equator, because it is not known whether the magnetic field at the time was normal or reversed. In either case, the terrane traveled a long distance to become part of present-day Alaska (◆Figure 2). It now appears that about 25 percent of the western edge of North America, from Alaska to Baja California, formed in this way; that is, by bits and pieces being grafted onto the core of the continent and thus enlarging it. In the distant future, part of California may become an exotic terrane of Alaska.

If you've deduced that terrane accretion onto the continents (also known as *docking*) is a characteristic of active continental margins, you are right (◆Figure 3). The east coast of North America is a rifted (pulled-apart), or *passive,* margin. Material is added to it by river sediment forming flat-lying sedimentary rocks that become part of the continent. Such rocks are not subject to the mountain-building forces of active margins and they accumulate in thick, undisturbed sequences.

*Terrane, as noted, is a geological term describing the area or surface over which a particular rock or group of rocks is prevalent. *Terrain* is a geographical term referring to the topography or physical features of a tract of land.

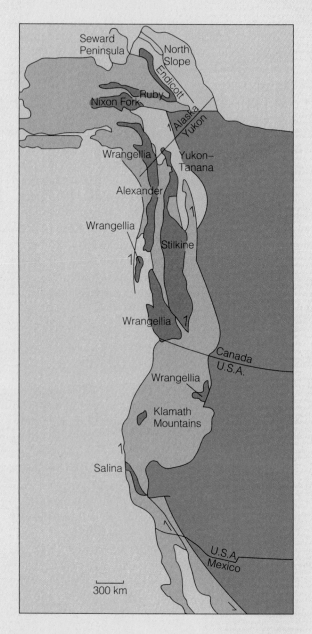

◆ FIGURE 1 Accreted terranes of the west coast of the United States and Canada. The Salina, Wrangellia, Alexander, Stikine, Klamath Mountains, Ruby, and Nixon Fork blocks (dark brown) were probably once parts of other continents and have been displaced long distances. The terranes shown as light-green areas are probably displaced parts of North America, the stable North American craton (darker green). The lighter-brown areas represent rocks that have not traveled great distances from their places of origin.

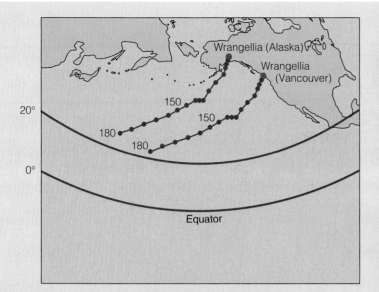

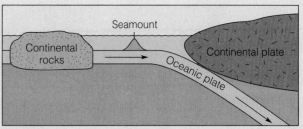

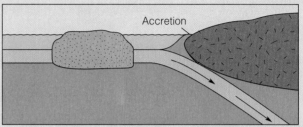

◆ FIGURE 2 An analysis of Wrangellia terrane trajectories for the period 180 Ma to 100 Ma, assuming initial position in the Northern Hemisphere. In this interpretation both Vancouver Island and the Alaskan terranes (see Figure 1) rode with the oceanic plate until they collided (docked) at their present positions on the North American continent about 100 million years ago. The data points (the dots) represent paleomagnetic evidence at the respective locations.

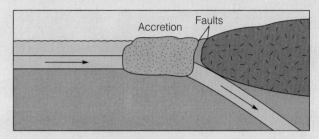

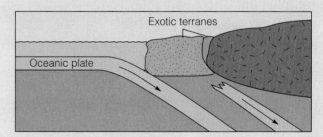

◆ FIGURE 3 Formation of exotic terranes by a small continental mass and a submarine volcano being scraped off from the subducting plate and onto the continental mass. Many portions of the west coasts of the United States and Canada have grown in this manner.

GALLERY

gallery

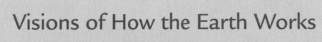

Visions of How the Earth Works

The evolution of modern plate-tectonic theory started with Alfred Wegener's 1912 hypothesis of Continental Drift. Almost fifty years later the idea of sea-floor spreading was proposed. Although scientists were skeptical at first, this far-reaching concept led to the unifying theory of plate tectonics.

BILDARCHIV PREUSSISCHER KULTURBESITZ

◆ FIGURE 1 Alfred Wegener formulated his hypothesis that continents move about on the face of the earth while sitting out a long Arctic winter in Greenland. His ideas were considered so outrageous that more than fifty years passed before scientists realized he was on the right track.

RODNEY SUPPLE

◆ FIGURE 2 Evidence of colliding and sliding tectonic plates. Collision of the Indian and Asian plates produced the great Himalaya Mountains. The rocks of Nepal's Dhaulāgiri I (8,172 m, 26,810 ft), one of the highest peaks in the world, were uplifted and contorted by the continent–continent collision. The steep-sided south wall is 15,000 feet high. This view shows folded strata in the mountain's lower slope.

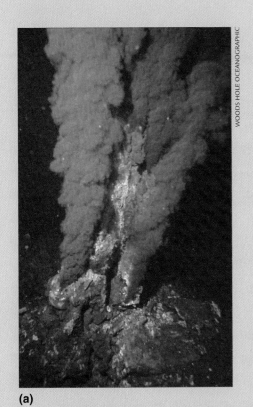

(a)

(b)

(c)

◆ FIGURE 3 (a) "Black smokers" of mid-oceanic ridge. Hot brine loaded with minerals leached from basalt emanates from vents on ridges and builds tall "chimneys." (b) Giant tube worms thrive around the "smokers" and live on bacteria in the Galapagos Rift (see Chapter 5). (c) Clams more than 25 centimeters (10 in) long inhabit the warm waters near the vents in the Galapagos Rift.

Summary

Continental Drift

Description

Alfred Wegener's theory that during the Permian Period there was a single supercontinent, Pangaea, which split apart during Jurassic time into individual continents that have since migrated to their present positions.

Cause

The theory provided no explanation for a driving force.

Evidence

1. Similar glacial deposits and plant fossils in Permian rocks (the Gondwana Succession) in Africa, South America, India, Australia, and Antarctica.
2. Similarities between fossils, ages of rocks, and geologic provinces on opposite sides of the Atlantic Ocean.
3. The jigsaw fit of the continents if the Atlantic and Indian Oceans were closed.

Sea-Floor Spreading

Description

Theory proposed by Harry Hess that new oceanic crust is created at mid-ocean ridges and moves laterally across the ocean basins to trenches, where it sinks and returns to the mantle.

Cause

Convection cells in the mantle generated by thermal and density differences carry the rigid overlying crust (lithosphere) away from the ridges until the crust cools and descends toward the mantle at oceanic trenches.

Evidence

1. Distribution of deep-focus earthquakes at trenches.
2. Lack of ancient sediments on the sea floor.
3. Increasing age of sea-floor sediments away from mid-ocean ridges.

Plate Tectonics

Description

Theory that unifies and incorporates the earlier theories of continental drift and sea-floor spreading. According to it, the lithosphere is composed of seven large plates and a number of smaller ones that move about on the surface of the earth. Plates interact in only three ways: they *collide* at convergent boundaries marked by trenches or mountain ranges; they *move apart*, forming new sea floor, at divergent boundaries marked by mid-ocean ridges; or they *slip by one another* along transform boundaries.

Cause

Convection currents in the asthenosphere (the plastic layer of the upper mantle) carry the plates as passive passengers from divergent boundaries to convergent and conservative boundaries. Gravity pulls plates down into the mantle at convergent boundaries.

Evidence

1. Land geology, ocean drilling, and geophysical data support past openings and closings of the ocean basins (Wilson cycle).
2. Magnetic anomaly patterns ("stripes") on both sides of mid-ocean ridges.
3. Chains of oceanic volcanoes and seamounts that systematically increase in age with distance from the present locations of active volcanism (hot spots) indicate that plates are moving over fixed plumes of lava coming from deep in the mantle.
4. The geological and geophysical evidence for continental drift and sea-floor spreading also support plate tectonics.

Explains

1. The thick salt deposits beneath the Mediterranean Sea and its desiccation 10 million years ago.
2. The existence of narrow seas and gulfs such as the Gulf of California and the Red Sea as spreading centers.
3. The location and geology of the Himalaya Mountains.
4. The location of violent earthquakes and volcanoes at convergent margins.

Key Terms

anomaly	crust	mantle	plate tectonics
asthenosphere	divergent boundary	mantle lid	sea-floor spreading
Benioff–Wadati zone	Gondwana succession	marine trench	subduction
continental drift	graben	mid-ocean ridge	terrane
convergent boundary	hot spot (plume)	Pangaea	transform boundary
core	lithosphere	passive margin	Wilson cycle
craton	magnetic polarity	plate	

Study Questions

1. What geologic and geographic evidence did Alfred Wegener cite for continental drift?

2. How do we know that earth's interior is layered?

3. Why does convection occur? Give some examples of convection from everyday life.

4. What are the major spheres, or shells, of the earth's interior based upon composition? What is meant by the term *lithosphere*? Describe the lithosphere and tell how it is separated from the deep mantle.

5. What is a tectonic "plate"? Give an example (exact geographic location) of each type of plate boundary. How fast do plates move?

6. What geophysical observations and measurements support the model of sea-floor spreading and plate tectonics?

7. How can the change in direction of a line of oceanic islands—such as that between the Hawaiian Islands chain and the Emperor Seamount chain—be explained?

8. Draw a cross section of a typical convergent plate boundary (a subduction zone) and label the following: trench, volcanic areas, Benioff-Wadati zone, lithosphere, asthenosphere.

9. How might the distribution of the Permian Gondwana Succession be explained without invoking plate tectonics (ancient landmass, land bridges, etc.)? Might an alternative explanation be more reasonable than drifting continents and plate tectonics?

10. How have the narrow bodies of the Red Sea and the Gulf of California formed? Study a map, seeking an obvious relationship between the Red Sea and the Dead Sea.

For Further Information

Books and Periodicals

Ballard, Robert. 1983. *Exploring our living planet.* Washington, D.C.: National Geographic Society.

Bonatti, Enrico. 1994. The earth's mantle below the oceans. *Scientific American,* March: 44–51.

Cox, Allan. 1986. *Plate tectonics: How it works.* Palo Alto, Calif.: Blackwell Scientific.

Dalziel, Ian W. D. 1995. Earth before Pangaea. *Scientific American,* January: 58.

Debiche, M. G., Allan Cox, and D. Engebretson. 1987. *The motion of allochthonous terranes across the north Pacific basin.* Geological Society of America Special Paper 207, 49 pp.

Hallam, A. 1973. *A revolution in the earth sciences.* Oxford, U.K.: Oxford University Press.

Hsu, Ken. 1983. *The Mediterranean was a desert.* Princeton, N.J.: Princeton University Press.

Kearey, P., and F. J. Vine. 1996. *Global tectonics,* 2d ed. Palo Alto, Calif.: Blackwell Scientific.

Larson, Robert. 1995. The mid-Cretaceous superplume episode. *Scientific American,* January: 82–86.

Moores, E. M., ed. 1990. Shaping the earth: tectonics of continents and oceans. *Scientific American readings.* New York: W. H. Freeman.

Murphy, J. Brendan, and R. Damian Nance. 1992. Mountain belts and the supercontinent cycle. *Scientific American,* April: 84–91.

Nance, R. Damian, T. R. Worsley, and J. B. Moody. 1988. The supercontinent cycle. *Scientific American* 259: 172–79.

Orestes, N., 1999, *The Rejection of Continental Drift,* Oxford University Press.

Schneider, David. 1997. Hot spotting. *Scientific American,* April: 22–23.

Sullivan, Walter. 1974. *Continents in motion: The new earth debate.* New York: McGraw-Hill.

Tarling, D., and Maureen Tarling. 1970. *Continental drift: A study of the earth's moving surface.* New York: Anchor Books.

Uyeda, Seiya. 1977. *The new view of the earth.* New York: W. H. Freeman.

Wegener, Alfred. 1929, 1966 translation. *The origin of continents and oceans.* New York: Dover Publications.

Wilson, J. Tuzo. 1972. Continents adrift. *Scientific American readings.* San Francisco: W. H. Freeman.

ENVIRONMENTAL
Geology⇌Now™

Assess your understanding of this chapter's topics with additional quizzing and comprehensive interactivities at http://earthscience.brookscole.com/pipkingeo4e as well as current and up-to-date Web links, additional readings, and InfoTrac College Edition exercises.

Earthquakes and Human Activities

ASSOCIATED PRESS, AUGUST 25, 1999 "TURKEY'S PRIME MINISTER has promised stricter building rules to prevent the shoddy construction blamed for the thousands of deaths from last week's massive earthquake. Eight days after the powerful 7.4-magnitude temblor reduced much of western Turkey to rubble, an estimated 200,000 survivors remained camped out in parks and on vacant lots."

The cause of the earthquake was a sudden rupture along northern Turkey's Anatolia fault. The severe ground motion that was generated collapsed hundreds of buildings, many of which had been built below earthquake-resistant standards. Interestingly, the Anatolia fault and California's infamous San Andreas fault have some striking geographic similarities; see ◆Figure 1. Let's hope that is where the likeness ends.

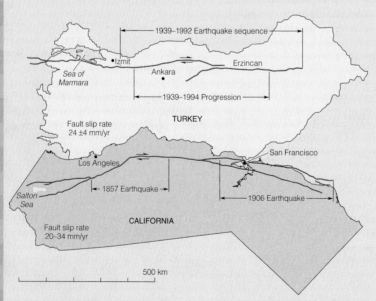

◆ FIGURE 1 *The North Anatolian fault in Turkey and the San Andreas fault in California. The faults have the same sense of motion, are about the same age and length, and are capable of similar-magnitude earthquakes. The earthquake sequences shown on the North Anatolian fault represent events that occurred, one after the other from east to west, along the fault.*

The Nature of Earthquakes

After experiencing an earthquake in Concepción, Chile, in 1835, Charles Darwin noted that "A bad earthquake at once destroys the oldest associations; the world, the very emblem of all that is solid, had moved beneath our feet like crust over a fluid." Darwin's reflections are vivid, and many of us have felt the same way when experiencing strong earthquake motion. What Darwin did not know was the cause of earthquakes and the resulting ground motion.

Earthquakes are the result of abrupt movements on **faults**—fractures in the earth's lithosphere. The types of faults and the earth forces that cause them are shown in ◆ Figure 4.1. The

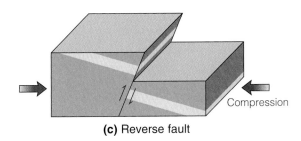

(a) Normal fault

(b)

(c) Reverse fault

(d)

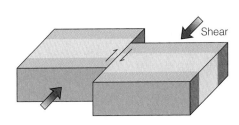

(e) Right-lateral strike-slip fault

(f)

◆ FIGURE 4.1 (a) Normal fault geometry and (b) examples in a road cut on I-40 near Kingman, Arizona. The normal fault at the right is in relatively soft sedimentary rocks. Fault at the left is also a normal fault with a small displacement of white sandstone between it and a small fault that shows reverse displacement. (c) Reverse fault geometry and (d) low-angle reverse fault displacing sandstone strata at Wasatch Plateau near Salina, Utah. (e) Right-lateral strike-slip fault geometry and (f) a plowed field displaced by a strike-slip fault in the Imperial Valley, California, in 1979.

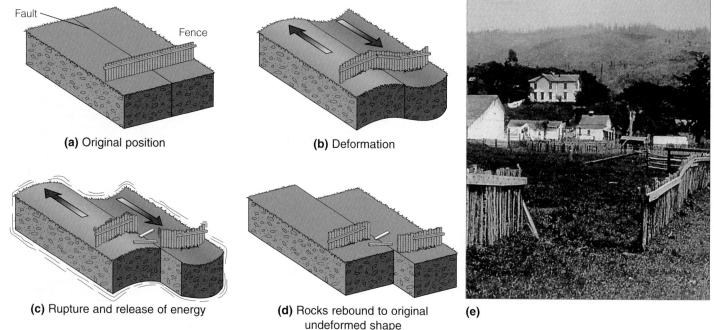

(a) Original position

(b) Deformation

(c) Rupture and release of energy

(d) Rocks rebound to original undeformed shape

(e)

◆ FIGURE 4.2 (a–d) The cycle of elastic-strain buildup and release for a right-lateral strike-slip fault according to Reid's elastic rebound theory of earthquakes. At the instant of rupture, (c), energy is released in the form of earthquake waves that radiate out in all directions. (e) Right-lateral offset of a fence by 2.5 meters (8 ft) by displacement on the San Andreas fault in 1906; Marin County, California.

movements occur as the earth's crustal plates slip past, under, and away from one another. Because the **stress** (force per unit area) that produces **strain** (deformation) can be transmitted long distances in rocks, active faults do not necessarily occur exactly on a plate boundary, but they generally occur in the vicinity of one. The mechanism by which stressed rocks store up strain energy along a fault to produce an earthquake was explained by Harold F. Reid after the great San Francisco earthquake of 1906. Reid proposed a mechanism to explain the shaking, which resulted from movement on the San Andreas fault, known as the **elastic rebound theory** (◆Figure 4.2). According to this theory, when sufficient strain energy has accumulated in rocks, they may rupture rapidly—just as a rubber band breaks when it is stretched too far—and the stored strain energy is released as vibrations that radiate outward in all directions (◆Figure 4.3).

Most earthquakes are generated by movements on faults within the crust and upper mantle that do not produce ruptures at the ground surface. We can thus recognize a point within the earth where the fault rupture starts, the **focus,** and the **epicenter,** the point on the earth surface directly above the focus (Figure 4.3). Elastic waves—vibrations—move out spherically in all directions from the focus and strike the surface of the earth. Damaging earthquake foci are generally within a few kilometers of the earth's surface. Deep-focus earthquakes, on the other hand, those whose foci are 300–700 kilometers (190–440 mi) below the surface, do little or no damage. Earthquake foci are not known below 700 kilometers. This indicates that the mantle at that depth

behaves *plastically* due to high temperatures and confining pressures, deforming continuously as ductile substances do, rather than storing up strain energy.

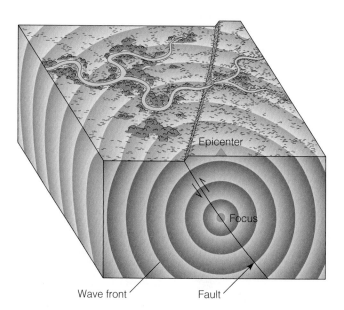

ENVIRONMENTAL
◆ **Geology⇌Now™** ACTIVE FIGURE 4.3 The focus of an earthquake is the point in the earth where fault rupture begins, and the epicenter is directly above the focus at the ground surface. Seismic-wave energy moves out in all directions from the focus. (From *Essentials of Geology,* 2nd ed., by R. Wicander and J. Monroe, p. 228, Brooks/Cole, 1999)

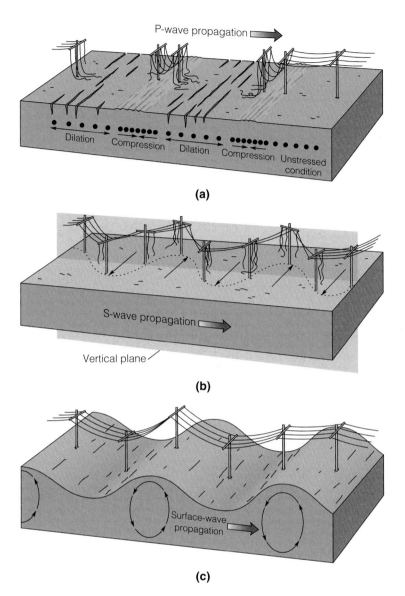

P-wave propagation →

Dilation   Compression   Dilation   Compression   Unstressed condition

(a)

S-wave propagation →

Vertical plane

(b)

Surface-wave propagation →

(c)

ENVIRONMENTAL
◆ Geology⇌Now™ ACTIVE FIGURE 4.4
Ground motion during passage of earthquake waves. (a) P-waves compress and expand the earth. (b) S-waves move in all directions perpendicular to the wave advance, but only the horizontal motion is shown in this diagram. (c) Surface waves create surface undulations that result from a combination of the retrograde elliptical motions of Rayleigh waves (shown) and Love waves, which move from side to side at right angles to the direction of wave propagation.

The vibration produced by an earthquake is complex, but it can be described as three distinctly different types of waves (◆Figure 4.4). Primary waves, **P-waves,** and secondary waves, **S-waves,** are generated at the focus and travel through the interior of the earth; thus, they are known as **body waves.** They are designated P- and S-waves because they are the first (*primary*) and second (*secondary*) waves to arrive from distant earthquakes. As these body waves strike the earth's surface, they generate surface waves that are analogous to water ripples on a pond.

P-waves (◆Figure 4.4a) are **longitudinal waves;** the solids, liquids, and gases through which they travel are alternately compressed and expanded in the same direction the waves move. Their velocity depends upon the resistance to change in volume (in compressibility) and shape of the material through which they travel. P-waves travel about 300 meters (1,000 ft) per second in air, 300–1,000 meters (1,000–3,000 ft) per second in soil, and faster than 5 kilometers (3 mi) per second in solid rock at the surface. P-wave velocity increases with depth, because the materials composing the mantle and core become less compressible with depth. This increase causes the waves' travel paths to be bowed downward as they move through the earth. P-waves speed through the earth with a velocity of about 10 kilometers (6 mi) per second. At the ground surface they have very small amplitudes (motion) and cause little property damage. Although they are physically identical to sound waves, they vibrate at frequencies below what the human ear can detect. The earthquake noise that has been reported could be P-waves of a slightly higher frequency or some other, related vibrations whose frequencies are in the audible range.

S-waves (◆Figure 4.4b) are **transverse (shear) waves;** they produce ground motion perpendicular to their direction of travel. This causes the rocks through which they travel to be twisted and sheared. These waves have the shape produced when one end of a garden hose or a loosely hanging rope is given a vigorous flip. They can travel only through material that resists shearing—that is, material that resists two forces acting in opposite directions in different planes. Thus, S-waves travel only through solids, because liquids and gases have no shear strength (try piling water in a mound). S-wave ground motion may be largely in the horizontal plane and can result in considerable property damage. S-wave velocity is approximately 5.2 kilometers (3.2 mi) per second from distant earthquakes; hence, they arrive after the P-waves.

Unexpended P- and S-wave energy bouncing off the earth's surface generates the complex surface waves (◆Figure 4.4c). These waves produce a rolling motion at the ground surface. In fact, slight motion sickness is a common response to surface waves of long duration. The waves are the result of a complex interaction of several wave types, the most important of which are *Love waves,* which exhibit horizontal motion normal to the direction of travel, and *Rayleigh waves,* which exhibit retrograde (opposite to the direction of travel) elliptical motion in a plane perpendicular to the ground surface. Surface waves have amplitudes that are greatest in near-surface unconsolidated layers and are the most destructive of the earthquake waves.

Locating the Epicenter

A **seismograph** is an instrument designed specifically to detect, measure, and record vibrations in the earth's crust (◆Figure 4.5). It is relatively simple in concept (◆Figure 4.6), but some very sophisticated electronic systems have been developed. Seismic data are recorded onto a **seismogram** (◆Figure 4.7). Because seismographs are extremely sensitive to vibrations of any kind, they are installed in quiet areas such as abandoned oil and water wells, cemeteries, and parks.

When an earthquake occurs, the distance to its epicenter can be approximated by computing the difference in P- and S-wave arrival times at various seismograph stations. Although the method actually used by seismologists today is more accurate and determines the depth and location of the quake's focus and its epicenter, what is described here serves to show how seismic-wave arrival times can be used to determine the distance to an epicenter. Because it is known that the two kinds of waves are generated simultaneously at the earthquake focus and that P-waves travel faster than S-waves, it is possible to calculate where the waves started.

Imagine that two trains leave a station at the same time, one traveling at 60 kilometers per hour and the other at 30 km/h, and that the second train passes your house one hour after the first train goes by. If you know their speeds, you can readily calculate your distance from the train station of origin as 60 kilometers. The relationship between distance to an epicenter and arrival times of P- and S-waves is illustrated in ◆Figure 4.8. The epicenter will lie somewhere on a circle whose center is the seismograph and whose radius is the distance from the seismograph to the epicenter. The problem is then to determine where on the circumference of the circle the epicenter is. To learn this, more data are needed: specifically, the distances to the epicenter from two other seismograph stations. Then the intersection of the circles drawn around each of the three stations—a method of map location called *triangulation*—specifies the epicenter of the earthquake (◆Figure 4.9).

> **CONSIDER THIS** *You are shopping for a house in earthquake country. How could an isoseismal map of an earthquake whose epicenter was near a house you think you would like to buy be useful? What patterns would you look for when examining the isoseismal map?*

Earthquake Measurement

Intensity Scales

The reactions of people (geologists included) to an earthquake typically range from mild curiosity to outright panic. However, a sampling of the reactions of people who have been subjected to an earthquake can be put to good use. Numerical values can be assigned to the individuals' perceptions of earthquake shaking and local damage, which can then be contoured upon a map. One **intensity scale** developed for measuring these perceptions is the 1931 **modified Mercalli scale*(MM).*** The scale's values range from *MM* = I (denoting not felt at all) to *MM* = XII (denoting widespread destruction), and they are keyed to specific U.S. architectural and building specifications. (See Appendix 4.) People's perceptions and responses are compiled from returned questionnaires, and then lines of earthquake intensity, called **isoseismals,** are plotted on maps. Isoseismals enclose areas of equal earthquake damage and can indicate areas of weak rock or soil as well as areas of substandard building construction (◆Figure 4.10a). Such maps have proved useful to planners and building officials in revising building codes and safe construction standards.

Recently a new method of plotting shaking intensities has evolved that is faster and that compares favorably with the questionnaire method. It is the Community Internet Intensity Maps (CIIM), an example of which is shown in Figure 4.10b. The CIIM takes advantage of the Internet, and the time to generate intensity maps drops from months to

◆ FIGURE 4.5 Seismometer *(left)* and seismogram *(right)* used to demonstrate method of detecting and recording earthquakes. The record on the seismogram is not an earthquake, but represents the seismometer's detection of vibrations from students' footsteps amplified about 2,000 times.

minutes. The responses, which can take place within 3 minutes of the event, are summarized by computer, and an intensity number is assigned to each ZIP code area. This method is particularly useful in areas with sparse seismograph coverage. Although perhaps not as colorful as the map generated by the traditional *MM* intensities obtained by using postal questionnaires, the CIIM values agree well and have actually proved more reliable in areas of low shaking. However, neither method allows comparison of earthquakes with widely spaced epicenters because of local differences in construction practices and local geology.

Richter Magnitude Scale—The Best-Known Scale

The best-known measure of earthquake strength is the **Richter magnitude scale,** which was introduced in 1935 by Charles Richter and Beno Gutenberg at the California Institute of Technology. It is a scale of the energy released by an earthquake and thus, in contrast to the intensity scale, may be used to compare earthquakes in widely separated geographic areas. The Richter value is calculated by measuring the maximum amplitude of the ground motion as shown on the seismogram using a specified seismic wave, usually the surface wave. Next, the seismologist "corrects" the measured amplitude (in microns) to what a "standard" seismograph would record at the station. After an additional correction for distance from the epicenter, the Richter **magnitude** is the common logarithm of that ground motion in microns. For example, a magnitude-4 earthquake is specified as having a corrected ground motion of 10,000 microns ($\log_{10}$ of 10,000 = 4) and thus can be compared to any other earthquake for which the same corrections have been made. It should be noted that because the scale is logarithmic, each whole number represents a ground shaking (at the seismograph site) 10 times greater than the next-lower number. Thus a magnitude-7 produces 10 times greater shaking than a magnitude-6, 100 times that of a magnitude-5, and 1,000 times that of a magnitude-4. Total energy released, on the other hand, varies logarithmically as some exponent of 30. Compared to the energy released by a magnitude-5 earthquake, a $M = 6$ releases 30 times (30^1) more energy; a $M = 7$

◆ FIGURE 4.6 The first earthquake detector, invented about A.D. 130 by Chinese scholar Chang Heng. Balls held in the dragons' mouths were aligned with a pendulum inside the vase. When an earthquake occurred the pendulum swung and pushed balls into the mouths of the frogs aligned with the pendulum's swing. If the frogs facing north and south contained balls, for example, Chang Heng could say that the earthquake epicenter was north or south of the instrument.

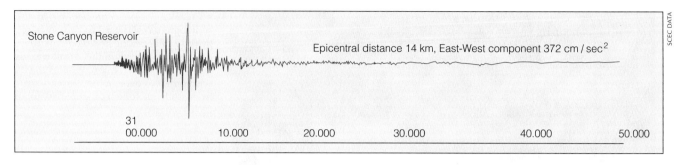

◆ FIGURE 4.7 Seismogram of the main shock of the Northridge earthquake, January 17, 1994. The time in minutes and seconds after 4:00 A.M. appears at the bottom. Distance from the epicenter and direction and ground acceleration at the recording site are also indicated. (Acceleration is explained in the next section.)

releases 900 times (30^2) more; and a $M = 8$ releases 27,000 (30^3) times more energy (◆Figure 4.11a).

The Richter scale is open-ended; that is, theoretically it has no upper limit. However, rocks in nature do have a limited ability to store strain energy without rupturing, and no earthquake has been observed with a Richter magnitude greater than 8.9—yet.

Moment Magnitude—The Most Widely Used Scale

Seismologists have abandoned Richter magnitudes in favor of **moment magnitudes** (M_w or M) for describing earthquakes. The reason is that Richter magnitudes do not accurately portray the energy released by large earthquakes on faults with great rupture lengths. The seismic waves used to determine the Richter magnitude come from only a small part of the fault rupture and, hence, cannot provide an accurate measure of the total seismic energy released by a very large event.

Moment magnitude is derived from seismic moment, M_0 (in dyne centimeters), which is proportional to the average displacement (slip) on the fault *times* the rupture area on the fault surface *times* the rigidity of the faulted rock. The amount of seismic energy (in ergs) released from the ruptured fault surface is linearly related to seismic moment by a simple factor, whereas Richter magnitude is logarithmically related to energy. Because of the linear relationship of seismic moment to energy released, the equivalent energy released by other natural and human-caused phenomena can be conveniently compared to earthquakes' moment magnitudes on a graph (◆Figure 4.11b).

Moment magnitudes (M_w) are derived from seismic moments (M_0) by the formula: $M_w = (\frac{2}{3} \log M_0 - 10.7)$. This table compares the two most commonly used scales for selected significant earthquakes:

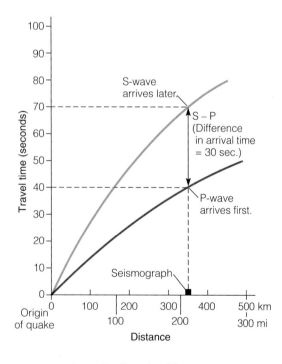

◆ FIGURE 4.8 Generalized graph of distance versus travel time for P- and S-waves. Note that the P-wave has traveled farther from the origin of the earthquake than the S-wave at any given elapsed time. The curvature of both wave paths is due to increases in velocity with depth (distance from epicenter).

Earthquake	Richter Magnitude	Moment Magnitude
Chile, 1960	8.3	9.5
Alaska, 1964	8.4	9.2
New Madrid, 1812	8.7 (est.)	8.1 (est.)
Mexico City, 1985	8.1	8.1
San Francisco, 1906	8.3 (est.)	7.7
Loma Prieta, 1989	7.1	7.0
San Fernando, 1971	6.4	6.7
Northridge, 1994	6.4	6.7
Kobe, Japan, 1995	7.2 JMA*	6.9

*The Richter scale is not used in Japan. The official earthquake scale there is the scale developed by the Japanese Meteorological Agency (JMA).

Confusion arises because some of the earthquake magnitudes are reported as Richter magnitude (M_L, teleseismic

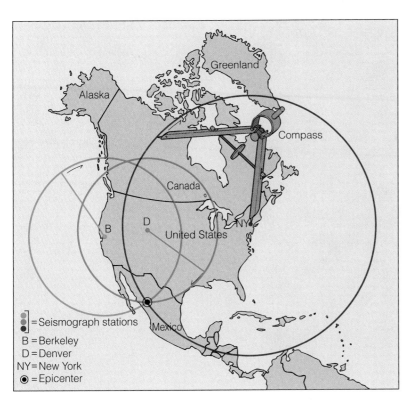

◆ **Geology ⇌ Now** ™ ACTIVE FIGURE 4.9 An earthquake epicenter can be closely approximated by triangulation from three seismic stations.

= Seismograph stations
B = Berkeley
D = Denver
NY = New York
◉ = Epicenter

body wave magnitude (m_b), duration magnitude (m_d), surface wave magnitude (M_s), or moment magnitude (M_w or M). This is simplified somewhat because most newspapers and the Internet report earthquakes as either Richter or moment magnitude.

At the close of the twentieth century the Chilean earthquake of 1960 ($M_w = 9.5$) still held the record for the greatest seismic moment and energy release ever measured; that is, it had the longest fault rupture and the greatest displacement. The Alaskan earthquake of 1964 had the highest Richter magnitude of the 1900s and theoretically more severe ground motion than the event in Chile. (See ❂ Case Study 4.2 on page 103).

CONSIDER THIS *"Then the Lord rained upon Sodom and Gomorrha brimstone and fire. . . . And he overthrew those cities, and all the plain about . . ." (Genesis 19: 24–25). Sodom and Gomorrha are believed to have been located near the modern industrial city of Sedom in Israel at the south end of the Dead Sea. Could some geological event have caused the destruction of these towns? If so, what events might have been possible?*

Click **Geology Interactive** to work through an activity on Earthquakes and Space and Time through Earthquakes and Tsunamis.

Fault Creep, the "Nonearthquake"

Some faults move almost continuously, or in short spurts, and do not produce detectable earthquakes. This type of movement, called **fault creep,** is well known but poorly understood. The Hayward fault, which is part of the San Andreas fault system, is an example of a creeping fault. It is just east of the San Andreas fault and runs through the cities of Hayward and Berkeley, California. (It's on the map in Figure 4.33). Creep along this fault causes displacements of millimeters per year, and is one way that this plate boundary accommodates motion between the Pacific and North American plates. An alternative "accommodation" of plate motion on the San Andreas system is the storage of the strain energy and periodic release as a large displacement on a fault that generates a damaging earthquake. The obvious benefit of a creeping fault does not come without a price, as the residents of Hollister, California, can testify. The "creeping" Calaveras fault runs through their town and, as in Hayward, it gradually displaces curbs, sidewalks, and even residences.

A good strategy in areas subject to fault creep is to map the surface trace of the fault so that it can be avoided in future construction. The University of California–Berkeley Memorial Stadium was built on the Hayward fault before the hazard of fault creep was recognized. The fault creep caused damage to the drainage system that requires periodic maintenance. It is ironic that creep damage occurs at an

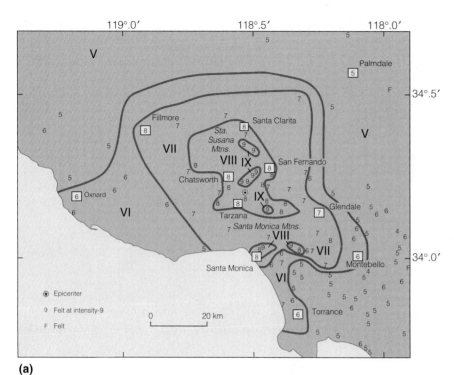

(a)

(b)

ZIP-code areas with no data shown in gray.

◆ FIGURE 4.10 (a) Isoseismal map of the near-field (close to the epicenter) modified Mercalli scale intensities of the 1994 Northridge, California, earthquake. Intensity-IX values are assigned where there are spectacular partial collapses of modern buildings, destroyed wood-frame buildings, and collapses of elevated freeways (see Appendix 4). (b) Community Internet Intensity Map (CIIM) for the same earthquake. The main difference in appearance is due to the fact that the intensities were not contoured but assigned to ZIP code areas of the respondents. (a) SCEC data (b) USGS

therefore excite more short-period seismic waves. This method requires extensive analysis of seismograms and is one of the tools that will be used to measure compliance by signers of the Comprehensive Test Ban Treaty. The following are select anomalous seismic events from 1995 to 2003, some of which were believed to be nuclear explosions. Note that the August 12, 2000 event marks the tragic explosion and sinking of the Russian submarine *Kursk,* and the February 1, 2003 event marks the *Columbia* space shuttle disintegration. Although not on this list of special studies compiled by Lynn Sykes of Lamont–Doherty Earth Observatory of Columbia University, the collapse of the Twin Towers on September 11, 2001 was also recorded on seismographs at Lamont–Doherty.

Location	Date	Type of Event
Novaya Zemlya	January 13, 1996	earthquake
Kola Peninsula	September 19, 1996	chemical explosion
Germany	September 11, 1996	mine collapse
East Kazakhstan	August 03, 1997	chemical explosion
India	May 13, 1998	alleged nuclear explosions
Novaya Zemlya	September 23, 1999	alleged nuclear explosion
Kursk, Barents	August 12, 2000	chemical explosion
Novaya Zemlya	February 23, 2002	earthquake
U.S. space shuttle *Columbia*	February 01, 2003	explosion 100 seconds long

institution that installed the first seismometer in the United States.

Forensic Use of Seismic Records

Seismic records can be used, in conjunction with other methods, to distinguish between earthquakes and nuclear explosions. The basis for the seismic method is that nuclear explosions are more concentrated in space and time and

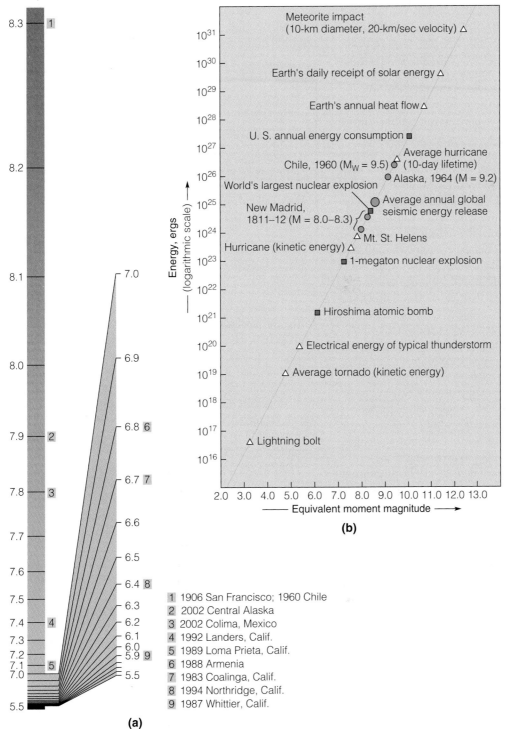

(a)

◆ FIGURE 4.11 Earthquake magnitude and energy. (a) Richter magnitudes of nine selected earthquakes, 1906–1994. (b) Equivalent moment magnitudes of energetic human-caused events (*squares*), natural events (*triangles*), and large earthquakes (*circles*). The erg is a unit of energy, or work, in the metric (cgs) system. To lift a one-pound weight one foot requires 1.4×10^7 ergs.

Seismic Design Considerations

Ground Shaking

Ground movements, particularly rapid horizontal displacements, are most damaging during an earthquake. Shear and surface waves are the culprits, and the potential for them must be evaluated when establishing design specifications. The design objective for earthquake-resistant buildings is relatively straightforward: structures should be designed to withstand the maximum potential horizontal ground acceleration expected in the particular region. Engineers call this acceleration *base shear*, and it is usually expressed as a percentage of the *acceleration of gravity* (g). On earth,

g is the acceleration of a falling object in a vacuum (9.8 m/sec², or 32 ft/sec²). In your car it is equivalent to accelerating from a dead stop through 100 meters in 4.5 seconds. An acceleration of 1 g downward produces weightlessness, and a fraction of 1 g in the horizontal direction can cause buildings to separate from their foundations or to collapse completely. An analogy is to imagine rapidly pulling a carpet on which a person is standing; most assuredly the person will topple.

The effect of high horizontal acceleration on poorly constructed buildings is twofold. Flexible-frame structures may be deformed from cube-shaped to rhomb-shaped, or they may be knocked off their foundations (◆Figure 4.12). More rigid multistory buildings may suffer "story shift" if floors and walls are not adequately tied together (◆Figure 4.13). The result is a shifting of floor levels and the collapse of one floor upon another like a stack of pancakes. Such structural failures are not survivable by inhabitants, and they clearly illustrate the adage that "earthquakes don't kill people; buildings do."

Damage due to shearing forces can be mitigated by bolting frame houses to their foundations and by *shear walls.* An example of a shear wall is plywood sheeting nailed in place over a wood frame, which makes the structure highly resistant to deformation. Wall framing, usually two-by-fours, should be nailed very securely to a wooden sill that is bolted to the foundation. Diagonal bracing and blocking also provide shear resistance (◆Figure 4.14). L-shaped structures may suffer damage where they join, as each wing of the structure vibrates independently. Such damage can be minimized by designing *seismic joints* between the building wings or between adjacent buildings of different heights. These joints are filled with a compressible substance that will accommodate movement between the structures.

Wave period is the time interval between arrivals of successive wave crests, or of equivalent points of waves, and it is expressed as T in seconds. It is an important consideration when assessing a structure's potential for seismic damage, because if a building's natural period of vibration is equal to that of seismic waves, a condition of resonance exists. **Resonance** occurs when a building sways in step with an oscillatory seismic wave. As a structure sways back and forth under resonant conditions, it gets a push in its direction of sway with the passage of each seismic wave. This causes the sway to increase, just as pushing a child's swing at the proper moments makes it go higher with each push. Resonance also may cause a wine glass to shatter when an operatic soprano sings just the right note or frequency.

Low-rise buildings have short natural wave periods (0.05–0.1 seconds), and high-rise buildings have long natural periods (1–2 seconds). Therefore, high-frequency (short-period) waves affect single-family dwellings and low-rise buildings, and long-period (low-frequency) waves affect tall structures. Close to an earthquake epicenter, high-frequency waves dominate, and thus more extensive home and low-rise

damage can be expected. With distance from the epicenter, the short-period wave energy is absorbed or dissipated, resulting in the domination of longer-period waves.

CONSIDER THIS   *Over time, tectonic stresses strain the rocks along a fault and increase the chance of sudden rupture and a potentially damaging earthquake. For this reason, it is generally believed that a region that has had many small earthquakes (magnitude <4.0) has a low probability for a large earthquake. Question: For a magnitude-7.0 earthquake with a 50-year recurrence interval, how many magnitude-4.0 earthquakes would be required to "balance the energy books"? How many small earthquakes a day is this? Is this belief realistic in a seismically active region based upon your calculations?*

Landslides

Thousands of landslides are triggered by earthquakes in mountainous or hilly terrain; there were an estimated 17,000 during the Northridge earthquake alone (see ◎ Case Study 4.1 on page 103). The 1989 Loma Prieta earthquake caused landsliding in the Santa Cruz Mountains and adjacent parts of the California Coast Ranges. Such areas are slide prone under the best of conditions, and even a small earthquake will trigger many slope failures. In the greater San Francisco Bay area there was an estimated $10 million damage to homes, utilities, and transportation systems because of landslides and surficial ground failures resulting from the Loma Prieta earthquake.

One of the worst earthquake-triggered tragedies in the United States occurred in August 1959, a short distance from Yellowstone Park. A landslide, in reality a massive rockslide (see Chapter 7), originated on Red Mountain above the Madison River in Montana. The rockslide was triggered by a moderate earthquake that caused the metamorphic rock (schist) composing the mountain to slide down its foliation planes, which were inclined parallel to the hill slopes (◆Figure 4.15). However, the rockslide occurred above a popular campground, and 26 people were killed by the falling rock. The slide generated a terrific shock wave of air that lifted cars, trees, and campers off the ground. In addition, the Madison River was dammed by the slide and a lake formed, later named "Quake Lake."

Ground or Foundation Failure

Liquefaction is the sudden loss of strength of water-saturated, sandy soils resulting from shaking during an earthquake. Sometimes called **spontaneous liquefaction,** it can cause large ground cracks to open, lending support to the ancient myth that the earth opens up to swallow people and animals during earthquakes. Shaking can cause saturated sands to consolidate and thus to occupy a smaller volume. If the water is slow in draining from the consolidated material, the overlying soil comes to be supported only by

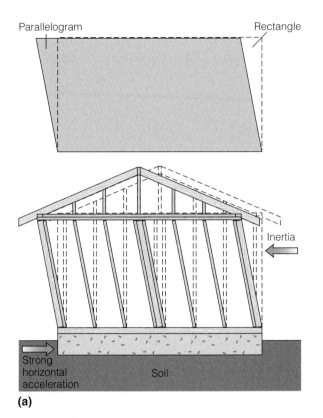

(a)

(b)

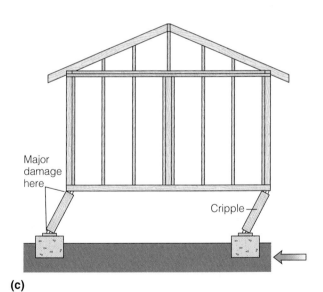

(c)

(d)

◆ FIGURE 4.12 (a) Strong horizontal motion may deform a house from a cube to a rhomboid or knock it from its foundation completely. (b) A Coalinga, California, frame house that was deformed by the magnitude-6.3 earthquake in 1983. (c) A cripple-wall consists of short vertical members that connect the floor of the house to the foundation. Cripple-walls are common in older construction. (d) A Watsonville, California, house that was knocked off its cripple-wall base during the Loma Prieta earthquake of 1989. Without exception, cripple-walls bent or folded over to the north relative to their foundations.

JAMES C. ANDERSON, USC

◆ FIGURE 4.13 (Total vertical collapse as a result of "story shift"; Mexico City, 1985. Such structural failures are not survivable.

pore water, which has no resistance to shearing. This may cause buildings to settle, earth dams to fail, and sand below the surface to blow out through openings at the ground surface (◆Figure 4.16a). Liquefaction at shallow depth may result in extensive lateral movement or spreading of the ground, leaving great cracks and openings.

Ground areas most susceptible to liquefaction are those that are underlain at shallow depth—usually less than 30 feet—by layers of water-saturated fine sand. With subsurface geologic data obtained from water wells and foundation borings, liquefaction-susceptibility maps have been prepared for many seismically active areas in the United States.

Similar failures occur in certain clays that lose their strength when they are shaken or remolded. Such clays are called *quick clays* and are natural aggregations of fine-grained clays and water. They have the peculiar property of turning from a solid (actually a gel-like state) to a liquid when they are

agitated by an earthquake, an explosion, or even vibrations from pile driving. They occur in deposits of glacial-marine or glacial-lake origin and are therefore found mostly in northern latitudes, particularly in Scandinavia, Canada, and the New England states. Failure of quick clays underlying Anchorage, Alaska, produced extensive lateral spreading throughout the city in the 1964 earthquake (◆Figure 4.16b).

The physics of failure in spontaneous liquefaction and in quick clays is similar. When the earth materials are water-saturated and the earth shakes, the loosely packed sand consolidates or the clay collapses like a house of cards. The pore-water pressure pushing the grains apart becomes greater than the grain-to-grain friction, and the material becomes "quick" or "liquifies" (◆Figure 4.17). The potential for such geologic conditions is not easily recognized. In many cases it can be determined only by information gained from bore holes drilled deeper than 30 meters (100 ft). Because of this expense, many site investigations do not include deep drilling, and the condition can be unsuspected until an earthquake occurs.

Ground Rupture and Changes in Ground Level

Structures that straddle an active fault may be destroyed by actual ground shifting and the formation of a fault scarp (◆Figure 4.18). By excavating trenches across fault zones, geologists can usually locate past rupture surfaces that may reactivate. In 1970 the California legislature enacted the Alquist–Priolo Special Study Zones Act, which was renamed the *Earthquake Fault Zones Act* in 1995. It mandates that all known active faults in the state be accurately mapped and zoned for seismic safety. The act provides funds for state and private geologists to locate the youngest fault ruptures within a zone and requires city and county governments to limit land use adjacent to identified faults within their jurisdic-

◆FIGURE 4.14 Methods of reinforcing structures against base shear. (a) Diagonal cross-members and blocks resist horizontal earthquake motion (shear). (b) Plywood sheeting forms a competent shear wall, and metal "L" braces and bolts tie the structure to the foundation.

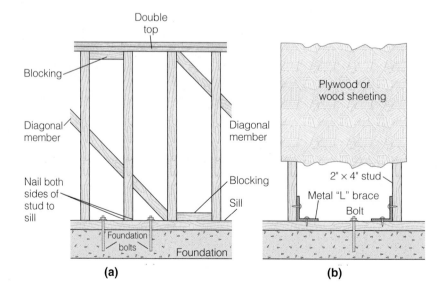

(a)   (b)

B. PIPKIN

◆ FIGURE 4.15 Pictured is the site of the earthquake-induced rockslide at Red Mountain. The Madison River is just off the picture in the foreground. The rockslide was triggered by an earthquake and took place in schist with foliation planes inclined parallel to the mountain slope (see Chapter 7).

tions. Ironically, the faults responsible for the 1992 Landers earthquake (M_w = 7.3) were designated as Earthquake Fault Zones just prior to the June 28 event.

Changes in ground level as a result of faulting may have an impact, particularly in coastal areas that are uplifted or down-dropped. For instance, during the 1964 Alaskan event, parts of the Gulf of Alaska thrust upward 11 meters (36 ft), exposing vast tracts of former tidelands on island and mainland coasts.

Fires

Fires caused by ruptured gas mains or fallen electric power lines can add considerably to the damage caused by an earthquake. In fact, most damage attributed to the San Francisco earthquake of 1906 and much of that in Kobe, Japan, in 1995 was due to the uncontrolled fires that followed the earth-

quakes. The Kobe quake hit at breakfast time. In neighborhoods crowded with wooden structures, fires erupted when natural-gas lines broke and falling debris tipped over kerosene stoves. Broken water mains made fire-fighting efforts futile. One of the principal "do's" for citizens immediately after a quake is to shut off the gas supply to homes and other buildings in order to prevent gas leaks into the structure. This in itself will save many lives (see Figure 4.41).

Tsunamis

The most myth-ridden hazard associated with earthquakes (and submarine volcanic eruptions and landslides) is tsunami (pronounced "soo-nah-mee"), or seismic sea waves. **Tsunami** is a Japanese word meaning "great wave in harbor" and it is appropriate because these waves are impulsively generated and most commonly wreak death and destruction inside bays and harbors. They have nothing to do with the tides, even though the term "tidal wave" is commonly used in the English-speaking world. The Japanese written record of tsunamis goes back 200 years and their power is dramatically displayed in the well-known print by Hokusai (◆Figure 4.19). Tsunamis, their causes, and effects are treated in detail in Chapter 10.

ENVIRONMENTAL
Geology⇌Now™

Click **Geology Interactive** to work through an activity on Tsunamis through Earthquakes and Tsunamis.

Four Earthquakes That Make a Point

Each year in the world, on average, there are at least two earthquakes of magnitude 8.0 or greater and 20 earthquakes in the magnitude 7.0 to 7.9 range. Release of seismic energy

NOAA

(a)

NGDC 97

(b)

◆ FIGURE 4.16 (a) These apartment buildings tilted as a result of soil liquefaction in Niigata, Japan, in 1964. Many residents of the building in the center exited by walking down the side of the structure. (b) "House of cards" collapse of quick clay structure in Turnagain Heights, Anchorage, Alaska in 1964. Total destruction occurred within the slide area, which is now called Earthquake Park.

◆ FIGURE 4.17 Liquefaction (lateral spreading) due to repacking of spheres (idealized grains of sand) during an earthquake. The earthquake's shaking causes the solids to become packed more efficiently and thus to occupy less volume. A part of the overburden load is supported by water, which has no resistance to lateral motion.

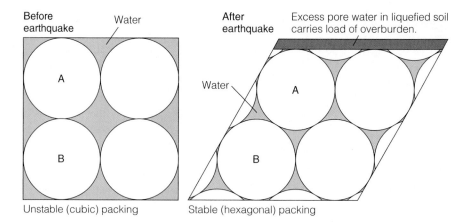

in the form of earthquakes has occurred throughout geologic time, and recorded history contains many references to strong earthquakes. More than 3,000 years of seismicity are documented in China, and Strabo's *Geography* mentions an earthquake in 373 B.C. in Greece. Thus, humans have probably always been subject to earthquakes (✳ Table 4.1).

◆ FIGURE 4.18 The 1992 Landers earthquake (M_w = 7.3) in the Colorado Desert of southern California was felt from Phoenix, Arizona, to Reno, Nevada. Right-lateral offset of 4.27 meters (14 ft) on the Johnson Valley fault created a 2-meter (6.5-ft) vertical scarp due to lateral offset of a ridge. The geologist is 185 cm tall.

This section describes four earthquakes that occurred between 1994 and 2003. All four had very strong magnitudes but had quite different results because of differences in local foundation conditions, building standards, population density, and societal factors. The following is a synopsis of these events.

- A magnitude-7.7 earthquake struck Gujarat, India at 8:46 A.M. on January 26, 2001. It was 60 times more powerful than the Northridge, California, earthquake of 1994 but the death toll was 400 times greater.

- A magnitude-7.9 earthquake struck central Alaska on November 3, 2002. It was on the Denali fault and was one of the largest strike-slip ruptures in Alaska of the past two centuries. One minor injury was reported and the dollar loss was minimal.

- A magnitude-7.8 earthquake struck the state of Colima, Mexico, on January 22, 2003. The destruction was widespread, but the death toll was only 29. The victims were mostly in dwellings that were poorly built and collapsed. Modern high-rise buildings faired well.

- A magnitude-6.7 struck the San Fernando Valley at the community of Northridge on January 17, 1994. It is recognized as the second most costly natural disaster in U.S. history, second only to Hurricane Andrew (see Chapter 10).

Gujarat, India, 2001

A headline in a well-known American newspaper stated "Bad quake, worse buildings." This just about sums up the impact of this strong event in the state of Gujarat (◆Figure 4.20). It was a reverse-fault earthquake and the closest plate boundary lies many hundreds of kilometers away. It was felt 2000 km away and over an area 16 times that of the M_w = 7.8 San Francisco earthquake of 1906. Field investigation in Gujarat revealed no ground rupture, which is unusual for an earthquake of this magnitude. Liquefaction was widespread because of the high water table and thickness of unconsolidated sediment in the

◆ FIGURE 4.19 "The Breaking Wave off Kanagawa," wood-block color print by Katsushika Hokusai (1760–1849) from the series "Thirty-six Views of Mount Fuji," 1826–33. Metropolitan Museum of Art, H. O. Havemeyer Collection (JP 1847)

Location	Year	Richter Magnitude**	Impact
TABLE 4.1 Selected Significant World Earthquakes, in Chronological Order			
San Francisco	1906	8.3	700 killed, $7 million damage, fire
Messina (Sicily)	1908	7.5	160,000 killed
Tokyo, Japan	1923	8.3	140,000+ killed, fire
Assam, India	1950	8.4	30,000 killed
Chile	1960	M_w 9.5	5,700 killed, 58,000 homes destroyed, tsunami
Alaska	1964	M_w 9.2	131 killed, tsunami
T'ang-shan, China	1976	7.9	330,000 killed
Mexico City	1985	8.1	10,000+ reported killed
Armenia	1988	6.8	55,000 killed
Loma Prieta (California)	1989	M_w 6.9	67 killed, $8 billion damage
Northridge, California	1994	M_w 6.7	60 killed, $25 billion damage
Kobe, Japan	1995	M_w 6.9	5,378 killed, $100 billion damage
Izmit (Kocaeli), Turkey	1999	M_w 7.4	15,370 killed, $10–$20 billion damage
Chi-Chi, Taiwan	1999	M_w 7.6	2,300+ killed
Gujarat, India	2001	M_w 7.7	Reported 20,000+
Alaska	2002	M_w 7.9	Little impact, no casualties
Colima, Mexico	2003	M_w 7.8	Widespread destruction, 27 fatalities
Bam, Iran	2003	M_w = 6.5	30,000 believed killed, 30,00 injured

*A "significant" earthquake is defined as one that registers a moment magnitude (M_w) of at least 6.5 or a lesser one that causes considerable damage or loss of life. The world averages 60 significant earthquakes per year.
**Estimated prior to 1935. M_w = Moment magnitude.
Sources: U.S. Geological Survey: *Earthquakes and Volcanoes;* California Division of Mines and Geology; *Academic American Encyclopedia* (1990, Grolier Electronic Publishing Company).

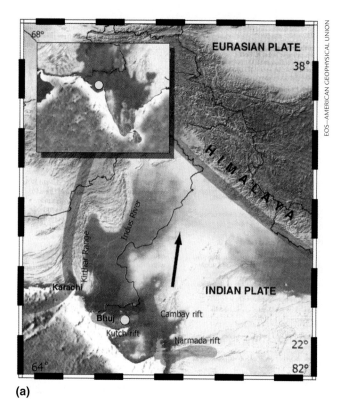

(a)

(b)

◆ FIGURE 4.20 (a) The regional geography and plate tectonic setting of the Gujarat (Bhuj) earthquake. The epicenter is shown by the yellow dot. Bhuj lies 400–500 km from a plate boundary and a greater distance from the Himalayan range. Also shown are buried rift basins that are seismically active and very similar to those of the central United States (see page 94 on New Madrid). (b) Collapsed houses in the town of Ratnal, in the epicentral region of the Bhuj earthquake.

region. The quake occurred in an old rift system, which is a well-known seismic zone that is very similar to parts of the central United States. One doesn't have to be a seismologist to know that an earthquake of this magnitude has real potential to cause devastation. Here, the lack of strict building codes governing construction methods, some of which date back to the British Colonial period, exacerbated the destruction that led to the loss of more than 20,000 lives. Here also, one can become an architect or building contractor without any special education or license. The damage covered an area over 500 km wide from the large city of Ahmedabad in the east to the Arabian Sea in the west. Severe damage was reported within an area 50 km by 70 km in which most high-rises (anything more than 3 stories) and low-rise cobble-stone structures collapsed.

What is surprising to geologists is that this earthquake occurred in an area of relative flat topography without the usual tectonic landforms suggestive of faulting and seismic activity. Strong earthquakes in India usually occur as the Indian plate moves northward against the Himalayas. However, a similar event occurred in 1819 in the Rann of Katchchh, also in the state of Gujarat, just to the southeast of the 2002 earthquake epicenter. This suggests a greater rate of tectonic (earthquake) activity than is indicated by the land forms there.

Geologists from the United States are particularly interested in this event because it may serve as an analog for mid-continent earthquakes and the New Madrid, Missouri, seismic zone (Figure 4.30). Major earthquakes are rare in intraplate regions like New Madrid and Gujarat, far from plate edges and the usual seismicity associated with the world's tectonic-plate boundaries. As pointed out, there is little evidence at the ground surface left by these earthquakes. Thus, each intraplate earthquake is an opportunity to study and better understand the hazards posed by these events.

Alaska, 2002

On November 3 a magnitude-7.9 earthquake occurred some 90 miles south of Fairbanks along the Denali fault (◆Figure 4.21). Denali is the native Tanana word for "high one," and the Denali fault is the most well-known and studied active fault in the state. This earthquake was among the strongest ever recorded in the United States. It shut down the Alaska pipeline, and caused lakes to ripple in Iowa and water to slosh out of swimming pools as far south as Louisiana. The earthquake was shallow and its energy went directly to the surface to produce the distant effects. One resident of Porcupine Creek is quoted as saying "A charging brown bear I can handle. This (the earthquake) scared the heck out of me." Alaska has a history of strong earthquakes, demonstrated in 1964 when the Denali fault ruptured and left 159 dead and many communities in ruins, mostly due to tsunami. This most recent event is a reminder that the fault has a great potential for damage.

Colima, Mexico, 2003

On January 22 a magnitude-7.8 earthquake struck near the village of Tecoman in the state of Colima (◆Figure 4.22). Twenty-seven people lost their lives in Colima and two more in the adjoining state of Jalisco. In this shallow-focus quake, the epicenter occurred near the junction of three tectonic

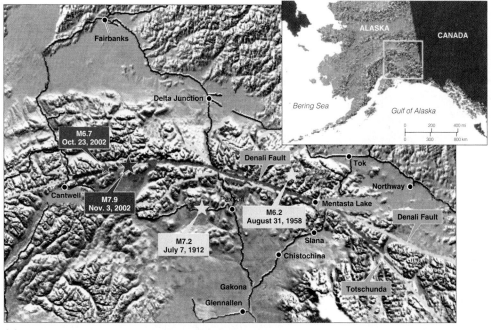

(a)

(b)

◆ FIGURE 4.21 (a) The magnitude-7.9 Denali fault earthquake of November 3, 2002, resulted from predominantly right-lateral offset along portions of the Denali and Totschunda fault systems in Alaska. Total length of the surface rupture was about 320 kilometers (200 mi). The western 49 km of the rupture shows mainly low-angle thrust offset as much as 1.5 m, with the northwest side up. Shown here are the epicenters of the November 3 event (red) and the magnitude-6.7 foreshock of October 23, 2002 (blue), as well as two previously recorded large shallow earthquakes (yellow) in the vicinity of the fault. (b) Aerial view of the Trans-Alaska Pipeline System (TAPS) line near the Denali fault, looking west. This is where the pipeline is supported by rails on which it can move freely in the event of fault offset. Here the line has slid toward the west end of the rails. Alyeska Pipeline Service Company reported no breaks to the line and therefore no loss of oil. Out of view to the left (south) is the 2.5 m (ft) right-lateral offset of the highway where it crosses the fault. Photo courtesy of Rod Combellick, Division of Geological and Geophysical Surveys, State of Alaska

plates: North American plate to the northeast, the Cocos plate to the south, and the Rivera plate to the northwest. Both the smaller plates are being subducted beneath the North American plate, producing significant seismic activity in the region. Twenty of the fatalities occurred in the collapse of adobe-brick houses. Over 150 houses were reported destroyed in the city of 200,000 that dates back to sixteenth-century Spanish colonial times. Mexican structural engineers are very aware of the seismic hazards that exist in this part of the country and are quite good at designing buildings to be earthquake-resistant. As a result, no damage occurred to modern high-rise buildings from this powerful earthquake. However, it was felt strongly and caused some to panic in Mexico City, where an earthquake in 1985 with its epicenter

in the same region took over 10,000 lives. Two 20-story buildings in the capital swayed so much that they actually collided, but did little damage.

Northridge, California, 1994

The largest earthquake in Los Angeles's short history occurred at 4:30 A.M. Monday, January 17, 1994, on a hidden fault below the San Fernando Valley. The $M_w = 6.7$ earthquake started at a depth of 18 kilometers (11 mi) and propagated upward in a matter of seconds to a depth of 5–8 kilometers (3–5 mi, ◆Figure 4.23). There were thousands of aftershocks, and their clustered pattern indicated the causative fault to be a reverse fault with a shallow dip of 35°

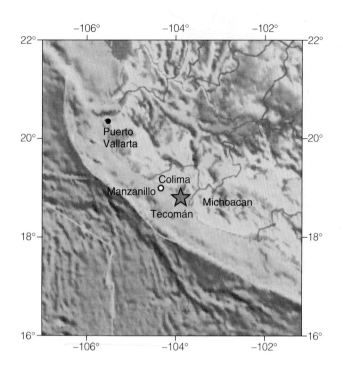

◆ FIGURE 4.22 Magnitude-7.8 earthquake near Tecomán, Colima, Mexico, on January 22, 2003. USGS

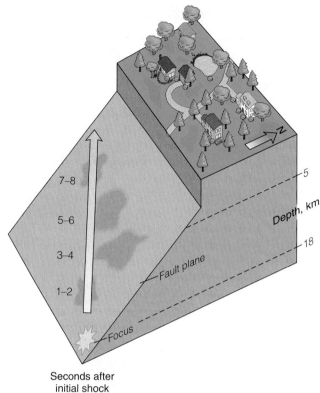

Seconds after initial shock

◆ FIGURE 4.23 In the 1994 Northridge earthquake, the fault rupture progressed up the fault plane from the focus at the lower right of the figure to the upper left in 8 seconds. Rather than rupturing smoothly, like a zipper opening, it moved in jerks along the fault plane, as shown by the pink patches. Total displacement was about 4 meters (12 ft).

to the south (Case Study 4.2 and Appendix 3). Such a low-angle reverse fault is called a *thrust fault,* and, because the thrust did not rupture the ground surface, it is described as "blind" (◆Figure 4.24). Blind thrusts were first recognized after the 1987 Whittier earthquake 40 kilometers (25 mi) southwest of Northridge, but they were not fully appreciated until after the Northridge event.

Blind thrust faults were a newly identified seismic hazard, and there is evidence that a belt of them underlies the northern Los Angeles basin. They are especially dangerous because they cannot be detected by traditional techniques such as trenching and field mapping (thus they do not fall within the Earthquake Fault Zones Act as it is written) and yet, as we know now, they can generate significant earthquakes. Just as the residents of earthquake country were beginning to feel confident that geologists knew the location and behavior of most active faults, blind thrusts made their presence known.

Field inspection of the epicentral region was a depressing experience. Thirteen thousand buildings were found to be severely damaged; 21,000 dwelling units had to be ordered evacuated; 240 mobile homes had been destroyed by fire; 11 major freeway overpasses were damaged at 8 locations (✪ Case Study 4.3 on page 104). The reason for the extensive damage was the high horizontal and vertical accelerations generated by the earthquake. Accelerations of more than 0.30 *g* are considered dangerous, and the vertical acceleration of 1.8 *g* measured by an instrument bolted to bedrock in nearby Tarzana probably set a world record. The high ground acceleration explains why many people were literally thrown out

of their beds and objects as heavy as television sets were projected several meters from their stands. Although the damage was not as visually spectacular as that at Mexico City in 1985, it was heartbreaking to many residents—brick chimneys, both reinforced with steel bars and unreinforced, came down; houses moved off their foundations; concrete-block walls crumbled; gaps broke open in plaster walls; the number of shattered storefront windows was beyond counting (◆Figure 4.25). The horizontal ground motion was directional; that is, it was strongest in the north–south direction. With few exceptions, block walls oriented east–west tipped over or fell apart, whereas those oriented north–south remained standing. Many two- and three-story apartment buildings built over open first-floor garages collapsed onto the residents' cars. These open parking areas' lack of shear resistance led to failure of the vertical supports, and everything above came down.

California State University at Northridge sustained almost "textbook" earthquake damage. In its library is "Leviathan II," recognized as one of the most advanced automated book-withdrawal systems in the country. On this day it recorded a record withdrawal of about 500,000 books—all of them onto the floor of the library. Bottled chemicals fell

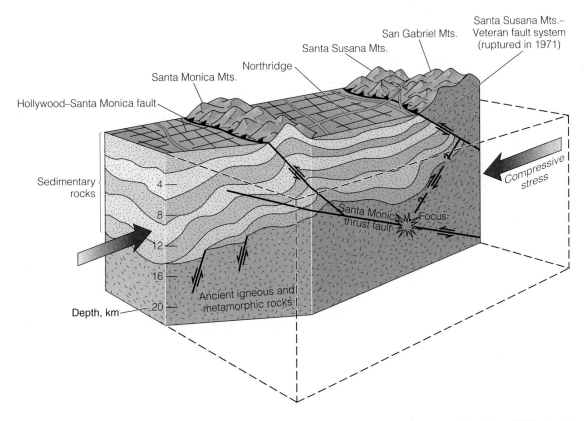

◆ FIGURE 4.24 Interpretation of the fault system in the San Fernando Valley and surrounding area of Southern California. Compressive stresses built up over a long period, causing the subsurface blind thrust fault to rupture. The ensuing Northridge earthquake impacted the entire valley, and extended from the Santa Susana Mountains on the north to the city of Santa Monica on the south. USGS DATA

off their storage shelves (a major concern at any university) and caused a large fire in the Chemistry Building. In addition, one wall of a large open parking structure collapsed inward to produce the stunning art deco architecture seen in ◆Figure 4.25c.

What We Can Learn

These four earthquakes reinforce the adage that "earthquakes don't kill people, falling buildings kill people." The magnitudes were in the range where destruction is certain if construction methods are not adequate. In Gujarat high-rise structures and stone dwellings alike collapsed due to inadequate design leading to a high death toll. In Alaska, where the population density is about one person per square mile, the strong temblor took no lives and really did minimal damage. On the other hand, the example in Mexico illustrates that even when high population areas are struck by a powerful earthquake, recognition of the importance of seismic design criteria can reduce the loss of life. The Northridge earthquake was an unusual case. There the combination of a blind thrust fault and very high ground accelerations, both vertical and horizontal, combined to cause the extensive damage in Southern California.

It should be noted that other factors beside construction, population density, and kind of fault movement can cause damage. Site effects are extremely important. During the 1989 Loma Prieta (San Francisco) earthquake the ground shaking around the bay area varied considerably. There was liquefaction and widespread destruction in the Marina District, where two-level Interstate 880 was built on bay muds. Near Oakland it shook violently and failed (◆Figure 4.26) causing fatalities, whereas structures in San Francisco built on bedrock suffered much less or no damage. The seismograms shown in ◆Figure 4.27 illustrate this point (also see Case Study 4.3).

Site effects were noted almost 200 years ago when an observer of the New Madrid sequence (next topic) noted "The convulsion was greater along the Mississippi, as well as along the Ohio, than in the uplands. The strata in both valleys are loose. The more tenacious layers of clay and loam spread over the adjoining hills . . . suffered but little derangement" (see Drake, 1815, in For Further Information).

Does Earthquake Country Include Idaho, Missouri, and New York?

Even though the vast majority of the world's earthquakes occur at plate boundaries, areas hundreds and even thousands

(a)

(b)

(c)

◆ FIGURE 4.25 (a) Two wings of an apartment building that collapsed toward one another. (b) A staircase that leads to nowhere. In reality, it was the second-floor breezeway that collapsed onto the ground floor. (c) Part of the collapsed parking structure of Fashion Center, Northridge. Such collapses instigated new design criteria for open structures. Note the outside staircase that separated from the main structure.

of kilometers away are not free of seismic activity. In the United States the five most seismically active states between 1980 and 1991 were

	Earthquakes 1980–1991	
State	**Number Recorded**	**Largest, M_w**
Alaska	10,253	9.2
California	6,732	7.2
Washington	615	5.5
Idaho	536	7.3
Nevada	398	5.6

Although Alaska and California continue to lead the United States in the number of shakers, the earthquakes that have been felt over the largest area occurred in Missouri in 1811 and 1812, and significant seismic hazards are recognized in 39 states. No state is earthquake proof, as the seismic-risk map of the United States shows (◆Figure 4.28). It appears that U.S. residents who want to be seismically safe should move to Texas, Florida, or Alabama.

Why do such strong earthquakes occur *intraplate*—that is, far from a plate boundary (✳ Table 4.2)? Intraplate earthquakes have several characteristics in common (see also the Gujarat earthquake, this chapter):

1. The faults causing them are deeply buried and have not broken the ground surface.

◆ **FIGURE 4.26** The I-880 freeway failed because it was built on bay muds that reacted to seismic waves like a bowl of gelatin. Pictured is the Cypress Street viaduct with the upper deck collapsed upon the lower one. There was loss of life here.

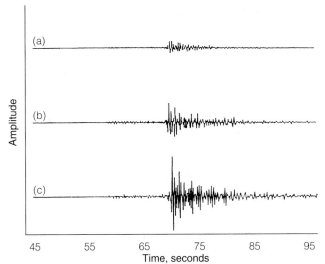

◆ **FIGURE 4.27** Seismograms for a magnitude-4.1 aftershock as recorded on (a) firm bedrock, (b) alluvium (stream-deposited sediment), and (c) areas of fill and bay muds. This is where the most damage occurred in the San Francisco earthquakes of 1906 and 1989. Data from EOS—American Geophysical Union

2. Because the rocks of the continental interior are stronger than those at plate boundaries, which are laced with faults, they transmit seismic waves better, causing ground motion over a huge area. In the United States this is usually many states (◆Figure 4.29).

3. They do not appear to be random events.

The Charleston, South Carolina, earthquake of 1886 was larger than the 1989 San Francisco shaker. It killed scores of people, ruined the city, and slowed the South's recovery from the Civil War. The New Madrid (pronounced "mad′rid") earthquakes of 1811 and 1812 caused extensive topographic changes, locally reversed the course of the Mississippi River, and may have been felt over a larger area than any other earthquake in recorded history. Ground motion was felt as far away as Washington, D.C., where it caused church bells to ring and scaffoldings on the Capitol to collapse.

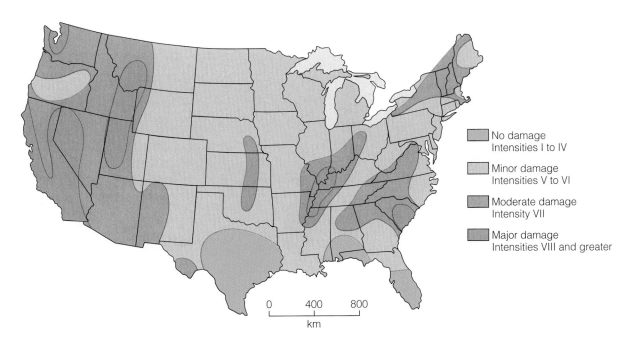

No damage
Intensities I to IV

Minor damage
Intensities V to VI

Moderate damage
Intensity VII

Major damage
Intensities VIII and greater

◆ **FIGURE 4.28** Seismic-risk map of the 48 contiguous states based upon historical records and intensities collected by the U.S. Coast and Geodetic Survey. The Coast and Geodetic Survey gathers intensity data from questionnaires after earthquakes. From M. L. Blair and W. W. Spangle, USGS prof. paper 941-B, 1979

✳TABLE 4.2 Selected North American Intraplate Earthquakes

Location	Year	Moment Magnitude*	Impact
New Madrid, Missouri	1811	8.2	Reelfoot Lake formed in NW Tennessee.
New Madrid	1812	8.3	Elevation changes caused Mississippi River to reverse its course locally.
New Madrid	1812	8.1	
Charleston, South Carolina	1886	7.6	Felt from New York to Chicago; 60 killed.
Charleston, Missouri	1895	6.8	Damage in 6 states (see Figure 4.28).
Grand Banks, Newfoundland	1929	7.4	Submarine landslides broke trans-Atlantic cable, disrupting communications.

*Moment magnitude is used in reconstructing the strength of "preinstrument" earthquakes, because we know something of fault length, rupture length, and area felt.

Recent studies suggest that intraplate earthquakes are concentrated in areas where normally stable crust has been stretched and faulted. Such zones form where continents have been split apart *(rifted)*, forming two continents, as occurs at divergent plate boundaries (see Chapter 3). When North America was separated from Africa at the Mid-Atlantic Ridge about 180–200 million years ago, the continental crust being rifted at the ridge was thinned, stretched, and faulted. The North American continent was transported westward with the plate, and the weakened part of the crust along its eastern edge was hidden by a thick cover of younger sedimentary rocks. Buried rifted crust under the Eastern seaboard states forms the intraplate earthquake zone from the Carolinas to Canada.

The Midwest earthquake zone through Arkansas, Missouri, and Illinois is believed to be where an ancient (pre-Pangaea) divergent boundary started to form but "failed" for some reason, leaving behind a significant buried fault zone. Known as the *Reelfoot Rift*, the buried block of crust is down-dropped between faults (the hachured lines on the map of ◆Figure 4.30). The rift is 60 kilometers wide and 300 kilometers long (roughly 40 mi × 190 mi) and formed at least 500 million years ago. The linear trend of earthquakes from Marked Tree, Arkansas, northeastward to Caruthersville, Missouri, reveals upwarped sedimentary rocks along the rift axis. Detailed analysis of the geology across the Reelfoot fault scarp by trenching revealed evidence of three large earthquakes within the past 2,000 years, which yields a recurrence-interval estimate of 600–900 years for the fault, but the recurring earthquakes are not necessarily of the same magnitude as the 1811–1812 events. In fact, scientists estimate that a magnitude-6–7 earthquake will occur in the New Madrid seismic zone between 2000 and 2050 with a 90 percent probability.

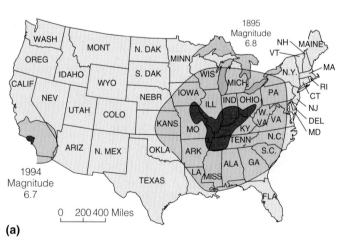

(a)

(b)

◆ **FIGURE 4.29** Mid-continent earthquakes. (a) Comparison of the effects of two earthquakes of similar magnitude. The 1895 Charleston, Missouri, earthquake impacted a relatively huge area in the central United States, whereas the 1994 Northridge earthquake's effects were limited to Southern California. Red indicates areas of minor-to-major damage to buildings and their contents. Yellow indicates areas where shaking was felt but there was little or no damage to objects, such as dishes. (b) Damage from the 1886 earthquake along East Bay Street in Charleston, South Carolina.

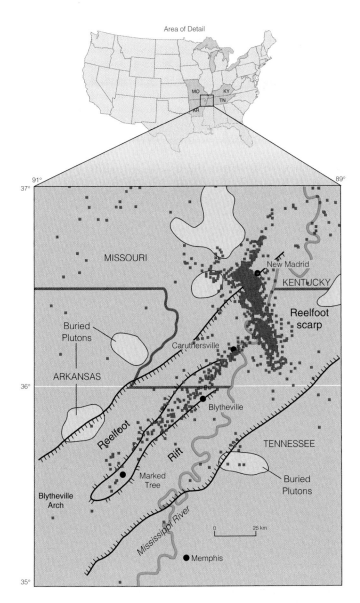

◆ FIGURE 4.30 The New Madrid seismic zone covers parts of four states and consists of an ancient, fault-bounded depression (rift) buried below 1.5 kilometers (almost a mile) of sedimentary rocks. An area of major seismic hazard, it is studied intensively. USGS DATA

Iben Browning, a scientist with a Ph.D. in physiology but who is best-known for his work on climate, predicted there would be a repeat of the 1811 New Madrid earthquake on December 3, 1990. His prediction was based upon alignment of the planets and consequent gravitational pull, which he believed would be sufficient to trigger an earthquake on that date. Although the prediction was discounted by scientists, it created considerable anxiety in Arkansas, Tennessee, Alabama, and Missouri. Months before "the day," earthquake insurance sales boomed, moving companies were booked up, and bottled-water sales rose dramatically—as did sales of quake-related souvenirs. One church sold "Eternity Preparedness Kits," and "Survival Revivals" were held. On the day of the scientifically discredited prediction nothing earthshaking occurred.

Many earthquakes with epicenters in weakened intraplate continental crust have shaken parts of the Midwest, eastern Canada, New England, New York, and the East Coast of the United States. New York does not have a major fault, but it has many faults. In the 1980s an earthquake in Westchester County ($M = 4.0$) toppled chimneys and caused enough panic for the state to implement new seismic codes that added 2–5 percent to building costs. In April 2002 the state experienced a magnitude-5.1 shaker that centered 15 miles southwest of Plattsburgh near the Canadian border. It rattled dwellings from Maine to Maryland and left cracks in foundations and chimneys. There were no injuries. The point here is that a relatively small intraplate earthquake can be felt over a huge area. By the mid 1990s New York and Massachusetts were the only Northeastern states that had earthquake building codes.

The Pacific Northwest

In years past, earthquake hazards were considered minor in Oregon and Washington. During the 1980s research changed the perception, however; it revealed geologic evidence that "major" (≥ 7.0 but < 8.0) or "great" (≥ 8.0 but < 9.0) subduction-zone earthquakes have occurred in the past and that they can occur in the future. The Cascadia subduction zone (◆Figure 4.31), extending 1,200 kilometers (740 mi) from northern California to Vancouver Island in Canada, has destructive earthquake potential. Geological evidence, consisting of carbon-dated tsunami deposits and drowned red-cedar forests, suggests that great earthquakes strike the Pacific Northwest roughly every 500 years, the last one in A.D. 1700, approximately 300 years ago. Minor earthquakes occur daily, however, and the cities of Portland, Seattle, and Vancouver, British Columbia, are in earthquake country (◆Figure 4.32). The states now are working to minimize damage should the "big one" occur on the subduction zone or on the Seattle fault, which runs through downtown Seattle.

The year 2000 marked the tricentennial of the last great (equal to or greater than $M = 8.0$) earthquake generated along the Cascadia Subduction Zone. To commemorate this, almost 100 earth scientists and public officials gathered at Seaside, Oregon, to assess the seismic hazard posed by the

Nobody knows how many people died in the 1811–1812 earthquakes. An amateur observer in Louisville, Kentucky, recorded almost 2,000 shocks with a homemade pendulum. Liquefaction was widespread along the Mississippi River, and the town of New Madrid, Missouri, was completely destroyed when it settled from 8 meters (25 ft) above sea level to only 4 meters (12 ft). Subsidence caused some swamps to drain and others to become lakes; an example is Reelfoot Lake in northwest Tennessee, which today is more than 50 feet deep. The New Madrid seismic zone is a major geologic hazard in the United States, and efforts are being made to reduce the impact of a future large earthquake there.

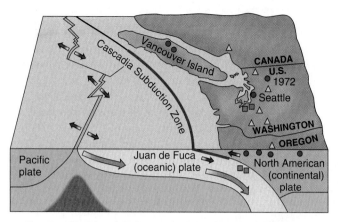

■ Deep earthquakes (>65 km or 40 mi deep)* occur when the oceanic plate descends beneath the continental plate. The largest deep earthquakes in recent times were in 1949 ($M = 7.1$) and in 1965 ($M = 6.5$).

● Shallow earthquakes (<17 km or 10 mi deep)* are caused by faults in the North American continent. Magnitude 7+ earthquakes have occurred along the Seattle fault in 1872, 1918, and 1946.

— Subduction earthquakes are huge quakes that occur when the subduction boundary ruptures. The most recent Cascadia Subduction Zone earthquake was 1700 and it sent a tsunami as far as Japan.

*Shallow and deep do not refer to focus of quake; rather, depth is subjective and is related to damage caused by faults in the Seattle area.

◆ FIGURE 4.31 Faults and epicenters in the Seattle, Washington and Vancouver, British Columbia area. USGS

Cascadia Subduction Zone. Besides ways to mitigate great loss of life, selected major points of agreement were:

■ Most of the 1100 km of the subduction zone ruptured on January 26, 1700. A correlative tsunami struck Japan and the characteristics of the wave suggest a M_w of 9.0.

■ Cascadia earthquakes generate tsunami, the most recent of which was about 10 meters high based on studies of the inundation zone.

■ Strong ground shaking from a M_w −9.0 earthquake will last three minutes or more and will be damaging as far inland as Vancouver, Portland, and Seattle.

■ The recurrence interval for an event equivalent to the 1700 earthquake is 500–600 years.

Twenty years ago scientists were debating whether great earthquakes occurred at subduction zones. Now there are few doubters, and the 2000 meeting spent a great deal of time and effort in discussing methods of mitigating loss of life and damage. It is ironic that the conference hotel lies within the inundation zone of the 1700 tsunami.

Prediction

A guy ought to be careful about making predictions. Particularly about the future.
Yogi Berra, baseball player

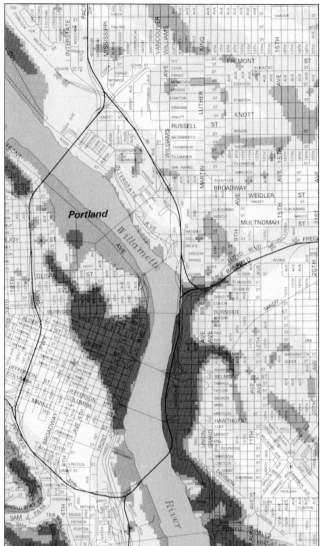

◆ FIGURE 4.32 Seismic-hazard map of downtown Portland, Oregon. The hazard zones are based upon liquefaction potential, landslide potential, and the probable degree of ground shaking. Red denotes greatest hazard; pale yellow, the lowest hazard. Oregon Dept. of Geology and Mineral Industries

Earthquake prediction has the great potential for saving lives and reducing property damage. A good prediction gives the location, time, and magnitude of a future earthquake with acceptable accuracy. Prediction was the hottest area of geophysical and geological research from the 1970s to the early 1990s, based upon the belief that measurable phenomena occurring before large and small earthquakes called **precursors** could be identified. Researchers studied earthquakes and seismograms where presumed precursors were seen. Such things as changes in the ratio of P-wave and S-wave velocities, ground tilt, water-well levels, and emission of noble gases in ground water were measured and piles of data were accumulated. Unfortunately, the hope of finding precursors that would lead to reliable predictions seems to have evapo-

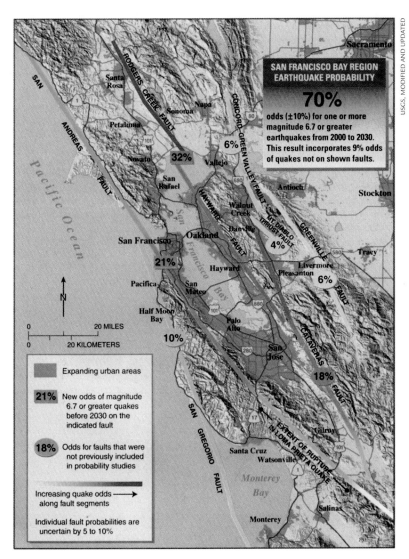

◆ FIGURE 4.33 Probabilities that one or more magnitude-6.7 or greater earthquakes will strike on specific faults in the San Francisco Bay region between 2000 and 2030. The San Andreas, Rogers Creek, and Hayward faults have the highest probabilities. Total probability for the region is computed as 70% (+ or − 10%). These probabilities were developed by scientists with the U.S. Geological Survey, part of the U.S. Department of the Interior, and thus constitute official long-term *forecasts*. The message is that all communities in the Bay region should keep preparing for earthquakes.

rated. In fact, the U.S. Geological Survey is on record as saying that the prospect of earthquake prediction is very dismal. In short, earthquakes cannot be predicted because the mechanics of earthquake generation are, with the present state of our knowledge, too complicated to predict.

Forecasts

We are familiar with weather forecasts and are aware that accuracy declines as the time span of the prediction becomes longer. We can usually rely on one-week forecasts, but a year ahead would be asking too much. The same can be said of earthquakes, only in reverse. There is an old saying in geology that "the longer it has been since the last earthquake, the sooner we can expect the next one."

A new approach is to evaluate the probability of a large earthquake occurring on an active fault during a given time period. This falls under the heading of long-term forecasting. An example of this is the U.S. Geological Survey and other

scientists' conclusion that there is a 70 percent probability of at least one magnitude-6.7 or greater earthquake striking the San Francisco Bay region between 2000 and 2030. Such an event would be capable of causing widespread damage (◆Figure 4.33). Such forecasts are made on the basis of measured plate tectonic motion and slip on faults. The inexorable movement of the Pacific plate past the North American plate loads strain on the San Andreas network. Periodically this strain is released on one of the faults in the system and an earthquake occurs.

Statistical Approach

By compiling statistical evidence pertaining to past earthquakes in a region, we acquire basic data for calculating the statistical probability for future events of given magnitudes. These calculations may be done on a worldwide scale or on a local scale, such as the example in ◆Figure 4.34. Analysis of the graph indicates that for the particular area in Southern

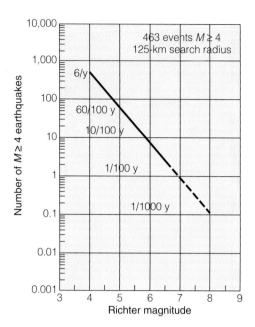

◆ FIGURE 4.34 A graphing of 463 earthquakes of magnitude 4 or greater in a small area near a nuclear reactor in Southern California over a period of 44 years. Statistically, the graph shows that 600 magnitude-4 earthquakes can be expected in 100 years, or 6 per year on average. The probability of a magnitude-8 earthquake in 100 years is 0.1, which would be 1 in 1,000 years. Such plots can be constructed for a region or for the world. The graphs are all similar in shape; only the numbers differ.

California, the statistical **recurrence interval**—that is, the length of time that can be expected between events of a given magnitude—is 1,000 years for a magnitude-8 earthquake (0.1/100 years), about 100 years for a magnitude-7 earthquake (1/100 years), and about 10 years for a magnitude-6 earthquake (almost 10/100 years). The probability of a magnitude-7 occurring in any one year is thus 1 percent, and of a magnitude-6, 10 percent. On an annual basis worldwide, we can expect at least two magnitude-8 earthquakes, 20 magnitude-7 earthquakes, and no less than 100 earthquakes of magnitude 6. Thus for seismically active regions, historical seismicity data can be used to calculate the *probability* of damaging earthquakes. Although these numbers are not really of predictive value as we defined it, they can be used by planners for making zoning recommendations, by architects and engineers for designing earthquake-resistant structures, and by others for formulating other life- and property-saving measures.

CONSIDER THIS *Humans have caused earthquakes by injecting fluids, such as oil-field waters and chemical warfare wastes, into deep wells. How does this open the possibility of controlling earthquake activity, and what might be a major deterrent to the effort?*

◆ FIGURE 4.35 (a) Historic seismic activity of about 450 kilometers (275 mi) of the San Andreas fault south of San Francisco before the 1989 Loma Prieta earthquake. The southern Santa Cruz Mountains seismic gap, where Loma Prieta is located, is highlighted. Other seismic gaps apparent in this historical record at the time were one between San Francisco and the Portola Valley and one southeast of Parkfield. (b) Seismicity record of the Southern Santa Cruz Mountains portion of the fault after the 1989 earthquake. The former seismic gap had been "filled" by the earthquake (*open circle*) and its aftershocks.

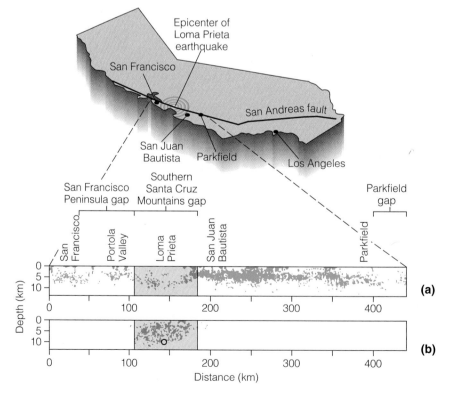

Geological Methods

Research suggests that active faults and segments of long, active faults tend to have recurring earthquakes of characteristic magnitude, rupture length, and displacement. For example, six earthquakes have occurred along the San Andreas fault at Parkfield since 1857 (Figure 4.35). They had Richter magnitudes of around 5.6, rupture lengths of 13–19 kilometers (8–12 mi), and displacements averaging 0.5 meter. These earthquakes have recurred about every 22 years since the earliest recorded earthquake. A 95 percent probability of an earthquake was predicted for this segment between 1988 and 1993, the first prediction in the United States to be endorsed by scientists and subsequently issued by the federal government. The prediction was a failure.

Geophysical and seismological precursors, as noted previously, have not proved to be reliable predictors of earthquakes. Seismic gaps, however, can indicate stretches of future fault activity. Seismic gaps are stretches along known active fault zones within which no significant earthquakes have been recorded. It is not always clear whether these fault sections are "locked" and thus building up strain energy, or if motion (creep) is taking place there that is relieving strain. Such gaps existed and were "filled" so to speak during the Mexico City (1985), Loma Prieta (1989; Figure 4.35), and Izmit (1999) earthquakes. Seismic gaps serve as alerts or warnings of possible future events and can be used as forecasting tools.

Since recorded history in North America is short, geologists need other means of collecting frequency data of large prehistoric earthquakes. One method is to dig trenches into marsh or river sediments that have been disrupted by faulting in an effort to decipher a region's **paleoseismicity,** its rock record of past earthquake events. Kerry Sieh of the California Institute of Technology has done this across the San Andreas in Southern California (◆Figure 4.36). Sieh found an intriguing history of seismicity recorded in disrupted marsh deposits at Pallett Creek and liquefaction effects extending from the seventh century to a great earthquake in 1857. Ten large events, extending from A.D. 650 to 1857, were dated using [14]C. The average recurrence interval for these ancient earthquakes is 132 years, but they are clustered in four groups. Within each cluster, the recurrence interval is less than 100 years, and the intervals between the clusters are two to three centuries in length. The last big one, also the last one of a cluster, was in 1857. Thus it appears that this section of the San Andreas may remain dormant until late in the twenty-first century or beyond.

Earlier in this chapter we quoted a geologist saying about earthquake prediction that "the longer it's been since the last earthquake, the sooner we expect the next one." Unfortunately, this is about the status of our present predictive ability. We remain uncertain that a quake will follow well-defined percursory phenomena, such as wave velocity changes or anomalous animal behavior (see ◎ Case Study 4.4 on page 106). It seems that

◆ FIGURE 4.36 Disrupted marsh and lake deposits along the San Andreas fault at Pallett Creek near Palmdale, California. Sediments range in age from about A.D. 200 at the lower left to A.D. 1910 at the ground surface. Several large earthquakes are represented here by broken layers and buried fault scarps.

no one phenomena is a predictor, and many changes will have to be monitored over a long period of time before our knowledge of fault behavior is refined enough for reliable prediction.

ENVIRONMENTAL
Geology⇌Now™

Click **Geology Interactive** to work through an activity on Seismic Risk USA through Earthquakes and Tsunamis.

Mitigation

Reducing earthquake risks is an admirable goal of scientists and lawmakers. Building codes provide the first line of defense against earthquake damage and help to ensure the public safety. It has been shown that strict building codes in seismic regions, such as the Pacific Rim, reduce damage and loss of life. Laws passed after the 1933 (Long Beach) and 1971 (Sylmar) California earthquakes have proven the effectiveness of strict earthquake-resistant design. A good example is the magnitude-6.2 Morgan Hill (California) earthquake that

◆ FIGURE 4.37 Seismic records (upper right) obtained during the 1984 Morgan Hill, California, earthquake led to an improvement in the Uniform Building Code (a set of standards used in many states). The center of the gym roof shook sideways three to four times as much as the edges. The code has since been revised to reduce the flexibility of such large-span roof systems and thereby improve their seismic resistance. USGS

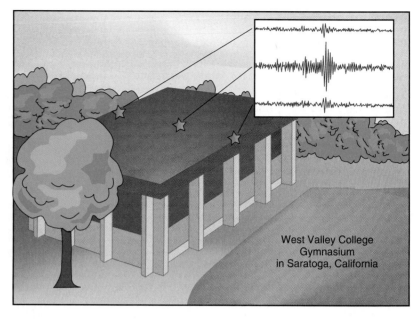

West Valley College
Gymnasium
in Saratoga, California

shook West Valley College 20 miles from the epicenter. Seismic instruments on the gymnasium showed that the roof was so flexible that in a strong seismic event it could collapse (◆Figure 4.37). Flexible roofs were permitted by the code at that time and many gyms and industrial buildings were built that way. As a result of the experience at West Valley College the Uniform Building Code was revised—this is the code used by hundreds if not thousands of municipalities across the country. The revision requires that the roofs be constructed to be less flexible and thus able to withstand nearby or strong distant earthquakes. Most large cities in strong earthquake zones have their own building codes, patterned after the Uniform Building Code, that require construction to modern seismic standards. For instance, ground response due to differing soil types is now more appreciated as contributing to quake damage. ◆Figure 4.38 shows how the code has evolved from 1955 as we obtained more information. For example, soft clays react more violently with earthquake waves than do, say, granites, and this difference in response is now taken into consideration.

The primary consideration in earthquake design is to incorporate resistance to horizontal ground acceleration, or "base shear." Strong horizontal motion tends to topple poorly built structures and to deform more flexible ones. In California, high-rise structures are built to withstand about 40 percent of the acceleration of gravity in the horizontal

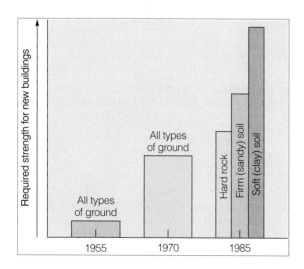

◆ FIGURE 4.38 Earthquake requirements in building codes have increased over time as scientists and engineers have obtained new information. Note that recent codes specify separate criteria for different ground types. USGS data

◆ FIGURE 4.39 Salt Lake City and County Building, Utah. Constructed in 1890, largely of brick, it is now seismically retrofitted to meet the modern building code.

◆ FIGURE 4.40 Basement of the Salt Lake City and County Building. The building has been seismically retrofitted by cutting the massive structure from its foundation pillars (one shown on the far left, another in the center of the photograph), and lifting the building onto 440 large rubber isolators (one left of center, another at the extreme right of the photograph).

direction (0.4 *g*), and single-family dwellings are built to withstand about 15 percent (0.15 *g*). *Base isolation* is now a popular design option. The structure, low- or high-rise, is placed upon Teflon plates, rubber blocks, seismic-energy dissipators that are similar to auto shock absorbers, or even springs, which allows the ground to move but minimizes building vibration and sway.

The ultimate earthquake resistance for large buildings is provided by "base isolation." The century-old Salt Lake City and County Building (◆Figure 4.39), scheduled at the time for demolition, was cut from its foundation and retrofitted with 440 rubber base isolators. These permit the structure to move as a single unit during an earthquake while the ground shakes in all directions (◆Figure 4.40). About seventy-five percent of Utah's population live near the Wasatch Range and the fault responsible for the uplift. The retrofit of older public buildings is an example of the awareness of earthquake hazards by government officials and the public. In this case it saved an architectural treasure for future generations.

Survival Tips

Knowing what to do before, during, and after an earthquake is of utmost importance to you and your family.

Before an Earthquake:

◆Figure 4.41 provides the Federal Emergency Management Agency's (FEMA) suggestions for minimizing the possibility of damage, fire, and injuries in the home.

During an Earthquake:

- Remain calm and consider the consequences of your actions.
- If you are indoors, stay indoors and get under a desk, bed, or a strong doorway.
- If you are outside, stay away from buildings, walls, power poles, and other objects that could fall. If driving, stop your car in an open area.
- Do not use elevators, and if you are in a crowded area, do not rush for a door.

After an Earthquake:

- Turn off the gas at the meter.
- Use portable radios for information.
- Check water supplies, remembering that there is water in water heaters, melted ice, and toilet tanks. Do not drink waterbed or pool water.
- Check your home for damage.
- Do not drive.

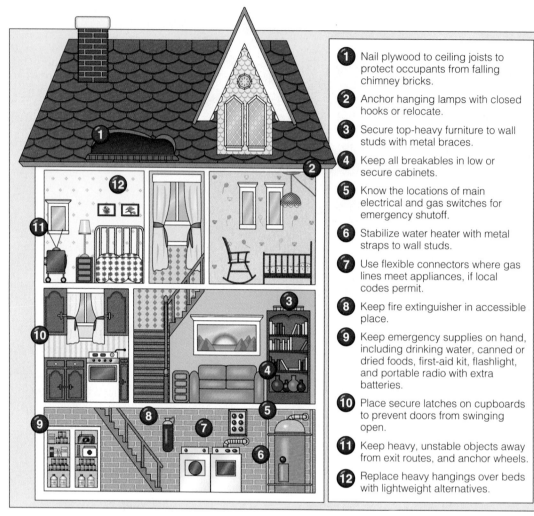

1. Nail plywood to ceiling joists to protect occupants from falling chimney bricks.
2. Anchor hanging lamps with closed hooks or relocate.
3. Secure top-heavy furniture to wall studs with metal braces.
4. Keep all breakables in low or secure cabinets.
5. Know the locations of main electrical and gas switches for emergency shutoff.
6. Stabilize water heater with metal straps to wall studs.
7. Use flexible connectors where gas lines meet appliances, if local codes permit.
8. Keep fire extinguisher in accessible place.
9. Keep emergency supplies on hand, including drinking water, canned or dried foods, first-aid kit, flashlight, and portable radio with extra batteries.
10. Place secure latches on cupboards to prevent doors from swinging open.
11. Keep heavy, unstable objects away from exit routes, and anchor wheels.
12. Replace heavy hangings over beds with lightweight alternatives.

(a)

Shut-off valve positions

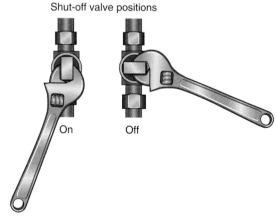

On   Off

(b)

◆ FIGURE 4.41 (a) How to minimize earthquake damage in the home in advance. (b) Keep a small crescent wrench at the gas meter. Turn off the gas by turning the valve end 90°. FEMA

Earthquakes, Landslides, and Disease

Dynamically induced landslides are not particularly newsworthy, even though the 1994 Northridge earthquake caused 17,000 of them. However, the soil dislodged during sliding caused an outbreak of *coccidioidomycosis* (CM), commonly known as "valley fever." Endemic to the Southwest, the disease causes persistent flu-like symptoms and in extreme cases can be fatal. The victims breathe in airborne *Coccidioides immitis* spores that are released from the solid during sliding. From January 24 to March 15, 166 people

were diagnosed with valley fever symptoms in Ventura County, up from only 53 cases in all of 1993. Most of the cases were reported from the Simi Valley, an area in which only 14 percent of the county's population resides (◆Figure 1).

Large clouds of dust hung over the Santa Susana Mountains for several days after the earthquake, promoted by the lack of winter rains preceding the quake. During this period, pressure-gradient winds known locally as "Santa Ana" winds of 10–15 knots (11–17 mph) blew

into the Simi Valley, carrying in spore-laden dust from the Santa Susana Mountains to the northeast. Researchers believe that the more "coherent" metamorphic rocks of the San Gabriel Mountains northeast of the San Fernando Valley probably account for the lack of cases reported in the epicentral region. This is the first report of an earthquake-associated outbreak of valley fever, even though many earthquakes have occurred in CM endemic areas.

◆ FIGURE 1 Histogram of the almost-epidemic outbreak of coccidioidomycosis (valley fever) in Ventura County in early 1994 following the Northridge earthquake.

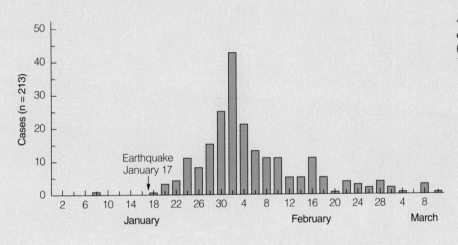

Predictable "Future Shocks"

A large earthquake is normally followed by thousands of smaller-magnitude earthquakes known as aftershocks (◆ Figure 1). If the main shock is small—say in the magnitude-4.0–5.0 range—aftershocks are small and nonintimidating. Following Northridge-size and bigger earthquakes, however, the strongest aftershocks

(M = 5.0–6.0 at Northridge) can cause buildings damaged by the main earthquake to collapse and, even worse, can greatly increase anxiety in the already damaged psyche of the local citizenry. Aftershocks are caused by small adjustments (slips) on the causative fault or on other faults close to the causative one. For example, try this:

push the eraser on the end of a lead pencil across a desktop. You'll find that it does not slide smoothly; it moves in jerky jumps and starts. This is called "stick–slip" and it is what happens along faults that are adjusting after a big earthquake.

Los Angeles experienced 2,500 aftershocks in the week following the

CONTINUED

Northridge earthquake. Three strong ones, magnitude 5 or greater, occurred the first day. The largest ($M = 5.6$) occurred eleven hours after the main shock, causing concern among rescuers digging for victims beneath the rubble and the already traumatized citizens. Aftershocks follow statistically predictable patterns, as exemplified by the Northridge sequence. On the first day there were 188 aftershocks of magnitude 3 or greater, and on the second day only 56 were recorded (◆Figure 2). By fitting an equation to the distribution of aftershocks during the first few weeks, seismologists were able to estimate the number of shocks to be expected in the future. Statistically, there was a 25 percent chance of another magnitude-5 or greater aftershock occurring within the following year, but it did not happen.

◆ FIGURE 1 Seismogram of early aftershocks following the Northridge earthquake, one of them a magnitude-5.6. Southern California Earthquake Center (SCEC)

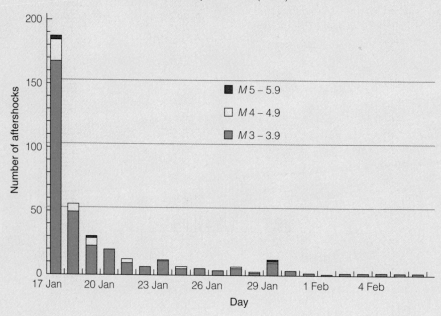

◆ FIGURE 2 Daily record of aftershocks of magnitude 3.0 to 5.9 during the three weeks following the main shock at Northridge. Note the sharp drop in aftershock frequency in the first four days. Redrawn from F. Harp and R. Jibson, USGS

Rx for Failed Freeways

Extensive damage to freeway bridges and overpasses typically accompanies earthquakes in large urban areas. Such damage has occurred in Alaska, California, and Japan. Overpass damage commonly is due to failure of the shorter columns, which lack the flexibility of longer ones. During an earthquake the tall columns supporting a bridge or overpass system bend and sway with the horizontal forces of the quake. Because the parts of the overpass system are tied together, the stresses are transferred through the structure to the short columns, which are designed to bend only a few centimeters (◆Figure 1). Failure causes the short columns to bulge just

CONTINUED

◆ FIGURE 1 The difference in long and short columns' flexibility results in failure of the short ones.

In a quake, long columns survive because they sway.

Short columns, unable to bend, absorb horizontal energy produced by longer columns and blow out.

above ground level, which breaks and pops off the exterior concrete, exposing the warped, "birdcaged" steel in the interior. In addition, high vertical accelerations can cause some columns to actually punch holes through the platform deck. With excessive horizontal motion some deck spans may slip off their column caps at one end and fall to the ground like tipped dominoes.

The California Department of Transportation in 1971 decided to retro- fit 122 overpasses to alleviate these problems. One aspect of the retrofitting was jacketing the short columns with steel or a composite substance and filling the space between the jacket and the original column with concrete (◆Figure 2a). This allows the columns to bend 12.5 centimeters (5 in), instead of 2.5 centimeters (1 in), without shattering. Another solution is to increase columns' horizontal strength by wrapping heavy steel rods around their vertical support bars, particularly on short columns. This allows the columns to bend but prevents "birdcaging" or permanent bending (◆Figure 2b). To prevent the decks from slipping off their supports and dropping to the ground, steel straps or cables are installed at the joints (◆Figure 2c). Of the 122 overpasses that CalTrans retrofitted, not one collapsed in the 1994 Northridge earthquake. Ten of the eleven that collapsed were slated for future retrofitting.

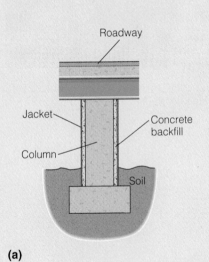

Roadway

Jacket

Concrete backfill

Column

Soil

(a)

(b)

B. PIPKIN

◆ FIGURE 2 Earthquake-resistant design for bridges and overpasses. (a) Short columns are retrofitted with steel or composite jackets that allow them to bend five times as much. (b) Old method of constructing highway support columns. The vertical steel supports have bent and "birdcaged." (c) Steel straps are used to hold deck sections together, preventing them from falling off their support columns. (a), (c) After *The Los Angeles Times*

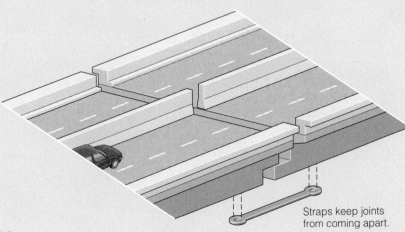

Straps keep joints from coming apart.

(c)

CASE STUDY 4.4 Depressed Tigers, Restless Turtles, and Earthquakes!

Anomalous animal behavior preceding earthquakes is well documented. Domesticated animals such as barnyard fowl, horses, cats, and dogs have been known to behave so peculiarly before big events that they attracted the notice of people not knowledgeable in what is normal or abnormal animal behavior (◆ Figure 1). Remember that although anomalous animal behavior precedes many earthquakes, it does not appear to precede every earthquake, and abnormal animal behavior is not always followed by an earthquake. Other natural phenomena, such as atmospheric disturbances, also can cause animals to behave strangely. Here are some reported incidents of unusual animal behavior noted before earthquakes:

- Tientsin Zoo, China, 1969: 2 hours before the magnitude-7.4 earthquake the tiger appeared depressed, pandas screamed, turtles were restless, and the yak would not eat.
- Haicheng, China, 1975: 1½ months before the magnitude-7.3 earthquake, snakes came out of hibernation; 1–2 days before, pigs would not eat and they climbed walls; 20 minutes before, turtles jumped out of the water and cried.
- Tokyo, Japan, 1855: 1 day before the magnitude-6.9 earthquake wild cats cried, and rats disappeared.
- Concepción, Chile, 1835: 1 hour and 40 minutes before the earthquake, flocks of sea birds flew inland, and dogs left the city.

- San Francisco, 1906: dogs barked all night before the magnitude-8.3 earthquake.
- Friuli, Italy, 1976: 2–3 hours before the magnitude-6.7 earthquake, cats left their houses and the village, mice and rats left their hiding places, and fowl refused to roost.

◆ FIGURE 1 Anomalous animal behavior. An earthquake may be in the offing if your dog dons a hard hat.

gallery

Earthquakes Are Hard on Cars

The sight of cars buried beneath a pile of bricks and rubble after an earthquake seem to be standard fare for the press. We captured a few of these scenes just by coming upon them and hope you feel, as we do, that there is a light side to most bad natural disasters. The other photos were just too unusual or amusing to pass up.

◆ **FIGURE 1** This luxury car seems to be pursuing the garage in which it is normally kept. The house and garage slid down the hillside, but the car was parked with its rear wheels on the driveway, which was on stable ground. Nobody was hurt; Northridge, 1994

◆ **FIGURE 2** A row of cars in need of body-and-fender work because their owners were asleep in the apartment building at the time of the earthquake; Northridge, 1994

◆ **FIGURE 3** Earthquakes have no respect for "No Parking" signs; State University–Los Angeles, California, 1971

YI-BEN TSAI, NATIONAL CENTRAL UNIV., TAIWAN

◆ FIGURE 4 One hundred yard . . . dash? The September 21, 1999 earthquake (M_w = 7.8) in Taiwan was caused by movement on the Chelungpu fault. This running track at Kuang Fu High School was built across the fault that here had a vertical offset of 2.5 meters (8.1 ft). It is obvious that no 100-meter-dash records will be broken here.

◆ FIGURE 5 The Hector Mine earthquake of October 16, 1999 was a real bomb! The magnitude-7.0 event occurred in California's remote Mojave Desert. The fault scarp, shown here, forms a gash across the desert for 40 kilometers (25 mi), but the earthquake did little damage. This area happens to be a U.S. Marine Corps bombing range, as you can see, which made geologists' field work interesting.

TED REEVES, CHAFFEY HIGH SCHOOL, UPLAND, CALIF.

Summary

Earthquakes

Cause

Movements on fractures in the crust known as *faults* that result in three types of wave motion: P- and S-waves, which are generated at the focus of the earthquake and travel through the earth, and surface waves.

Distribution

Most (but not all) large earthquakes occur near plate boundaries, such as the San Andreas fault, and represent the release of stored elastic strain energy as plates slip past, over, or under each other. Intraplate earthquakes can occur at locations far from plate boundaries where deep crust has been faulted, probably at "failed" continental margins.

Measurement Scales

Some of the scales used to measure earthquakes are the modified Mercalli intensity scale (based on damage), the Richter magnitude scale (based on energy released as measured by maximum wave amplitude on a seismograph), and the moment magnitude (based on the total seismic energy released as measured by the rigidity of the faulted rock, the area of rupture on the fault plane, and displacement).

Earthquake-Related Hazards and Mitigation

Ground Shaking

Damaging motion caused by shear and surface waves.

Ways to reduce effects—seismic zoning; building codes; construction techniques such as shear walls, seismic joints, and bolting frames to foundations.

Landslides

Hundreds of landslides may be triggered by an earthquake in a slide-prone area.

Ways to reduce effects—proper zoning in high-risk areas.

Ground Failure (Spontaneous Liquefaction)

Horizontal (lateral) movements caused by loss of strength of water-saturated sandy soils during shaking and by liquefaction of quick clays.

Ways to reduce effects—building codes that require deep-drilling to locate liquefiable soils or layers.

Ground Rupture/Changes in Ground Level

Fault rupture and uplift or subsidence of land as a result of fault displacement.

Ways to reduce effects—geologic mapping to locate fault zones, trenching across fault zones, implementation of effective seismic zonation like the Earthquake Fault Zones Act in California.

Fire

In some large earthquakes fire has been the biggest source of damage.

Ways to reduce effects—public education on what to do after a quake, such as shutting off gas and other utilities.

Tsunamis

Multidirectional sea waves generated by disruption of the underlying sea floor (see Chapter 10).

Earthquake Prediction

It is not feasible at this time to predict within useful limits the time, magnitude, and location of an earthquake. However, seismologists continue to search for clues to accurate prediction, which is what scientists do and what, we might add, science is.

Upside/downside—no proven large earthquakes predicted. Inaccuracies in short-term prediction have made the public wary.

Earthquake Forecasts

Based on plate movement and known fault movement there is a 70 percent chance there will be a magnitude-6.7 earthquake in the San Francisco Bay area between 2000 and 2030. This kind of forecast is the research thrust today and forecasts will vary for each seismic region.

Upside/downside—good for planning purposes but not for short-term warnings.

Statistical Methods

Tell us that worldwide there will be two $M = 8.0+$ and at least 20 $M = 7.0+$ earthquakes each year. Statistics can be applied for a smaller area in earthquake country.

Upside/downside—good for planning purposes but not for short-term warnings.

Geological Methods

Active faults are studied to determine the characteristic earthquake magnitudes and recurrence intervals of particular fault segments; sediments exposed in trenches may disclose historic large fault displacements (earthquakes) and, if they contain datable C-14 material, their recurrence intervals.

Upside/downside—useful for long-range forecasting along fault segments and for identifying seismic gaps; not useful for short-term warnings.

Key Terms

base isolation	epicenter	focus (pl. *foci*)	isoseismal
body wave	fault	forecasts	liquefaction
elastic rebound theory	fault creep	intensity scale	longitudinal wave

magnitude	P-wave	seismograph	transverse (shear) wave
modified Mercalli scale (*MM*)	recurrence interval	spontaneous liquefaction	tsunami (pl. *tsunamis*)
moment magnitude (M_w or *M*)	resonance	strain	wave period (*T*)
paleoseismicity	Richter magnitude scale	stress	
precursor	seismogram	S-wave	

Study Questions

1. What is "elastic rebound," and how does it relate to earthquake motion?

2. Distinguish among earthquake intensity, Richter magnitude, and moment magnitude. Which magnitude scale is most favored by seismologists today? Why?

3. Why should one be more concerned about the likelihood of an earthquake in Alaska than of one in Texas?

4. In light of plate tectonic theory, explain why devastating shallow-focus earthquakes occur in some areas and only moderate shallow-focus activity takes place in other areas.

5. What should people who live in earthquake country do before, during, and after an earthquake (the minimum)?

6. Describe the motion of the three types of earthquake waves discussed in the chapter and their effects on structures.

7. Why do wood-frame structures suffer less damage than unreinforced brick buildings in an earthquake?

For Further Information

Books and Periodicals

Atwater, Bryan, and others. 1999. *Surviving a tsunami: Lessons from Chile, Hawaii, and Japan.* U.S. Geological Survey circular 1187, 19 pp.

Bolt, Bruce. 1998. *Earthquakes, newly revised and expanded.* New York: W. H. Freeman.

Brocher, Thomas M., 2000, Urban seismic experiments investigate Seattle fault and basis, *EOS,* Transactions of the American Geophysical Union, vol. 81, no. 46, November 14.

Celebi, M., P. Spudich, Robert Page, and Peter Stauffer. 1995. *Saving lives through better design standards.* U.S. fact sheet 176–95.*

Drake, D. 1815. *Natural and statistical view, or picture of Cincinnati and the Miami County, illustrated by maps.* Cincinnati, Ohio: Looker and Wallace.

Ellis, Michael, Joan Gomberg, and E. Schweig. 2001. Indian earthquake may serve as an analog for New Madrid earthquakes, *EOS,* Transactions of the American Geophysical Union, vol. 82, no. 32, August 7.

Field, Edward H. 2000. Accounting for site effects in probabilistic seismic hazard analysis of Southern California. *Bulletin of the Seismological Society of America,* vol. 90, no. 6b, December.

Gomberg, Joan, and Eugene Schweig. 2002. *East meets Midwest: An earthquake in India helps hazard assessment in the central United States.* U.S. Geological Survey fact sheet 007-02.

González, Frank I. 1999. Tsunami. *Scientific American,* May: 56–65.

Gore, Rick. 1995. Living with California's faults. *National geographic,* April:2–34.

Hickman, Steve, and John Langbein. 2000. *The Parkfield experiment—Capturing what happens in an earthquake,* U.S. Geological Survey fact sheet 049-02.

Iacopi, Robert. 1971. *Earthquake country.* Menlo Park, Calif.: Lane Books.

Knopoff, L. 1996. Earthquake prediction: The scientific challenge. In *Earthquake prediction: Proceedings of the National Academy of Sciences,* pp. 3719–3720.

Kockelman, William J. 1984. *Reducing losses from earthquakes through personal preparedness.* U.S. Geological Survey open file report 84-765.

Koper, Keith D., and others. 2001. Forensic seismology and the sinking of the *Kursk, EOS,* Transactions of the American Geophysical Union, vol. 82, no. 4, January 23.

Kovachs, Robert. 1995. Earth's fury: An introduction to natural hazards and disasters. Englewood Cliffs, N.J.: Prentice-Hall.

Michael, Andrew J., and others. 1995. *Major quake likely to strike between 2000 and 2030.* U.S. Geological Survey fact sheet 151–99.

Mori, James J. 1994. Overview: The Northridge earthquake: Damage to an urban environment. *Earthquakes and volcanoes* 25, no. 1 (special issue).

Renwald, Marie, Tammy Baldwin, and Terry C. Wallace. 2003. Seismic analysis of the space shuttle *Columbia* disaster (abstract). Geodaze Geoscience Symposium, University of Arizona, Tucson, p. 75.

Richter, C. 1958. *Elementary seismology.* New York: W. H. Freeman.

Wuethrich, Bernice. 1995. Cascadia countdown. *Earth: The science of our planet,* October:24–31.

Yeats, Robert S. 1998. *Living with earthquakes in the Pacific Northwest.* Corvallis, Oreg.: Oregon State University Press.

*Fact sheets are one- or two-page condensations of a geological problem and are free of charge. Some are found on the Web at http://quake.usgs.gov. Hard copies may be obtained from the U.S. Geological Survey, Mail Stop 977, 345 Middlefield Road, Menlo Park, CA 94025.

ENVIRONMENTAL
Geology⇌Now™

Assess your understanding of this chapter's topics with additional quizzing and comprehensive interactivities at http://earthscience.brookscole.com/pipkingeo4e as well as current and up-to-date Web links, additional readings, and InfoTrac College Edition exercises.

Volcanoes

"Ashes, Ashes, All Fall Down"

ONE NIGHT IN 1982 A BRITISH AIRWAYS BOEING 747 on a routine flight from Kuala Lumpur, Malaysia, to Australia cruised at an altitude of 12,000 meters (37,000 ft). Just before midnight, sleeping passengers were awakened by a pungent odor. Through the windows, they saw the huge plane's wings lit by an eerie blue glow. Suddenly engine number 4 flamed out, followed almost immediately by the other three. The plane glided silently for an agonizing 13 minutes. At 4,500 meters (14,500 ft), engine number 4 was restarted, then numbers 2, 1, and 3. Nonetheless, an emergency was declared, and the plane landed in Jakarta, Indonesia, with only three engines operating.

This near-death experience gained the attention of the world's airline passengers and pilots. Before this, such a failure seemed virtually impossible in modern aircraft, which have backup systems for almost every contingency. Flying air-gulping jet engines through clouds of volcanic ash, however, is not one of them.

In December 1989, a KLM Boeing 747 encountered airborne ash from Redoubt Volcano at about 8,500 meters (28,000 ft) during a descent to land at Anchorage, Alaska (◆Figure 1). All four engines flamed out, and the large aircraft suddenly became a glider. After a 4,100-meter fall (13,300 ft, more than 2 mi), the engines were restarted, and the plane went on to make what was described as an "uneventful" landing. All four engines required replacement, as did the windshield and the leading edges of the wings, flaps, and vertical stabilizer, which had all been "sandblasted." One may wonder what it takes to make a landing "eventful." The plane's interior was filled with ash, and the seats and avionic equipment had to be removed and cleaned. It cost $80 million to return the aircraft to service.

Ash clouds are difficult to distinguish from rain clouds, both visually and with radar. The ash cloud from Mount Pinatubo's 1991 eruption traveled westward more than 5,000 miles in three days from the Philippines to the east coast of Africa. Twenty aircraft were damaged by the cloud, most of them while flying more than 1,000 kilometers (600 mi) from the eruption.

North Pacific air routes are some of the busiest in the world. Since 1980, flying through

◆ FIGURE 1

> *Earth exists by geologic consent, subject to change without notice.*
>
> Will Durant, Historian (1885–1981)

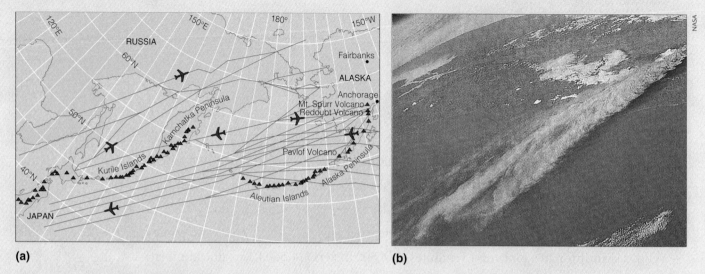

(a)　　　　　　　　　　　　　　　　　　(b)

◆ FIGURE 2 *More than 10,000 passengers and millions of dollars in cargo are flown daily on these North Pacific and Far East air routes. (a) The routings (red lines) pass over or near more than 100 active volcanoes (red triangles) on this subduction-zone complex. (b) Air traffic to the Far East and Asia was disrupted in September 1994 when Mount Kliuchevskoi (4,750 m, 15,580 ft) on the Kamchatka Peninsula erupted, spewing ash into high-altitude airlanes. This photo was taken by astronauts on board the* Endeavor.

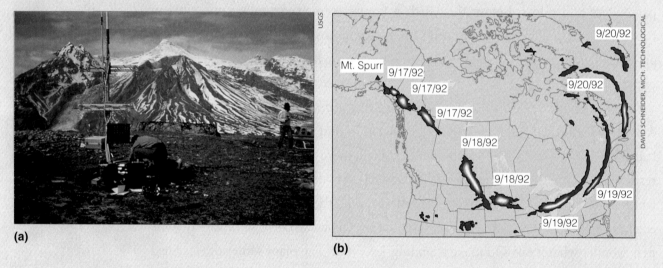

(a)　　　　　　　　　　　　　　　　　　(b)

◆ FIGURE 3 *Mount Spurr Volcano, a potential hazard to aviation. (a) Scientists from the Alaskan Volcano Observatory and the University of Alaska install seismometers at Mount Spurr to detect magma activity below the volcano. (b) The ash cloud from Mount Spurr's 1992 eruption traveled across Canada and the United States on prevailing westerly winds.*

ash clouds on those routes has caused damage to at least 15 aircraft (including the KLM flight), and worldwide there have been more than 100 inadvertent ash-cloud entries. The Alaskan Peninsula and the Aleutian Islands have 40 historically active volcanoes, and the Kamchatka Peninsula has even more (◆Figure 2). In 1992 Mount Spurr Volcano on the Alaskan Peninsula disrupted air traffic several times in the United States and Canada (◆Figure 3).

Historically, there are an average of 5 eruptions per year along the 2,400-nautical-mile great circle from Alaska to northern Japan. On average, volcanic ash is present 4 days each year above 9,300 meters (30,000 ft), where passenger jets fly. A massive effort in place to accurately forecast eruptions—the best way to address this potential hazard—involves the Alaskan Volcano Observatory (AVO), the U.S. Geological Survey (U.S.G.S.), University of Alaska, Federal Aviation Administration (FAA), National

Oceanic and Atmospheric Administration (NOAA) Weather Service, Michigan Technological University, and the Russian Institute of Volcanic Geology and Geochemistry in Petropavlovsk-Kamchatskii. Forecasting data are collected by monitoring seismic activity in the volcanic arc, reviewing satellite imagery to detect ash plumes, and making overflights of the 40 active volcanoes to measure gas emissions, which precede many eruptions—all to help make "flying the friendly skies" of the North Pacific and Far East safe for the traveler. (For more information see http://www.avo.alaska.edu/)

Although people commonly regard volcanoes and volcanic activity as environmentally harmful and life-threatening, volcanoes are not all bad. Many of them, such as Japan's Mount Fujiyama, are esthetically beautiful, some offer outstanding skiing and hiking opportunities, as does Oregon's Mount Hood and California's Mammoth Mountain; and volcanic soils, such as those on the Island of Hawaii, are highly productive. The eruption of Italy's Mount Vesuvius in A.D. 79 illustrated why volcanoes present a problem to humans (⊙ Case Study 5.1, page 137). An active volcano may be dormant for hundreds or even thousands of years, and human settlement may occur nearer and nearer to the volcanic cone during that time. Like a hibernating bear, the sleeping giant may suddenly awaken and vent an explosive eruption that wreaks havoc upon the surroundings, as did Mount St. Helens in 1980. Today, however, we can forecast an eruption with a good deal of accuracy, as explained later in the chapter. Knowing that a potential volcanic hazard exists allows policy makers and emergency managers to plan for various eruption and evacuation scenarios, thus reducing casualties when an eruption does occur.

Who Should Worry

Most geologic activity that is dangerous to humans occurs along plate boundaries. This is where we find earthquakes and active volcanoes. Explosive volcanic activity, which presents the greatest challenge to life and property, occurs mostly at convergent plate boundaries. Inspection of the Pacific Ocean basin shows that it is surrounded by trenches, which form at convergent plate boundaries. Associated with these subduction zones is the so-called Ring of Fire, the location of two thirds of the world's active violent volcanoes. A recent catalog of the 1,350 volcanoes that have erupted within the last 10,000 years (*Holocene time* to the geologist) shows 900 of them around the Pacific Rim. The largest number are found in New Zealand, Japan, Alaska, Mexico, Central America, and Chile (◆Figure 5.1). Mount Erebus in Antarctica is the southernmost active volcano in this belt. Some of the most famous active volcanoes are found in the Mediterranean, including Vesuvius, Etna, and Stromboli. Stromboli has been called the "lighthouse of the Mediterranean" because of the nearly continuous activity in its crater since the time of Greek colonization more than 2,000 years ago.

The most frequently active volcanoes tend to produce mild eruptions of fluid lava and occur along divergent plate boundaries at mid-ocean ridges. Hot-spot volcanism may be explosive or mild and may occur on continents or in ocean basins (see also Chapter 3).

There are twice as many dry land volcanoes north of the equator as there are south of it. This is particularly interesting to meteorologists, because large amounts of volcanic ash in the stratosphere result in a net heat loss for the earth, which affects climate. Any concentration of active volcanoes at a particular latitude must be considered as a potential agent of change in global climate. Although most climatic variability is related to causes other than volcanism, historical examples of volcanic eruptions affecting weather are those of Tambora (Indonesia, 1815), El Chichón (Mexico, 1982), and most recently, Mount Pinatubo (the Philippines, 1991). Because two thirds of the world's active volcanoes and land area are north of the equator, the Northern Hemisphere is more vulnerable to climatic impact by volcanic eruptions than is the southern half of the world.

Mapping of the potential volcanic risk in the United States, such as ◆Figure 5.2, reveals that there is virtually no threat of volcanic outbursts east of New Mexico. This is understandable in light of plate tectonics. The odd-shaped projections from the high-risk area on the map are due to the estimated transport of ash on the prevailing winds. The northeast–southwest trending high-risk zone through Idaho and into the corner of Wyoming is the trace of the southwestward drift of North America over a stationary mantle hot spot. A zone of volcanic vents, ash deposits, and lava flows mark the drift path, with Yellowstone being the currently active center over the hot-spot plume.

The Cascadia subduction zone is responsible for the many potentially or recently active volcanoes in Washington (6), Oregon (21), and California (16; see ◆Figure 5.3). The zone of active volcanoes extends 1,100 kilometers (700 mi) from Mount Lassen in California northward into British Columbia. Alaska ranks as the second-most active volcanic region in the world, and Hawaii is not far behind. Thus the study of volcanoes is important to citizens of the United States.

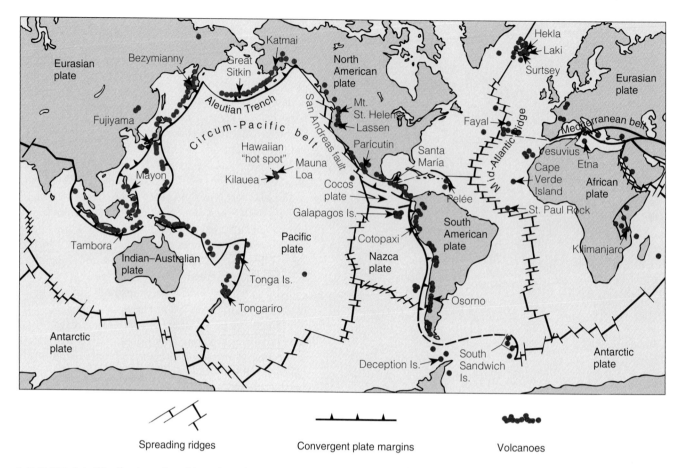

◆ **FIGURE 5.1** Distribution of earth's active volcanoes at plate boundaries and hot spots. Note the prominent "Ring of Fire" around the Pacific Ocean, which contains 900 (66 percent) of the world's potentially active volcanoes. The remaining 450 are in the Mediterranean belt (subduction zones) and at mid-ocean-ridge spreading centers (divergent boundaries). A few important volcanic centers are related to hot spots, such as those in the Hawaiian and Galápagos Islands. After R. I. Tilling, C. Heliker, and T. L. Wright, *Eruptions of Hawaiian Volcanoes: Past, Present, and Future* (USGS, 1987)

◆ **FIGURE 5.2** Map of potential volcanic hazards in the 48 contiguous states. The red areas are subject to lava flows, lateral blasts, ash falls, and mudflows. The severity of risk decreases from red (greatest), to green, to blue (least). After Hays, USGS prof. paper 1240-B, 1981

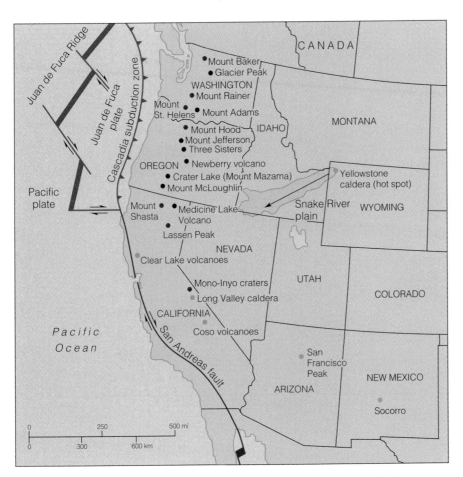

◆ FIGURE 5.3 Active volcanoes and volcano clusters in the continental United States. The volcanoes of the Cascades are geologically related to the Cascadia subduction zone.

- ● Volcanoes that have short-term eruption periodicities (100–200 years or less) or that have erupted in the past 200–300 years, or both.
- ● Volcanoes that appear to have eruption periodicities of 1,000 years or greater and that last erupted 1,000 years or more ago.
- ● Volcanic centers that are greater than 10,000 years old, but beneath which exist large, shallow bodies of magma that are capable of producing exceedingly destructive eruptions. The Snake River plain is the track of the Yellowstone hot spot.

ENVIRONMENTAL
Geology⇌Now™

Click **Geology Interactive** to work through activities on Distribution of Volcanism, Volcano Watch, and Seismic Case Study, Alaska, 1964, USA through Volcanism.

The Nature of the Problem

How a volcano erupts determines its impact on humankind. A cataclysmic eruption, such as that of Mount St. Helens in 1980, has serious results. Simple outpourings of lava such as at Kilauea in Hawaii, on the other hand, can be good for tourism and business. Unfortunately, there is no scale for describing the "bigness" of volcanic eruptions as there is for earthquakes. ◆Figure 5.4 shows some of the criteria commonly used to describe an eruption's "explosivity." **Volcanic Explosivity Index (VEI)** values range from 0 to 8 according to the volume of material ejected, the height to which the material rises, and the duration of the eruption. Its "Classification" scale associ-

ates the particular eruption with a well-known volcano that exhibited the same kind of activity. The "Description" scale employs adjectives such as those used in newspaper headlines to describe the eruption. The 1980 eruption of Mount St. Helens, for example, could be indexed as a *4* and appropriately described as an *explosive-to-cataclysmic* eruption.

In many (though not all cases) a volcano's potential explosivity has been found to increase with time since the start of its previous eruption (◆Figure 5.5). This fact allows a region's risk of volcanic eruptions to be assessed when the date of the volcano's last eruption is known. Some cataclysmic eruptions, such as the one that took place at Yellowstone hot spot 600,000 years ago, are considered a minor threat because of their rarity. Thus, both magnitude and frequency must be considered in evaluating volcanic hazards.

A magma's *viscosity*—its resistance to flow—is another determinant of a volcano's explosivity, and viscosity is a function of lava's temperature and composition. Viscosity varies inversely with temperature for most fluids, including

◆ FIGURE 5.4 Graphic representation of the Volcanic Explosivity Index (VEI). As an example, a VEI 5 would be a very large eruption, described as *cataclysmic, Vulcanian* in its type of eruption, and have an ash plume up to 25 kilometers (15 mi) high. There have been 19 historic VEI-5 eruptions. After T. Simpkin and L. Siebert, *Volcanoes of the World,* 2nd ed. (Tucson, Ariz.: Geoscience Press, 1994)

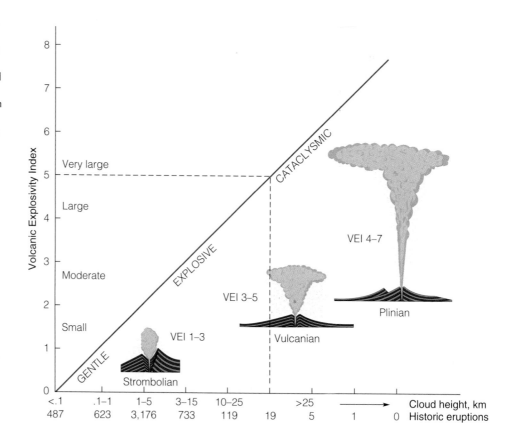

◆ FIGURE 5.5 The longer the time interval since an eruption, the greater is the next eruption's potential explosivity. These data for 4,320 historic eruptions relate known intervals between eruptions to the Volcanic Explosivity Index (Figure 5.4). Also shown is the percentage of eruptions in each VEI that have caused fatalities. Data from T. Simpkin and others, *Volcanoes of the World* (Dowden, Hutchinson, and Ross)

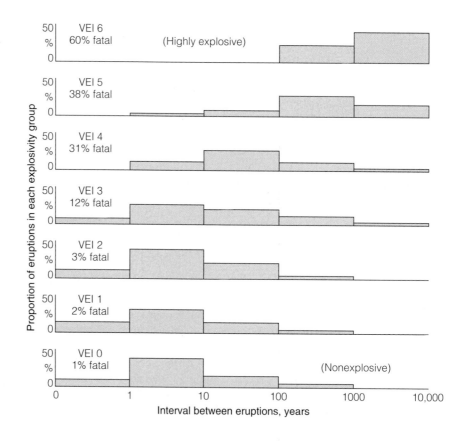

magma: the higher the temperature, the less viscous; the lower the temperature, the more viscous, other factors being equal. A magma's composition, most importantly its silica (SiO_2) content, also influences its viscosity. The silicon-oxygen bond is very strong; considerable heat energy is required to break it. The fluidity of siliceous (high-silica) magmas depends on the continual breaking and remaking of this bond. Thus, the higher the SiO_2 content, the more viscous is the lava—again, other factors being equal.

Viscous magmas with a high gas content can build up such high gas pressures as to be explosive, whereas gases escape readily from fluid lavas and hence are much less dangerous. This fact gives us another method of categorizing large volcanoes and their activity: the nature of the magmas they tap. Magmas form at depths of 10 to 250 kilometers (6–150 mi), where temperatures are sufficiently high to melt rocks completely or partially. At mid-ocean ridges and hot spots, volcanoes draw from magmas in the upper mantle that contain abundant iron and magnesium and a relatively low amount of silica (50 percent or less). These are **mafic** magmas (*ma-* for magnesium and *-f-* for the Latin word for iron, *ferrum,* plus *-ic*), which yield fluid (low-viscosity) lavas that retain little gas and hence do not erupt violently. Volcanoes that are adjacent to subduction zones, on the other hand, tap magmas that derive from oceanic crust and sediment, upper-mantle material, and melted continental rocks. These are **felsic** magmas (*fel-* for feldspar and *-s-* for silica, plus *-ic*), which yield thick, pasty lavas with SiO_2 contents up to 70 percent. Even though hot lavas have lower viscosities and flow more readily than do cool ones, just as with honey and various other common fluids, the major determinant of magma fluidity is silica content.

High-silica, high-viscosity magmas retain gases, which leads to violent, explosive-type eruptions. The geologically important boundary between mild oceanic eruptions and the more explosive continental ones of the Pacific basin is called the **andesite line** (after the intermediate rock andesite from the Andes Mountains). The line is generally drawn southward from Alaska to east of New Zealand by way of Japan, and along the west coasts of North and South America. The andesite line is also a petrologic boundary between mostly mafic magmas, which yield basalt, and felsic magmas, which yield andesite or some other high-silica rock like rhyolite. Thus, explosive volcanoes are always found on the continental side of the andesite line, and gentle volcanoes are found within the Pacific Ocean basin.

Types of Eruptions and Volcanic Cones

The type of volcanic eruption determines the shape of the structure or cone that is built, and by appearance alone, one can get an idea of a volcano's hazard potential (◆Figure 5.6).

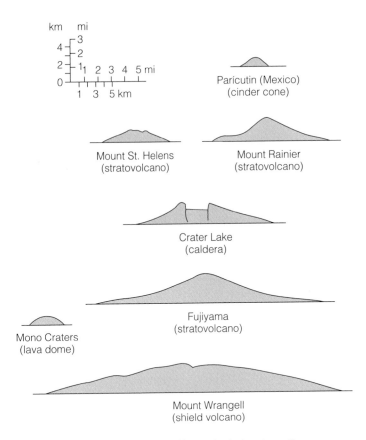

◆ FIGURE 5.6 Comparative profiles and relative sizes of some Pacific rim volcanoes and the Mono Craters in the western interior of the United States.

Effusive Eruptions

Shield Volcanoes

Shield volcanoes are built by gentle outpourings of low-silica fluid lavas from a central vent or conduit. The lavas cool to form basalt, the most common volcanic rock. A shield volcano's profile is gently convex upward like that of a shield laid on the ground. Mostly oceanic in origin, shield volcanoes are found in Iceland (where the *shield* name was first applied), the Galápagos Islands, and the Hawaiian Islands (◆Figure 5.7). They are built up from the sea floor, layer upon layer, to elevations thousands of meters above sea level. Mauna Loa ("long mountain") and Mauna Kea ("white mountain") volcanoes on the Island of Hawaii are shield volcanoes that project 4.5 kilometers (2.8 mi) above sea level, and their bases are in water about 5 kilometers (3 mi) deep. With a total height of 9.5 kilometers (31,000 ft), these are the highest mountains on earth, exceeding Mount Everest by about 650 meters (2,150 ft). (If one counts the fact that these volcanoes have also caused the sea floor to sink beneath them as they've grown, then each is more than 15,000 meters or 50,000 feet high!)

B. PIPKIN

◆ FIGURE 5.7 Shield volcano Fernandina in the Galápagos Islands. This active volcano is related to the Galápagos "hot spot." Its convex profile is similar to those of the volcanoes of Hawaii.

The Island of Hawaii (the "big" island to locals) is composed of five separate volcanoes that have built the island over the past 700,000 years (◆Figure 5.8). Kilauea (Hawaiian, "much spreading") is the most easterly of the group and is the most active. This is exactly what one would predict as sea floor moves over the fixed Hawaiian hot spot.

A new volcano, Loihi ("the long one"), is growing on the flanks of Kilauea 28 kilometers southeast, providing further

evidence of the northwest motion of the Pacific plate over a hot spot. The summit of Loihi is about 1,000 meters below sea level, and it rises 3,500 meters from the Pacific floor (◆Figure 5.9). In 1996 it gave rise to the largest swarm of earthquakes ever recorded near any Hawaiian volcano— more than 4300 earthquakes in a month. Loihi is believed to be fed by the same 50-km-wide hot spot that is the source of Mauna Loa and Kilauea volcanic activity. Loihi will probably pop its head above water in about 50,000 years and become a tourist attraction.

It is not uncommon for a shield volcano to erupt from a *fissure* (crack or "rift"), on its flanks, rather than from a central vent. This is typical of Kilauea's east rift zone, where the countryside has been flooded under a sea of lava. Pu'u O'o ("hill of the o'o bird") is a large crater built upon a vent in the east fissure zone. Pu'u O'o's crater has reached impressive proportions as a result of scores of eruptions since 1983. During Pu'u O'o's eruptions, Kilauea's summit deflates slightly, rising again or reinflating between eruptive episodes. This indicates that the rift zone and Kilauea's vent plumbing are connected with the main magma reservoir beneath the summit. Lava flows associated with Pu'u O'o caused significant damage in the Royal Gardens subdivision near Kalapana about 8 kilometers (5 mi) from the vent. Flows destroyed at least 75 homes and covered 10 kilometers (6 mi) of residential streets (◆Figure 5.10). In addition, lava has built out hundreds of meters into the sea, adding over 200 hectares (500 acres) of new land to the island.

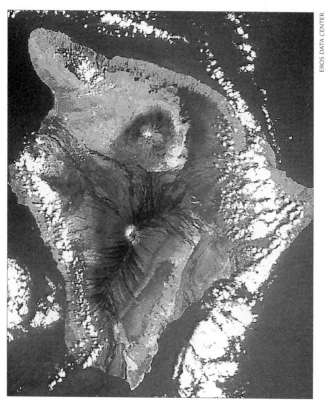

EROS DATA CENTER

◆ FIGURE 5.8 The Island of Hawaii. (a) Satellite photograph with north at the top. Mauna Kea and Mauna Loa are clearly visible. (b) The volcanoes that form the island have grown over 700,000 years in this order: Kohala (indicated by K), Mauna Kea (MK), Hualalai (H), Mauna Loa (ML), Kilauea (KI), and Loihi (L; submarine). Solid stars denote vigorous growth; open stars, waning growth; open circles, little activity. The island to the northwest is Maui. Its Haleakala Volcano is shown.

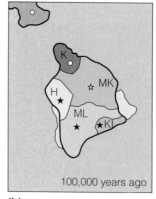

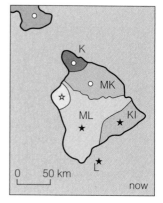

(a)

(b)

(a)

(b)

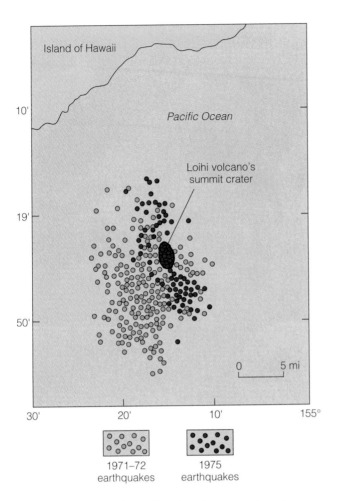

◆ FIGURE 5.9 Locations of earthquakes near submarine Loihi Volcano, 1971–1972 and 1975. Similar earthquake swarms occurred in 1984–1985, providing further evidence that Loihi is an active volcano. USGS

Spectacular lava fountains 400 meters (1,300 ft) high have occurred with fissure eruptions at Pu'u O'o (◆Figure 5.11). Some of the more fluid droplets set into glassy, tear-shaped blobs or tiny threads, known, respectively, as "Pele's tears" and "Pele's hair," named after the Hawaiian goddess of volcanoes (◆Figure 5.12). Birds have been known to make nests from some of the more pliable threads of volcanic glass.

Continental Flood Basalts

The landscapes of some areas of the world are dominated by thick, flat-lying basalt flows that form immense plateaus: the Columbia River plain in Washington, Oregon, and Idaho, and the Deccan plateau in northwestern India to name two (◆Figure 5.13). In scale they cover many times the area of the large volcanoes. When deeply eroded, these "flood basalts" form stairstep topography consisting of flat "treads" (soft rock) and vertical "risers" (hard rock). This geomorphic expression led early geologists to call the rocks collectively *traps*, the Swedish word for "staircase." When you hear or

(c)

◆ FIGURE 5.10 (a) A lava flow ignores a stop sign in the Royal Gardens subdivision below the east rift zone. (b) Lava advances through the Kalapana community, leveling everything in its path. (c) Firefighters arrest the speed of an advancing flow by dousing its front to create a resistant crust. This gives homeowners time to finish evacuating belongings.

◆ FIGURE 5.11 Lava fountaining at Pu'u O'o, Kilauea Volcano east rift zone, Island of Hawaii.

(a)

(b)

◆ FIGURE 5.12 (a) Artist's rendition of the revered Hawaiian fire goddess Pele. Her traditional home is at the summit of Kilauea Volcano. (b) "Pele's hair," thin filaments of glassy lava that have been carried by wind.

read terms like "the Deccan Traps" or "the Siberian Traps," this is the reason.

The Deccan Traps are 1–2 kilometers (3,000–6,000 ft) thick and cover a half-million square kilometers. What is the explanation for these great outpourings of lava? The process is believed to start as a giant plume of hot rock (see Chapter 3) originating at the mantle–core boundary that is several hundred kilometers in diameter. When the superplume rises to the base of the lithosphere it is 200°–300° C hotter than the surrounding mantle, and the rock at the base of the lithosphere rapidly melts. Over a period of just a few million years a basalt plateau grows on the surface above. Some plateaus can be matched across continents—the Parana of Brazil and the Etendeka of Namibia, for instance—giving more credence to Wegener's original ideas on drifting continents.

The Deccan Traps have been linked by some to the extinction of the dinosaurs 65 million years ago (end of the Cretaceous Period). They argue that the sulfurous fumes from the massive basaltic eruptions could have severely altered climates and poisoned ecosystems worldwide. Others cite even more compelling evidence that a meteorite impact in Central America at about the same time caused the die-off. The balance of fossil evidence favors the meteorite theory. However, the two events could be linked. Some geophysicists suggest that a large meteorite impact on one side of the earth might trigger an upwelling or superplume on the opposite side, causing vast outpourings of lava there. India

and Central America, are (perhaps not coincidentally) positioned opposite one another on the globe.

Explosive Eruptions

Stratovolcanoes

A combination of effusive and explosive volcanic activity builds **stratovolcanoes,** sometimes referred to as *composite cones,* because they are composed of layers of both pyroclastic material and lava. Typically thousands of feet high and 10–20 kilometers (6–12 mi), across at the base, they have a concave upward profile and a central vent. They are stratified, thus the name *strato*volcano, consisting of alternating layers of ash, cinders, and lava.

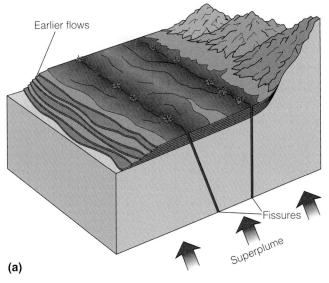

(a)

(b)

◆ FIGURE 5.13 (a) Fissure eruptions forming a lava plain. (b) Antoja caves, Buddhist shrines excavated into the basalts of the Deccan lava plateau, India.

The upper, steep slopes are formed mostly of pyroclastic volcanic ejecta, and the less steep, lower slopes are composed of alternating layers of lava and pyroclastics. These cones are some of the most beautiful tourist attractions in the world. Noteworthy examples are Mount Vesuvius in Italy, Mount Fujiyama in Japan, Mount Hood in Oregon, Mount Rainier in Washington, and Volcán San Cristobal in Nicaragua (◆Figure 5.14). They occur on the landward side of subduction zones, where melting of descending oceanic crust and mantle forms magma. The magmas rise because, as they melt, they expand, becoming more buoyant than the surrounding lithosphere. As the magma penetrates the continental rocks, partial melting occurs, and the magma becomes enriched in potassium and silica. This mixing and partial melting changes the volcanism from undramatic basalt to explosive andesitic and high-silica rhyolitic volcanism. As the

(b)

◆ FIGURE 5.14 Three examples of stratovolcanoes. (a) Mt. Shasta, California, with Shastina, a parasitic stratocone on Shasta's northwest flank. Shasta, mantled in snow and ice, is deeply eroded, since the volcano's eruption frequency is waning. Shastina grew after the end of the last ice age, and preserves a youthful conical shape. (b) Volcán San Cristóbal, Nicaragua, shown here erupting in 1976. The dark cauliflower-shaped cloud is volcanic ash. The white cloud is volcanic gas and steam.

◆ FIGURE 5.15 Crater Lake is a caldera, resulting from the explosion–collapse of a huge stratovolcano known as Mount Mazama, about 6,900 years ago.

magma rises it becomes gas-charged, and the high-silica lavas, rhyolites, dacites, and the like, are very viscous and thus do not flow easily. They may congeal in the volcano's central conduit (vent), which can cause gas pressures to build to explosive proportions. For this reason, stratovolcanoes present the most immediate threat to humans. Some eruptions have been so great that large cones have simply disappeared in the explosion. Crater Lake in Oregon represents the stump of the former Mount Mazama, a very large stratovolcano; it was probably the size of Mount Rainier. About 6,900 years ago, 70 cubic kilometers (17 mi³) of Mount Mazama disappeared, due in part to explosive activity, but mostly due to the collapse of the remaining cone into the deflated magma chamber beneath it. The "crater" of Crater Lake is a **caldera,** defined as a volcanic crater that is many times larger than the vent that feeds it (◆Figure 5.15). Calderas may form by explosive disintegration of the top of a volcano, by collapse into the magma chamber, or by both mechanisms (see Case Study 5.1). Yellowstone National Park includes several huge calderas, as does Long Valley in the eastern Sierra Nevada, both of them lying above shallow bodies of magma that are capable of producing destructive eruptions. The Yellowstone calderas cover an area of 40 by 65 kilometers (25 by 40 miles).

Lava Domes

Lava domes are formed when bulbous masses of lava pile up around the vent because the lava is too thick and viscous to flow any significant distance from its source. Sometimes

called *volcanic domes,* they usually grow by expansion from within. As the outer surface cools, the brittle crust breaks and tumbles down the sides, as at the Mono Craters on the eastern Sierra Nevada of California (◆Figure 5.16). Relatively small domes may form in the crater of a larger composite cone, such as the one that formed in the crater of Mount St. Helens after the 1980 eruption (◆Figure 5.17). The source vents of large domes may be substantially plugged, which offers the potential for explosive eruptions, particularly

◆ FIGURE 5.16 Lava domes; Mono Craters, east-central California. The Sierra Nevada range is in the background.

(a)

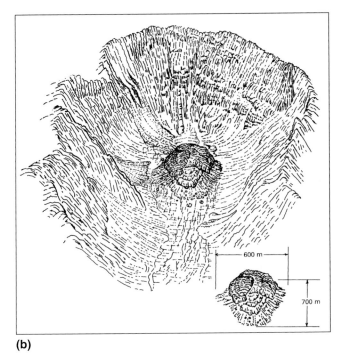

(b)

◆ FIGURE 5.17 (a) Steaming lava dome in the crater of Mount St. Helens, October 1981. (b) Oblique map of the same lava dome, October 8, 1981. Tao Rho Alpha

where the lavas have access to underground water or seawater. Mount Pelée on Martinique in the West Indies and Mount Lassen and Mammoth Mountain in California are all dormant lava domes. They extrude lavas such as rhyolite with silica (SiO_2) contents of 65–75 percent, and they also extrude glassy rocks that cool quickly, such as obsidian and pumice.

Cinder Cones

Cinder cones are the smallest and most numerous of volcanic cones (◆Figure 5.18). They are built of pyroclastic material of all sizes—from blocks and bombs to the finest ash (✳ Table 5.1). Pyroclastic material of all shapes and sizes is collectively known as **tephra**. *Bombs* are blobs of still-molten lava that assume an aerodynamically induced spindle shape and solidify before striking the ground (◆Figure 5.19). They can be found in great numbers around certain cinder

◆ FIGURE 5.18 Parícutin Volcano, Michoacán, Mexico, in 1943. The only totally new volcano in North American history, it began as a fuming crack in a farmer's field. In a short time it was 400 meters (1,312 ft) high, and lava from it covered two nearby villages, leaving only a church steeple protruding through the lava. Volcanic activity decreased rapidly, and the volcano was inactive by 1952.

cones, and collectors have impoverished many landscapes of their specimens. Cinder cone activity is local, within a few kilometers of the source vent, and is usually short-lived. The only new volcano to form in historic time in North America erupted in a farm field in 1943 near the village of Parícutin

✳TABLE 5.1	Classification of Pyroclastic Ejecta (Tephra)	
Name	**Size**	**Condition When Ejected**
Blocks	>32 mm	Cold, solid
Bombs	>32 mm	Hot, plastic
Lapilli (cinders)	4–32 mm	Molten or solid
Ash	¼–4 mm	Molten or solid
Dust	<¼ mm	Molten or solid

B. PIPKIN

◆ FIGURE 5.19 A volcanic bomb; Galápagos Islands, Ecuador. Note the spindle shape that the bomb acquired as it was flung through the air.

in the state of Michoacán, Mexico (Figure 5.18). Eruptions caused by frothing gases hurled blobs of lava into the air that became cinders and then fell back around the vent, eventually building a cone nearly 400 meters (1,300 ft) high. An observatory was established on a nearby hill, and the volcano's every burp and belch was recorded for nine years. After the initial cone-building stage had passed, basaltic lava flowed from the base of the cone, inundating the nearby pueblo of San Juan and burying about 160 square kilometers (100 mi²) of agricultural land. Fortunately, no lives were lost. Cinder cones are generally thought to be one-shot events; that is, once the eruption sequence ends, it is rarely reactivated. Nonetheless, exceptions are known, and one usually finds many cinder cones clustered together.

ENVIRONMENTAL
Geology⇌Now™

Click **Geology Interactive** to work through activities on Magma Chemistry and Explosivity and Volcanic handforms through Volcanism.

Benefits of Volcanic Action

Most of the atmosphere and hydrosphere that we know on the earth today has come from the interior of the earth via volcanic activity. The earth's primitive atmosphere was probably composed of hydrogen and helium, the two most abundant gases found in the universe. Because the young planet was still very hot, these gases most likely escaped to outer space. A second atmosphere came into being, sweated out from the earth's interior by volcanic eruptions and fumeroles (steam vents, see Figure 2 in the Gallery on page 144). Working on the principle that "the present is the key to the past," we assume that volcanic action spewed out the same gases then as it does today:

water vapor (80%), carbon dioxide (10%), and the remainder mostly nitrogen and some rare gases. The predominantly water and carbon-dioxide atmosphere was continually modified by volcanic **outgassing** until sufficient water was supplied to form clouds, from which rain fell to form the oceans. (There is some evidence that cometary impacts also supplied large amounts of water to the early earth.) From this start, plants evolved two to three billion years ago, and they consumed much of the carbon in the atmospheric CO_2. As the proportion of CO_2 decreased, the atmosphere became richer in oxygen and nitrogen. So we can credit volcanoes for our oceans and atmosphere. In fact, volcanic outgassing remains a vital player in maintaining earth's habitability, even at present.

Volcanic activity also provides products that we utilize in many ways. The final polish your teeth receive when you get them cleaned is an example. Dental pumice is refined and flavored volcanic pumice with a hardness just slightly less than tooth enamel. "Lava" brand hand soap, a rough, abrasive bar soap, contains the same powdered rock. Light-weight bricks, cinder blocks, and many road-foundation and decorative stone products originated deep within the earth. Where volcanic cinders are mined for road base, they may also be mixed with oil to form asphalt pavement. The reddish pavement of state highways in Nevada and Arizona contains basalt cinders that are rich in oxidized iron, which gives it the red color. Pieces of rock pumice are sold in drugstores, supermarkets, and hardware stores for use as a mild abrasive for removing skin calluses and unsightly mineral deposits from sinks and toilets. Powdered, it is used in abrasive cleaners and in furniture finishing.

Glassy volcanic rock, such as obsidian, is easy to chip and form. Hence, Native Americans and stone-age people in many parts of the world used it to make tools and arrowheads. Today obsidian and many similar volcanic rocks are the raw material of "rock hounds" and artisans for producing polished pieces and decorative art.

Volcanoes provide opportunities for recreational activities, ranging from mountain climbing and skiing to more passive activities such as photography and birdwatching. Of course, skiing on an active volcano poses some measure of risk, as skiers at Mount Ruapehu in New Zealand found out in 1995 (see ✪ Case Study 5.1 on page 137).

CONSIDER THIS *Although a few areas on the Island of Hawaii are regarded as relatively free from geologic hazards, you might not want to live in one of those "safe" places. Select an area on the island that appeals to you (for its view, beach access, culture, or some other amenity) and determine what geologic hazards you could encounter there.*

Geothermal Energy

By far the most important and beneficial volcanic product is also the most hazardous; it is heat. The same heat energy that causes eruptions also drives geysers and hot springs and, when controlled, it can be converted to other uses. Not surprisingly, the prospects for geothermal energy are best at or near plate boundaries, where active volcanoes and high heat flow are found. The Pacific Rim (Ring of Fire), Iceland on the Mid-Atlantic Ridge, and the Mediterranean belt offer the most promise. Energy from earth heat is discussed at length in Chapter 14.

Volcanic Hazards

Sicily's Mount Etna (from the Greek *aitho,* "I burn") exhibits almost continuous activity in its crater. In 1992 eruptions and lava flows were threatening several villages. Empedocles (circa 490–430 B.C.), Greek philosopher and statesman, is perhaps most notable for throwing himself into the crater of Mount Etna to convince his followers of his divinity. His dramatic suicide inspired Matthew Arnold's epic poem "Empedocles on Etna," which bears no resemblance to the little rhyme attributed to Bertrand Russell:

> Empedocles that ardent soul—
> Fell into Etna and was roasted whole!

Volcanoes comprise the third most dangerous natural hazard globally in terms of loss of life, after coastal flooding (hurricanes and typhoons) and earthquakes. Following the Mount St. Helens eruption of 1980, U.S. civil defense agencies published suggestions for what to do when a nearby volcano erupts. The Federal Emergency Management Agency's suggestions appear later in the chapter. For now, however, let us examine the various types of hazards associated with eruptions, several of which might occur during a single eruptive phase (◆Figure 5.20).

Lava Flows

Most hazards, natural and human-made, decrease in severity with distance from the point of origin. This is true of earthquakes generally, tornadoes, and falls of volcanic ash, for example. Lava flows may be the exception to this, because they generally burn or bury everything in their path, even to their farthest limit. Holocene lava flows in Queensland, Australia, traveled 100 kilometers (62 mi) from their vents down riverbeds with very low gradients. Knowledge of the paths that flows might take from given vents makes it possible

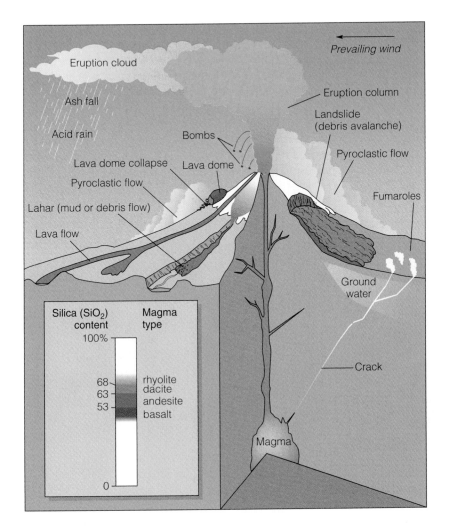

◆ FIGURE 5.20 Volcanic hazards typical of those found in the western United States and Alaska. Some hazards, such as lahars and landslides, can occur even when a volcano is not erupting. Table 5.1 provides descriptions of pyroclastic ejecta portrayed here. USGS image, fact sheet 002-97, "What Are Volcano Hazards?"

to delineate low- and high-risk volcanic-hazard areas. Basaltic lavas, such as those from Mount Etna in 1992, may emerge from a vent or fissure at temperatures in excess of 1,100° C and flow rapidly downslope like a river because of their low viscosity. As a flow cools, its viscosity increases, and then it may only "chug" along slowly. In either case, the end product is a layer of rock of varying thickness that covers everything in its path.

Two kinds of basalt flows are recognized. **Pahoehoe** ("pa-ho´-e-ho´e") forms when a skin develops on a flow and buckles up as the flow moves, making a smooth or ropy-looking surface (◆Figure 5.21a). Other basalt flows develop upper and lower surfaces that are rough and blocky. This rough, angular lava is called **aa** ("ah´-ah") and it is characterized by a very slow advance in which the top, cold rubble rolls over the front of the flow, moving much like the track of a Caterpillar tractor (◆Figure 5.21b).

A volcanic eruption cannot be prevented, but in some cases, it is possible to divert or chill a flow so as to keep it from encroaching onto croplands or structures. Flows have been diverted in Hawaii and Italy by constructing earthen dikes. Icelandic fire fighters have successfully chilled flow fronts by spraying them with large volumes of seawater from fire hoses. In the 1930s and 1940s the U.S. Army Air Corps tried, unsuccessfully, to divert lava flows by bombing them.

Ash Falls

Roman naturalist and historian, Gaius Plinius (Pliny the Elder, born A.D. 23) died during the eruption of Vesuvius in A.D. 79 from complications brought on by inhaling noxious fumes and by overexertion. He is revered as geology's first martyr, having been quoted by his nephew as saying, "Fortune favors the brave." Bravery was to be his undoing, however, as corpulent Pliny died in an area of heavy fallout from the eruption. Fortunately, however, Pliny had enjoyed life: he is credited with coining the phrase *In vino veritas* ("In wine there is truth"). Before he died, he witnessed a great vertical plume of ash with a mushroom- or anvil-shaped head rising from Vesuvius's summit. This roiling, high-speed column of steam, gas, and fragmented laval is called a **Plinian column** in commemoration of Pliny the Elder. Plinian columns typically penetrate the stratosphere as high as 15–25 kilometers (approximately 10–15 mi), and inject huge amounts of climate-cooling, sunset-altering dust and sulfur dioxide into the upper atmosphere. Plinian columns formed at the onset of the Mount St. Helens eruption in 1980 and at Mount Pinatubo (◆Figure 5.22) in 1992. Acidic gases from the 1912 Katmai, Alaska Plinian eruption—the largest eruption in the past century—ate clothes hanging on the line in Seattle, Washington, two thousand miles downwind! As a Plinian cloud spreads, its ash and pumice fall out, with the larger, heavier particles of pumice falling closest to the vent. An **ash-fall** from the A.D. 19 eruption fell across Pompeii as people fled. Many carried mattresses and pillows on their heads as protection against the marble- and golf-ball-sized debris dropping from the sky. Farther downwind, the finer dust-like ash falls out. Some particles are so small that they cannot be seen, but certainly can be tasted. During the Mount St. Helens eruption, westerly winds carried ash eastward, causing difficulties in Yakima and Spokane, where several inches of ash were deposited. At Butte, Montana, about 750 kilometers (465 mi) east, the fine ash and dust in the atmosphere forced the closure of the airport. People in Yakima wore masks to filter out the choking dust. Bankers politely requested that all patrons remove their masks before entering their banks because of potential identification problems.

(a)

(b)

◆ FIGURE 5.21 (a) Smooth, ropy pahoehoe lava; Galápagos Islands, Ecuador. (b) Rough, blocky aa flow in Hawaii. Human approach is possible because the flow moves very slowly.

◆ FIGURE 5.22 *Ascending Plinian ash cloud over Mount Pinatubo.*

◆ FIGURE 5.23 A pyroclastic flow (nuée ardente) hurling down the slope of Augustine Volcano in Alaska, April 1986.

Pyroclastic Flows

Pyroclastic flows are turbulent mixtures of hot gases and pyroclastic material that travel across the landscape with great velocity. They are generated quickly and with such force that obstructions in their path may be blown down or carried away (◆Figure 5.23). The term **nuée ardente,** French for "glowing cloud," is commonly applied to this kind of eruption because of the intense heat in the cloud's interior. A nuée ardente caused most of the fatalities at Mount St. Helens and devastated an area in excess of 160 square kilometers (◆Figure 5.24).

One of the most infamous of all historic eruptions is that of Mount Pelée on the Island of Martinique in the West Indies on May 8, 1902. It caused the deaths of approximately 30,000 people in Saint-Pierre at the foot of the volcano and initiated the study of nuée ardentes and the unique volcanic deposits it forms. Pelée rumbled for some time before its main eruption. Because of this precursor activity, the population of Saint-Pierre swelled from 19,700 to at least 30,000 as refugees fled there from outlying villages. Local officials encouraged people to stay to participate in an upcoming election, claiming that Mount Pelée posed no more of a threat to St. Pierre than Vesuvius does to Naples(!). The concentration of sulfurous gases in the air became so great that citizens were forced to cover their faces with wet cloths. On the fateful day there was a tremendous explosion and a lateral blast from Pelée that boiled and rolled down the slopes of the mountain at a velocity estimated at 100 to 150 kilometers per hour (60–90 mph). The searing cloud engulfed the town in seconds and burned or suffocated all its inhabitants except for two survivors, one of them a condemned prisoner who had been in a dungeon and later went on circus tours to recount his horrible ordeal.

Such eruptions are common in the Lesser Antilles and similar subduction-related island arcs around the world (see ◌ Case Study 5.2 on page 140).

The old Roman city of Pompeii was buried under six meters (20 ft) of volcanic ash over a period of several days when Mount Vesuvius erupted in A.D. 79 (◆Figure 5.25).

(a)

(b)

◆ FIGURE 5.24 Mount St. Helens. (a) A blast erupted from the volcano on Sunday morning, May 18, 1980, and culminated with a powerful pyroclastic flow. (b) The force of the pyroclastic flow was evidenced by the downed trees stripped of their bark in the "blow-down" area. This is the area where most of the fatalities occurred.

Uncovering the city and preserving its treasures marked the beginning of modern systematic archeology. Less well known, but more dramatic, was the eruption's impact on Pompeii's neighboring city Herculaneum, settled by the Greeks and named after Hercules. Herculaneum was overcome in a matter of minutes by a pyroclastic flow that buried the city to a depth of 20 meters (65 ft). The pyroclastic deposits are extremely hard, and a modern city, Ercolano, is now situated on the old city's grave. For this reason, only the equivalent of eight city blocks there have been excavated. Until the 1980s it was thought that Herculaneum's residents had escaped, because only 30 bodies had been recovered there, compared to 3,000 at Pompeii. Excavations since then along the former waterfront have uncovered hundreds of skeletons huddled together in cavelike openings in buildings used to store fishing boats. Apparently, the victims were overcome by the hot flows after running to the beach seeking refuge. More of this sad attempt of humans to escape the overwhelming forces of a volcanic eruption will be revealed in the continuing archeological excavations at Herculaneum.

Suffocating pyroclastic flows are a major volcanic hazard and have occurred in historic times in Alaska, Washington, and California. Prehistoric but geologically young pyroclastic-flow deposits of truly enormous size are found in all the Western states. They are known as ash flow or **welded tuffs,** because in certain layers the minerals and glass composing them were fused together by intense heat.

Lahars

Water from heavy precipitation, melting snow, a lake, or a river may mobilize debris on the flanks of a volcano and cause it to move a great distance downslope as a thick mush of rock, ash, and cinders. **Lahar** is the Indonesian word for such a fast-moving volcanic debris flow. A lahar accompanied the eruption of Mount St. Helens and filled the north fork of the Toutle River, a pristine mountain valley and stream before the eruption (◆Figure 5.26). Lahars are basically identical to nonvolcanic debris flows except for their volcanic source (see Chapter 7). Lahar-generating volcanoes may show little activity when such flows occur, and there may be little warning.

Another example of a destructive lahar is the one caused by the mild eruption of Colombia's Nevado del Ruiz, 5,432 meters (17,800 ft) high, in November 1985. Although, as one observer put it, the volcano simply "ran a fever and cleared its throat," a lahar generated by melting snow and ice cover covered the town of Armero 50 kilometers (30 mi) to the east in just a few minutes. A wall of mud 40 meters high (130 ft) careened down the narrow canyon of the Lagunilla River and overwhelmed the city at the canyon's mouth, leaving more than twenty thousand victims entombed in mud. Because the disaster struck at 11 o'clock at night, many of the victims were sleeping and unable to run to higher ground.

The potential for destructive debris flows is predictable. The map of the Mount Rainier area in ◆Figure 5.27 specifies the various identified potential volcanic hazards from this stratovolcano. Because Mount Rainier is covered with ice and has some areas of escaping steam and hot water, it is considered capable of generating lahars that could reach the suburbs of Seattle and Tacoma. Even a small eruption would subject the area at its base to inundation. Similar maps have been developed for most of the active volcanoes near urban areas along the west coast of the United States.

Tsunamis

Destructive tsunamis associated with volcanic activity are rare and they occur mainly in the western Pacific Ocean and around Indonesia. Of the 405 tsunamis recorded since 1900, 12 were caused by submarine volcanic eruptions, and only two of those resulted in significant damage. The most famous tsunami of all time, however, was caused by Krakatoa's eruption in 1883. Krakatoa Volcano lies in the

(a)

(c)

(b)

◆ FIGURE 5.25 (a) The excavated wall of Terzigno quarry at the southeast base of Vesuvius. Find the dark bed in the lower center of the wall with the rippled upper contact. The ripples are furrows from an ancient Roman farm now buried under almost 8 meters (25 ft) of younger volcanic deposits. (b) Plaster casts of a person and a dog whose body molds were preserved in the A.D. 79 ash covering Pompeii. More than 2,000 body molds have been discovered. (c) A fully excavated Pompeii avenue. The wood bracing is to strengthen ancient walls weakened by modern earthquakes. (d) A fresco preserved on the wall of a Pompeiian home. The red color ("Pompeii Red") of these frescoes may not be original, but baked in by the heat of the eruption in A.D. 79.

Sunda Strait between the islands of Java and Sumatra. In 1883 an eruption destroyed the volcano and formed a 7-kilometer-wide (4 mi) caldera. The islands of Krakatoa, Verlaten, and Lang are remnants of this volcano. The eruption was so enormous (VEI = 6) that its explosion was heard 3,000 miles away, and it spewed a dust cloud 80 kilometers (50 mi) into the atmosphere. About 30 minutes after the climactic phase of the eruption began, a huge tsunami swept the west coasts of Java and Sumatra, rising to a height of 35 meters (115 ft) above sea level and killing an estimated 36,000 people. The wave traveled across the Pacific, still a meter high when it

(d)

reached South America, and was noticed as far away as the English Channel. Since 1927 small eruptions in the caldera have been frequent, and a new volcanic island has appeared, Anak Krakatoa ("Son of Krakatoa"; ◆Figure 5.28). It is characterized by explosive eruptions, with turbulent ash clouds rising 1,200 meters (4,000 ft) above the crater.

◆ FIGURE 5.26 Lahar-filled north fork of the Toutle River at Mount St. Helens, formerly a pristine V-shaped stream valley. This photograph was taken in July 1980, well after the main eruption on May 18, and the lahar was still steaming.

Weather and Climate

Benjamin Franklin is regarded as the first person to recognize the connection between cold weather and volcanic eruptions while he was serving as U.S. ambassador to France in 1783–84. Laki Fissure in Iceland erupted in June 1783, putting sufficient ash into the atmosphere that the following winter was extremely cold in Europe. The summer of 1783

was bleak in Iceland, resulting in famine and the loss of a fifth of its population and more than half its sheep, cattle, and horses. The eruption of Tambora Volcano in Indonesia in April 1815 hurled so much fine ash into the upper atmosphere that the following year became known as the "year without a summer." This led to a famine that killed no less than 80,000 humans, produced the coldest three years in British history, and brought snow to New England in July. Crop failures caused the price of flour to double in Great Britain during the two crop years of cold weather.

The reason for these drastic climate changes is that the tiny particles blown into the upper atmosphere by a volcano are very effective in reducing incoming solar radiation. The average length of time volcanic dust particles (0.0001–0.005 mm) remain in the upper atmosphere, the *residence time,* has been found to be 1 to 2 years. Sulfur dioxide gas emitted by volcanic action produces white coatings on the particles, which makes them superreflectors of solar energy. This loading of the upper atmosphere with volcanic dust reduces the amount of incoming radiation relative to outgoing radiation, resulting in a net heat loss for the earth. An attempt has been made to quantify these so-called dust veils into a Dust Veil Index (DVI) that can be used to relate a specific eruption to its impact on climate. Indexes for selected volcanoes are shown in ✳ Table 5.2. As apparent in the table, Tambora and Krakatoa have had potentially the most significant effects on climate in recent history, whereas Mount St. Helens, with a DVI of 1, had little effect. Dust veils are obviously sporadic

◆ FIGURE 5.27 Volcanic hazard map of Mount Rainier area in Washington. USGS data

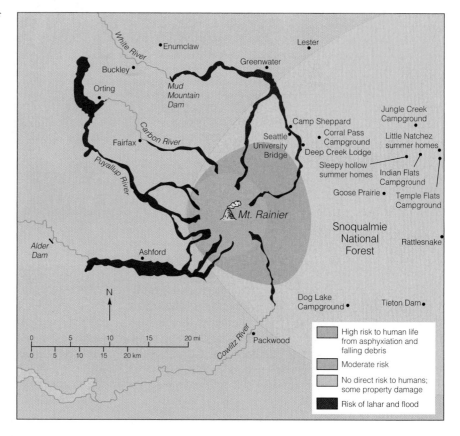

*TABLE 5.2 Dust Veil Indexes of Selected Volcanoes	
Volcano	**DVI**
Tambora, Indonesia, 1815	1,500
Krakatoa, Indonesia, 1883	1,000
Mount Pelée, Martinique, 1902	500
Mount St. Helens, Washington, 1980	1

SOURCE: Lamb, H. H. Volcanic Dust in the Atmosphere (Phil. Trans. Royal Society of London, Ser. A, 266, 425–533).

and transient, and many volcanoes would have to erupt in concert to produce significant long-term effects on climate.

> **CONSIDER THIS** *Skiers flock to Washington, Oregon, and California's many popular ski resorts on dormant volcanoes but are seldom given instructions about what to do if one of them erupts. What evacuation instructions should skiers on the slopes of a volcano that "warms up" to the point that the snowpack begins to melt rapidly have? (This has occurred in New Zealand and is not at all improbable in the western United States.)*

Gases

Denver, Mexico City, and Los Angeles have their smog; the Island of Hawaii, the "big island," has its *vog*—short for "*v*olcanic gas." Since 1986 Kilaeua Volcano near the island's southeast coast has been working almost nonstop producing lava and, as a byproduct, 1,000 tons of sulfur dioxide (SO_2) per day. This sulfurous gas quickly combines with water to form sulfuric acid, which is hazardous to health and corrosive to metallic components of structures and machinery. There are reports of the nails and hinges of houses being eaten away

to such an extent that the structures have almost collapsed and of damage to plants and crops by the acidic gases. During most of the year trade winds blow the vog toward the west side of the island, where it swirls behind massive Mauna Loa and gets trapped over the island's main tourist area, the city of Kailua Kona and the Kona coast (◆Figure 5.29). Health officials and citizens are concerned, and there is much debate about vog as a health hazard. However, the volcano itself will close the debate when this activity cycle ends.

Tree-killing emissions of CO_2 were discovered in 1990 at Mammoth Mountain, California. These emissions continue to this day and pose a serious hazard to campers and hikers (☉ Case Study 5.3 on page 141). The intimate association of lethal carbon dioxide gas and volcanism yielded tragic results in 1986. A cloud of carbon dioxide gas emitted from a crater lake in Cameroon, West Africa, killed more than 1,700 people (see ☉ Case Study 5.4 on page 143).

Selected historic volcanic eruptions and prehistoric eruptions indicated by thick, widespread tephra deposits of Holocene age are listed in * Table 5.3. Although other significant eruptions are known, those in the table were chosen to illustrate the volcanic hazards and impacts discussed in this chapter.

ENVIRONMENTAL
Geology⟲Now™

Click **Geology Interactive** to work through an activity on Magma Chemistry and Explosivity through Volcanism.

The Mitigation and Prediction

Diversion

As mentioned earlier in the chapter, various tactics have been used in attempts to prevent lava flows from overwhelming the land. At various times, people have dammed them, diverted

◆ FIGURE 5.28 Anak Krakatoa, "Son of Krakatoa," in Indonesia. Krakatoa's 1883 eruption caused one of the largest tsunami in history and left a huge caldera. This "son" volcano formed in the caldera. It had nine episodes of explosive activity between 1963 and 2000. This view (1993) shows a rising ash cloud and dark-colored ash falling into the sea.

B. PIPKIN

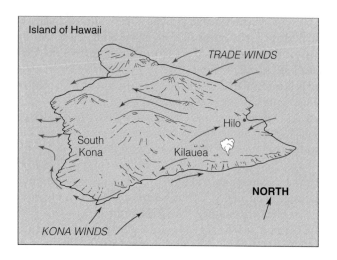

◆ FIGURE 5.29 During trade-wind conditions the almost continuous stream of vog produced by Kilauea (white plume in figure) is blown to the west side of the Island of Hawaii. Traces have been found on Johnston Island 1,600 kilometers (1,000 mi) to the southwest. Onshore (daytime) and offshore (nighttime) winds, indicated by double-headed arrows, trap vog on the island's popular Kona Coast. When the light Kona winds blow (red arrows) the vog is concentrated on the east side of the island. Modified from USGS fact sheet 169-97, 1997

TABLE 5.3   Selected Notable Worldwide Volcanic Eruptions

Year	Volcano Name and Location	VEI	Comments
4895 B.C.±	Crater Lake, Oregon	7	Posteruption collapse formed caldera
1390 B.C.±	Santorini (Thera), Greece	6	Late Minoan civilization devastated
79	Vesuvius, Italy	5	Pompeii and Herculaneum buried; at least 3,000 killed
186	Taupo, New Zealand	7	16,000 km^2 devastated; largest eruption in last 5,000 years
1631	Vesuvius, Italy	4	Modern Vesuvius eruptive cycle begins
1783	Laki, Iceland	4	Largest historic lava flows; 9,350 killed
1792	Unzen, Japan	2	Debris avalanche and tsunamis killed 14,500
1815	Tambora, Indonesia	7	Most explosive eruption in history; 92,000 killed; weather changed
1883	Krakatoa, Indonesia	6	Caldera collapse; 36,000 killed, mostly by tsunamis
1902	Mount Pelée, Martinique, West Indies	4	Saint-Pierre destroyed; 30,000–40,000 killed; spine extruded from lava dome
1912	Katmai, Alaska	6	Perhaps largest 20th-century eruption; 33 km^3 of tephra ejected
1914–17	Lassen Peak, California	3	California's last historic eruption
1943	Parícutin, Mexico	3	New cone formed; event observed and documented from first eruption
1959	Kilauea, Hawaii	2	Lava lake formed that is still cooling
1963	Agung, Bali	4	1,100 killed; climatic effects
1968	Fernandina, Galápagos	4	Caldera floor drops 350 meters
1980	Mount St. Helens, Washington	4–5	Ash flow, 600 km^2 devastated
1982	El Chichón, Mexico	4	Ash flows kill 1,877; climatic effects
1991	Mount Pinatubo, Philippines	5–6**	Probably the second largest eruption of the 20th century; huge volume of SO$_2$ emitted
1991	Unzen, Japan	4**	Pyroclastic flows killed 41 people including 3 volcanologists; lava dome
1991–1993	Etna, Italy	1–2**	Longest continuous activity (473 days) in 300 years ended; 300 million m^3 lava extruded
1983–	Kilauea, Hawaii	1–2**	Longest continuing eruption, with more than 50 eruption events
1995–	Montserrat, West Indies	1–2**	Pyroclastic flows and ash falls that have devastated this idyllic island

*Historic lava or tephra volume ≥2 km^3, Holocene >100 km^3, fatalities ≥1,500.
**Author's estimate.
SOURCE: Data adapted from Smithsonian Institution, Global Volcanism Program.

USGS

◆ FIGURE 5.30 Lava flows on Heimaey Island off the south coast of Iceland were successfully chilled by spraying them with seawater in 1973.

them, and even bombed them. Diversion barriers, high earth and rock embankments piled up by bulldozers, have been constructed on Mauna Loa to protect the city of Hilo, and also to divert potential flows away from the National Oceanographic and Atmospheric Administration (NOAA) observatory at Mauna Loa. Practically, diversion can be considered only if the topography allows the diverted lava to flow onto unimproved lands whose owners do not object to it. Flows have been slowed and even stopped by spraying them with seawater at Heimaey Island, Iceland (◆Figure 5.30). This was done in 1973 at the fishing village of Vestmannaeyjar. Both damming and diversion were used in Sicily in 1992 when eruptions of Mount Etna generated extensive lava flows.

Volcano Hazard Zones and Risk

The purpose of a hazard-zone map of any type is to give accurate information on the frequency and severity of the particular hazard. In the 1950s, the area below Kilauea's east rift zone on the Island of Hawaii was developed. The volcano had been dormant for 17 years and it had not erupted on the lower east rift zone since 1840. Kilauea became active in 1952, however, and by 1996 there had been 13 separate rift-zone eruptions, the last of which has been almost continuous since 1983 (◆Figure 5.31). In 1990 lava-flow hazard maps were prepared for the island, delineating zones ranging from low risk (9) for Kohala at the north end of the island to high risk (1 and 2) within and below the east rift zone (◆Figure 5.32). The Royal Gardens residential subdivision, a development project of the 1950s, has been almost totally buried under lava (see Figure 5.10), and soon after the hazard-zone maps were published, the town of Kalapana was covered by aa flows. Had the developers known that 90 percent of the land surface of Kilauea had been covered by lava since the arrival of the first Hawaiians 1,500 years earlier, they may have reconsidered building there. May have, that is!

◆ FIGURE 5.31 Lava flows on Kilauea Volcano's east rift zone. By late 1992 more than 83 square kilometers (33 mi^2) had been covered by the 1983 to 1992 eruptive sequence shown here, and activity continued into the twenty-first century. USGS data

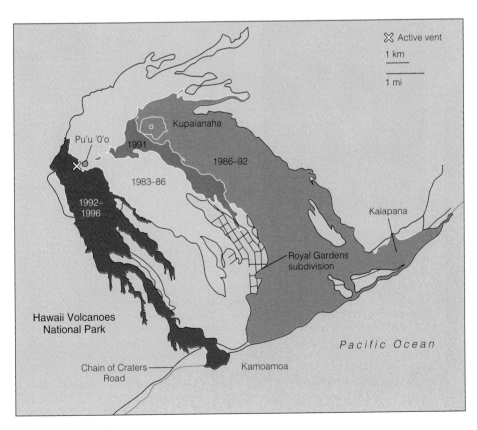

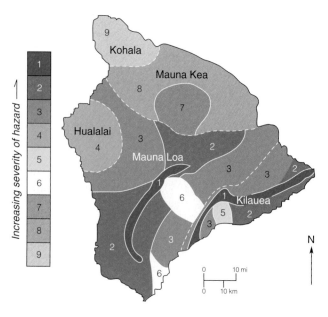

◆ FIGURE 5.32 Lava-flow hazard zones for the Island of Hawaii. The zones range from low risk (9) to high risk (1). Volcano boundaries are denoted by dashed lines. USGS data

Eruption Forecasting

A handful of volcano observatories exist in particular regions around the world to study volcanoes and forecast their eruptions. The oldest is the Vesuvius Volcano Observatory, established by the King of Naples in 1845. Researchers at the Hawaiian Volcano Observatory on the rim of Kilauea Caldera, established in 1912, developed many of the ground-based technologies used in forecasting eruptions (◆Figure 5.33). More recently scientists at the Alaska Volcano Observatory working with laboratories in Kamchatka, Japan, and the U.S. Pacific Northwest have applied satellites very effectively to study eruptions across thousands of kilometers of little inhabited territory.

Prediction is the ability to identify when and where something will happen. Although volcano scientists have gotten very close to actually predicting some eruptions within a few days of their occurrence, it is more accurate to say that they are *forecasters*—that is, that they state in general terms the *probabilities* of eruptions taking place. Like hurricane forecasters, they do this by looking for certain precursory warning

◆ FIGURE 5.33 Volcanologists in the field. (a) Historical photo of scientist taking measurement using a wet-tube tiltmeter, one of the first kinds of tilt-meters invented for volcano studies. As the slope of a volcano changes, so too does the water level in three precisely spaced metal water pots connected by hoses that are set on that slope. (b) Hawaiian geologist Gary Puniwai with a proton-precession magnetometer, used to detect shallow magma intrusions. (c) U.S.G.S. researcher Donald Peterson measures the depth to a lava pool in a vent on the Kilauea east rift zone using a range-finder. (d) A Dartmouth College undergraduate collects fumarole gas samples using a gas chromatograph in the crater of Casita Volcano, Nicaragua.

signs. The ability to forecast depends upon how well a volcano is monitored with instruments, and to a certain extent by how well its history of eruptions is known. Almost every volcano shows its own unique "personality" or eruptive behavior. Some show patterns of behavior that can be very useful in forecasting. Others are more chaotic, exhibiting great changes in their eruptive styles over time.

Scientists at the Hawaiian Volcano Observatory closely monitor the behavior of Kilauea and Mauna Loa, two among the world's most active volcanoes. Each shield contains a magma reservoir several kilometers below the surface that gradually fills with molten rock from weeks to years before an eruption. The whole mountain swells up just as if air were being blown into a balloon. The uplift causes the slopes of the mountain to steepen very gradually, a change that can be precisely measured using a *tiltmeter* (◆Figure 5.34). Gradually increasing earthquake activity typically accompanies the tilting. The locations of earthquake epicenters and foci indicate where the outburst is likely to occur. In the final hours leading up to an eruption, the reservoir suddenly deflates as molten rock works its way toward the surface. Simultaneously a continuous shuddering of the ground, called *volcanic tremor,* takes place. The scientists alert civil defense authorities that an eruption may be imminent: Take no chances; it is time to evacuate!

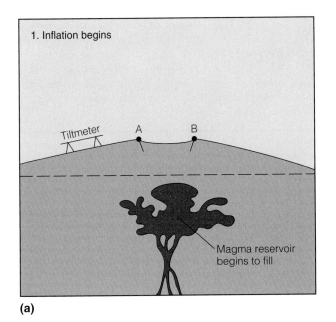

(a)

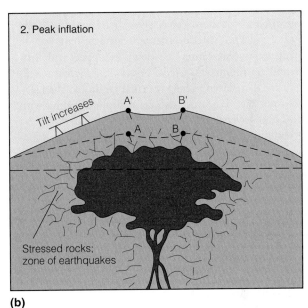

(b)

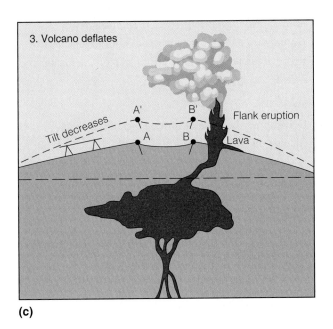

(c)

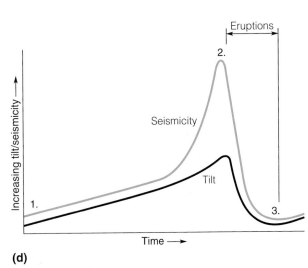

(d)

◆ FIGURE 5.34 Three stages are apparent in Hawaiian volcanoes' typical eruption sequence. (a–c) Inflation and deflation of the cone are accompanied by (d) measurable changes in slope and seismicity. Eruptions occur after peak inflation and continue until magma is exhausted. USGS data

Eruptions do not always follow these warnings. Magma may simply intrude at a shallow depth without breaching the surface. Nor do the warnings provide any clue as to the magnitude and duration of an eruption, should one occur. But it is better to be safe than sorry, and thousands of lives have been saved by successful eruption forecasting.

Other tools have also proved useful in studying volcanic activity. The ascent of magma into shallow reservoirs usually causes a change in the earth's gravity field, which can be measured using *gravimeters*. Change in the makeup of gases, especially in concentrations of chlorine relative to sulfur issuing from fumaroles, often accompanies the ascent of fresh magma. Tell-tale swellings may be detected using *satellite interferometry*—a method of comparing images from space at different times to detect even slight changes in the shape of a volcano. Interferometry is an excellent way to study volcanoes that cannot be easily equipped with tiltmeters and seismometers. *Global Positioning System* (GPS) receivers are also placed to help measure landform changes.

Volcano scientists must also be skillful at dealing with the press and government officials, a craft that can only be learned on the job. Scientists must be careful to convey information accurately so that they do not cause panic or unnecessary and costly evacuations. At the same time, they must judge the precise moment when it is mandatory to urge people to leave their homes. These are not easy calls to make—and throughout a period of crisis the experts and authorities wrestle with life-threatening probabilities.

The ability to save lives from volcanic activity also depends upon the level of public education. Volcano scientists working through the United Nations and other organizations are eager to broadcast information about warning signs, potential hazards, and what people can do to protect themselves during an eruption. There are many troublesome misconceptions that people have about volcanoes.

One source of confusion is the terms "active," "dormant," and "extinct." Many people think of active volcanoes as ones that are *presently* in a state of eruption. Scientifically, however, an active volcano is one that has erupted frequently in historic time, and will almost certainly continue to do so in the future whether or not it is presently erupting. (Just what "frequently" means is a subjective term, however. Also, notice that the length of "history" differs from one region to another around the world.) An active volcano has to be regarded as threatening, whether it is actually erupting or not.

As a rule of thumb, a dormant volcano is one that shows signs of having erupted within the past few hundred or few thousand years, but which has not erupted in living memory. Many dormant volcanoes have only slightly weathered lava flows, appear not to be greatly eroded, and have little vegetation relative to the surrounding area. However, some dormant volcanoes are so heavily vegetated and eroded that people don't even know that they exist without a careful geologic study! Because a volcano is dormant does not mean that it is safe to live near. We simply don't know if or when it will erupt again. Moreover, if an eruption does occur it is

likely to be dangerously explosive, due to the cap of old hardened rock that plugs the throat of the volcano. Examples of dormant volcanoes that have returned to life catastrophically include El Chinchon (Mexico), Mount Lamington (New Guinea), Mount Pinatubo, Vesuvius, and Mount St. Helens. Geologists do not know exactly how many dormant volcanoes there are in the world at present.

An extinct volcano is typically deeply eroded because it has not erupted in thousands if not millions of years. Cinder cones, however, are examples of extinct volcanoes that may have fresh, young appearances, because they tend to erupt only once in their lives. As you might guess, there are many more extinct volcanoes than dormant and active ones in the world today.

CONSIDER THIS *Volcanic activity is perhaps the most spectacular of all geologic hazards, and the destructive potential of many volcanoes is immense. Fortunately, hazards associated with volcanoes are not spread equally around the earth's surface. What areas in the United States are most likely to be affected by future volcanic eruptions, and how have they been identified?*

What to Do When an Eruption Occurs

Based upon the experience of people at Mount St. Helens, the Federal Emergency Management Agency (FEMA) made the following suggestions to ease the trauma of a volcanic eruption.

At Home

- Stay calm and get children and pets indoors.
- If outside, keep eyes closed or use a mask and seek shelter.
- If indoors, stay there until heavy ash has settled.
- Close doors and windows, and place damp towels at door thresholds and other draft openings.
- Do not run exhaust fans or use clothes dryers.
- Remove ash accumulations from low-pitched roofs and gutters.

In Your Car

- Drive slowly because of limited visibility.
- Change oil and air filters every 50–100 miles (80–160 km) in heavy dust, or every 500–1,000 miles (800–1,600 km) in light dust (visibility up to 200 feet, or 60 meters).
- Use both windshield washer and wipers.
- Volcanic ash is abrasive and can ruin engines and paint.

Pets

- Keep pets indoors.
- Brush or vacuum them if they are covered with ash.
- Do not let them get wet, and do not try to bathe them.

ENVIRONMENTAL
Geology ⇌ Now ™

Click **Geology Interactive** to work through an activity on Seismic Risk USA through Earthquakes and Tsunamis.

...ast in the Past, a Skifield, and Fluidization

...han 50 kilometers (30 mi) high, which spread downwind, forming light-yellow ash-fall deposits more than 5 meters (8 ft) thick in some places. At 100 kilometers (60 mi) distant the deposits are 25 centimeters (10 in) thick. Toward the end of the eruption sequence the hot column became so heavy that it collapsed and flowed outward in all directions as a dense mixture of expanding gases and volcanic fragments called a pyroclastic flow. Flow velocity is estimated to have been 600 kilometers per hour (375 mph), and the material traveled more than 80 kilometers (50 mi) from its source, overwhelming all the topography and vegetation in its path. Fortunately, the native New Zealanders, the Maori, had not yet settled the island when the eruption occurred. When they did come to the island about A.D. 1000, they regarded the Tongariro–Taupo area as sacred. They believed that their gods and demons had brought volcanic fire with them from Hawaiki, the Maori's ancestral homeland.

The aftermath of this eruption was a huge collapse basin or caldera, which is now filled with water and known as Lake Taupo. The area is still hot, and north of the lake are the Rotorua thermal springs and geysers, which attract many tourists, and the Wairaki geothermal energy complex (see Chapter 14).

How can a pyroclastic material move at such a phenomenal speed and travel such a great distance, seemingly defying friction and gravity? Some pyroclastic flows are **fluidized,** a process by which solids are transported that is well known in chemical engineering. A "fluidized bed" is a body of fine-grained solids through which a stream of high-pressure gas flows so that the grains separate and move as fluid. Many chemical engineering professors demonstrate this by forcing air through a powder on the bottom of a see-through box. As the air pressure is increased, the powder begins to quiver; then particles start dancing around, and eventually the whole mass is in motion. When the professor has it just right, he or she places a plastic duck (professor's choice) into the container, and it literally floats on the air–solid mixture. If "ducky" is pushed down, it immediately pops up, proving that it is floating on a denser medium. Part of the weight of the solids is taken by the air, reducing the interparticle friction to such a degree that the particles will not stand in a heap. The particles are not *entrained* (carried away bodily) like snowflakes in a blizzard; rather, the mass behaves as a fluid substance would. The point of this is that a well-understood process demonstrates how pyroclastic flows can fluidize and travel far from their source, destroying everything in their path.

At the south end of the Taupo Volcanic Zone is a cluster of stratovolcanoes in the Tongariro National Park. Two of the volcanoes are active and are popular playgrounds for New Zealanders, primarily for skiing and tramping (hiking). Mount Ruapehu, with a summit elevation of 2,797 meters (9,100 ft), is the highest point on North Island and has a lake in its summit crater. It is also the location of the Whakapapa Skifield, a very popular, well-equipped ski area. The terrane at Whakapapa (Maori, pronounced "whack-a-papa") is bare ragged lava and requires a lot of snow to be skiable (◆Figure 2). Throughout the skifield (or lava field, depending on the time of year) signs alert skiers and trampers to stay on high ground and out of the canyons should Ruapehu erupt, which it did in 1969, 1971, 1982, and sensationally in 1995. The eruptions are *phreatic;* that is, superheated water at the bottom of the lake flashes to steam and overflows the crater. The lake's surface-water temperature, which ranges from 20°C to 60°C (68°–140° F), is monitored as a potential eruption predictor. When the water heats up, some kind of activity can be expected. A lahar warning system at Whakapapa is based on detection of earthquake swarms, which precede phreatic and magmatic eruptions. Tongariro (Maori *tonga-,* "south wind," and *-riro,* "carried away") National Park is another case of humans putting nature to good use.

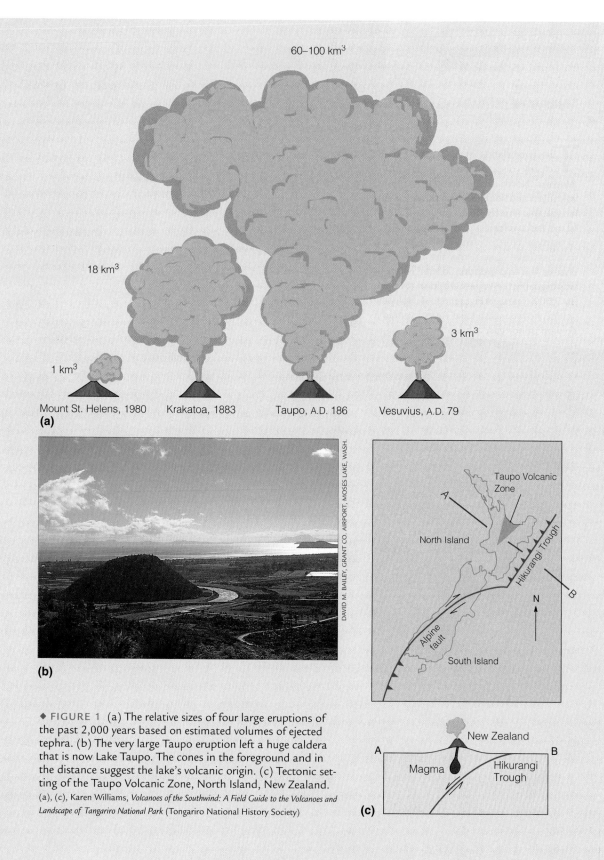

60–100 km³

18 km³

3 km³

1 km³

Mount St. Helens, 1980   Krakatoa, 1883   Taupo, A.D. 186   Vesuvius, A.D. 79

(a)

DAVID M. BAILEY, GRANT CO. AIRPORT, MOSES LAKE, WASH.

(b)

Taupo Volcanic Zone

North Island

Hikurangi Trough

Alpine fault

South Island

A

B

N

New Zealand

Magma

Hikurangi Trough

A   B

(c)

◆ **FIGURE 1** (a) The relative sizes of four large eruptions of the past 2,000 years based on estimated volumes of ejected tephra. (b) The very large Taupo eruption left a huge caldera that is now Lake Taupo. The cones in the foreground and in the distance suggest the lake's volcanic origin. (c) Tectonic setting of the Taupo Volcanic Zone, North Island, New Zealand.

(a), (c), Karen Williams, *Volcanoes of the Southwind: A Field Guide to the Volcanoes and Landscape of Tangariro National Park* (Tongariro National History Society)

(a)

(b)

◆ FIGURE 2 (a) Whakapapa Skifield is on volcanic Mount Ruapehu, the highest point on New Zealand's North Island and the site of a beautiful summit crater lake. Signs on the slopes warn skiers to stay out of canyons and low spots should a phreatic eruption create a dangerous lahar. (b) On September 25, 1995 a combination of hot magma and cold water produced a violent steam eruption that hurled boulders over the upper slopes of the Whakapapa Skifield. The eruption did little damage, but projectiles as big as cars were thrown over the rim. The Maori name for Ruapehu means "exploding pit."

CASE STUDY 5.2

Montserrat, Paradise Lost

The Soufrière Hills Volcano on Montserrat, a tidy little island and British crown colony in the eastern Caribbean Sea, began a series of eruptions in 1995 that continued episodically into 2000. The eruptions began with a dome of andesitic lava in the central crater that periodically collapsed, generating pyroclastic flows (◆Figure 1). Ash falls followed the flows and smothered most of Plymouth, the resort capital, and St. Patricks, a resort town. The world followed the eruptions on television and on the Internet, particularly the evacuation of half of Montserrat's 12,000 inhabitants away from the shadow of the smoldering peak. Volcanic hazard zones were established in September 1997 by the British Geological Survey's Montserrat Volcano Observatory (◆Figure 2). To add to Montserrat's problems, it was also subjected to the extreme hurricane season of 1999.

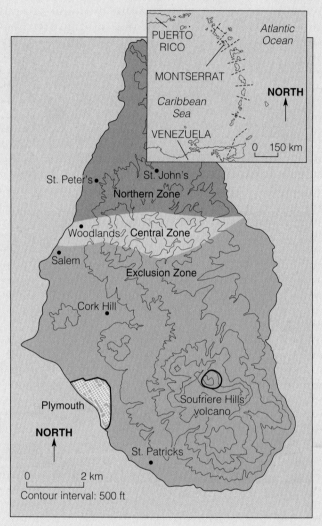

◆ FIGURE 2 Map of Montserrat Island showing Soufrière Hills Volcano and the volcanic hazard zones. *Exclusion Zone:* no admittance except for scientific monitoring and national security matters. *Central Zone:* residents only on a heightened state of alert; all residents to have a rapid means of exit 24 hours a day and to wear hard hats and dust masks when outside. *Northern Zone:* significantly lower risk; suitable for residential and commercial occupation. Montserrat Volcano Observatory Web site map, Jan. 1998, British Geological Survey

SIMON R. YOUNG, MONTSERRAT VOLCANO OBSERVATORY

◆ FIGURE 1 Pyroclastic flow moving down the slope of Soufrière Hills Volcano on the island of Montserrat, British West Indies. Seen here is one of many life-endangering flows that occurred during July 1997 and that threatened the city of Plymouth.

Carbon Dioxide, Earthquakes, and the Los Angeles Water Supply

The Eastern Sierra Nevada is a mecca for tourists who want beautiful mountains and outstanding recreation. Much of the area's appeal results from volcanic action. About 760,000 years ago a huge eruption blew out 617 cubic kilometers (150 mi³) of magma that spread ash over nine states and as far to the east as present-day Nebraska (◆Figure 1). The earth surface above the magma chamber sank 2 kilometers, forming the huge, oval caldera (16 km × 33 km) that is now called *Long Valley*. The eruption was 2,000 times the size of the 1980 Mount St. Helens eruption and it left hard welded tuff all over today's east-central California. Mammoth Mountain and Mono Craters, a linear chain of glassy domes (see Figure 5.16), formed within the last 400,000 years, and the people of Long Valley are living with the caldera's continuing restlessness.

The restlessness is manifest in hot springs, carbon dioxide emissions, and swarms of earthquakes, the most intense of which began in 1980 after two decades of quiescence. There were four strong, magnitude-6 earthquakes, three of them occurring on the same day. These quakes prompted studies by the U.S. Geological Survey that found uplift of 60 centimeters (about 2 ft) in the caldera center. Uplift, earthquake swarms, and gas emissions leave little doubt that magma is rising and getting closer to the ground surface. Late in 1997, just after the volcano movie *Dante's Peak* was released, the real volcano here acted up with more than 8,000 earthquakes larger than magnitude 1.2, and a couple of rattlers approached magnitude 5.0. More than 1,000 earthquakes occurred one day, and many scientists believe the volcano was waving a red flag signaling "danger."

Much of the water for the City of Los Angeles comes from Sierra Nevada snowmelt and travels from Mammoth Lakes to the city via open aqueduct. Should the volcano erupt, it would jeopardize or totally cut off this invaluable resource. Furthermore, the area's main highway access is aligned in the direction of the prevailing wind from the mountains. For this reason, an "escape road" was built in the opposite direction, toward the north, to provide an alternate route for the area's residents and visitors.

Mammoth Mountain, a large rhyolitic volcano, is still hot. Tree-killing emissions of CO_2 were discovered at the base of the mountain in 1990. This indicated magma activity at shallow depth (◆Figure 2). By 1995 about 30 hectares (75 acres) of pine and fir trees had suffocated, and CO_2 concentrations of up to 90 percent of total gas were found in the trees' soil and root systems. Where CO_2 concentrations exceeded 30 percent, most of the trees were dead. One large campground had to be closed because park rangers found high levels of CO_2 in restrooms and cabins after campers exhibited symptoms of asphyxia (◆Figure 3). It appears that carbon dioxide degasses from the magma and migrates into a deep gas reservoir. When earthquake swarms cause fractures to develop in the reservoir rock, the gas finds its way to the surface. It is estimated that between 200 and 1,000 tons of carbon dioxide flows into the soil and atmosphere per day. The good news is that gas emissions have been declining, which may indicate that the magma has exhausted its gas supply.

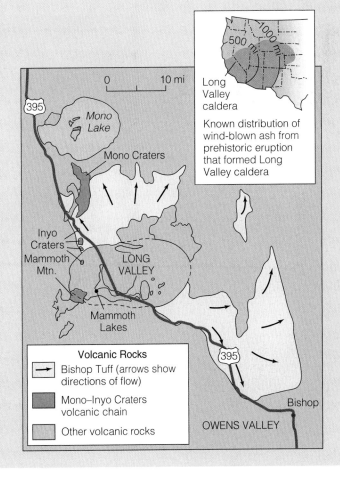

◆ **FIGURE 1** Mammoth Mountain and the Mono–Inyo Craters volcanic chain (red). The cataclysmic eruption 760,000 years ago that formed Long Valley ejected hot, glowing pyroclastic flows that cooled to form the Bishop Tuff (yellow). The inset map shows the distribution of Bishop Tuff and ash. USGS data

(a)   (b)

◆ FIGURE 2   The Horseshoe Lake tree-kill area, Mammoth Lakes, California. (a) Aerial view looking north toward Mammoth Mountain. Inyo–Mono Craters and Mono Lake are seen in the right background. The light colored area (foreground) at the base of the Mammoth Mountain and on the shore of Horseshoe Lake is the tree-kill area. (b) Close-up of trees killed by leaking volcanogenic carbon dioxide gas.

◆ FIGURE 3   Looking south through the Horseshoe Lake tree-kill area on the south side of Mammoth Mountain. The trees have died from high levels of CO_2 in the root zone.

Volcanic CO₂ and Fountaining Lakes

Lake Nyos is contained within a volcanic crater, one of 40 such craters that are roughly aligned southwest to northeast in Cameroon, West Africa. Specifically, the lake is in a **maar,** a low-relief crater produced by explosive interaction of magma with underground water. Such a high-pressure steam explosion, called a **phreatic eruption,** does not produce much new magmatic material, which is why the resulting crater is relatively shallow. Lake Nyos and several nearby crater lakes and springs are known to be highly charged with carbon dioxide, CO_2, just as is a bottle of soda water.

On the evening of August 21, 1986, almost noiselessly and without warning, the lake erupted, discharging 80 million cubic meters (100 million yd^3) of CO_2 dissolved in its waters into the atmosphere. So much gas escaped from the lake that its level dropped 1 meter. Because carbon dioxide weighs about one and a half times as much as air, the suffocating gas descended downslope, following stream valleys and drainages for 16 kilometers (10 mi). The CO_2 cloud was 50 meters (160 ft) thick and it killed all animal life in its path, including 1,200 residents of Nyos village and 500 persons in nearby communities. The entire event took less than an hour (◆Figure 1).

What happened the evening of the tragedy that caused the gas in the lake to explode catastrophically? Was it a small volcanic disturbance, or did some other physical phenomenon upset the lake? It is known that lakes periodically *overturn;* that is, their surface water sinks to the bottom and is replaced by bottom water rising to the surface. This usually occurs in temperate climates in the fall, because as the surface water cools, it becomes denser than the bottom water and displaces it. Overturn usually requires a trigger, such as strong winds, a landslide, or a drastic change in atmospheric pressure. The best hypothesis appears to be that winds associated with the rainy season blew the lake's surface water to one side of the crater. This reduced the hydrostatic (confining) pressure on the bottom waters, which allowed the highly charged gas to come out of solution and form bubbles. The formation of bubbles further reduced the density of the water, and the gas in solution frothed and rose with dramatic speed. That this sequence of events indeed happened is supported by stripped vegetation along the lake shore. The leaves and other parts of plants were most likely removed by a surge of water associated with the explosion—just as a fountain of foam is associated with popping the top of a warm can of soda.

Bantu folklore is filled with stories of exploding lakes and "rains" of dead fish. According to legend, there have been at least three earlier episodes of catastrophic lake explosions. Dead fish found floating on the water surface during minor degassing events are believed by the local people to be gifts from their ancestors who live in the lake.

Since 1986 the gas content of the lake has increased by additions from subterranean springs and magmatic sources. Engineering studies indicate that Lake Nyos and other dangerous crater lakes in the region could be defused by installing pipes in them to bring charged water from the lake bottom to the surface as fountains. Once started, the fountains would be driven entirely by the lifting force of the released gas.

◆ FIGURE 1 (a) Deadly carbon dioxide belched from the bottom of volcanic Lake Nyos, which stood muddy and full of debris for several days. The gas settled into the valleys below, tragically suffocating 1,700 people. (b) Cattle suffocated by carbon dioxide at Lake Nyos.

(a)

(b)

gallery

Volcanic Wonders

Volcanism creates landscapes that are attractive to humans. One of the most eye-catching volcanic features is *columnar jointing*, which forms as cooling volcanic rock contracts around many equally spaced centers, causing the lava to split into polygonal columns. Six-sided columns are most common, but four-, five-, seven-, and eight-sided columns also occur. An excellent example is seen at the Devil's Tower, Wyoming, a shallow intrusion (◆Figure 1).

JOHN MEAD, SCIENCE PHOTO LIBRARY/PHOTO RESEARCHERS

◆ FIGURE 1 Devil's Tower National Monument in northeastern Wyoming is an exposed volcanic neck with spectacular columnar jointing. Also known as "Grizzly Bear Lodge" and "Mato Tepee," it rises 865 feet above the surrounding landscape.

JACK GREEN, CALIF. STATE UNIV., LONG BEACH

◆ FIGURE 2 Nature's foot-warmer; Sarangson fumarole field, Minhasa, Celebes (Sulawesi). Fumarole temperatures range from 30° to 90° C.

◆ FIGURE 3 Call the automobile club!

◆ FIGURE 4 A DC-10 pointed skyward due to Mount Pinatubo ash on its tail; Cubi Point Naval Air Station, Philippines.

The columns form perpendicular to the main cooling surfaces of the flows, which is why they are sometimes curved. The 264-meter-tall (865 ft) Devil's Tower, called "Grizzly Bear Lodge" and "Mato Tepee" by early Native Americans, is surrounded by colorful Cheyenne and Lakota Sioux legends. According to one legend, a number of children were being chased by a grizzly bear. The Great Spirit saw their plight and raised the ground they were on. The bear scratched deep grooves into the rocks in his attempts to reach the frightened children. Imagine the tribal elders passing on this legend to wide-eyed children. It is a much more memorable explanation for the origin of the columns than the scientific one.

Fumaroles represent the dying stages of volcanism. They emit heat, steam, and sulfurous gases and build brightly colored surface mounds. They also have other uses (◆Figure 2). The other gallery photos speak for themselves.

◆ FIGURE 5 A mud volcano formed by volcanic gases escaping through fine-grained clay deposits; Imperial Valley, California.

◆ FIGURE 6 Surrealistic lava tree molds form when fluid lava surrounds a tree trunk and cools, leaving a cylinder of solidified rock; east rift zone, Kilauea Volcano, Hawaii.

Summary

Volcanoes

Defined

A vent or series of vents that issue lava and pyroclastic material.

Distribution

Adjacent to convergent plate boundaries around the Pacific Ocean (the Ring of Fire); in a west–east belt from the Mediterranean region to Asia; along the mid-ocean ridges; and in the interior of tectonic plates above hot spots in the mantle. In the U.S., a 1,100-kilometer (680-mi) belt extends northward from northern California through Oregon and Washington.

Measurement

The Volcanic Explosivity Index assigns values of 0 to 8 to eruptions of varying size.

Explosivity

Depends upon the lava's viscosity and gas content. Viscosity, a function of temperature and composition, determines the type of eruption: felsic (high-SiO_2) lavas are viscous and potentially explosive; mafic (low in SiO_2, high in magnesium and iron) lavas are more fluid and less explosive.

Types of Volcanoes and Volcanic Landforms

Shield (Hawaiian or "Effusive" Type)

Gentle outpourings of lava produce a convex-upward edifice resembling a shield.

Continental Flood Basalts

Lava erupts from long cracks, building up broad lava plateaus such as those found in the Columbia River Plateau (Oregon–Washington–Idaho), Iceland, and India. They are fed by mantle plumes.

Stratovolcanoes (Explosive Type)

Characterized by concave-upward cone thousands of meters high. Vesuvius (Italy), Fujiyama (Japan), and Mount Hood (Oregon) are examples. Crater Lake (Oregon) is the stump of a stratovolcano that exploded and collapsed inward, leaving a wide crater known as a *caldera*.

Lava Domes

Bulbous masses of high-silica, glassy lavas, such as found at Mono Craters (California). Domes may form in the crater of a composite cone after an eruption, such as at Mount St. Helens (Washington).

Cinder Cones

The smallest and most numerous of volcanic cones, composed almost entirely of tephra.

Benefits of Volcanic Activity

Atmosphere and Hydrosphere

Outgassed from the earth's interior.

Building Materials

Cinder blocks, road-base materials, pumice, light-weight concrete, decorative stone.

Geothermal Energy

Cooling magma at depth heats water that can be converted to steam to drive electrical generators.

Hazards

Lava Flows

Destroy or burn everything in their path, can travel a distance of 100 kilometers or more.

Ash Falls

Heavy falls of volcanic ash from Mount Vesuvius buried Pompeii in A.D. 79. There were heavy ash falls from Mount Pinatubo, the Philippines, 1991. Vertical plumes of ash may rise many kilometers from a vent, which are then carried by the wind and blanket the terrain. Such plumes rising from a vent are known as *Plinian eruptions*.

Pyroclastic Flows

Pyroclastic flows are hot, fluid masses of steam and pyroclastic material that travel down the flanks of volcanoes at high speeds, blowing down or suffocating everything in their path. Such flows have killed thousands of people.

Lahars

Catastrophic mud flows down the flanks of a volcano. They cause the most volcanic fatalities, more than pyroclastic flows.

Tsunamis

Submarine volcanic eruptions (such as that of Krakatoa in 1883) create enormous sea waves that may travel thousands of miles and still do damage along a shoreline.

Weather

Sulfur dioxide coatings on ash and dust particles increase their reflectivity and cause cooling of weather, such as occurred after the eruption of Mount Pinatubo, 1991–1993. The effects of ash and dust in the stratosphere are short-term—a few years.

Gases

Corrosive gases emitted from a volcano can be injurious to health, structures, and crops. Such gases are called *vog* (*v*olcanic *g*as) on the island of Hawaii. Carbon dioxide emitted from the bottom of volcanic Lake Nyos in Cameroon formed a cloud that sank to the ground and suffocated 1,700 people in 1986. Tree kills due to CO_2 have occurred at Mammoth Mountain, California.

Mitigation and Prediction

Diversion and Chilling

Flows have been diverted in Hawaii and elsewhere and have been chilled with seawater in Iceland.

Prediction

Prediction in Hawaii, based upon seismic activity (earthquake swarms and harmonic tremors) and the tilt of the cone as magma works its way upward, has become quite accurate. Similar phenomena were observed at Mount St. Helens and Mount Pinatubo.

Key Terms

aa	felsic	maar	phreatic eruption	tephra
andesite line	fluidization	mafic	Plinian column	Volcanic Explosivity
ash fall	lahar	nuée ardente	pyroclastic flow	Index (VEI)
caldera	lava dome	outgassing	shield volcano	welded tuff
cinder cone	lava plateau	pahoehoe	stratovolcano	

Study Questions

1. What are the four kinds of volcanic cones, and what type of eruption can be expected from each kind? What are flood basalts, and how have they shaped the earth's landscape?

2. How would you expect the eruption of a volcano that taps mafic magma to differ from the eruption of one that taps felsic magma? What kind of rocks will result from each eruption?

3. Explain the distribution of the world's 1,350 active volcanoes in terms of plate tectonic theory.

4. List the six geologic hazards connected with volcanic activity in decreasing order of their threat to human life and limb.

5. Where in the United States are the most dangerous volcanic hazards encountered? Which geologic hazard related to volcanic activity presents the greatest danger to humans, and how can this hazard be mitigated, or at least minimized?

6. Why can eruptions on the Island of Hawaii be predicted more accurately than volcanic eruptions at subduction zones?

7. How may calderas be formed? What geologic evidence is found at Crater Lake, Oregon, that indicates it was formed by a special set of circumstances?

8. In what ways are volcanic activity and its products useful to humankind?

9. Name five volcanic areas of great scenic wonder and the geologic feature found there; for example, Crater Lake, caldera.

10. How do volcanic eruptions affect climate and weather? What impact does this have on populations, particularly Third World peoples?

11. In what ways do gases, including water vapor, emitted by a volcano impact the nearby area and people living there?

12. Define or sketch these volcanic features: aa, pahoehoe, lahar, Dust Veil Index, lapilli, tephra, welded tuff.

For Further Information

Books and Periodicals

Decker, Robert, and Barbara Decker. 1981 and 1997. *Volcanoes.* New York: W. H. Freeman.

Duennebier, F. K. (contact person) and the Loihi Science Team. 1997. Researchers rapidly respond to submarine activity at Loihi Volcano, Hawaii. EOS 78, no. 22 (3 June).

Evans, W. C., L. D. White, M. L. Tuttle, G. W. Kling, G. Tanyileke, and R. L. Michel. 1994. Six years of change at Lake Nyos, Cameroon, yield clues to the past and cautions for the future. *Geochemistry* 28: 139–162.

Fierstein, Judy, and others. 1997. *Can another great volcanic eruption happen in Alaska?* U.S. Geological Survey fact sheet 039-97.

Francis, Peter. 1993. *Volcanoes.* New York: Oxford University Press.

Hazlett, Richard W. 1987. *Kilauea volcano: Geological field guide.* Hilo, Hawaii: Hawaiian Natural History Association.

Hill, David, and others. 1996. *Living with a restless caldera, Long Valley, California.* U.S. Geological Survey fact sheet 108-96.

——————. 1997. *Future eruptions in California's Long Valley area: What's likely.* U.S. Geological Survey fact sheet 073-97.

Kling, George W., and others. 1987. The 1986 Lake Nyos gas disaster in Cameroon, West Africa. *Science* 236: 169–174.

McDonald, G. A., A. T. Abbott, and F. L. Peterson. 1983. *Volcanoes in the sea: The geology of Hawaii,* 2d ed. Honolulu: University of Hawaii Press.

McGee, Kenneth A., and others. 1997. *Impacts of volcanic gases on climate, the environment, and people.* U.S. Geological Survey open file report 97-262.

McPhee, John. 1989. *The control of nature.* New York: Farrar, Straus, and Giroux.

Myers, Bobbie, and others. 1997. *What are volcano hazards?* U.S. Geological Survey fact sheet 002-97.

Neal, Christina A., and others. 1997. *Volcanic ash: Danger to aircraft in the North Pacific.* U.S. Geological Survey fact sheet 039-97.

Schuster, Robert L., and J. P. Lockwood. 1991. Geologic hazards at Lake Nyos, Cameroon, West Africa. *Association of Engineering Geologists News,* April.

Sigvaldason, G. E. 1989. International conference on Lake Nyos disaster, Yaounde, Cameroon, 16–20 March 1987: Conclusions and recommendations. *Journal of volcanology and geothermal research* 39: 97–109.

Simkin, Tom, and others. 1981. *Volcanoes of the world,* Stroudsberg, Penn.: Hutchinson Ross Publishing Co.

Smith, A. L., and M. J. Roobol. 1991. *Mt. Peleé, Martinique: An active island-arc volcano.* Geological Society of America memoir 175.

Sorey, Mike, Bill Evans, Mack Kennedy, John Rogie, and Andrea Cook. 1999. Magmatic gas emissions from Mammoth Mountain. *California geology,* September/October.

Stager, Curt. 1987. Killer lake. *National geographic,* September: 404–420.

Sutton, Jeff, and others. 1997. *Volcanic air pollution: A hazard in Hawaii.* U.S. Geological Survey fact sheet 169-97.

Tilling, Robert I. 1984. *Monitoring active volcanoes.* Denver, Colo.: U.S. Geological Survey.

Williams, A. R. 1997. Montserrat: Under the volcano. *National geographic,* July.

Wright, Thomas L., and T. C. Pierson. 1992. *Living with volcanoes.* U.S. Geological Survey circular 1073. U.S. Geological Survey, Box 25425, Federal Center, Denver, CO 80225.

ENVIRONMENTAL
Geology ⇌ Now™

Assess your understanding of this chapter's topics with additional quizzing and comprehensive interactivities at http://earthscience.brookscole.com/pipkingeo4e

as well as current and up-to-date Web links, additional readings, and InfoTrac College Edition exercises.

6

Soils, Weathering, and Erosion

The Wind Blew and the Soil Flew

> *A few inches between humanity and starvation . . .*
>
> Anonymous

THE GREAT DEPRESSION OF THE 1930S WAS exacerbated by a protracted drought in the Great Plains of Oklahoma, Colorado, Texas, New Mexico, and Kansas. This part of the Great Plains was totally dependent upon rainfall for crop production and, because the government had guaranteed wheat prices, farmers tilled thousands of acres. A longstanding agricultural practice was to plow the fields after the fall harvest, crop stubble and all, and let the land lie fallow all winter. This was good practice as long as it didn't rain too much—or too little. If too much rain falls the soil is subject to extreme sheet and rill erosion, and if there is drought and the soil dries out the result is the same—extreme soil erosion, only by wind.

There was drought and the wind did blow in the southern Great Plains in the mid 1930s. Huge quantities of topsoil were simply blown away (◆Figure 1). Fine particles were lifted 5 kilometers (16,000 ft) in the air and carried eastward as far as Washington, D.C., and beyond. Five states that were the source of this airborne dust became known as the "Dust Bowl." Farming was impossible and farm families, burdened by debt for equipment, seed, and supplies, left their farms to become migrant workers. Many traveled to California and became derisively known as "Okies" (◆Figure 2). These unfortunate victims of severe soil erosion were the subject of John Steinbeck's touching novel *The Grapes of Wrath.* Even today wind erosion exceeds water erosion in many parts of the "Dust Bowl" states.

UPI/BETTMANN

◆ FIGURE 1 *A dust storm hits a southwestern Great Plains village in October 1937. Storms like this one destroyed crops and buried pasturelands. Minutes after this photograph was taken the village was completely engulfed in choking dust.*

BETTMANN ARCHIVE

◆ FIGURE 2 *Dust Bowl America. In 1938 Dorothea Lange, who worked for the Farm Security Administration, took this historic photograph of a homeless migrant family of seven on a highway in Pittsburgh County, Oklahoma. By 1939, half of the families in Oklahoma were on relief.*

Soil is the solid earth's most fundamental resource. Through it we produce much of our food, and it supports forest growth, which gives us essential products. Soil material, organisms within the soil, and vegetation constitute an ecological system in which the four ingredients needed for plant growth are recycled. These ingredients are water, air, humus (organic matter), and mineral matter (◆Figure 6.1).

The multiple uses of **soil** are reflected in the many definitions and classification schemes that have been developed for soils. To the engineer, a soil is the loose material at the earth's surface; that is, material that can be moved about without first being dynamited and upon which structures can be built. The geologist and the soil scientist see soil as weathered rock and mineral grains that are capable of supporting plant life. A farmer, on the other hand, is mostly interested in which crops a soil can grow and in whether the soil is rich or depleted with respect to humus and minerals. Soils form by the weathering of **regolith,** the fragmental rock material at the earth's surface. By studying soils we can make inferences about their parent materials and the climate under which they formed and can determine their approximate age.

The carrying capacity of our planet—that is, the number of people the earth can sustain—depends on the availability and productivity of soil. This is why reducing soil erosion is so important. For all practical purposes, once productive topsoil is removed, it is lost to human use forever.

Five environmental factors determine the development of soils: (1) climate, (2) organic activity, (3) relief of the land, (4) parent material, and (5) the length of time that soil-forming processes have been acting. These factors can be entered into the "CLORPT equation," a memory aid for these five factors:

$$\text{Soil} = f(Cl, O, R, P, T)$$

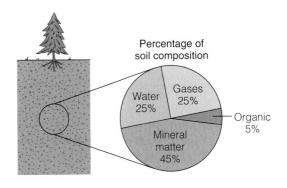

◆ FIGURE 6.1 Composition of a typical soil. The organic component includes humus (decomposed plant and animal material), partially decomposed plant and animal matter, and bacteria.

where f is read "a function of" and Cl, O, R, P, and T represent climate, organic activity, relief, parent material, and time, respectively. The rate at which a soil weathers is a function of climate. In order to understand how soils form, change with time, and are distributed on the surface of the earth, we must first understand how rocks, minerals, and even the structures we place on the earth are broken down by weathering—their interaction with water and air.

Weathering

Weathering is a destructive process by which rocks and minerals are broken down by exposure to atmospheric agents. Specifically, **weathering** is the physical disintegration and chemical decomposition of earth materials at or near the earth's surface. **Erosion,** on the other hand, is the removal *and transportation* of weathered or unweathered materials by wind, running water, waves, glaciers, underground water, and even gravity (◆Figure 6.2).

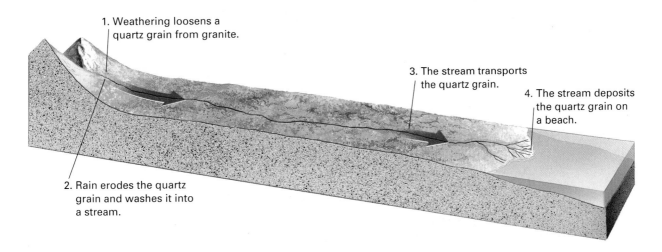

1. Weathering loosens a quartz grain from granite.

2. Rain erodes the quartz grain and washes it into a stream.

3. The stream transports the quartz grain.

4. The stream deposits the quartz grain on a beach.

◆ FIGURE 6.2 Weathering occurs by water (rain) and the atmosphere, which loosen the quartz grains in the host rock, granite. The loosened grains are then transported, in this example by a stream, to a beach, which may be at a lake or at the ocean.

Physical Weathering

Physical weathering makes little rocks out of big rocks by many processes. Rocks are wedged apart along planes of weakness by thermal expansion and contraction and by the activities of organisms. **Frost wedging** occurs when water freezes in joints or other rock openings and expands. This expansion, which amounts to almost ten percent in volume, can exert tremendous pressures in irregular openings and joints. As one would suspect, this process operates only in temperate or cold regions that have seasonal or daily freeze-and-thaw cycles. Roots of even the smallest plants can act as wedges in rock in any climate (♦Figure 6.3). The most dramatic example of this is probably the damage to paving and structures caused by tree roots. Although this process is independent of climate, it operates in tandem with the effects of climate.

Thermal expansion and contraction may loosen mineral grains and contribute to rock disintegration in extreme climates. The quantitative importance of this process is difficult to evaluate, and experiments suggest it is probably not great. Salt-crystal growth is known to pry loose individual minerals in a rock, even granites. It is particularly effective in porous, granular rock such as sandstone. It occurs mainly in arid regions.

Animals, on the other hand, contribute significantly to weathering and soil formation by aerating the ground and mixing loose material. Insects, worms, and burrowing mammals, such as ground squirrels and gophers, move and mix soil material and carry organic litter below. Some tropical ants' nests are tunnels hundreds of meters long, for example, and common earthworms can thoroughly mix soils to depths of 1.3 meters (4¼ ft). Earthworms are found in all soils where there is enough moisture and organic matter to sustain them,

♦ FIGURE 6.3 Tree roots and their trunk wedge openings in granite; Yosemite Valley, California.

and it is estimated that they eat and pass their own weight in food and soil minerals each day, which amounts to about ten tons per acre per year on average. An Australian earthworm is known that grows to a length of 3.3 meters (11 ft) and is a real earth mover!

Chemical Weathering

Chemical reactions also work on rock and mineral debris at the earth's surface. These reactions are complex and they involve many steps, but the fundamental processes are reactions between earth materials and atmospheric constituents such as water, oxygen, and carbon dioxide. The products of chemical weathering are new minerals and/or dissolved minerals. The specific reactions are *solution,* the dissolving of minerals; *oxidation,* the "rusting" of minerals; *hydration,* the combining of minerals with water; and *hydrolysis,* the complex reaction that forms clay minerals and the economically important aluminum oxides. ✳ Table 6.1 shows the formulas for the typical chemical-weathering reactions that are explained in the following discussion.

Solution

Carbon dioxide released from decaying organic matter and from the atmosphere combines with water to form weak carbonic acid. This natural weak acid attacks solid limestone, dissolving it and yielding an aqueous solution of calcium and bicarbonate ions. Solution in limestone terrane creates caverns, such as the Mammoth Caves in Kentucky and Carlsbad Caverns in New Mexico. (Chapter 7 explains this.) Acid rain, which forms when sulfur dioxide or nitrogen oxides combine with water droplets in the atmosphere to form sulfuric and nitric acids, takes its toll on rock monuments (♦Figure 6.4).

Oxidation and Hydration

Oxidation produces iron-oxide minerals (hematite and limonite) in well-aerated soils, usually in the presence of water. It is the "rusting" commonly seen on metal objects that are left outdoors. Pyroxene, amphibole, magnetite, pyrite, and olivine are most susceptible to oxidation because they have high iron contents. It is the oxidation of these and other iron-rich minerals that provides the bright red and yellow colors seen in many of the rocks of the Grand Canyon and the Colorado Plateau.

Hydrolysis

The most complex weathering reaction, hydrolysis, is responsible for the formation of clays, the most important minerals in soil. A typical hydrolytic reaction occurs when orthoclase feldspar, a common mineral in granites and some sedimentary rocks, reacts with slightly acidic carbonated water to form clay minerals, potassium ions, and silica in solution.

The ions released from silicate minerals in the weathering process are salts of sodium, potassium, calcium, iron, and

∗TABLE 6.1 Examples of Chemical-Weathering Reactions

Reaction	Example
Solution	
Step 1: A natural acid forms.	$CO_2 + H_2O \rightarrow H_2CO_3$ carbon dioxide + water → carbonic acid
Step 2. Acid dissolves limestone, forming calcium ions and bicarbonate ions in solution.	$CaCO_3 + H_2CO_3 \rightarrow Ca^{++} + 2(HCO_3)^-$ limestone + carbonic acid → calcium ions + bicarbonate ions
Oxidation and Hydration	
Oxygen and water combine with iron, producing hydrated iron oxide ("rust").	$4Fe^{++} + 3O_2 + 6H_2O \rightarrow 2(Fe_2O_3 \cdot 3H_2O)$ iron minerals + oxygen + water → limonite (rust)
Hydrolysis	
Orthoclase feldspar combines with acid and water, forming clay minerals and potassium ions.	$2KAlSi_3O_8 + 2H^+ + 9H_2O \rightarrow$ orthoclase feldspar + acid ions + water $Al_2Si_2O_5(OH)_4 + 2K^+ + 4H_4SiO_4$ clay minerals + potassium ions + soluble silica

magnesium, which become important soil nutrients. Clay minerals are important in soils, because their extremely small grain size—less than 4 microns (0.004 mm)—gives them a large surface area per unit of weight. Soil clays can adsorb significant amounts of water on their surfaces, where it stays within reach of plant roots. Also, because the clay surfaces have a slight negative electrical charge, they attract positively charged ions, such as those of potassium and magnesium, making these important nutrients available to plants. Other soil components being equal, the amounts of clay and humus in a soil are what determine its suitability for sustained agriculture. Sandy soils, for example, have much less water-holding capacity, less humus, and fewer available nutrients than do clayey soils. But clay-rich soils drain poorly and are hard to work.

The Rate of Weathering

The rate of chemical weathering is controlled by the surface environment, grain size (that is, the surface area), and climate. The stability of the individual minerals in the parent material is determined by the pressure and temperature conditions under which they formed. Quartz is the last mineral to crystallize from a silicate magma, and the temperature at which it does so is lower than the temperatures at which olivine, pyroxene, and the feldspars crystallize. For this reason, quartz

(a)

(b)

◆ FIGURE 6.4 Comparative weathering of granite obelisks, both believed to have been shaped 3,500 years ago (1500 B.C.). (a) Obelisk amid the ruins of Karnak near Luxor, Egypt. The hieroglyphs are sharp, and the polished surface is still visible. (b) The red-granite obelisk known as "Cleopatra's Needle" that stands in New York's Central Park was a gift from the government of Egypt in the late 19th century. New York City's humidity and other environmental conditions have caused chemical weathering of the incised hieroglyphs that had endured several thousand years in Egypt's arid climate.

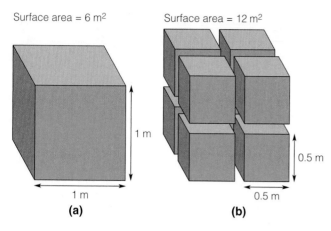

Surface area = 6 m²   Surface area = 12 m²

1 m

1 m

0.5 m

0.5 m

(a)   **(b)**

◆ FIGURE 6.5 Surface area increases with decreasing particle size. A cube that is 1 meter on each side has a surface area of 6 m². If the cube is cut in half through each face, 8 cubes result that have a total surface area of 12 m².

has greater chemical stability under the physical conditions at the earth's surface. The relative resistances to chemical weathering of minerals in igneous rocks thus reflect the opposite order of their crystallization sequence in a magma. Most stable to least stable, the sequence is

quartz (most stable; crystallizes at lowest temp.)

  orthoclase feldspar

    amphibole

      pyroxene

        olivine (least stable; crystallizes at highest temp.)

Micas and plagioclase feldspars form over a range of temperatures and thus vary in weathering stability. Note that ferromagnesian (iron and magnesium) minerals—olivine, pyroxene, and amphibole—are least stable and most subject to chemical weathering. Igneous rocks made up of iron-rich minerals weather readily in humid climates.

A monolithic mass of rock is less susceptible to weathering than an equal mass of jointed and fractured rock. This is because the monolith has less surface area on which atmospheric processes can act. A hypothetical cubic meter of solid rock would have a surface area of 6 square meters. If we were to cut each face in half, however, we would have 8 cubes with a total surface area of 12 square meters (◆Figure 6.5). Thus smaller fragments have more area upon which chemical and physical weathering can act. For example, whereas soils have developed on volcanic ash in the West Indies after only 30 years, some nearly 200-year-old solid basalt flows in Hawaii have yet to grow a radish. This reflects differences in particle size and surface area.

The effect of climate has already been discussed. It is important to remember that chemical weathering dominates in humid areas, and physical weathering is most active in arid or dry alpine regions. (◆Figure 6.4 may help you remember this.) However, humans have accelerated chemical weathering by polluting the atmosphere with acid-forming compounds. Limestone monuments in English cemeteries are reported to require 250–500 years to weather to a depth of 2.5 centimeters and thus to obliterate inscriptions. Buildings in northern England are known to weather more on the side that receives sulfur-laden winds from a nearby industrial area. A sixteenth-century monastery in Mexico that was built of volcanic tuff has been ravaged by chemical weathering to the extent that some of the carved statues on its exterior are barely recognizable (◆Figure 6.6a). A rock wall protecting Chicago's Lakeshore Drive was built of Indiana limestone. Chemical weathering by water overtopping the wall has dissolved clearly visible pits in a pattern known as "honeycomb" weathering (◆Figure 6.6b). You can see examples of weathering in your own community by visiting a cemetery. Compare the legibility of the inscriptions on monuments of various ages and materials. Slate and granite monuments last much longer than those of limestone or marble, as you might suspect.

(a)

(b)

◆ FIGURE 6.6 (a) A 16th-century monastery in Mexico shows the ravages of weathering mostly from wind and wind-driven rain. The rock is volcanic tuff. (b) A Lake Michigan seawall built of Indiana limestone is chemically weathered by waves overtopping the structure. Solution of the limestone has caused "honeycomb" weathering.

Geologic Features of Weathering

Spheroidal weathering is chemical weathering in which concentric shells of decayed rock are separated from a block bounded by joints or other fractures (◆Figure 6.7). It occurs when a rectangular block is weathered from three sides at the corners and from two sides along its edges, while the block's plane faces are weathered uniformly. It is also called "onion-skin" weathering, since the layers of an onion provide a good analogy. **Exfoliation domes** (Latin *folium*, "leaf," and *ex-*, "off") are the most spectacular examples of physical weathering. These domes result from the unloading of rocks that had been deeply buried. As erosion removes the overburden, it reduces pressure on the underlying rocks, which in turn begin to expand. As the rocks expand, they crack and fracture along **sheet joints** parallel to the erosion surface. Slabs of rock then begin to slip, slide, or spall (break) off the host rock, revealing a large, rounded domelike feature. Sheeting along joints occurs most commonly in plutonic rocks like granite and in massive sandstone. Half Dome in Yosemite National Park, California, and Stone Mountain in Georgia are granite exfoliation domes (◆Figure 6.8).

CONSIDER THIS *Tennis is a sport that is played throughout the world on a variety of court surfaces. For example, the five most prestigious professional tennis tournaments are played on grass (in England), clay (in France and Italy), and concrete (in the United States and Australia). Why do you suppose clay-surface courts are more common in the eastern United States, and in tropical and humid countries, than in the southwestern United States, where concrete is the preferred surface?*

Soils

Soil Profile

As we have noted, loose rock and mineral fragments at the surface of the earth are referred to as *regolith*. Regolith may consist of sediment that has been transported and deposited by rivers or wind, for example, or it may be rock that has decomposed in place. Where weathering has been sufficient, a surface layer that can support plant life—soil—has developed. Climate is the most important factor in soil development; given enough time, soils formed on different rocks under the same climate will look much the same. The influence of the parent material is most apparent in young soils such as decomposed granite, which contains quartz and feldspar grains, some of them having been altered to clay minerals.

A mature **soil profile** has several recognizable **soil horizons,** layers roughly parallel with the ground surface that are products of soil-forming processes (◆Figure 6.9). The uppermost soil horizon, the "A horizon," is the *zone of leaching*. Here, mineral matter is most strongly dissolved by downward-percolating water. It may be capped by a zone of organic matter (*humus*) of variable thickness called the "O horizon," which provides CO_2 and organic compounds that make the percolating water slightly acidic. As the dissolved chemicals move downward, some of them are redeposited as new minerals and compounds in the "B horizon," called the *zone of accumulation*. This horizon is usually lighter in color and harder than the A horizon. Below this is the "C horizon," the weathered transition zone that grades downward into fresh parent material.

◆ FIGURE 6.7 Spheroidal weathering in jointed granite; Alabama Hills, California. Mount Whitney, just right of the twin spires, and the Sierra Nevada are in the background.

◆ FIGURE 6.8 Sheet jointing has produced these two spectacular domes in granite. North dome and Basket Dome; Yosemite Valley, California.

Soil horizons

O Humus

A Zone of leaching of soluble salts (topsoil)

B Zone of accumulation of salts (subsoil)

C Weathered parent material (bedrock)

Gradational contact

Fresh parent material (bedrock)

◆ FIGURE 6.9 An ideal soil profile. Heavily irrigated soils and very young ones exhibit variations of this profile.

The thickness and development of the soil profile are profoundly influenced by climate and time. In desert regions of the Southwest, the A horizon may be thin or nonexistent, and the B horizon may be firmly cemented by white, crusty calcium carbonate known as **caliche.** Caliche forms when downward-percolating water evaporates and deposits calcium carbonate that has been leached from above. Also, if the water table is close to the ground surface, upward-moving waters may evaporate into the layers, leaving deposits of caliche. Some of these deposits are so thick and hard that unusual measures are required to till or excavate through them (◆Figure 6.10).

Residual Soils

Residual soils are soils that have developed in place on the underlying bedrock. For example, in temperate climates granite decomposes into soils that contain quartz, weathered feldspar, and clay minerals formed from the weathering of feldspar. This soil is usually light brown or gray in color and can be mistaken easily for its parent rock. Engineers and geologists call slightly weathered granite *decomposed granite,* or simply *d.g.* It is an excellent foundation material for structures and roads. Heavy clay soils may develop on shale bedrock. The expansive nature of some of these clays makes them troublesome. (This is discussed in the next section of this chapter.) Residual soils may take hundreds or even thousands of years to develop and they are always subject to erosion and removal by one of many geologic processes.

CONSIDER THIS *Decomposed granite (referred to as d.g. in the building trades) is the preferred sandy soil material for making durable, attractive garden pathways, trails, and even driveways. What is the mineralogy of this immature soil, and what mineral makes it cohesive and firm when it is rolled?*

Transported Soils

Some soils have formed on transported regolith. These transported soils are designated by the geologic agent that is responsible for their transportation and deposition: *alluvial* soils by rivers, *eolian* soils by the wind, *glacial* soils, *volcanic* soils, and so on. ◆Figure 6.11 illustrates a soil that developed on volcanic ash overlying an older basalt on which little or no soil had developed. Known as the *Pahala ash,* this soil supports such exotic crops as macadamia nuts, sugar cane, and papaya on the Island of Hawaii.

As Pleistocene glaciers expanded and moved over northern North America, they scraped off Canada's soils and rede-

◆ FIGURE 6.10 Caliche exposed in an arroyo near Tucson, Arizona.

◆ FIGURE 6.11 Volcanic ash (horizontal layers) overlying weathered basalt; Pahala, Hawaii. This volcanic soil supports rich crops of sugar cane, pineapple, and macadamia nuts.

duced especially rich agricultural lands in the Palouse region of eastern Oregon and Washington, and in the central Great Plains, sometimes called the "breadbasket" of the United States. The fertility of the Ukraine, formerly part of the Soviet Union but now a republic, is due to a favorable climate and its loessal soils. So highly regarded is this soil that Adolph Hitler had large quantities of it moved to Germany during World War II.

Without doubt, the largest loess deposits are in China, where they cover 800,000 square kilometers (312,500 mi²) and are as thick as several hundred meters. These deposits were derived from the Gobi and Takla Makan deserts to the north and west. On windy days airborne silt from these deserts is noticeable in Beijing, hundreds of miles away. Similarly, the loess in Africa's eastern Sudan was carried there by winds from the Sahara Desert.

Loess has several noteworthy physical properties. It is tan to almost yellow in color and rather loosely packed (low density), which makes it easily excavated and eroded by running water. Its interlocking angular grains give it cohesive strength, however, and it is capable of standing in almost-vertical slopes without falling (◆Figure 6.13). These unusual properties served the Union and the Confederacy well in Mississippi during the Civil War. Both sides excavated tunnels, trenches, revetments, and other earthworks in the loess soil prior to the siege and battle at Vicksburg, which ultimately was won by U. S. Grant's Union forces. In northern China's Shansi Province in the 1500s, elaborate caves scraped out of loess provided refuge for more than a million people. Unfortunately for the cave residents, loess is subject to collapse under dynamic stresses such as those of an earthquake, and an estimated 800,000 people were buried in their homes during the earthquake of 1556. The load of loess carried by the Huang Ho River as it flows through this area gave the river its other name, the *Yellow River*—which empties into the Yellow *Sea*.

Soil Color and Texture

Color is a good indicator of a soil's *humus* (organic material) and iron content. As the proportion of humus increases, soil color darkens from brown to nearly black. Oxides of iron impart a reddish or yellowish color found mostly in semitropical and tropical areas (see Figure 6.16a). Bright red soils occur where there is aeration and good drainage, whereas yellow hydrated-iron compounds form where there is poor drainage. In arid climates white B horizons and white-coated soil fragments indicate calcium carbonate or other salts. Because these compounds are water-soluble, they do not collect where rainfall is high.

A soil's texture—size of the individual grains (particles) composing it—is a major determinant of its utility. A sample of any particular soil, excluding any gravel or boulders in it, consists of three basic grain sizes—sand, silt, and clay. The

posited them in a band across the United States from Montana to New Jersey. Soils formed on these widespread deposits—collectively known as *glacial drift*—are extremely variable; they may be thin or thick, bouldery or fine-grained, fertile or relatively barren. Some areas of drift are so bouldery that they are impossible to farm. This is one of the reasons that pastures for grazing dairy cattle, rather than cultivated grain crops, are the agricultural mainstay of such states as Wisconsin and New Hampshire (see Gallery Figure 2).

Loess (approximately "luss," to rhyme with cuss) are wind-blown silt deposits composed of angular grains of feldspar, quartz, calcite, and mica. Covering about 20 percent of the United States and about 10 percent of the earth's land area (◆Figure 6.12), loess is arguably the earth's most fertile soil. The sources of loess are glacial deposits and deserts. Strong winds blowing off Ice Age glaciers that covered most of Canada and the northern United States swept up nearby meltwater deposits and redeposited them far away as loess in valleys and on ridge tops. In the United States loess has pro-

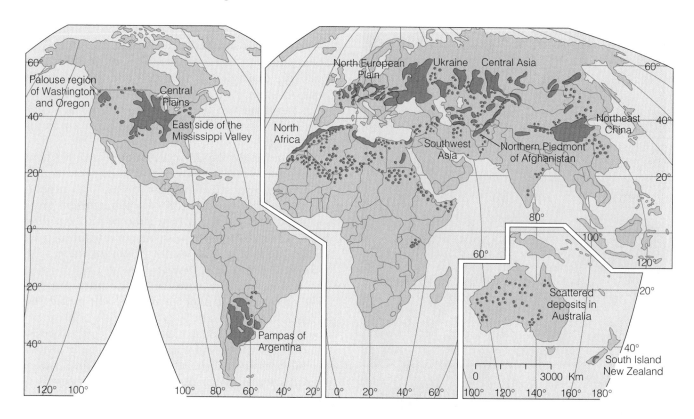

◆ FIGURE 6.12 Loess-covered regions of the world. Note that the largest areas are in China, central and western Asia, and North America.

grain-size range used by the U. S. Department of Agriculture for each texture category is shown in ◆Figure 6.14a. Based upon the percentages of sand, silt, and clay in a soil sample, the soil can be classified as loam, sandy loam, and so forth (◆Figure 6.14b).

The texture of a soil has practical significance, as it determines the soil's properties—its looseness, workability, drainage, and so on. For example, coarse-textured soils are eas-

ily worked, readily penetrated by plant roots, and easily eroded. Fine-textured (clayey) soils tend to be heavy, harder to work, and sticky when wet. Water retention and infiltration are also functions of soil texture. Loose, sandy soils tend to allow water to pass through them readily, but they have low water-retention capacity. The opposite is true of fine-textured soils; they have large water-retention capacity but low filtration rates. This condition leads to poor drainage and excessive

◆ FIGURE 6.13 Villagers in Askole, Pakistan, have dug caves in these vertical loess cliffs.

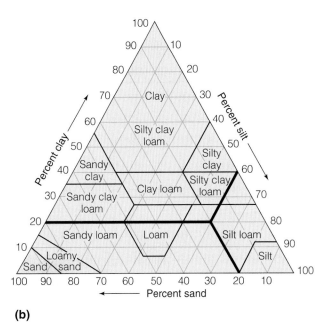

(a)

(b)

◆ FIGURE 6.14 (a) Grain-size scale adopted by the U.S. Department of Agriculture and used by most soil scientists. (b) Soil texture classification triangle. Each side of the triangle is a scale for the percentage content of one of the three textural grades—sand, silt, or clay. A hypothetical soil sample has been found to be 60% silt, 20% sand, and 20% clay. Using this triangle, lines are followed inward from the respective percentages to learn the textural class of the soil, "silt loam." USDA data

water runoff. Loamy soils are generally considered most favorable for agriculture, because they provide a balance between the coarse- and fine-textured soils, as shown in ◆Figure 6.14b.

Soil Classification

Zonal Classification

Pedologists (Greek *pedon*, "soil," and *logos*, "knowledge of") have long known that soils that form in tropical regions are different from those that form in arid or cold climates. Where rainfall is heavy, soils are deep, acidic, and dominated by chemical weathering. Soluble salts and minerals are leached (removed) from the soil, iron and aluminum compounds accumulate in the B horizon, and the soil supports abundant plant life. In arid regions, on the other hand, soils are usually alkaline and thinner, coarse-textured or rocky, and dominated by physical weathering. Because of low precipitation there, salts are not leached from soils, and when soil moisture evaporates, it leaves additional salts behind (⊙ Case Study 6.1 on page 168). These salts may form crusts and lenses of caliche ($CaCO_3$ and other salts), most commonly in the B horizon.

These understandings led to the development of a "zonal" classification of soils based upon climate, which then evolved into a system of Great Soil Groups. These groups may be simplified into four major types (◆Figure 6.15):

- **pedalfers,** soils that are high in aluminum (Al) and iron (Fe) and that are characteristic of humid regions (rainfall >50 cm/year, or >20 in/y);
- **pedocals,** soils that are high in calcium and found in desert or semiarid regions (rainfall <50 cm/y);
- **laterites,** the brick-red soils of the tropics that are enriched in hydrated iron oxides (rust); and
- **tundra soils,** the soils of severe polar climates, found mostly in the Northern Hemisphere.

Pedalfers are the rich soils of the prairies and steppes that support forest and grassland growth. Pedocals form where there is little rainfall, resulting in desert or scrub vegetation. Pedocals and pedalfers are mid-latitude, temperate-climate soils. In contrast, laterites (Latin *later*, "brick, tile") are the residual soils of the tropics, where heavy rainfall leaches out all soluble materials, leaving behind the clay minerals and hydrated aluminum and iron oxides that impart a

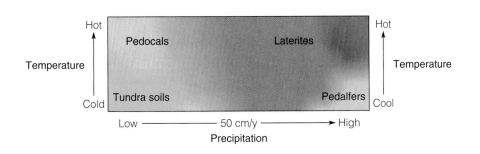

◆ FIGURE 6.15 The Great Soil Groups and their climatic environments.

◆ FIGURE 6.16 (a) Laterite, shown here in Madagascar, is a deep, red soil that forms in response to intense chemical weathering in the tropics. (b) Bauxite, the ore of aluminum, forms in horizon B of laterites derived from aluminum-rich parent materials.

(a)

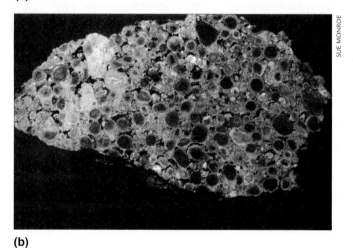

(b)

red color (◆Figure 6.16a). Laterite is not uniquely identified with any particular rock type, but it does require an iron-containing parent, abundant rainfall, and well-drained terrain. Laterites are currently forming in humid tropical and subtropical regions. Leaching of these soils may be so extensive that with the right parent rock, economically valuable aluminum-rich *bauxite* deposits are the end products of the soil-forming process (◆Figure 6.16b). Finally, tundra soils form in arctic and subarctic regions and support only such vegetation as mosses, sedges, lichens, and dwarf shrubs, making for bleak, forestless landscapes (◆Figure 6.17).

U.S. Comprehensive Soil Classification System

Because of dissatisfaction with existing classification schemes, the U.S. Soil Conservation Service instituted the first of seven trial classification systems in 1952. Each trial system was sent to a select group of international pedologists for their comments. After incorporating the suggestions and findings of these experts, the "Seventh Approximation," as it is called, was officially adopted by the Soil Conservation Service in 1965. The distinguishing features of this system are that:

1. Soils are classified by their physical characteristics, rather than their origin; that is, the system is nongenetic.

2. Soils that have been modified by human activities are classified along with natural soils.

3. The soils' names convey information about their physical characteristics.

◆ FIGURE 6.17 Treeless tundra landscape near Kotzebue, Alaska. Most of the soil in tundra regions is frozen much of the year.

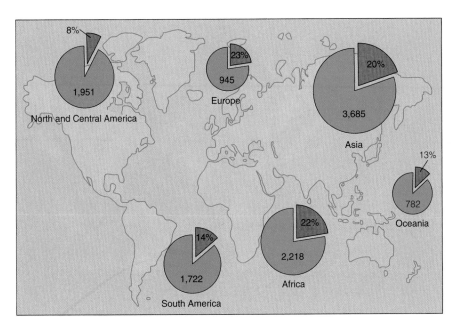

◆ FIGURE 6.18 World soil degradation, 1945-1991. The areas degraded are given both in percentage of the arable area (pie-shaped areas in orange) and in hectares (a hectare is 10,000 square meters = 2.47 acres). Table 6.3 (on page 162) summarizes the various causes of soil degradation in the world.

The system subdivides soils into a hierarchy of six levels—orders, families, series, and so forth, with ten orders at the highest level in the system. With a practiced eye, each soil order can be easily identified. Because the number of soil entries increases astronomically at the lower levels (illustrated by the fact that 14,000 soil series are recognized in the U.S.), this classification is for the professional, not the layperson.

Soil Problems

Soil Erosion

Although soil is continually being formed, it is for practical purposes a nonrenewable resource, because hundreds or even thousands of years may be necessary for soil to develop. Ample soil is essential for providing food for the earth's burgeoning population, but throughout the world it is rapidly disappearing. Although the United States has one of the world's most advanced soil conservation programs, soil erosion remains a problem after 60 years of conservation efforts and expenditures of billions of dollars. Erosion removed an estimated two billion tons of topsoil from U.S. farmlands in 1992—down from slightly more than three billion tons a decade earlier. With the average annual soil loss estimated as 10–12 tons per hectare (4–4.8 tons/acre) and an assumed annual soil-formation rate of 2–4 tons per hectare (0.8–1.6 tons/acre), three to five times as much soil is being lost as is being formed in the United States. Not only does soil erosion destroy fertility, but also tons of soil settle in lakes and clog waterways and drainages with sediment, pesticides, and nutrients each year. Losses of only 2–3 centimeters (1 in) of topsoil represent about 200 tons per hectare. Such erosion cuts crop yields and may render the land unproductive.

On a global basis, an area the size of China and India combined has suffered *irreparable degradation* from agricultural activities and overgrazing—mostly in Asia, Africa, and Central and South America (◆Figure 6.18). Lightly damaged areas can still be farmed using the modern conservation techniques explained later in this chapter, but "fixing" serious erosion problems is beyond the resources of most Third World farmers and their governments.

Erosion Processes

The agents of soil erosion are wind and running water. Heavy rainfall and melting snow create running water, which removes soil by sheet, rill, and gully erosion. **Sheet erosion** is the removal of soil particles in thin layers more or less evenly from an area of gently sloping land. It goes almost unnoticed. **Rill erosion,** on the other hand, is quite visible in discrete streamlets carved into the soil (◆Figure 6.19). If these rills become deeper than about 25–35 centimeters (10–14 in), they cannot be removed by plowing, and gullies form. Gullies are created in unconsolidated material by widening, deepening, and headward erosion of rills. **Gullying** is a big problem in the easily eroded loess soils of the Palouse region of eastern Washington. Snow melting there saturates the loess soil, causing it to become fluid and then to flow like wet concrete across the underlying still-frozen ground. Estimated annual U.S. cropland soil loss from sheet and rill erosion is shown by state in ◆Figure 6.20a. Tennessee headed the list in 1980, followed by Hawaii, Missouri, Mississippi, and Iowa, as shown in ＊Table 6.2.

Wind erodes soils when tilled land lies fallow and dries out without vegetation and root systems to hold it together. The dry soil breaks apart, and wind removes the lighter particles. If the wind is strong, a dust storm may occur, as it did in the 1930s, creating the "Dust Bowl" (described at the beginning of the chapter). Much of the reason for the Dust Bowl was

(a)

(b)

◆ FIGURE 6.19 (a) Rill erosion in cultivated soil. The rill will probably be plowed under. (b) Rills left unattended have led to extreme gullying in tilled land near Lumpkin, Georgia (width of area in photo about 11/4 kilometers).

the widespread conversion of marginal grasslands, which receive only 25–30 centimeters (10–12 in) of rainfall per year, to cropland. Unfortunately, this is still occurring in some places in the United States and in the rest of the world. ◆Figure 6.19b presents estimates of the amount of soil lost each year to wind erosion in the Great Plains states. When the 50 states' estimated wind and water erosion losses are combined (Table 6.2), the state with the greatest annual loss is Texas, followed by Tennessee, Hawaii, and New Mexico, respectively. ✳ Table 6.3 summarizes the recognized causes of accelerated soil erosion worldwide and assesses their relative impact.

In addition to the soil erosion caused by overgrazing, deforestation, and bad agricultural practices, some recreational pursuits also cause soil erosion, particularly those that involve offroad vehicles (ORVs). A 1974 study indicated that recreational pressure on semiarid lands of Southern California by ORVs was "almost completely uncontrolled" and that those desert areas had experienced greater degradation than had any other arid region in the United States (◆Figure 6.21). In the two decades since that study, much has been accomplished in controlling offroad traffic in California and other states, but ORVs continue to exert severe impacts upon areas established for their use.

The Impact of Cropland Loss on Humans

The main concern when addressing soil loss—whether the loss is due to erosion or to conversion of cropland to nona-gricultural uses—is that soils are biological and chemical factories that produce food. Increasingly, the U.S. population is moving into urban areas, and to accommodate them, cities are expanding outward onto agricultural land. About 20 percent of the United States was under cultivation in 1992 (◆Figure 6.22), and each year some more of that land is urbanized or converted to forestland, particularly in the northeastern and southeastern states. This trend is also seen in India and other nations with high growth rates. Between 1950 and 1998 the amount of grain-growing land per person worldwide declined from .23 hectare to .012 hectare (.57A and .03A). World cropland decreases had not been regarded as a major problem till the late 1900s, because increases in productivity had been great. Increases in productivity leveled off in 1990, however, and it is doubtful that future increases will keep pace with increasing population and decreasing cropland soils.

Some of the many ways soils are important to humans are shown in ◆Figure 6.23. When soil is lost, our beneficial use of the resource is reduced.

Mitigation of Soil Erosion

Thomas Jefferson was among the first people in the United States to comment publicly on the seriousness of soil erosion. He called for such soil conservation measures as crop rotation, contour plowing, and planting grasses to provide soil cover during the fallow season. These practices are now common in industrialized nations but, unfortunately, only minimally employed in Third World countries. Proper choices of where to plant, what to plant, and how to plant are

TABLE 6.2	Significant U.S. Soil Erosion by State, 1980	
Rank	**State**	**Tons/Acre/Year**
	Sheet and Rill Erosion	
1	Tennessee	14.12
2	Hawaii	13.71
3	Missouri	11.30
4	Mississippi	10.92
5	Iowa	9.91
	Wind Erosion	
1	Texas	14.9
2	New Mexico	11.5
3	Colorado	8.9
4	Montana	3.8
5	Oklahoma	3.0
	Total Soil Erosion, Water and Wind*	
1	Texas	18.40
2	Tennessee	14.12
3	Hawaii	13.71
4	New Mexico	13.50
5	Colorado	11.40

*The U.S. average annual rate of total soil erosion is estimated at 4.0–4.8 tons/acre (10–12 tons/hectare).
Source: *Environmental Trends* (Washington, D.C.: Council on Environmental Quality, 1981), cited in Sandra Batie, *Soil Erosion: A Crisis in America's Cropland?* (Washington, D.C.: The Conservation Foundation, 1983).

the components of soil conservation. The major erosion-mitigation practices are described here.

- *Terracing*—creating flat areas, terraces, on sloping ground—is one of the oldest and most efficient means of saving soil and water (◆Figure 6.24).

- *Strip-cropping*, by which close-growing plants are alternated with widely spaced ones, efficiently traps soil that is washed from bare areas and also supplies some wind protection. An example of this is alternating strips of corn and alfalfa.

- *Crop rotation* is the yearly alternation of soil-depleting crops with soil-enriching crops. Rotating corn and wheat with clover in Missouri, for example, has been found to reduce soil runoff from 19.7 tons/acre for corn to 2.7 tons/acre, a huge loss reduction. Groundcovers such as grasses, clover, and alfalfa tend to be soil-conserving, whereas rowcrops like corn and soybeans can leave soil vulnerable to runoff losses. Retaining crop residues, such as the stubble of corn or wheat, on the soil surface after harvest has been found to reduce erosion and to increase water retention by more than half.

- *Conservation-tillage* practices minimize plowing in the fall and encourage the contour plowing of furrows perpendicular to the field's slope so that they will catch runoff, increasing water infiltration.

- *No-till* and *minimum-till* practices also conserve soil. With no-till farming, seeds are planted into the soil through the previous crop's residue, and weeds are controlled solely with chemicals. Specialized no-till

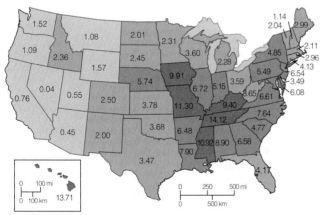

(a) Sheet and rill erosion losses in tons per acre per year

- 9.0 – 15.0
- 4.8 – 8.9
- 2.0 – 4.7
- Less than 2.0

National average: 4.8

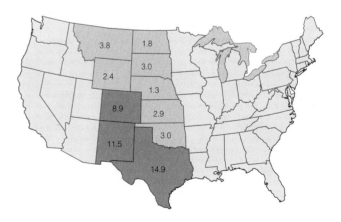

(b) Wind erosion losses in tons per acre per year

- 8.0 – 15.0
- 1.0 – 7.9

Great Plains average: 5.3

◆ FIGURE 6.20 Estimated annual rates of cropland soil loss in the United States, 1980. (a) Sheet and rill erosion losses. (b) Wind erosion losses in the Great Plains. (Light green represents <1 ton/acre per year.)

*TABLE 6.3	Causes of World Soil Erosion and Deterioration	
Cause	**Estimated % of Total**	
Overgrazing	35	
Deforestation	30	
Bad agricultural practices	28	
Other causes, natural and human	7	

Source: World Resources Institute, United Nations Environment and Development Program, *World Resources: A Guide to the Global Environment* (New York: Oxford University Press, 1992).

equipment is required. "No-till" has been especially successful in soybean farming. In 1993 U.S. farmers left 30 percent of their crop debris on 37 million cultivated hectares (92 million acres), nearly a third of all cultivated land in the United States. Ten percent of the U.S. planted acreage was not plowed at all. No-till agriculture is most beneficial during times of drought, when plowed fields dry out and topsoil is more vulnerable to wind erosion. No-till practices saved much of the agricultural soil that would have otherwise been eroded by floodwaters of the Great Flood of 1993 in the Mississippi River valley.

Wind-caused soil losses can also be reduced by planting windbreaks of trees and shrubs near fields and by planting row crops perpendicular to the prevailing wind direction. Wind losses are almost zero with cover crops such as grasses and clover.

◆ FIGURE 6.21 Hill-climbing motorcyclists have devastated this slope in Jawbone Canyon in the Mojave Desert, California. The bikers have compacted the soil surface, causing reductions in vegetation and soil permeability. The reduced soil moisture content has seriously harmed the local plant and animal ecology. Desert rains have eroded the tire tracks, transforming them into gullies, which increases water runoff and sediment yield (erosion).

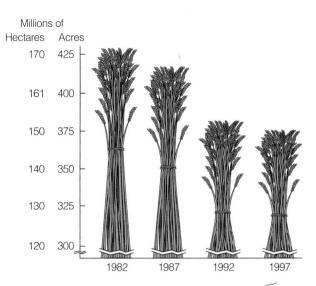

◆ FIGURE 6.22 Total U.S. cropland, 1982 to 1997. Reductions were due to urbanization, conversion to other uses, and soil erosion. American Geological Institute

Expansive Soils

Certain clay minerals have a layered structure that allows water molecules to be absorbed between the layers, which causes the soil to expand. (This process differs from the *adsorption* of water to the surface of nonexpansive soil clays.) Soils that are rich in these minerals are said to be **expansive soils.** Although this reaction is somewhat reversible, as the clays contract when they dry, soil expansion can exert extraordinary uplift pressures on foundations and concrete slabs, with resulting structural and cosmetic damage (◆Figure 6.25). Damage caused by expansive soils costs about $6 billion per year in the United States, mostly in the Rocky Mountain

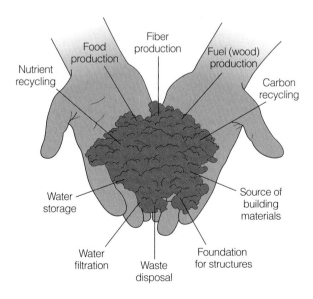

◆ FIGURE 6.23 Some of soil's beneficial uses. American Geological Institute

◆ FIGURE 6.24 Terracing steep hillsides is an ancient soil conservation practice; China.

◆ FIGURE 6.25 This well-built concrete-block wall in Colorado was uplifted and tilted by stresses exerted by the underlying expansive soil.

states, the Southwest, and Texas and the Gulf Coast states (◆Figure 6.26). The swelling potential of a soil can be identified by several standard tests, usually performed in a soils engineering laboratory. An obvious test is to compact a soil in a cylindrical container, soak it with water, and measure how much it swells against a certain load. Clays that expand more than 6 percent are considered highly expansive; those that expand 10 percent are considered critical. Treatments for expansive soils include (1) removing them, (2) mixing them with nonexpansive material or with chemicals that change the way the clay reacts with water, (3) keeping the soil moisture constant, and (4) using reinforced foundations that are designed to withstand soil volume changes. Identification and mitigation of damage from expansive soils is now standard practice for U.S. soil engineers.

CONSIDER THIS *Expansive soil is the soil problem most likely to be encountered in maintaining a home. Assuming that you, as a homeowner, found that your newly purchased dream house is built on expansive soil, what could you do to minimize its impact on your home?*

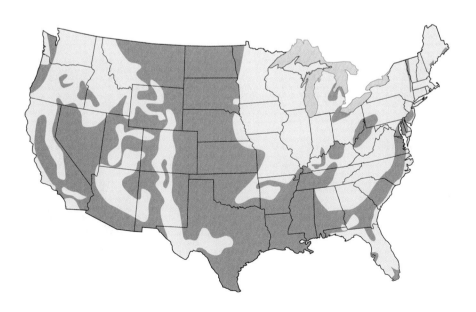

◆ FIGURE 6.26 Distribution of expansive soils (shown in brown) in the United States.

Permafrost

The term **permafrost,** a contraction of *permanent* and *frozen,* was coined by Siemon Muller of the U.S. Geological Survey in 1943 to denote soil or other surficial deposits in which temperatures below freezing are maintained for several years. More than 20 percent of the earth's land surface is underlain by permafrost, and it is not surprising that most of our knowledge about frozen ground has been derived from construction problems encountered in Siberia, Alaska, and Canada. Permafrost becomes a problem for us humans when we change the surface environment by our activities in ways that thaw the near-surface soil ice, since this results in soil flows, landslides, subsidence, and related phenomena. Permafrost forms where the depth of freezing in the winter exceeds the depth of thawing in the summer. If cold ground temperatures continue for many years, the frozen layer thickens until the penetration of surface cold is balanced by the flow of heat from the earth's interior. This equilibrium between cold and heat determines the thickness of the permafrost layer. Thicknesses of 1,500 meters and 600 meters (almost 5,000 and 2,000 ft) have been reported in Siberia and Alaska, respectively. Such thick layers are very old; they must have formed during the Pleistocene Epoch of geologic time, the time known as the Great Ice Age.

The top of the permanently frozen layer is called the **permafrost table** (◆Figure 6.27). Above this is the **active layer,** which is subject to seasonal freezing and thawing and is always unstable during summer. When winter freezing does not penetrate entirely to the permafrost table, water may be trapped between the frozen active layer and the permafrost table. These unfrozen layers and lenses, called *talik* (a Russian word), are of environmental concern, because the unfrozen ground water in the talik may be under pressure. Occasionally, pressurized talik water bursts through to the surface, forming a dome-shaped mound 30–50 meters

◆ FIGURE 6.28 Utility poles extruded from permafrost ground; Alaska. One solution has been to support the poles with tripods founded on the ground surface.

(100–160 ft) high called a *pingo.* The presence of ice-rich permafrost becomes evident when insulating vegetation, usually tundra, is stripped away for a road, runway, or building. Thawing of the underlying soil ice results in subsidence, soil flows, and other gravity-induced mass movements (◆Figure 6.28). Even the casual crossing of tundra on foot or in a jeep can upset the thermal balance and cause thawing. Once melting starts, it is impossible to control, and a permanent scar is left on the landscape (◆Figure 6.29).

Structures that are built directly on or into permafrost settle as the frozen layers thaw (◆Figure 6.30). Both active and passive construction practices are used to control the problem. Whereas active methods involve complete removal or thawing of the frozen ground before construction, passive methods build in such manners that the existing thermal regime is not disturbed. For example,

■ a house can be built with an open space beneath the floor so that the ground can remain frozen (◆Figure 6.31a),

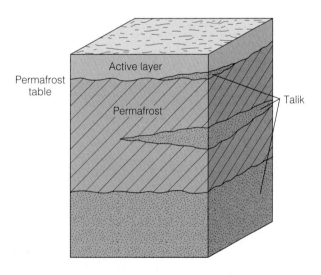

◆ FIGURE 6.27 Permafrost zones. Talik, unfrozen layers and "lenses," may occur in the active layer and in the permanently frozen layer. USGS

◆ FIGURE 6.29 Scars made by vehicles on tundra soil remain for many years; Alaska.

LARRY HINZMAN, UNIV. OF ALASKA

◆ FIGURE 6.30 Differential melting of permafrost caused this cabin to subside at one end. Note that part of the foundation of the cabin is open to the atmosphere, which prevented the permafrost from melting.

- a building can be constructed on foundation piers whose temperature is held constant by heat dissipators or coolant (◆Figure 6.31b), and
- roads are constructed on top of packed coarse-gravel fill to allow cold air to penetrate beneath the roadway surface and keep the ground frozen.

The most ambitious construction project on permafrost terrain to date is the 800-mile oil pipeline from Prudhoe Bay on Alaska's North Slope to Valdez on the Gulf of Alaska. About half the route is across frozen ground, and because oil is hot when it is pumped from depth and also during transport, it was decided that the pipe should be above ground so as to avoid the thawing of permafrost soil (◆Figure 6.32). The four-foot-diameter pipeline also crosses several active faults, three major mountain ranges, and a large river. The pipeline's above-ground placement allows for some fault displacement and for easier maintenance if it should be damaged by faulting. This placement has the added advantage of accommodating the longstanding migratory paths of herding animals such as caribou and elk.

(a)

(b)

◆ FIGURE 6.31 Two methods of mitigating permafrost damage: (a) Raised WW-II Quonset hut; Nome, Alaska. (b) Heat-dissipating foundation piers on a modern structure; Kotzebue, Alaska

(a)

◆ FIGURE 6.32 The 800-mile Alaskan oil pipeline was constructed to accommodate permafrost and wildlife. (a) The meandering pattern allows for expansion and contraction of the pipe and for potential ground movement along the Denali fault, which it crosses (see Ch. 4). The pipeline supports feature heat-dissipating radiators. (b) Moose and caribou are able to pass beneath the elevated pipeline.

(b)

Settlement

Settlement occurs when a structure is placed upon a soil, rock, or other material that lacks sufficient strength to support it. Settlement is an engineering problem, not a geological one. It is treated here, however, because settlement and poor soil seem inextricably associated. All structures settle, but if settlement is uniform—say, two or three centimeters over the entire foundation—there is usually no problem. When differential settlement occurs, however, cracks appear or the structure tilts.

The Leaning Tower of Pisa, Italy, is probably the best-known, most-loved example of poor foundation design. It may have scientific importance as well, as Galileo is said to have conducted gravity experiments at the 54.5-meter (163 ft) structure. Begun in 1174, the tower started to tilt when the third of its eight stories was completed, sinking into a two-meter-thick layer of soft clay just below the ground surface. Initially the tower leaned north, but after the initial settlement and throughout the rest of its history it has leaned south. To compensate for the lean, the engineer in charge had the fourth through eighth stories made taller on the leaning side. The added weight caused the structure to sink even further. At its maximum lean, it tilted toward the south about 5.2 meters (17 ft) from the vertical, about 5.5°, which gives one an ominous feeling when standing on the south side of the tower (◆Figure 6.33).

In 1989, a similarly constructed leaning bell tower at the cathedral in Pavia, Italy, collapsed, causing officials to close the tower at Pisa to visitors. In 1990 the Italian government established a special commission of structural engineers,

◆ FIGURE 6.33 Inadequate foundation investigation and design; Pisa, Italy.

soils engineers, and restoration experts to determine new ways to save the monument at Pisa. The commission's goal was to reverse the tilt 10–20 centimeters (4–8 in), which they believed would add 100 years to the lifetime of the 800-year-old structure. After considering many plans it was decided to remove soft sediment from the north side of the tower, causing it to reverse its lean to the south. The method of soil extraction was proposed by a commission member and professor of soil mechanics, John Burland of London's Imperial College. This method was simple in concept but difficult to implement. It was done with drills 200 mm (about 8 in) in diameter, which remove cores of soil about 15–20 liters (4.8–5.2 gal) in volume. The weak sediment under the north-side foundation quickly filled the void left by the drill and the tower started to straighten—that is, tilt back toward the north. The concept was simple compared to installing stabilizing cables or constructing enlarged and stronger foundations. After being shut down for eleven years, the tower was reopened, still leaning, but 40 cm (16 in) and 0.5° less. It is estimated that the realignment will add 300 years to the life of the tower. The project cost $30 million, but that is dwarfed by the enthusiasm of the four million visitors who come to visit the tower each year.

It should be noted that tilting towers are not exceptional in Italy. Few of the 170 campaniles (bell towers) still standing in Venice are vertical, for example, and many other towers throughout Italy and Europe tilt. These graceful towers are certainly testimony to past architects' ability and talent, but also to their failure to understand that a structure can be truly beautiful only if local geology permits.

> **CONSIDER THIS** *It is jokingly said that the twelfth-century architect of the Leaning Tower of Pisa saved money by not having a foundation investigation. Had engineers and geologists with today's knowledge and tools existed at that time, what routine tests would they have performed on borehole soil samples that would show settlement was inevitable? What recommendations might they have made to the client?*

Other Soil Problems

Laterite soils, the red soils of the tropics, are soft when they are moist but become brick-hard and durable when they dry out (♦Figure 6.34). The Buddhist temple complex at Angkor Wat in Cambodia is partly constructed of laterite bricks that have endured since the thirteenth century. Deforestation of areas underlain by laterites is an invitation to this hardening, as U.S. forces discovered in Vietnam when they attempted to grade roads and runways on lands that had been defoliated for military reasons. Deforested laterites make for swampy conditions during the rainy season, because they lack permeability; there is little infiltration of the water. This can raise havoc with wheeled vehicles and restrict mobility in such regions.

Although laterite soils support some of the densest rain forests in the world, they are relatively infertile because of

◆ FIGURE 6.34 Dried and hardened laterite; Brazil. This soil is so hard that soil scientists call it "ironstone."

extensive leaching of minerals. The nutrients that support the rain forest are in the vegetation, and they are recycled when the plants die and decompose into the soil. Deforested laterites used for agricultural purposes must be fertilized intensively after a few harvests in order to support productive crops.

In some areas of extensive irrigation, agricultural soils form impervious clayey layers called **hardpans** in the B horizon. Hardpans inhibit waters from percolating and thus reduce crop yields. They must be broken up by deep plowing using large, heavy tractors pulling long, single-tooth rippers that penetrate the subsoil. Hardpans and claypans are the scourge of farmers, because they can render their farmland unproductive.

The presence or absence of beneficial elements in soils can result in various dietary benefits or deficiencies in humans. This is because the soil is the actual beginning of the food chain. Soil elements are taken up by plants (vegetables, fruits, and grains), which are then eaten by animals. Some examples of soil elements and health are discussed in ❂ Case Study 6.2 on page 168. The presence of pollutants in the soil now requires expensive cleanup (❂ Case Study 6.3 on page 170).

Salinization and Waterlogging

Saline soils contain sufficient soluble salts to interfere with plant growth. Salinization is the oldest soil problem known to humans, dating back at least to the fourth century B.C. At that time a highly civilized culture was dependent upon irrigation agriculture in the southern Tigris–Euphrates Valley of Iraq, the "Cradle of Civilization." By the second century B.C., the soils there were so saline that the area had to be abandoned. Soils in parts of California, Pakistan, the Ukraine, Australia, and Egypt are now suffering the same fate.

Salinization can be human-induced by intensive irrigation, which raises underground water levels very close to the soil. Capillary action then causes the underground water, containing dissolved salts, to rise in the soil, just as water is drawn into a paper towel or sugar cube. The soil moisture is then subject to surface evaporation, and as the water in the soil evaporates, it leaves behind salts that render the soil less productive, and eventually barren (see ◆Figure 1). The productivity of 15 percent of all U.S. cropland is totally dependent upon irrigation, and 30 percent of U.S. crops are produced on this land.

For millennia, annual flooding of the Nile River added new layers of silt to the already rich soil and removed the salts. After completion of the Aswan Dam in 1970, however, the yearly flooding and flushing action stopped, and salts have now accumulated in the soils of the floodplain. Salinization is particularly evident in the arid Imperial Valley of California. Soils of the Colorado Desert, or any desert for that matter, can be made productive if enough water is applied to them. Because deserts have low rainfall, irrigation water must be imported, usually from a nearby river. The All-American Canal brings relatively high salinity water from the Colorado River to the Imperial Valley, where it is used to irrigate all manner of crops from cotton to lettuce. In places the water table has risen to such a degree that the cropland lies barren, its surface covered by white crusts of salt. The Imperial Valley's Salton Sea is one of the largest saltwater lakes in the United States, and it grows saltier and larger every year by the addition of saline irrigation water. The return of degraded irrigation water in canals to the Colorado River in southern Arizona has required that a desalinization facility be established near Yuma to make the water acceptable for agricultural use in Mexico.

The good news is that soil salinization can be reversed. It can be done by lowering the water table by pumping and then applying heavy irrigation to flush the salts out of the soil. Salinization can also be mitigated, or at least delayed, by installing perforated plastic drain pipes in the soil to intercept excess irrigation water and carry it offsite. Neither remedy is an easy or inexpensive solution to this age-old problem.

◆ FIGURE 1 Accumulation of salts at the soil surface due to evaporation of irrigation and ground water; Imperial Valley, California.

Soils, Health, and Locoweed

Humans need trace elements in their systems in order to function properly. **Trace elements** occur in small amounts in the body, usually a few parts per million, as opposed to the bulk elements, which are found in large quantities. The body obtains these elements naturally by this sequence: rocks to soils to plants and finally to animals. Humans get trace elements by eating plants or by eating animals that ate plants that grew on soils containing them or by using dietary supplements of them. Some water-soluble trace elements, such as fluoride, are sometimes added to municipal drinking-water supplies.

CONTINUED

The best-known relationship between soils and human health is the one between soil iodine deficiency and the enlargement of the thyroid gland known as goiter. Besides being cosmetically demoralizing, goiter is debilitating in other ways. Iodine deficiency during pregnancy can lead to "cretinism" in newborns, a severe mental and physical disorder. Endemic goiter has been firmly established in the northern United States in the so-called goiter belt (see ◆Figure 1). Because of the wide availability of iodized table salt, this disease has practically disappeared except among the poorly educated. Additionally, little of the U.S. diet is "home grown" anymore. This is another reason goiter is no longer an endemic problem. Soil-iodine thyroid diseases are also known in England, Thailand, Mexico, the Netherlands, and Switzerland.

Several trace elements—zinc, copper, iron, cobalt, manganese, and molybdenum—are associated with and stimulate enzymes, the catalysts of biochemical reactions. These enter the body mostly in plant and animal foods. Their importance might be better appreciated when we realize that silver, mercury, and lead are toxic because they are enzyme *inhibitors*. Some of the beneficial effects of trace elements are that zinc aids in healing wounds and burns, iron transports oxygen in the blood, and cobalt is a vital component of vitamin B_{12}. Deficiencies of these may result in anemia, slowed growth, vitamin-B_{12} deficiency, and other ailments.

A most interesting trace element is selenium. It is poisonous at high dosages, but deficiencies cause abnormalities in plants and animals. Selenium is abundant in soils of the high plains and western United States, and certain plants there concentrate it in their tissue. According to John McPhee in *Annals of the Former World* (page 284; see For Further Information), "A hundred million years ago stratovolcanoes stood in Idaho, and sent up ash that fell eastward in the sea. The selenium went into lime muds, and now these alien asters were drawing it out of the limestone and spreading the poison across the surface world, as few other plants can do. Most plants ignore selenium. Woody asters and a few others require selenium to germinate. After they take it up from the rock they convert it into a form that nearly all plants will, in turn, take up, too. Selenium-contaminated plants are eaten by sheep and cattle, which are served to people as chops and burgers."

Eating these plants, known as *locoweeds,* causes cattle to get the "blind staggers" and appear drugged. Selenium intake can affect people with heart, liver, and kidney problems. Selenium is a cumulative poison like lead and arsenic, and it can cause birth defects. Figure 1 shows areas where mineral-nutritional diseases are known to occur in animals. These are all localities where soils have deficiencies or excesses of vital or toxic trace elements.

Although molybdenum is probably best known as a metal that strengthens steel, it is also essential to plant and possibly animal health. Many plant enzymes contain molybdenum, two of which are important in metabolizing nitrogen, a major plant nutrient. In large concentrations, molybdenum interferes with the

CONTINUED

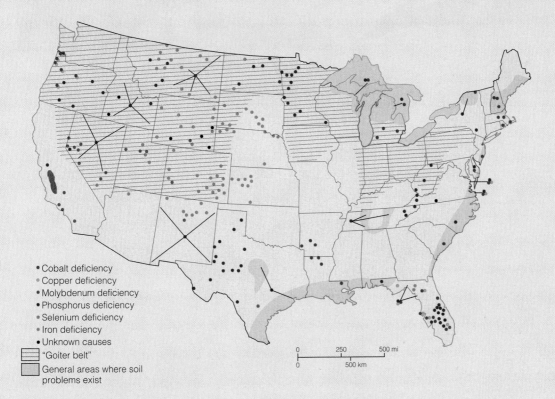

- • Cobalt deficiency
- • Copper deficiency
- • Molybdenum deficiency
- • Phosphorus deficiency
- • Selenium deficiency
- • Iron deficiency
- • Unknown causes
- "Goiter belt"
- General areas where soil problems exist

◆ FIGURE 1 Distribution of mineral-nutritional diseases of animals in the United States. Dots show specific locations where problems have been reported. J. M. Spencer, *Texas Journal of Science* 21 (1970)

absorption of copper, an essential element, and may be detrimental to certain livestock. *Molybdenosis* can cause anemia, depressed growth, poor reproduction, bone malformation, and loss of normal hair color (♦Figure 2). Black Angus cattle with molybdenosis turn a mouse-gray color, for example. If treated early enough, the symptoms of the disease can be reversed by corrective nutrition.

Lignite coal (see Chapter 14) is known to be rich in molybdenum. Ranchers in South Dakota reported problems resembling molybdenosis in cattle pastured near an abandoned open-pit lignite mine. Chemical analysis revealed that grass growing near the exposed lignite contained high levels of molybdenum. This led to the discovery that molybdenum was being leached from the coal and transported by surface-water runoff to nearby pastures, where it was taken up by range grass and thereby entering the food chain.

Molybdenum-related disease has not been found in humans in such countries as the United States where food-production and distribution are widespread. In a region in southern India, however, some individuals developed *genu valgun* (knock-knee), a crippling bone deformity, with a devastating impact on the local population. The outbreak was traced to a high level of molybdenum in sorghum, a staple in the residents' diet. High levels of the metal were then found in the ground water. Ultimately it was discovered that construction of a large dam and reservoir nearby had raised the water table, making the high-molybdenum ground water accessible to the sorghum crop.

(a) (b)

♦ FIGURE 2 Molybdenosis results from an imbalance of copper to molybdenum in the diet of cattle. The visible symptoms are hair discoloration and loss and weight loss. (a) A healthy Hereford steer. (b) A steer that is suffering from molybdenosis.

CASE STUDY 6.3

Rx for Contaminated Soils

Although spills of hazardous materials frequently contaminate soils, the Environmental Protection Agency (EPA) did not include a soil component in their system for ranking Superfund cleanup sites until 1990. Soils become contaminated when they are forced to absorb a wide variety of contaminants, the most common being oil-field brines (salty water), heavy metals, pesticides, cleaning agents, and petroleum products. Fortunately, this contamination by absorption is reversible, and soils can be "cleaned," *remediated,* by a variety of methods. A large soil remediation industry has sprung up in response to EPA requirements.

Various soil remediation techniques are used, depending upon the thickness of the soil to be cleaned and the chemical nature of the contaminant. *Dilution,* mixing contaminated soil with clean soil, is one technique. It works, but it is expensive and it contaminates an even larger volume of soil. The old saying "the solution to pollution is dilution" is relevant to many pollution problems, but it is not practical when large volumes of soil are contaminated.

CONTINUED

Bioremediation, a natural process, uses microorganisms to degrade and transform organic contaminants. Bacteria and fungi flourish naturally in soils where oil, sewage sludge, or food-production waste has been discarded. Biodegradation can be stimulated by increasing the amount of air in the soil through cultivation, vapor-extraction pumping (see below), or stockpiling the soil and forcing air into it. Adding easily decomposed organic matter such as hay, compost, or cattle manure also speeds up degradation of organic contaminants. Bioremediation is relatively inexpensive, and some contaminated soils simply require turning (disking or plowing) to speed up oxidation. A successful bioremediation project at Mobil Oil Company's natural-gas refinery at Liberal, Kansas, is shown in ◆ Figure 1. The soil contained up to 26,000 parts per million (2.6%) of hydrocarbons. It was decided that the contaminated soil should be spread in a layer less than 20 centimeters (8 in) thick and that a nutrient and

microbial amendment (cow manure) should be applied to it at a rate of about 60 tons per hectare (15 tons/acre). The soil was periodically disked to promote biological activity and was subsequently reclaimed and planted (◆Figure 1b). Under careful management even heavy metals such as selenium and arsenic can be converted to gaseous forms and released to the atmosphere. The three-year project was closely coordinated with the Kansas Department of Health and Environment.

Vapor extraction is used for removing volatile (easily evaporated) organic compounds such as cleaning solvents from soils. Air is drawn through the soil by means of dry wells connected to vacuum pumps. The air passing through the soil evaporates the compounds and exhausts them to the atmosphere. This technique is especially desirable for cleaning soils beneath buildings or roads, because no excavation is required except for the drilling of shallow wells. The down side is

that it may take 6 to 12 months to sanitize the soil adequately.

Phytoremediation, the use of plants to remove contaminants, is used mainly for extracting salts. This has been used in the Netherlands, where periodic seawater flooding contaminates soils. Plant scientists are developing *hyperaccumulator* plants that clear soils of heavy metals such as cobalt, nickel, and zinc by concentrating them in their tissues. The cost of phytoremediation materials, equipment, and labor is relatively low, but up to a decade can be required to remediate soil to the point that it can be used again.

The justification for soil remediation is obvious. It allows vegetation to be established; it decreases soil erosion, groundwater contamination, and air pollution; and it can restore soils' agricultural productivity. The soil remediation industry is in its infancy, and a considerable amount of research is under way for developing faster and more economical means of "cleaning dirt."

(a)

(b)

◆ FIGURE 1 A successful bioremediation project. (a) Oil-contaminated soil (2.6% hydrocarbons) at a Kansas refinery is removed for bioremediation by a bulldozer. Note the patchy distribution of the oil contamination. The soil is spread in a layer about 20 centimeters (8 in) thick. Nutrient amendment (cow manure) is added to stimulate biological degradation of the hydrocarbons. (b) Remediated soil (<.01% hydrocarbons) is spread and planted with grass. Note the clods of manure and the white chunks of caliche in the redistributed soil. The structure in the background is a grain silo in which wheat is stored.

gallery

A Few Inches between Humanity and Starvation

One of the factors that determines our planet's carrying capacity is soil, and the other is water. Soil is being lost much faster than weathering processes can form it. Poor agricultural practices and attempting to make marginal soils productive are major problems (◆Figure 1). Some soils defy conversion to agricultural use because of their origin (◆Figure 2).

◆ FIGURE 1 The "soil" of the Sahel of Africa is little more than dry, wind-blown sand. A region of severe drought and famine in recent years, the Sahel encompasses an east–west belt that separates the arid Sahara Desert from tropical Africa. It includes parts of Chad, Niger, Mali, Mauritania, Senegal, and Gambia.

◆ FIGURE 2 Remnants of a boulder fence in an area that was once cleared for agriculture; Attleboro, Massachusetts. Rocky glacial soils are difficult to farm, because freezing and thawing cycles continually lift boulders up into the soil root zone. This produces a new "crop" of rocks each spring in some areas. Boulder fences and walls are seen in fields and forests throughout the northeastern United States.

Summary

Soil

Defined

Many definitions depending upon use (engineering, agriculture, geology, etc.). For our purposes, however, soil is defined as weathered rock and mineral grains that can support plant life.

Environmental Factors of Soil Formation

Climate, organic activity, relief (topography), parent material, and time. Soils form from the weathering of regolith.

Weathering

Physical and chemical breakdown of bedrock and regolith.

1. *Physical weathering* is caused by temperature differences and disruption by plants and animals.
2. *Chemical weathering* processes include solution, oxidation, hydration, and hydrolysis. It is important in the formation of clays, the most important mineral group in soils. Clays are important because they hold moisture and exchange nutrients with plants.

Rate of Weathering

Depends upon the rock or regolith's mineral composition and surface area. Smaller particles weather faster than large masses, and ferromagnesian minerals weather faster than quartz or feldspar.

Features and Landforms

Spheroidal weathering and exfoliation domes are common features.

Soil Profile

Defined

Distinguishable layers, or horizons, in mature soils composed of the A horizon, or zone of leaching (where minerals are dissolved), and a lower B horizon known as the zone of accumulation, where salts and clays accumulate. Below this is the C horizon of weathered parent material. The thickness of the horizons and the development rate of soil are profoundly influenced by climate.

Residual Soil

Developed in place on underlying bedrock. Laterite, for example, is the brick-red soil of humid (subtropical and tropical) regions that represents intensive weathering of parent rock.

Transported Soil

Developed on regolith transported and deposited by wind (eolian), glaciers, rivers, or volcanic action. Loess, for example, is an eolian soil associated with ice-age continental glaciation. It is highly productive and is found throughout the central U.S. "breadbasket."

Soil Classification

Zonal

A climatic classification based upon rainfall and temperature. The four main zonal soils are pedalfers (rich brown soils of temperate regions), pedocals (alkaline soils of arid regions), laterites (tropical and subtropical soils), and tundra soils of polar regions.

U.S. Comprehensive Soil Classification System

Known as the *Seventh Approximation*, a classification of soils by their physical characteristics.

Soil Problems

Erosion

A major problem, because soil determines the earth's carrying capacity. Globally, irreparable soil degradation has impacted an area the size of China and India combined. U.S. soil losses vary greatly with area.

Processes

Sheet erosion by running water (thin layer is removed) can become rill erosion (defined streamlets are carved in the soil), which leads to gullying (the most destructive). Sheet and rill erosion can be halted, but once gullying starts, it is extremely difficult and costly to mitigate. Wind erodes soil on marginal lands that were once cultivated and then dried out. Without plant root systems to hold the soil, wind erosion can lead to "dust bowl" conditions. Off-road vehicles also seriously degrade soils.

Mitigation

Good agricultural practices such as terracing, no-till and minimum tillage, strip cropping, and crop rotation. Wind losses can be minimized by planting windbreaks, strip crops, and cover crops such as clover.

Other Soil Problems

Expansive Soils

Certain clay minerals absorb water and expand, exerting high uplift pressures that can crack or even destroy structures. Mitigated by removal of soil, mixing, or keeping the moisture content constant.

Permafrost

Permanently frozen ground (soil or rock) underly more than 20 percent of the earth's surface. Structures built directly on soil permafrost will settle, so mitigation involves building with an open space below the structure so as not to disturb the thermal regime in the soil and melt the permafrost. Removal of permafrost zones is sometimes possible.

Settlement

Occurs when the applied load is greater than the bearing strength of the soil; e.g., Leaning Tower of Pisa. An engineering problem that can be mitigated by proper design or by strengthening of the underlying soil.

Miscellaneous Soil Problems

Salinization and waterlogging (high water table), hardening of laterites due to deforestation and desiccation, hardpans (cemented B horizon), and presence or absence of trace elements essential to human health.

Contamination

Brines, petroleum products, crude oil, and chemicals of all sorts may contaminate soils. Common methods of cleansing the soil are bioremediation (bacteria), phytoremediation (plants), and vapor extraction (oxidation).

Key Terms

active layer	hardpan	phytoremediation	soil profile
bioremediation	laterite	regolith	spheroidal weathering
caliche	loess	residual soil	trace elements
erosion	pedalfer	rill erosion	tundra soils
exfoliation dome	pedocal	sheet erosion	vapor extraction
expansive soils	pedologist	sheet joint	weathering
frost wedging	permafrost	soil	
gullying (erosion)	permafrost table	soil horizon	

Study Questions

1. Sketch a soil profile and describe each soil horizon.
2. Distinguish chemical weathering, physical weathering, and erosion.
3. What kinds of soils would you expect to develop on basalt in Hawaii, New Jersey, and Arizona? Explain the differences in terms of the CLORPT equation.
4. What is permafrost, and where is it found?
5. What soil problem does permafrost present to engineering works, and what techniques are used to prevent damage from this problem?
6. What is expansive soil, and what problems and challenges does it pose to soils engineers? What can be done to prevent distress due to this condition?
7. Soil erosion and degradation are two of humankind's oldest problems. How has human-induced soil loss come about? What agricultural practices have led to salinization and water-logging of soils?
8. An inscribed rock obelisk dating from 1500 B.C. was resurrected from the ruins at Karnak, Egypt. Its inscriptions were crystal clear when it was brought to New York's Central Park as a gift of the Egyptian government in 1880, but today they are totally illegible. Explain the reason for this.
9. Laterite is one of the most interesting of the zonal soil groups. What special problems have laterites presented in deforested areas of Brazil and war-torn Vietnam?
10. Compare the rates of soil erosion and soil formation in the United States.

For Further Information

Books and Periodicals

Batie, Sandra S. 1983. *Soil erosion: A crisis in America's croplands?* Washington, D.C.: The Conservation Foundation.

Brady, Nyle C. 1990. *The nature and properties of soils.* New York: Macmillan Publishing Co.

Brevik, Eric. 2002. Problems and suggestions related to soil classification as presented in introduction to physical geology textbooks, *Journal of Geoscience Education,* vol. 50, no. 5, November.

Brown, K. W. 1994. New horizons in soil remediation. *Geotimes* (American Geological Institute), September.

Brown, Lester. 1999. *State of the world.* New York: W. W. Norton & Co.

Ferrians, O. J., and others. 1969. *Permafrost and related engineering problems,* U.S. Geological Survey professional paper 678.

Heiniger, Paolo. 1995. The leaning tower of Pisa. *Scientific American* 273, no. 6: 62–67.

Loynachan, Thomas E., Kirk W. Brown, Terence H. Cooper, and Murray Milford. 1999. *Sustaining our soils and society.* Alexandria, Virginia: American Geological Institute.

McPhee, John. 2000. *Annals of the Former World.* New York: Farrar, Straus, and Giroux.

Singer, M. J., and D. N. Munns. 2002. *Soils: An Introduction,* 5th ed. Upper Saddle River, New Jersey: Prentice Hall.

Winkler, Erhard M. 1998. The complexity of urban stone decay. *Geotimes* (American Geological Institute), September.

ENVIRONMENTAL
Geology⇌Now™

Assess your understanding of this chapter's topics with additional quizzing and comprehensive interactivities at http://earthscience.brookscole.com/pipkingeo4e as well as current and up-to-date Web links, additional readings, and InfoTrac College Edition exercises.

Mass Wasting and Subsidence

A Bad Slice on the Golf Course

Nature to be commanded, must be obeyed.

Francis Bacon, philosopher (1561–1626)

Golf's explosion in popularity in the late 1990s led to many new golf courses. Some are in resort communities, but others are in urban and suburban areas, like the beautiful Ocean Trails Golf Course, south of Los Angeles on the Palos Verdes Peninsula (◆Figure 1). Geologists know the peninsula well for its many landslides, due in part because it is uplifted 370 meters (1,300 ft) above sea level. Its stratified sedimentary rock inclines gently to the sea, in 30- to 50-meter cliffs where the layers are exposed. Just as a slice of bread slides down a breadboard when tilted, some weak layers have slid toward the sea in the past, carrying houses and roads with them. In other cases, constant battering by waves at the cliffs' base has created unstable conditions.

Ocean Trails Golf Course was almost ready for its early summer 1999 opening when, without warning, a fissure opened parallel to the cliff, and 300 meters of the 18th fairway slid into the ocean. Local resident Tony Baker and his dog were temporarily stranded on a precarious 215-meter-(700 ft)-long island between

B. PIPKIN

◆ FIGURE 1 *Ocean Trails golf course, Palos Verdes Peninsula, California, shortly after the "slice" occurred, disrupting grand opening plans.*

fissures. "I heard crumbling earth and started seeing dust rising," Baker said later. "The trail started cracking up. I was doing a little running around, jumping over big cracks . . . I just found a place and hunkered down. I wasn't sure if I was going to make it back."

After three years of wrangling the developers and lenders reached an impasse and the landslide stabilization work stopped. The beautiful golf course, now only 15 holes long, was blemished by the scar of the landslide and stabilization work. Enter Donald Trump, the New York entrepreneur, real estate magnate, and enthusiastic golfer. He bought the troubled project (at a bargain price) in 2002 and vowed to finish the project by the summer of 2003 (not fulfilled). However, the facility is on its way to completion and the buttress fill designed to stabilize the "slice" on the course will be buried under the last two holes.

Mass wasting is the general term that denotes any downslope movement of soil and rock under the direct influence of gravity. Mass-wasting processes include landslides, rapidly moving debris flows, slow-moving soil creep, and rockfalls of all kinds. Annual damage from landsliding alone in the United States is estimated at between one and two billion dollars. If we include other ground failures, such as subsidence, expansive soils, and construction-induced slides and flows, total losses are many times greater than the annual combined losses from earthquakes, volcanic eruptions, floods, hurricanes, and tornadoes.

Mass wasting is also a global environmental problem. Earthquake-triggered landslides in Kansu Province in China killed an estimated 200,000 people in 1920, and debris flows left 600 dead and destroyed 100,000 homes near Kobe, Japan, in 1938. The largest loss of life from a single landslide in U.S. history, 129 fatalities, occurred at Mameyes, Puerto Rico, in 1985.

The term **landslide** encompasses all moderately rapid falls, slides, and flows that have well-defined boundaries and that move downward and outward from a natural or artificial slope. They occur in all the states and are an economically significant factor in more than 25 of them (◆Figure

7.1). Wherever unstable slopes exist, some form of mass wasting usually occurs. Areas with the greatest topographic relief are at highest risk, because gravity acts to smooth out topography by reducing the high areas through mass wasting and filling in the low areas with slide debris. For example, more than two million mappable slides exist in the Appalachian Mountains from New England to Alabama. Almost ten percent of the land area of Colorado is landslide terrane, and landslides and debris flows in Utah caused $300 million worth of damage in 1983–84. In Southern California, eight wet years between 1950 and 1993 averaged $500 million in landslide damage. Slope stability problems have such impact that a national landslide-loss-reduction program was implemented by the U.S. Geological Survey in the mid-1980s. A high priority of the program is to identify and map landslide-prone lands.

The classification of slope movements used in this book is similar to the one widely used by engineering geologists and soil engineers (◆Figure 7.2). The bases of the classification are

- the type of material involved, such as rock or soil;
- how the material moves, such as by sliding, flowing, or falling;
- the moisture content of the material; and
- how fast the material moves, its velocity.

For instance, soil creep occurs at imperceptibly slow rates and is relatively dry; debris flows are water-saturated and move swiftly; and slides are coherent masses of rock or soil that move along one or more discrete failure surfaces, or **slide planes.** Failure occurs when the force that is

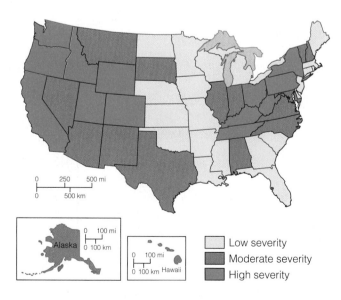

◆ FIGURE 7.1 Severity of landsliding in the United States. Note that the most serious problems are found in the Appalachian Mountains, the Rocky Mountains, and the coastal mountain ranges of the Pacific Rim. From J. P. Krohn and J. E. Slosson, "Landslide Potential in the U.S.," *California Geology 29*, no. 10 (1976)

MECHANISM		MATERIAL: Rock	MATERIAL: Fine-grained Soil	MATERIAL: Coarse-grained Soil	VELOCITY
SLIDE		Slump	Earth slump	Debris slump	Slow
SLIDE		Block glide	Earth slide	Debris slide	Rapid
FLOW		Rock avalanche	Mudflow, avalanche	Debris flow, avalanche	Very Rapid
FLOW		Creep	Creep	Creep	Extremely slow
FALL		Rockfall	Earthfall	Debrisfall	Extremely rapid

◆ FIGURE 7.2 Classification of landslides by mechanism, material, and velocity.

pulling the slope downward (gravity) exceeds the strength of the earth materials that compose the slope. Because gravity acting on a given slope is essentially constant, either the earth materials' resistance to sliding must decrease or the gradient of the slope must increase in order for sliding to occur.

Flows

Types of Flows

Creep is the slow (a few millimeters per year), essentially continuous downslope movement of soil and rock on steep slopes. It involves either freezing and thawing or alternate wetting and drying of a hill slope, which causes upward expansion of the ground surface perpendicular to the face of the slope. As the slope dries out or thaws, the soil surface drops vertically, resulting in a net downslope movement of the soil (◆Figure 7.3). Burrowing animals and other biological processes that produce openings in the soil also contribute to soil creep. Bent trees, leaning fence posts and telephone poles, and bending of tilted rock layers downslope are all evidence of soil creep. Homes built with conventional foundations 18–24 inches deep may develop cracks due to soil creep. This is seldom catastrophic; typically it is a cosmetic and maintenance problem. The influence of soil creep can be overcome by placing foundations through the creeping soil into bedrock.

Debris flows are dense, fluid mixtures of rock, sand, mud, and water. They may be generated quickly during heavy rainfall or snowmelt where there is an abundant supply of loose soil and rock. Moving with the consistency of wet concrete, they are very destructive, with velocities up to many meters per second. Because of their high density, commonly 1.5–2.0 times the density of water, debris flows are capable of transporting large boulders, automobiles, and even houses in their mass (◆Figure 7.4).

The areas most subject to debris flows are characterized by sparse vegetation and intense seasonal rainfall, or are in regions that are subject to drenching rains (see ✪ Case Study 7.1 on page 197). Debris flows are relatively common in Canada, South America in the Andes, and alpine and desert environments worldwide. Debris flows associated with volcanic eruptions (lahars) are potential hazards in all volcanic zones of the world. They pose a particular danger in populated areas near volcanoes in Washington, Oregon, and California, and are the most significant threat to life of all the landslide hazards (see Chapter 5). An *earthflow* is a sliding mass of soil and weathered bedrock over a discrete basal surface, usually fresh bedrock. Earthflows terminate in lobe-like forms and grade into debris flows as water content increases. They may be thought of as embryonic debris flows (see Figure 7.9b).

A **debris avalanche** is simply a fast-moving (4 m/sec) debris flow (◆Figure 7.5). A structure in the path of one of these flows can be severely damaged or completely flattened with attendant injury or loss of life. During the winter of 1982, debris flows and avalanches in the San Francisco Bay area killed 25 people and resulted in $66 million worth of damage. In Mill Valley, a small town north of San Francisco, a debris flow ripped a house from its foundations and deposited it 45 meters (150 ft) downslope—without serious injury to the residents (passengers). The house effectively dammed the canyon and prevented debris-flow damage farther downslope.

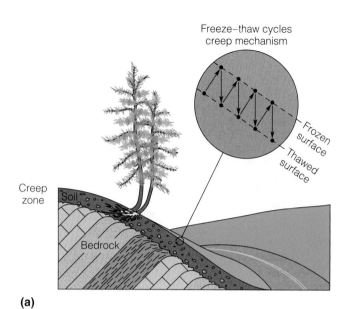

(a)

(b)

◆ FIGURE 7.3 (a) Diagram and (b) photograph of soil and rock creep; east slope of the Sierra Nevada, California.

Causes of Debris Flows and Debris Avalanches

Debris flows need steep slopes to start, but once started they may flow even upon gently sloping ground. Most hazardous to lives and property are areas in canyon bottoms, near the mouths of canyons that act as conduits for debris flows, and below slopes manufactured for homesites or roads. Once started, debris flows gain moisture as they travel, becoming more fluid, and may accelerate to speeds in excess of 50 kilometers per hour (30 mi/hr). Several flows may merge and travel many miles from their source area.

Studies of debris flows indicate that two rainfall conditions are necessary to cause them: (1) an initial period of rainfall, known as *antecedent rainfall,* that saturates the soil and (2) a subsequent period of intense rainfall that puts even more water into the soil and initiates flowage. Thus the onset of debris flows may be estimated during the rainy season if we have data on the amount of rainfall needed to initiate flowage. Such data have been compiled for flows and slides in the San Francisco Bay area of northern California (◆Figure 7.6). They indicate that a short period of intense rainfall or less-intense rainfall over a longer period can produce debris flows. Specifically, the curves of Figure 7.6 show that for a rainfall intensity of 0.5 inch per hour, the threshold time for the onset of debris flows is 8 hours in Marin County and 14 hours in Contra Costa County. The differing rainfall thresholds in these relatively close areas are due to the variability of geologic materials and topography. The lower curve of Figure 7.6 shows that less-intense rainfall will produce debris flows in semiarid and arid areas of California. This is because

these dry areas have little vegetation with root systems that retain soils and, therefore, have abundant loose surface debris. Although these curves are only preliminary, they will serve to alert residents in critical areas once the threshold conditions for flows have been attained, and they will be refined as more data become available.

Heavy rains associated with El Niño events have resulted in exceptional landslide and debris-flow activity, particularly along the west coasts of North and South America. The major El Niño of 1982–1983 was marked by widespread landsliding (slumps, slides, and flows) in various parts of the Western Hemisphere. The 1997–1998 El Niño, possibly the largest of the twentieth century, also had exceptional flows and slides. Hard hit was the San Francisco Bay area, where some of the 85,000 landslides shown on the thematic maps of the region were reactivated (◆Figure 7.7). This is an excellent example of interaction between the atmosphere, hydrosphere, and solid earth systems.

Mountain slopes burned by range and forest fires are also susceptible to debris flows during the wet season. The loss of active root systems to bind soil particles can result in an extremely dangerous condition. In addition, debris flows from burned slopes have been found to have longer runout distances into foothill areas than those from vegetated slopes.

◆ FIGURE 7.4 Two debris flows in 1.5 hours hit and moved the farmhouse and destroyed several other buildings. The farm structures were on an alluvial fan (see Figure 9.4) without a well-defined main channel. The flow occurred in Madison County, Virginia, which shows that debris flows are not limited to the southwestern United States.

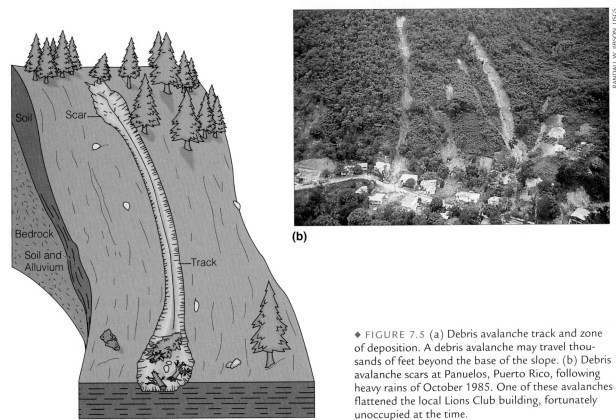

(b)

◆ FIGURE 7.5 (a) Debris avalanche track and zone of deposition. A debris avalanche may travel thousands of feet beyond the base of the slope. (b) Debris avalanche scars at Panuelos, Puerto Rico, following heavy rains of October 1985. One of these avalanches flattened the local Lions Club building, fortunately unoccupied at the time.

(a)

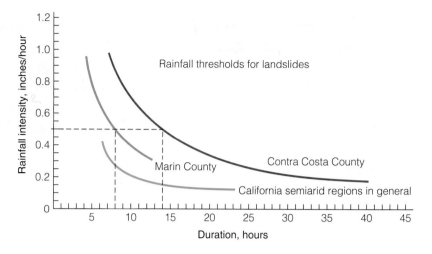

◆ FIGURE 7.6 Rainfall thresholds for debris avalanches and landslides for two central California counties and California semiarid regions in general. For example, a rainfall intensity of 0.5 inch/hour for 8 hours is sufficient to initiate flows in Marin County. The same intensity for 14 hours is the threshold for Contra Costa County. *California Geology* data from CDMG

Extreme forest fires occurred in Southern California in October 2003. In December of that year, 13 people lost their lives in a campground at the mouth of one of the burned canyons in the San Gabriel Mountains.

Not all debris flows and avalanches are triggered by intense rainfall. Slide Mountain, about halfway between Reno and Carson City, Nevada (and between Washoe Lake and Lake Tahoe), is not named for the ski area on its slopes; it is the site of many flows and slides. On May 30, 1983, a debris flow was generated on Slide Mountain that killed one person, injured many more, and destroyed a number of homes and vehicles, all in less than 15 minutes (◆Figure 7.8). About 720,000 cubic meters of weathered granite gave way from the mountain's steep flank and slid into Upper Price Lake. This mass displaced the water in the lake, causing it to overflow into a lower lake, which in turn overflowed into Ophir Creek gorge as a water flood. Picking up sediment as it went, it became a debris flow that emerged from the gorge, spread out, destroyed homes, and covered a major highway. Because this is a popular recreational area and debris flows can be sudden and hazardous to one's health, geologic hazard warning signs are posted on trails throughout the area.

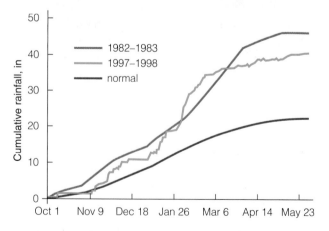

◆ FIGURE 7.7 Rainfall in excess of normal in the San Francisco Bay area during El Niño years 1982–1983 and 1997–1998. National Weather Service and NOAA

Debris flows are not just limited to the arid and semiarid Southwest and West. In Madison County, Virginia, in 1995 a storm caused debris flows that devastated a small area of the Blue Ridge. Livestock were killed and buildings, bridges, roads, and crops were damaged. There was one loss of life directly attributable to the debris flow (see Figure 7.4). Although this was an infrequent event for Madison County, it is estimated that such damaging debris flows may occur every three years somewhere in the Blue Ridge and southern Appalachian Mountains. As briefly mentioned previously, among the most destructive debris flows are those that accompany volcanic eruptions (see Chapter 5).

Landslides That Move as a Unit

Types

Two kinds of coherent slide masses are recognized, distinguished by the shape of the slide surface. They are **slumps,** or **rotational slides,** which move on curved, concave-upward slide surfaces and are self-stabilizing, and **block glides,** or **translational slides,** which move on inclined slide planes (see Figure 7.2). A block glide moves as a unit until it meets an obstacle or until the slope of the slide plane changes.

Slumps

Slumps are the most common kind of landslide and range in size from small features a few meters wide to huge failures that can damage structures and transportation systems. They are spoon-shaped, having a slide surface that is curved concave-upward and exhibiting a backward rotation (◆Figure 7.9). They occur in geologic material that is fairly homogeneous, such as soil or badly weathered or fractured bedrock, and the slide surface cuts across geologic boundaries. Slumps move about a center of rotation; that is, as the toe of the slide rotates upward, its mass eventually counterbalances the downward force, causing the slide to stop (◆Figure 7.9b). Slumping produces repeated uniform depressions and flat areas on otherwise sloping ground surfaces. Slumps grow

(c)

(a)

(b)

◆ FIGURE 7.8 Slide Mountain in western Nevada, site of the 1983 disaster. (a) Looking up Ophir Creek and the runout area of the debris flow. The scar of the debris-avalanche source area is the treeless area on the mountain. (b) Abraded bark on tree records the thickness of the flow. (c) A warning to be taken seriously.

headward (upslope). This is because as one slump mass forms, it removes support from the slope above it. Thus, a stairstep surface is common in landslide terrain. Slump-type landslides are the scourge of highway builders, and repairing or removing them costs millions of dollars every year.

Block Glides

Block glides are coherent masses of rock or soil that move along relatively planar sliding surfaces (failure planes), which may be sedimentary bedding planes, metamorphic foliation planes, faults, or fracture surfaces. For a block glide to occur, it is necessary that the failure plane be inclined *less steeply* than the inclination of the natural or manufactured hill slope. Slopes may be stable with respect to block glides until they are steepened by excavating for building subdivisions or roads, thus leading to landsliding (◆Figure 7.10).

A classic block glide of about ten acres is seen along a sea cliff undercut by wave action at Point Fermin in San Pedro, California. Movement was first detected there in January 1929, and by 1930 the landslide had moved two meters seaward. It has been intermittently active ever since (◆Figure 7.11). The rock of the slide mass is coarse sandstone (not the type of earth material usually involved in block glides), but a thin layer of **bentonite** dipping 15° seaward forms the slide plane. Bentonite is volcanic ash that has chemically weathered to clay minerals, which become plastic and slippery when wet (see Case Study 2.3). Bentonite is very commonly involved in slope failures; addition of water is all that is needed to initiate a landslide where dips are as slight as 5 degrees.

Some of these translational slides are large and move with devastating speed and tragic results. In 1985, tropical storm Isabel dumped a near-record 24-hour rainfall that averaged almost 470 millimeters (18.5 in) on a mountainous region near the city of Ponce on the south coast of Puerto Rico. Rainfall intensities peaked in the early morning hours of October 7, reaching 70 millimeters (2.8 in) per hour in some places. At 3:30 that Monday morning, much of the Mameyes residential district of the city was destroyed by a rock block

◆ FIGURE 7.9 A rotational landslide. (a) Oblique stylized view showing ideal spoon shape and crown and toe of landslide. (b) Oblique view of rotational landslide (slump) showing toe with earth flow and potential headward growth of landslide.

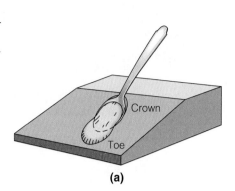

(a)

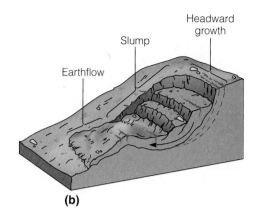

(b)

glide initiated during the most intense period of rainfall. This resulted in the worst loss of life from a landslide in U.S. history—129 deaths. The landslide was in a sandstone whose stratification parallels the natural slope of the slide mass, a condition called a *dip slope*. It moved at least 50 meters (165 ft), probably on a clay layer in the sandstone, before breaking up into the large blocks. The scarp at the top of the slide is 10 meters high (32 ft), and the maximum thickness observed at the toe of the landslide is 15 meters (49 ft).

The importance of water in initiating landslides and debris flows cannot be overemphasized. South-central Puerto Rico is the "dry" side of the island, where the average annual rainfall is about 1,000 millimeters (39 in). Ponce received almost half of that amount in the 24 hours preceding the landslide.

A similar landslide occurred in the Alps of northeastern Italy in 1963 at the site of Vaiont Dam, the highest thin-arch dam in the world at the time (275 m; 900 ft). The geology at

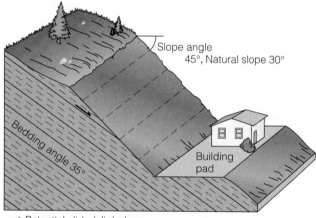

(a)

◆ FIGURE 7.10 Cutting a natural slope for homesites can lead to potential landslides on bedding or foliation planes. (a) After cutting a natural slope for a building pad the manufactured slope exposes potential slide planes. (b) A slide plane (smooth surface) and moving mass of shale after toe was cut; Santa Monica Mountains, California. (c) Striations gouged in clay on the slide plane by the sliding mass above it as shown in (b).

(b)

(c)

◆ FIGURE 7.11 The Point Fermin landslide at San Pedro, California, is a rock block glide that has moved intermittently since 1929. Damaged homes have been removed from the slide area in the foreground. The slide plane emerges just in front of the houses in the background, which are on relatively stable ground.

the reservoir consists of a sedimentary-rock structure that is bowed downward into a concave-upward **syncline** with its axis parallel to the Vaiont River canyon. Limestones containing clay layers dip toward the river and reservoir from both sides of the canyon due to this downfolded structure (◆Figure 7.12). Slope movements and slippage along clayey bedding planes above the reservoir had been observed before the dam was constructed. This condition gave engineers and geologists sufficient concern that they placed survey monuments on the slope above the dam for monitoring such movement. Heavy rains fell for two weeks before the disaster, and slope movements as large as 80 centimeters (31 in) per day were recorded. On the night of October 9, without warning, a huge mass of limestone slid into the reservoir so fast it generated a wave 100 meters (330 ft) high. The wave burst over the top of the dam and flowed into the Piave River valley, destroying villages in its path and leaving 2,500 people dead. The landslide velocity into the reservoir was so speedy that the slide mass almost emptied the lake. The dam did not fail and it is still standing today—a monument to excellent engineering but poor site selection.

Slump and block glides are the most common types of landslides, but there are also **complex landslides,** combinations of the two types. These landslides have elements of both rotational and translational failure, and there are many examples, most of them large landslides such as the Slumgullion landslide in Colorado (✪ Case Study 7.2 on page 198).

> **CONSIDER THIS** *You must choose between two hillside homesites. One site is underlain by schist, the other by gneiss, both with foliation dipping out of the hill ("daylighting"). Which site would you choose, and why? (Hint: Examine Figure 7.10, and refer to the discussion of sedimentary and metamorphic rocks in Chapter 2).*

Falls

Rockfalls occur very rapidly and are among the most common of mass-wasting processes. Signs alerting motorists to watch for falling rocks are a common sight. Chunks of rock of all sizes simply fall from vertical or very steep cliffs, having separated from the main mass along joints, faults, foliation planes, and other rock defects. The rocks are loosened by root growth, frost wedging, and heavy precipitation. A pile of rock debris may build up at the base of a cliff, forming a slope called **talus** (◆Figure 7.13). Two huge rockfalls in Yosemite Valley in 1996 killed one visitor and injured 14 others. The force of the shock-wave from the rockfall toppled nearby trees and stripped bark from more distant ones. The geology of this catastrophic event has been studied intensively by Radford University and the U.S. Geological Survey in an attempt to pinpoint areas of future rockfall hazard.

The Mechanics of Slides

Landslides do not just happen. They are explainable as a change in the balance between **driving forces,** the gravity forces that pull a slope downward, and **resisting forces,** the cohesive and frictional forces that hold a slope in place. A change in the balance can result from water seepage into slope material, an oversteepening of the slope by natural erosion or artificial cutting, or the addition of weight at the top of a slope. If any of these things happen, the driving forces may eventually exceed the resisting forces, which will cause a landslide to occur.

This can be demonstrated using a simple model of a sliding block on an inclined plane (◆Figure 7.14). The driving force, d, is the component of gravity acting parallel to the inclined plane at an angle α with the horizontal. The coefficient of friction between the block and the plane is f. Static

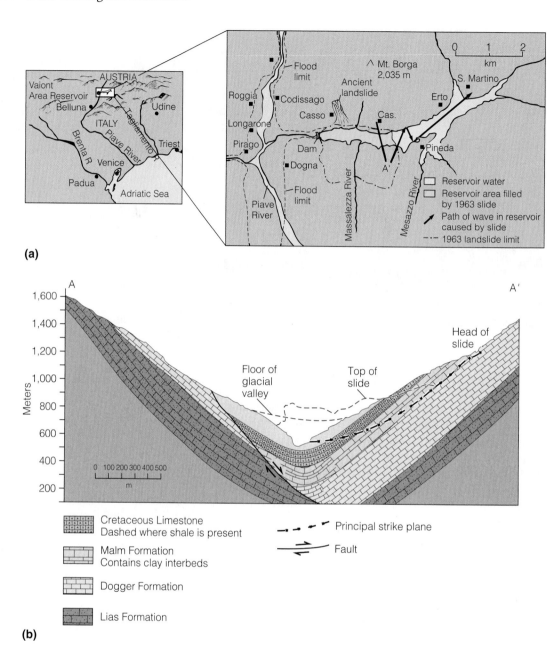

(a)

(b)

◆ FIGURE 7.12 The 1963 Vaiont Dam landslide. (a) Map of the slide mass that hurtled into the reservoir and the area that was impacted by the giant wave and flooding. Longarone and several other villages along the Piave River below the dam were devastated. (b) Cross-section through the Vaiont River valley showing a syncline with the sedimentary layers dipping toward the valley axis. The principal slide plane and resultant slide mass are indicated. The cross-section line is shown in (a).

friction between the block and slide plane increases as the normal force, n, acting across the slide plane increases. Friction between the block and the plane resists sliding. Thus,

resisting forces = friction coefficient × normal force

or

$$r = f \times n$$

A delicate balance exists, and sliding is imminent when

$$d(\text{driving force}) = r(\text{resisting force})$$

that is, when

$$r/d = 1.0$$

The Factor of Safety (F. S.) relates resisting force to driving force. When the two forces are equal, the F. S. equals 1.0. When resisting forces are greater than driving forces, the F. S. is greater than 1.0; and when resisting forces are less than driving forces, the F. S. is less than 1.0. Most modern building codes require that manufactured or natural slopes have a Factor of Safety of 1.5 or greater. In practice, one simply

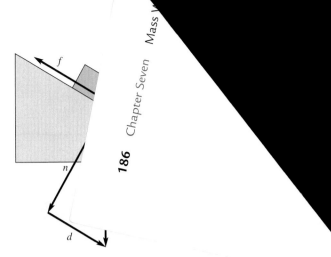

◆ FIGURE 7.14 Resolution of driving forces (*d*) and res.. forces (*f* × *n*) acting on a slide plane inclined at an angle α.

◆ FIGURE 7.13 Talus slopes caused by frost wedging and other physical weathering processes; Banff National Park in the Canadian Rocky Mountains.

sums up all the driving forces and resisting forces acting along a potential slide plane. From this it can be calculated whether the F. S. of the particular slope is greater than 1.0 (relatively stable) or less than 1.0 (relatively unstable).

If water pressure builds up in the pore spaces between sediment grains, it pushes them apart and reduces the soil's effective grain-to-grain friction—referred to as its **effective stress.** This is analogous to injecting water under pressure beneath the block on an inclined plane. The water pressure will support a portion of the block's weight, thereby reducing the normal force *n* and the sliding resistance of the block. Water in soil or sediment pore space supports part of the grain-to-grain pressure, thereby reducing the frictional resistance across the grain contacts. The effective stress increases as the normal force increases and decreases as the pore-water pressure increases. A "quick" condition exists when the pore pressure is equal to intergranular friction; in other words, when the net effective stress is reduced to zero. At this point, sand becomes a dense sand-water mixture known as *quicksand.* It is often said that water "lubricates" a failure plane, making it easier to slip. It has been demonstrated, however, that a certain amount of moisture actually increases the strength of some materials. Moist sand, for example, builds a better castle and stands on a steeper

slope than does dry sand (see Figure 7.16b). Water itself has no lubricating qualities. Rather, it is pore-water pressure that reduces grain-to-grain friction and therefore soil strength.

Lateral Spreading

Horizontal movement of a mass of soil overlying a liquefied or plastic layer characterizes a form of mass wasting called **lateral spreading.** Such slides are complex, as they involve elements of translation, rotation, and flow. Typically triggered by earthquakes, lateral spreading may result in the spontaneous liquefaction of water-saturated sand layers or in the collapse of **sensitive clays**—also known as **quick clays** (see Chapter 4). Spreading failures in the United States occur mostly in sand layers; however, areas underlain by glacial sediments may spread because of quick clay layers within the deposits.

Much of the damage in the Marina district of San Francisco in the 1906 and 1989 earthquakes was due to lateral spreading caused by liquefaction. Most of the damage in Anchorage, Alaska, during the 1964 quake was due to quick-clay-induced lateral spreads in the Turnagain Heights residential neighborhood and the downtown area. The spreading was so extensive that two houses that had been more than 200 meters (more than 2 football fields) apart collided within the slide mass. The Turnagain Heights area is now a tourist attraction known as "Earthquake Park" (◆Figure 7.15).

CONSIDER THIS *The geologists' report on a building site that you are considering buying states that the sedimentary strata—shale and sandstone—have a "bad attitude." Does this mean they are not nice? Is there some geological interpretation of this term?*

Factors That Lead to Landslides

Seldom can a landslide be attributed to a single cause; rather, landslides result from series of events that lead to failure.

◆ FIGURE 7.15 Turnagain Heights as it appeared after the "quick clay" lateral spread that was triggered by the 1964 Alaskan earthquake.

Nonetheless, the weakening of slope materials due to the addition of water is the most important causative factor of all slides and flows. Thus heavy rainfall, rapid snowmelt, leaking water mains, private sewage-disposal (cesspool) inflow, and poor building-pad drainage can all lead to landsliding. Excess water causes a buildup of pore-water pressure, which weakens the materials supporting the slope. This is why mass wasting is closely correlated with a series of heavy-rain years in semiarid Southern California and with torrential flooding in the eastern United States.

Recall that when driving forces exceed resisting forces, a failure is imminent. Factors that increase driving force are

- an increase in the slope angle,
- removal of lateral support at the toe of a slope, and
- added weight at the top of a slope.

The effect of increasing slope angle is best illustrated by the natural **angle of repose** of granular material such as dry sand or gravel. The angle of repose of such material is the maximum slope angle at which it can remain stable. For example, no matter how steeply you try to pile dry sand, it will always form a slope of about 32°–34°, its angle of repose (◆Figure 7.16). Moist sand has a greater angle of repose because of the temporary cohesion (added strength) imparted to it by moisture. This is why a castle built of moist sand collapses when the sand dries and loses its temporary cohesion. The angle of repose represents the point at which the frictional resistance between sand grains (shear strength) and the downslope component of gravity are in balance.

Theoretically, mass movements will not occur as long as the angle of repose of the particular slope material is not exceeded. Because undercutting the toe of a slope is equiva-

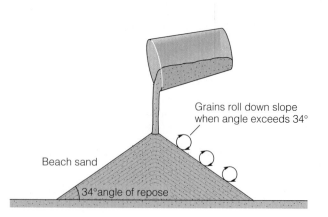

Grains roll down slope when angle exceeds 34°

Beach sand

34° angle of repose

◆ FIGURE 7.16 The angle of repose of granular material. For dry sand the angle is 34°. Coarser material, such as that found on talus slopes, has a slightly steeper angle of repose. The angle of repose depends not only upon the size of the material but also upon its moisture content. If one moistens beach sand, for example, one may build steep towers and walls with it.

lent to increasing the slope angle, it causes instability. Adding weight to the slope, as with fill material or an unusually heavy structure, increases the pore pressure and can thus induce failure. Factors that reduce a slope's strength and, therefore, its resistance to sliding are

- infiltration of water underground,
- weathering and breakdown of minerals, especially when clays are formed, and
- burrowing by animals.

In addition, it should be noted that sliding is also facilitated when a slope is cut in such a way that its bedding planes are exposed or inclined (dipping) out of the slope face.

A slope with a Factor of Safety less than 1 is ready to slide, but it might not do so unless it is "triggered" by an earthquake, heavy traffic, sonic boom, detonation of explosives, or another source of energy. Such triggers are split-second stresses that overcome *static friction,* the force resisting sliding between two surfaces at rest. Static friction is always greater than *sliding (kinetic) friction,* the friction existing between two surfaces in relative motion. This explains why landslides are very hard to stop once they start.

CONSIDER THIS *Suppose you wish to build a structure at the top of a vertical cut that is 3.1 meters (10 ft) high. The site is underlain by poorly cemented river sand. How far back from the top of the slope would you have to build to achieve an F. S. of 1.0? How far back for an F. S. of 1.5? (Hint: Consider the material's angle of repose.)*

Reducing Losses from Mass Wasting

Landslide Hazard Zonation

The first step in reducing losses due to mass-wasting over the long term is to make a careful inventory or map of known landslide areas. Ancient landslides have a distinct topography characterized by hummocky (bumpy) terrain with knobs and closed depressions formed by backward-rotated slump blocks. On topographic maps, slide terrain can be evidenced by curved, closely spaced, amphitheater-shaped contour lines indicating scarps at the heads of landslides. In the field, one may see scarps, trees, or structures that are tilted uphill; unexpected anomalous flat areas; and water-loving vegetation such as cattails where water issues at the toe of a landslide.

With an adequate data base provided by maps and field observations, it is possible to define landslide hazard zones. Maps can be generated that show the *landslide risk* in an area,

and then these areas can be avoided or work can be done to stabilize them (◆Figure 7.17). Mapping programs do not prevent slides, but they offer a means for minimizing slides' impact on humans. The human tragedy of landslides in urban areas is that they cause the loss of land as well as of homes.

Building Codes and Regulations

Building codes dictate which site investigations must be performed by geologists and engineers and the way structures must be built. Chapter 70 of the *Uniform Building Code,* which deals with slopes and alteration of the landscape, has been widely adopted by cities and counties in the United

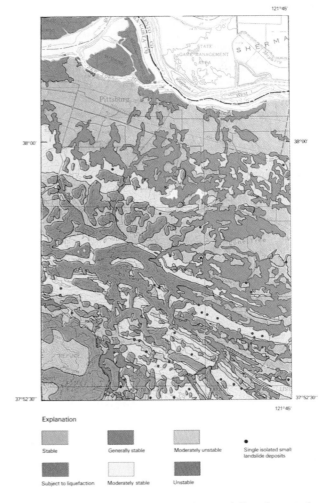

Explanation

Stable · Generally stable · Moderately unstable · Single isolated small landslide deposits

Subject to liquefaction · Moderately stable · Unstable

◆ FIGURE 7.17 Map of the relative slope stability of parts of Contra Costa County and adjacent counties derived by combining a slope map, a landslide-deposit map, and a map of susceptible geologic units. The area shown here is 15 kilometers (9 mi) wide. USGS

◆ FIGURE 7.18 Part of a typical grading ordinance (code) that shows the limitation of cut-slope steepness, the required benches on slopes for collecting rainfall and reducing slope-face erosion, and an approved method of constructing a fill slope with bedrock benches for increased stability.

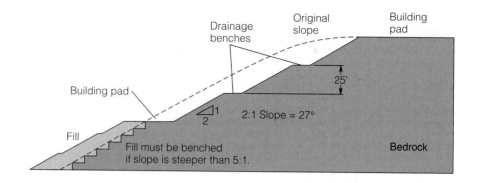

States. It specifies compaction and surface-drainage requirements and the relationships between such planar elements as bedding, foliation, faults, and slope orientation. Code requirements for manufactured slopes in Los Angeles resemble the specifications found in codes elsewhere (◆Figure 7.18). The maximum slope angle allowed by code in landslide-prone country is 2:1; that is, 2 feet horizontally for each foot vertically (27°). The rationale for the 2:1 (27°) slope standard is that granular earth materials have a natural angle of repose of about 34°; 2:1 slopes allow for a factor of safety.

Losses have been cut dramatically by enforcement of grading codes, and they will continue to decrease as codes become more restrictive in hillside areas. For instance, before grading codes were established in the City of Los Angeles in 1952, 1,040 building sites were damaged or lost by slope failures out of each 10,000 constructed—a loss rate of 10.4 percent. With a new code in effect between 1952 and 1962 that required minimal geologic and soil-engineering investigation, losses were reduced to 1.3 percent for new construction. In 1963, the city enacted a revised code requiring extensive geological and soils investigations. Losses were reduced to 0.15 percent of new construction in the following six years, which included 1969, the largest rain year in recent history. A new international code that is applicable anywhere in the world is in the works.

Control and Stabilization

The basic principles behind all methods of preventing or correcting landslides relate to strengthening the earth materials (increasing the resisting forces) and reducing the stresses within the system (decreasing the driving forces). For a given slide mass, this requires halting or reversing the factors that promote instability, which will probably be accomplished by one or more of the following: draining water from the slide area, excavating and redistributing the slide mass, and installing retaining devices. Debris flows require special devices that divert the path of the flow away from structures.

Water Drainage and Control

Water is the major culprit in land instability, and surface water must be prevented from infiltrating the potential slide mass. Water within the slide mass can be removed by drilling horizontal drains, called *hydrauger* (water bore) holes, and lining them with perforated plastic pipe. This is commonly done on hillside terrain where underground water presents a problem. In some areas the land surface of a building site above a suspect slope is sealed with compacted fill to help keep surface water from percolating underground. Also, plastic sheeting is widely used to cover cracks and prevent water infiltration and erosion of the slide mass (◆Figure 7.19). The efficacy of drainage control and planting is shown in the figure.

Excavation and Redistribution

Recontouring is a textbook method of stabilizing a slide mass. By this method, material is removed from the top of the landslide and placed at the toe. Compacted-earth structures called *buttress fills* are often designed to retain large active landslides and known inactive ones believed to be vulnerable. They are constructed by removing the toe of a slide and replacing it, layer by layer, with compacted soil. Such fills can "buttress" huge slide masses, thus reclaiming otherwise unusable land. The finished product is required to have concrete terraces for intercepting rainwater and preventing erosion and downdrains for conveying the intercepted water off the site (◆Figure 7.20). Buttressing unstable land has aesthetic and economic advantages, in that it can be landscaped and built upon. The cost of the additional work needed to stabilize unstable land is usually offset by a relatively low purchase price, which can make it economically "buildable." Of course, the ultimate mitigating measure is total removal of the slide mass and reshaping of the land to buildable contours.

(a)

(b)

◆ FIGURE 7.19 Excess water is the culprit in initiating many landslides. (a) Plastic sheeting placed over the head scarp of a landslide. This is, perhaps, closing the barn door after the cow has left but is common practice in landslide prone areas. (b) Surface-water erosion prevention measures on a graded slope; San Clemente, California. Whereas slope planting and terracing have kept the middle slope in good condition, the slope on the left has gullied badly and will need maintenance in the future.

Retaining Devices

Many slopes are oversteepened by cutting back at the toe, usually for the purpose of obtaining more flat building area. The vertical cut can be supported by constructing steel-reinforced concrete-block retaining walls with drain (weep) holes for alleviating water-pressure buildup behind the structure (◆Figure 7.21). Retaining devices are constructed of a variety of materials, including rock, timber, metal, wire-mesh fencing, and known as shotcrete.

Steep or vertical rock slopes that are jointed or very seamy are often strengthened by inserting long rock bolts into holes drilled perpendicular to planes of weakness. This binds the planes together much as a beam is bound. Rock bolts are used extensively to support tunnel and mine openings. They add considerably to the safety of these operations by preventing sudden rock "popouts." They are also installed on steep roadcuts to prevent rockfalls onto highways (◆Figure 7.22). Retaining devices also are effective for mitigating rockfalls and debrisfalls.

Diversion Techniques

Debris flows present a different challenge, because they can originate some distance away from the site of interest. It is simply not good practice to build at the bottom or the mouth of a steep ravine or gully, especially if loose soils are present on the higher slopes. This is particularly true in areas of high mean annual rainfall, areas that are subject to sudden cloudbursts, and areas that have been burned, such as in the tragic October 1993 firestorms in Southern California. As long as people continue to build houses in the "barrels" of canyons, as is done in the foothills of Southern California, debris flows will continue to take their toll. Debris-flow insurance is expensive, but the cost can be less if *deflecting walls* are designed into the house plans (◆Figure 7.23). These walls have proven to be effective in protecting dwellings, but they could have adverse consequences if they divert debris onto neighbors' structures, in which case neither an architect nor a geologist would be as helpful as a lawyer. ✳ Table 7.1 summarizes mitigating measures that can slow or stop landsliding and other mass-wasting processes.

> **CONSIDER THIS** *In many parts of the world, large numbers of people live on or close to potentially unstable hillslopes. How do grading codes help to reduce the risk of landslides? Do you think they are effective if they are enforced?*

Snow Avalanches

The mechanics of snow avalanches are similar to those of landsliding, the differences being in the material and the velocity. Although snow avalanches have been around as long as there have been mountains and snow, the relatively recent emergence of the recreational skiing industry as big business has transformed avalanche control from an art to a science. Unfortunately, it's too late for Hannibal. His crossing of the Alps in 218 B.C. was plagued by avalanches. Purportedly 18,000 of his troops and who knows how many of his elephants were killed by them. It's also too late for others. In World War I 6,000 troops were buried in one day by avalanches in the Dolomite Mountains of northern Italy. Avalanches in high mountain regions have wiped out villages, disturbed railroad track alignments, blocked roads, and otherwise made life difficult if not impossible for millennia.

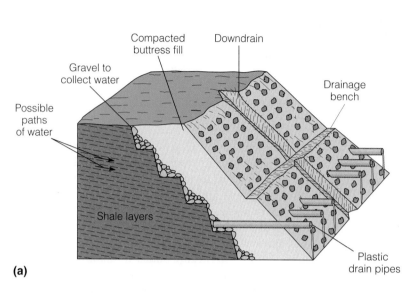

(a)

(b)

◆ FIGURE 7.20 (a) Buttress fill with horizontal drains for removing water and surface drains for preventing erosion. The fill acts as a retaining wall to hold up unstable slopes such as those that result from excavation. (b) A slope with benches and surface drains built according to accepted standards.

Avalanches are caused by an addition of heavy new snowfall (an increase in driving force) or a weakening of older snow (reduction of resisting force). A **slab avalanche** is a coherent mass of snow and ice that hurtles down a slope of more than 30° as an entity—much like a magazine sliding off a tilted coffee table (◆Figure 7.24). It may be a thick, wet snow accumulation sliding on ice, or a coherent slab of snow

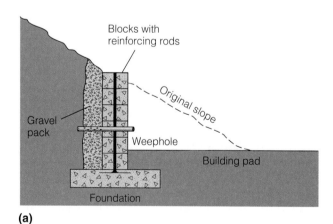

(a)

(b)

◆ FIGURE 7.21 (a)Typical retaining wall with drain (weep) holes to prevent water retention and the buildup of hydrostatic pressure behind the wall. The wall is constructed of concrete blocks with steel reinforcing rods and concrete in their hollow centers. (b) A steel cage is about to be placed in an 80-foot hole, which will then be filled with concrete. Thirty nine of these "caissons" were installed and tied together with a horizontal beam to retain ten homes at the top of a landslide.

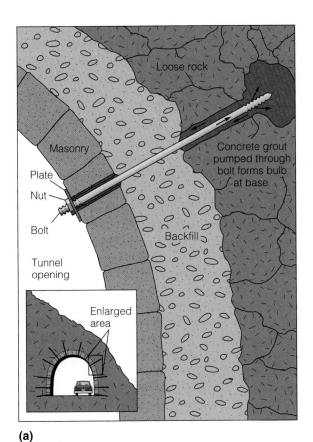

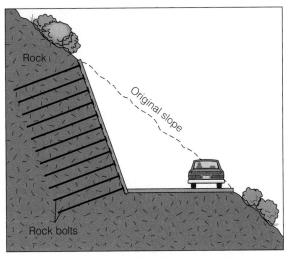

◆ FIGURE 7.22 (a) Typical rock bolt installation in a tunnel or mine opening. (b) Rock bolts support jointed rocks above a highway.

◆ FIGURE 7.23 The outlet of a debris-flow diversion structure; San Fernando Valley, California. See Figure 7.4 to view what lack of a debris-flow diversion structure can produce.

that detached along a *depth hoar* layer. Depth hoar begins to form at the surface when snow particles evaporate in very cold, calm weather and form a loose, open array of snow crystals. As the layer of crystals is buried, evaporation continues, resulting in a very weak, porous layer at depth. This hoar layer may collapse when it is disturbed and there forms a compressed-air layer upon which the overlying snow can move with great velocity. These avalanches account for the majority of avalanche deaths because of their sheer weight and their tendency to solidify when they stop. About 25 avalanche fatalities occur each year in North America, mostly in backcountry areas, and the victims tend to be cross-country skiers, snowmobilers, climbers, and hikers.

Powder-snow avalanches are more common in the Sierra Nevada and Coast Ranges, where depth hoar is rare and heavy snowfall results in deep snow packs. These "dry" avalanches are dangerous because they entrain a great amount of air and behave like a fluid. Furthermore, the strong shock wave that precedes them can uproot trees and flatten structures. Victims of dry avalanches tend to be buried in light snow close to the surface, which makes their chances of being rescued alive better than those of wet-avalanche victims.

Avalanches can be triggered by skiers, rockfalls, loud noises, or explosives. Avalanche control is thus achieved by

✳TABLE 7.1 Summary of Mass-Wasting Processes and Their Mitigation

Landslides

Cause	Effect	Mitigation
Excess water	Decreased friction (effective stress)	Horizontal drains, surface sealing
Added weight at top	Increased driving force	Buttress fill, retaining walls, decrease slope angle
Undercut toe of slope	Increased slope angle and driving force	Retaining walls, buttress fill
"Daylighted" bedding	Exposure of unsupported bedding planes in cut or natural slope	Buttress fill, retaining walls, decrease slope angle

Other Mass-Wasting Processes

Process	Mitigation
Rockfall	Rock bolts and wire mesh on the slope, concrete or wooden cribbing at bottom of the slope, cover with "shotcrete"
Debris flow	Diversion walls or fences
Soil creep	Deep foundations
Lateral spreading	Dewater, buttress, retain (difficult to mitigate)

triggering an avalanche with explosives or cannon fire when dangerous snow conditions exist in an area frequented by humans. The explosions collapse the depth hoar or detach powder snow, causing a "controlled" avalanche.

Various means have been developed for mitigating the destructive might of avalanches. Where structures already exist, fences can be built to divert avalanches away from them. In some areas in the Alps, the uphill sides of chalets are shaped similar to a ship's prow so that they will divert avalanching snow. Railroads and highways can be protected by building snowsheds along stretches adjacent to steep slopes.

Subsidence

Two million people live below the high-tide level in Tokyo. The canals of Venice overflow periodically and flood its beloved tourist attractions (✪ Case Study 7.3 on page 200). The Houston suburb of Baytown subsided almost three meters in the 1900s, and 80 square kilometers (31 mi^2) of this coastal region is permanently under water. Earthen dikes prevent the sea from flooding the land in some places in California, and Mexico City's famous Palace of Fine Arts rests in a large depression in the middle of the city (✳ Table 7.2). These areas suffer from natural and human-induced **subsidence**, a sinking or downward settling of the earth's surface.

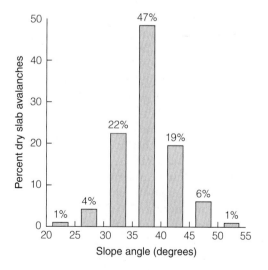

◆ FIGURE 7.24 Percentage of avalanches versus slope angle in degrees. Data from Knox Williams, *Avalanche Formation and Release*, *Rock Talk*, Colo. Geological Survey 3, no. 4, Oct. 2000

✳TABLE 7.2 Subsiding Cities, 1986

World		United States	
City	Maximum Subsidence, cm*	City	Maximum Subsidence, cm
Mexico City	850	Long Beach	900
Tokyo	450	(San Joaquin Valley)	880
Osaka	300	San Jose	390
Shanghai	263	Houston	270
Niigata, Japan	250	Las Vegas	66
Bangkok	100	Denver	30
Taipei, Taiwan	190	New Orleans	22
London	30	Savannah	20

*Since measurement began
Source: R. Dolan and H. Grant Goodell, "Sinking Cities," *American Scientist* 74 (1986), no. 1.

Subsidence is not usually catastrophic, and loss of life due to it is rare, but land subsidence is currently observed in 45 states and is estimated to cost $125 million annually. In the United States the area of human-induced subsidence is estimated as at least 44,000 square kilometers (17,000 mi^2), about the size of Maryland and New Jersey combined, but the actual area is probably even greater. Planners and decision makers need to be aware of the causes and impacts of subsidence in order to assess the risks and reduce material losses.

Human-Induced Subsidence

Human-induced subsidence occurs when humans extract underground water, engage in mining or oil and gas production, and when they cause loose sediments at the ground surface to consolidate or compress. The effects may be local or regional in scale. In the United States and Mexico alone, an area the size of Vermont has slowly subsided 30 centimeters (1 ft) due to withdrawal of underground water. Sinking of the land changes drainage paths, and it is particularly damaging to coastal areas and lands adjacent to rivers because it increases flood potential. Subsidence can also result from compaction of sediments rich in organic matter. Parts of the city of New Orleans, built upon cypress swamps of the delta of the Mississippi River, are below river level and sea level (see Chapter 9).

Annual losses from subsidence due to fluid withdrawal along the Texas Gulf Coast, for example, were estimated at $109 million between 1943 and 1973. Lowering of the water table in areas of cavernous limestone and gypsum has caused surface subsidence and collapse costing many millions of dollars each year (see the Chapter 9 discussion of hydrologic hazards). Each year subsidence of the land surface into abandoned coal mines costs the United States $30 million. About a quarter of the seven million acres that have been mined for coal has subsided, and some of this area underlies cities.

Natural Subsidence

Natural subsidence is caused by earthquakes, volcanic activity, and the solution of limestone, dolomite, and gypsum. Earthquake-related subsidence occurs rapidly. It is best known in Alaska and California but also occurs in a number of other states. Displacement along large faults can raise or lower the land surface over a large area. For instance, subsidence of one meter over 180,000 square kilometers (70,000 mi^2) took place during the Alaskan earthquake of 1964. Much of the subsidence was along the coast, and the subsided area is now flooded at high tide. In fact, the land actually tilted, and a large area of the Gulf of Alaska was uplifted several meters.

Severe ground shaking that leads to liquefaction may also lower the ground surface. This happened along the val-ley of the Mississippi River during the New Madrid earthquakes of 1811 and 1812 and in the Marina district of San Francisco in 1989. A very small-scale subsidence problem is the collapse of roofs of shallow lava tunnels and tubes in volcanic areas. Volcanic activity that empties magma chambers can cause collapse and subsidence over much larger areas when it forms calderas (see Chapter 5). Subsidence caused by tectonic (mountain-building) processes occurs so very slowly that it is not considered an environmental hazard.

To put this geologic hazard into perspective, most subsidence that adversely affects humans is caused by human activity and is irreversible. Once the land surface sinks by any significant amount, it cannot be "jacked" back up to its original level. As noted, most subsidence occurs so slowly that it poses little danger to people. There are some exceptions, however, such as sudden collapses into underground openings and subsidence-related dam failures. Subsidence that has well-identified connections with human activities is emphasized in this chapter.

A Classification of Subsidence

One method of classifying land subsidence is according to the depth at which the subsidence is initiated. *Deep subsidence* is initiated at considerable depth below the surface when water, oil, or gas is removed from there. *Shallow subsidence,* on the other hand, takes place nearer the ground surface when underground water or solid material is removed by natural processes or by humans. In addition, many poorly consolidated shallow deposits such as peat are subject to compaction by overburden pressures, ground-water withdrawal, and in some soils, simply by being saturated with water.

Settlement differs from subsidence. It occurs when an applied load—such as that of a structure—is greater than the bearing capacity of the soil onto which it is placed. Settlement is totally human-induced and it is a soils- and foundation-engineering problem. The Leaning Tower of Pisa is a classic example of foundation settlement; its interesting construction history is discussed in Chapter 6.

Deep Subsidence

Removal of fluids—water, oil, or gas—confined in the pore spaces in rock or sediment causes deep subsidence. Pore-water pressure—that is, the hydrostatic pressure of water in the pores between sediment grains—helps to support the overlying material. As pressure is reduced by extraction, the weight of the overburden gradually transfers to mineral-and-rock-grain boundaries. If the sediment was originally deposited with an open structure, the grains will reorient into a closer-packed arrangement, thus occupying less space, and subsidence will ensue (◆Figure 7.25). Because clays are

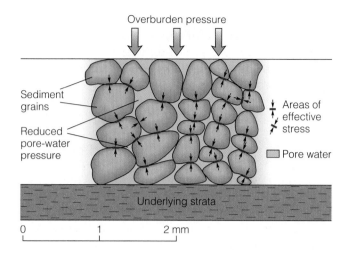

◆ FIGURE 7.25 A reduction of pore pressure causes increased effective stress and rearrangement and compaction of sediment particles.

more compressible than sands, most compaction takes place in clay strata.

At least 22 oil fields in California have subsidence problems, as do many fields in Texas, Louisiana, and other oil-producing states. A near world record for subsidence is held by the Wilmington Oil Field in Long Beach, California, where the ground has dropped 9 meters (about 30 ft). This field is the largest producer in the state, with more than 2,000 wells tapping oil in an upward-arched geologic structure called an *anti-cline* (see Chapter 14). The arch of the Wilmington field's anticline gradually sagged as oil and water were removed, causing the land surface to sink below sea level. Dikes were constructed to prevent the ocean from flooding the adjacent Port of Los Angeles facilities and the naval shipyard at Terminal Island. People who have their boats moored on Terminal Island must actually walk *up* to board them at sea level. When it rains it is necessary to pump water out of low spots upward to the sea, and oil-well casings and underground pipes have "risen" out of the ground as the land has sunk. To remedy the situation, in 1958 the City of Long Beach initiated a program of injecting water into the ground to replace the oil being withdrawn. Subsidence is no longer a problem, and the surface has even rebounded slightly in places.

Visible effects related to ground-water withdrawals are water-well damage, cracking of long structures such as canals, and development of tension cracks and fissures. In the Antelope Valley, California, home of Edwards Air Force Base, water levels beneath the dry lake have been lowered by more than 60 meters (200 ft). The dry lakebed at Edwards is the landing strip for the space shuttle and other high-performance aircraft. A ground fissure appeared in the lakebed that, initially only a few centimeters wide, became greatly enlarged by surface-water runoff and erosion. This halted operations on the lakebed until repairs could be made. Similar fissuring has occurred in Arizona, New Mexico, Texas, and Nevada. In Arizona ground-water withdrawal in some places has caused subsidence of 5 meters (more than 15 ft) and the development of cracks and earth fissures, particularly along the edges of

◆ FIGURE 7.26 Earth fissure due to subsidence caused by withdrawal of underground water near Picacho, Arizona.

T. L. HOLZER, USGS

some basins (◆Figure 7.26). The subsidence has changed natural drainage patterns and caused highways to settle and crack. It is influencing the siting of waste-treatment and solid-waste disposal sites and generally impacting all development activity in the state. The problem is so serious that the Arizona Department of Water Resources and the Arizona Geological Survey have established the Center for Land-Subsidence and Earth-Fissure Information, with the catchy acronym and "keyword" *CLASEFI*, to provide information and assistance to the public.

Coincidentally, the magnitude of the subsidence due to ground-water withdrawal is similar to oil-field subsidence. In the San Joaquin Valley of California, for example, 8.9 meters (29 ft) of subsidence resulted from 50 years' water extraction for agricultural irrigation (◆Figure 7.27).

Mexico City sits in a fault valley that has accumulated nearly 2,000 meters (6,600 ft) of sediments, mostly of pyroclastic origin. In 1925 it was demonstrated that subsidence was occurring here due to the withdrawal of water from these sediments. Total subsidence varies throughout the city, but almost 6 meters (20 ft) of subsidence had occurred by the 1970s in the northeast part of the city, mostly in the water-bearing top 50 meters (160 ft) of sediment. In 1951 an aqueduct was completed that brought water to the city and, as a result, a number of wells were closed. The effect on subsidence rates was dramatic, and by 1970 subsidence had decreased to less than 5 centimeters per year. The famous Our Lady of Guadalupe (the "brown Madonna") Cathedral tilts, because its foundation is partly on lava flows and partly on the subsiding lake sediments.

CONSIDER THIS *Subsidence of coastal cities such as Venice and Houston is particularly dangerous because it threatens them with reclamation by the sea. The subsidence problems in land-locked areas such as Mexico City and the San Joaquin Valley are a major geological concern. However, these areas do not face catastrophic inundation by the sea as do the major coastal cities of Tokyo, Venice, Houston, and New Orleans, to name but a few. What can be done to prevent or slow such potential catastrophes?*

Shallow Subsidence

Collapsing Soils

Subsidence that results from heavy application of irrigation water on certain loose, dry soils is called **hydrocompaction.** The in-place densities of such "collapsible" soils are low, usually less than 1.3 grams per cubic centimeter, about half the density of solid rock. Upon initial saturation with water, such as for crop irrigation or the construction of a water canal, the open fabric between the grains collapses and the soil compacts. Subsidence of monumental proportions has occurred because of hydrocompaction. In the United States, it is most common in arid or semiarid regions of the West and Midwest where soils are dry and moisture seldom penetrates below the root zone. Any sediment that was deposited rapidly and that has an open, granular structure is subject to compaction. These are most commonly debris flows or wind-blown (eolian) deposits. Known areas of hydrocompaction are the Heart Mountain and Riverton areas of Wyoming; Denver, Colorado; the Columbia Basin in Washington; southwest and central Utah; and the San Joaquin Valley of California.

Collapsing soils were a major problem for designers of the California Aqueduct in the San Joaquin Valley. In order to maintain a constant ground slope for the flow of water, it was necessary to identify those areas of loose soil that could not be avoided. Sediments were field-tested by saturating large plots and measuring subsidence at various levels

◆ FIGURE 7.27 Subsidence due to excessive ground-water withdrawal for agriculture in the San Joaquin Valley, California. The numbers on the pole indicate the ground level for the years indicated. The area is still subsiding, but at a much slower rate.

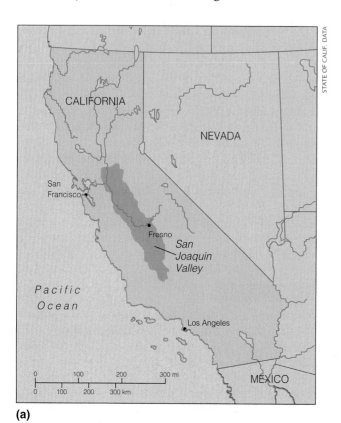

(a)

◆ FIGURE 7.28 (a) Location of the San Joaquin Valley. Collapsible soils occur almost entirely along the western border of the valley, whereas subsidence due to underground-water withdrawal is found throughout the valley. (b) Water-soaked test plot in the valley's collapsible soils. Total subsidence at this plot is on the order of 3 meters (10 ft).

(b)

beneath the surface. Subsidence was found to average 4.2 meters (close to 14 ft) over a period of 484 days, during which thousands of gallons of water had been applied to the test plots (◆Figure 7.28). On the basis of the field tests, 180 kilometers (112 mi) of the aqueduct alignment in the valley was "presubsided" by soaking the ground three to six months before construction. This assured that the gradient would be maintained after construction. Twenty-two million dollars was added to the cost of the project. Water transport systems, unlike highways or utility lines, are totally determined by the slope of the land beneath them. Sediments deposited on floodplains or on the deltas of large rivers, such as the Mississippi River delta, also are subject to compaction and subsidence. (See Chapter 9.)

> **CONSIDER THIS** *When a delta complex is drained for urban development, as in New Orleans, subsidence accelerates in the region. Which of these foundation types—(a) salty and sandy levee sediments, (b) backswamp clays, and (c) peat deposits—possesses the greatest and the least, subsidence potential?*

Mitigation

The best strategy for minimizing losses is to restrict human activities in areas that are most susceptible to subsidence and collapse. Unfortunately, most instances of land subsidence due to fluid withdrawals have not been anticipated up to

now. Knowledge of the causes, mechanics, and treatment of deep subsidence has increased with each damaging occurrence, so that we can now implement measures for reducing the incidence and impact of these losses.

Some of the measures now taken to minimize subsidence are controlling regional water-table levels by monitoring and limiting water extraction, replacing fluids by water-injection, and importing surface water for domestic use in order to avoid drawing on underground supplies.

Underground-mining operators are now required to leave pillars to support mined-out coal beds, and mine openings are being backfilled with compacted mine waste to support the overburden. This latter procedure, however, may introduce chemicals that could contaminate underground water unless the backfill is first stripped of pollutants. Pennsylvania's Mine Subsidence Insurance Act of 1962 allows some property owners to buy subsidence insurance after inspection of their property. Another Pennsylvania bill allows homeowners to buy the bituminous coal beneath their property at a fair price and therefore exercise control over mining operations.

Collapsible soils along canals are now routinely precollapsed by soaking, thereby anticipating in advance any adjustments of the canal gradient. Relatively inexpensive geophysical techniques are available for locating shallow limestone caverns beneath potential building sites. Subsidence problems will continue, but we should employ every means possible to foresee the problems and minimize their impact.

Dynamic Real Estate in Oregon

By volume, the Columbia River is the largest river flowing into the Pacific Ocean and the second-largest river in the United States. Its narrow gorge is a site of great scenic beauty, with towering basalt cliffs, beautiful river views, and boreal forests beyond description. No wonder people want to live there. There is a problem, however. On a stretch of the gorge between the towns of Warrendale and Dodson, Oregon, massive debris flows have occurred every 50 to 100 years since the gorge formed 13,000 years ago. The record-setting 1996 snowfall of 4 meters (12 ft) and then 32 centimeters (13 in) of rain on the 2,000–4,000-foot cliffs above the towns changed the idyllic environment to an agonizing one. About 3:00 A.M. on February 6, Dodson resident Mark Chandler went outside to call the family dog. He lived on Tumult Creek and could not help but notice it when the creek's rampaging waters suddenly stopped. He stared unbelievingly at the creek and then heard an ominous rumbling in the distance, like "two Union Pacific trains colliding" on the nearby tracks. Realizing that something was amiss, he dashed back to the house with the dog, awakened his family, and got them all out of town immediately.

A giant landslide had dammed Tumult Creek and was mobilized by the water to form a debris flow. With boulders the size of minivans, the flow broke off 20-meter (65-ft) pine trees, filled houses with mud, and blocked I-84 and the railroad. When it was all over, geologists calculated that there was enough debris to cover a football field with mud 215 meters (700 ft) thick (◆Figures 1 and 2).

In all, seven landslides struck the gorge between the two towns. County officials quickly placed restrictions on building along this stretch of the gorge and designated the towns as "hazardous areas." This means residents will have to hire a geological consultant for any building whatsoever. Property values have plummeted, and some owners think the government should buy them out. Others want to rebuild, but with government help. The last major slide in the area occurred in about 1920, and the lessons learned then apparently have been forgotten. Remarkably, no one was killed, only one person was injured, and a horse and a dog that were caught in the muck were rescued.

◆ FIGURE 1 This home was submerged to its second floor in the Tumult Creek debris flow. The beautiful high cliffs of basalt in the background were the source area of the landslide/rockfall and ensuing debris flow.

◆ FIGURE 2 The thickness and size of boulders in the flow are apparent in this photo. Again, note the precipitous basalt cliffs in the background.

Colorado's Slumgullion Landslide—
A Moving Story 300 Years Old

The Rocky Mountain states, and particularly Colorado, have some of the nation's highest levels of landslide hazard. Each year landslides in the Rockies take several lives and cause millions of dollars in damage to forests, roads, pipe and electrical-transmission lines, and buildings. What may be the largest active landslide in the United States is between Gunnison and Durango, Colorado, and bears the unusual name *Slumgullion*, a name that was probably bestowed upon it by early prospectors (◆Figure 1).

> **slumgullion** (slum-gul´yən), *n.* 1 *Slang.* a meat stew with vegetables, as potatoes and onion. 2. *Mining.* a muddy red residue in the sluice. *(Webster's New International Dictionary)*

Slumgullion, the stew, was a standard of mining-camp cooks, and slumgullion, the red residue, was what miners usually found instead of gold in the sluice box.

The Slumgullion landslide originated 700 years ago when highly weathered and altered Tertiary volcanic rocks gave way on a ridge above the Lake Fork tributary of the Gunnison River. It is a few kilometers upstream from Lake City, a historic mining town in the San Juan Mountains. The landslide dammed Lake Fork River, forming the largest natural lake in Colorado, Lake San Cristobal. The landslide is huge—6 kilometers (3.6 mi) long, 1 kilometer (0.6 mi) wide, and an average of 40 meters (132 ft) thick. As ◆Figure 2 shows, the Slumgullion is a landslide within a landslide; its 300-year-old active portion is flowing within the larger, 700-year-old mass. The total drop from headscarp to toe is about 762 meters (2,500 ft), and the slide moves at about 6 meters (20 ft) per year. The Slumgullion exhibits three distinct regions of deformation (◆Figure 3):

■ The top of the slide is in tension (extension). Normal faulting dominates this region.

■ The middle region flows as a rigid block, called *plug flow.* Nearly vertical strike-slip faults are found along plug's lateral edges.

■ The toe is a region of compression. It exhibits thrust (reverse) faults.

Within the active landslide the terrane is jumbled, and trees are either highly tilted or dead.

The landslide poses several geologic hazards. If it overruns Colorado Highway 149, it will isolate recreational development upstream. Downstream flooding will result if the lake level rises and overtops the natural dam. For these reasons, the U.S. Geological Survey is conducting multidisciplinary studies of the area. In addition to 30 years' worth of observation data, the USGS is evaluating:

■ precise measurements of the landslide's rate of movement using conventional survey techniques as well as Global Positioning System (GPS) technology.

ROBERT SCHUSTER, USGS

◆ FIGURE 1 The Slumgullion landslide in southwestern Colorado has a tremendous length and a sinewy path. The head scarp is more than 500 meters (1,600 ft) high.

old photographs of the area. These enable the researchers to estimate the slide's rate of headward growth. Because a slide's retrogression feeds rock debris and mass to the head of the slide, it serves to stimulate movement and growth of the slide.

■ detailed studies of Lake San Cristobal and its landslide dam. These will provide information that is applicable to other landslide dams in Colorado and the Rocky Mountains.

■ three-dimensional models of the active and inactive portions of the slide

based upon the accumulated data. These facilitate prediction of the Slumgullion's future behavior.

It is hoped that the results of this study will be applicable to large landslides in other states.

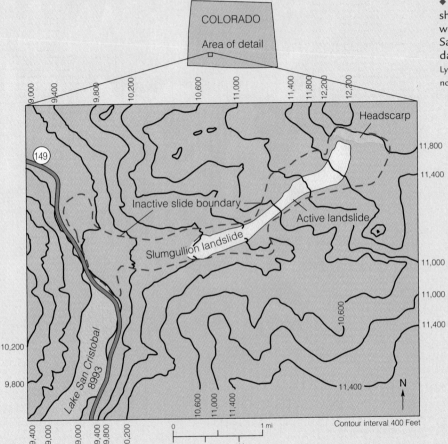

◆ FIGURE 2 Generalized contour map showing the 300-year-old active landslide within an inactive older slide mass. Lake San Cristobal formed when the landslide dammed the Lake Fork River. Redrawn from Lynn Highland, *Earthquakes and Volcanoes* 24, no. 3 (1993)

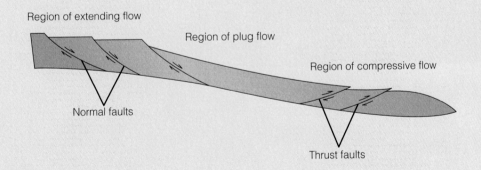

◆ FIGURE 3 Flow regions within the landslide. Extension (pull apart) at the top of the landslide leads to slumping there. The coherent plug flow in the middle of the landslide moves with lateral slip at the sides of the plug. Thrusting takes place at the toe of the landslide as the coherent plug attempts to override the shattered but stable toe. Redrawn from Lynn Highland, USGS

CASE STUDY 7.3

A Rescue Plan for the Merchants of Venice

Called "the city where streets are paved with water," Venice has special interest because of its historical importance, priceless antiquities, and beauty, as well as its charming canals. The city has a major flooding problem, however. St. Mark's Square and other squares are flooded during periods of extremely high tides the locals call the *aqua alta* (Italian, "high water"). It is said that the first Venetians settled here in A.D. 421 "to escape the barbarian Goths" that were sweeping down from the north. Sitting on a low mud bank in a lagoon sheltered by barrier islands (see Chapter 10), the city has subsided about three meters (10 ft) relative to sea level since it was founded. The sinking is due in part to the existence of 20,000 water wells in the region (now water is brought in by aqueduct), settlement into the soft sediments on which it was built, and rising sea level. The traditional building practice was to drive wooden piles into the soft sediment to form level foundations on which to build structures and, of course, timbers rot and settle. According to records, one cathedral is founded on more than a million timbers.

During periods of *aqua alta* the avenues and squares of Venice become flooded and gondola traffic comes to a screeching halt (◆Figure 1). The *aqua alta* occurs when an onshore flow of water (a storm surge) from the Adriatic Sea coincides with exceptionally high tides. There were an average of 30 high waters per year between 1957 and 1967, and the number had increased to 40–60 floods annually by 1990 (◆Figure 2). Dan Stanley of the Smithsonian Institution in Washington, D.C., likens Venice to a man standing in a swimming pool: His feet are embedded in lead, and the water level is rising. It's already to his nose. He'd better not make any waves.

The solution? Install movable floodgates at the entrances to the 90-mile-perimeter lagoon that can temporarily cut the lagoon off from the sea. The cost? Two billion dollars. The idea is to raise the floodgates like dams in anticipation of high water, and then to reopen them when the water level drops. There are environmental reasons as well as financial ones for not embarking on the project right away. Assuming a sea level rise of 5 cen-

◆ FIGURE 1 When the *aqua alta* floods St. Mark's Square in Venice, raised boardwalks are provided for pedestrians.

timeters over the next century, a conservative estimate, the gates would have to be raised 70 times a year, and that would interfere with boat traffic and tourism. According to British expert Edmund Penning-Rowsell, "The problem is that as sea level rises, they will have to be closed off so often that they will be quite ridiculous." The Italians love their antiquities and do a superior job of preserving them, but if the money is to be raised by taxing the city's merchants and citizens you may hear *"basta!"*

◆ FIGURE 2 Frequency of *aqua alta* flooding in Venice since 1910. The increase in frequency is due to subsidence, settlement, and rising sea level.
Adapted from P. Gatto and L. Carbognin, "The Lagoon of Venice," *Hydrological Sciences Bulletin* 26, no. 4 (1981)

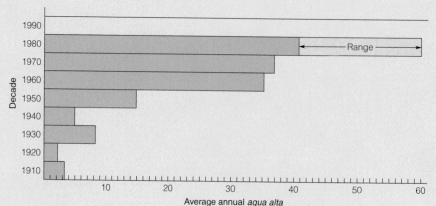

Slip, Slide, and Fall

Undeniably, the results of mass-wasting processes can be tragic and devastating, as explained throughout this chapter. In some cases, however, the results are whimsical. As examples, the subsidence at the Wilmington Oil Field in California left a fire hydrant high and dry (◆Figure 1), and a landslide occurred along Highway 89 that necessitated severing a telephone pole at its base. The pole now hangs from the wires it formerly supported and the base of the pole has moved some 70 feet downslope (◆Figure 2). A beautifully developed landslide on the planet Mars shows that geology also operates on this rocky planet (◆Figure 3).

◆ FIGURE 2 An active block glide in Montana. The person standing downslope is near the "stump" of the pole.

◆ FIGURE 1 Subsidence of the land but not the hydrant

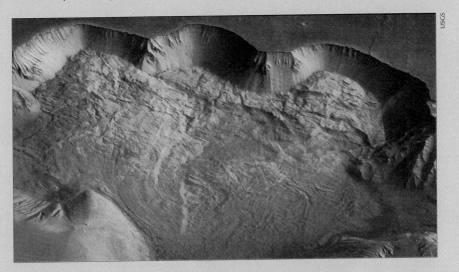

◆ FIGURE 3 My favorite Martian landslide. Oblique view from the south toward the Olympus Mons landslide complex, Mars. The arcuate landslide crown scarp in the background is about 25 kilometers (15 mi) wide and 4 kilometers (2.4 mi) high. The landslide debris flowed as far as 65 kilometers (39 mi) from the scarp toward the observer. Mass wasting is widespread on this rocky planet.

Summary

Mass Wasting

Defined

Downslope movement of rock and soil under the direct influence of gravity. The term *landslide* includes falls, flows, and slides that have well-defined boundaries.

Classification by

1. Type of material—rock or earth.
2. Type of movement—flow, slide, or fall.
3. Moisture content.
4. Velocity.

Falls

Defined

Relatively free fall of earth materials from a steep cliff or slope.

Types

1. Rockfall—involves hard bedrock of any size.
1. Debris fall—the collapse of weathered rock material and/or soil from a steep slope or cliff.

Flows

Defined

Mass movement of unconsolidated material in the plastic or semi-fluid state.

Types

1. Creep—slow, imperceptible downslope movement of rock and soil particles.
2. Debris flow—dense fluid mixture of rock, sand, mud, and water that is fast-moving and destructive.
3. Debris avalanche—fast-moving (75 km/h) debris flow.

Causes

Heavy rainfall on steep slopes with loose soil and sparse vegetation. Most common in arid, semiarid, and alpine climates and where dry-season forest fires expose bare soil.

Slides

Defined

Movement of rock or soil (landslide) as a unit on a discrete failure surface or slide plane.

Mechanics

Driving force (gravity) exceeds resisting forces (friction and force normal to the slide surface).

Types

1. Rotational slide (slump) on a curved, concave-upward slide surface.
2. Translational slide (block glide) on an inclined plane surface.
3. Complex landslide, a combination of (1) and (2).
4. Lateral spreads of water-saturated ground (liquefaction of sand and quick clays).

Causes

1. Weakening of the slope material by saturation with water, weathering of rock or soil minerals, and burrowing animals.
2. Steepening of the slope by artificial or natural undercutting at toe.
3. Added weight at the top of the slope such as by a massive earth fill.
4. A trigger such as an earthquake, vibrations from explosives, heavy traffic, or sonic booms.

Reduction of Losses

Geologic Mapping

To identify active and ancient landslide areas.

Building Codes

To limit the steepness of manufactured slopes and specify minimum soil and fill conditions and surface-water drainage from the site.

Control and Stabilization

Drain Insert Pipes

To drain subsurface water.

Excavation

To redistribute soils and rock from the head of the potential slide to the toe. This decreases driving forces and increases resisting force.

Buttress Fills and Retaining Devices

To retain active slides and potentially unstable slopes.

Rock Bolts

To stabilize slopes or jointed rocks.

Deflection Devices

To divert debris flows around existing structures.

Snow Avalanches

Types

Slab avalanche is old snow and ice that moves as a coherent mass. Powder snow avalanche is composed of new snow that moves like a fluid at high velocities.

Causes

Weak boundary layer between unstable layer above and stable mass below.

Mitigation

Controlled avalanching by shooting with light cannons, diversion structures, and snow sheds.

Subsidence

Defined

A sinking of the earth's surface caused by natural geologic processes or by human activity such as mining or the pumping of oil or water from underground.

Types

1. Deep subsidence caused by the removal of water, oil, or gas from depth.
2. Shallow subsidence of ground surface caused by underground mining or by hydrocompaction, the consolidation of rapidly deposited sediment by soaking with water.
3. Collapse of ground surface into shallow underground mines or into natural limestone caverns, forming sinkholes.
4. Settlement occurs when the load applied by a structure exceeds the bearing capacity of its geologic foundation.

Causes of Deep Subsidence

Removal of fluids confined in pore spaces at depth transfers the weight of the overburden to the grain boundaries, causing the grains to reorient into a closer-packed arrangement.

Causes of Shallow Subsidence and Collapse

The ground surface may collapse into shallow mines (usually coal). Sinkhole collapse is due to a lowering of the water table, which reduces the buoyant forces of underground water. Most sinkholes result from the collapse of soil and sediment cover into an underground limestone or dolomite cavern (see Chapter 9). Heavy irrigation causes some loose, low-density sediments and soils to consolidate (hydrocompaction). Compaction due to dewatering of peat and cypress-swamp (delta) clays and oxidation of the highly organic sediments causes subsidence.

Subsidence Problems

Ground cracking, deranged drainage, flooding of coastal areas, and damage to structures, pipelines, sewer systems, and canals. Whole buildings may disappear into sinkholes.

Mitigation

Underground water can be monitored and managed. Extracted crude oil can be replaced with other fluids. Mine operators are now required to leave pillars to support the mine roof, and rock spoil can be returned to the mine to provide support. Collapsible soils along canals can be precollapsed by soaking. Foundations on organic deltaic sediments should be on piles seated on firm geologic material. This protects structures from damage if the surrounding land subsides.

Key Terms

angle of repose	driving force	resisting force	slump
bentonite	effective stress	rotational landslide	subsidence
block glide	hydrocompaction	quick clay	syncline
complex landslide	landslide	sensitive clay	talus
creep	lateral spreading	settlement	translational landslide
debris avalanche	mass wasting	slab avalanche	
debris flow	powder-snow avalanche	slide plane	

Study Questions

1. What grading and land-use practices can reduce the likelihood of landslides?
2. How can vegetation, topography, and sediment characteristics aid in identifying old or inactive slumps, slides, and debris flows?
3. Name the processes that move earth materials downslope (gravity is the *cause*, not the *process*). On what factors does each depend?
4. Modern building codes require a Factor of Safety of 1.5 or greater in order to build on or at the top of a cut slope. What is the Factor of Safety, and why should it be 1.5?
5. What are some common, cost-effective methods of landslide control and prevention?
6. How did geologic structure factor into the disaster at Vaiont Dam?
7. What is the difference between settlement and subsidence? What can be done in advance to prevent settlement?
8. Distinguish between subsidence and collapse that are due to natural causes and due to human causes.
9. What can be done to slow or stop subsidence that is caused by withdrawal of water or petroleum?
10. What engineering problems make constructing canals and aqueducts in areas of hydrocompaction such an insidious geologic hazard?
11. What surface manifestations of subsidence due to groundwater withdrawal make it a serious problem, particularly in California and Arizona?
12. What hazard does subsidence present in coastal areas?

For Further Information

Books and Periodicals

California Department of Conservation. 1998. Hazards from mudslide, debris avalanches and debris flows in hillside and wildfire areas, California Div. of Mines and Geology, note 33, February.

Dolan, R., and H. Grant Goodell, 1986. Sinking cities. *American scientist* 74, no. 1:38–47.

Fellows, Larry D. 1995. Land-subsidence and earth fissure information. *Arizona geology* 25, no. 2.

Fleming, R. W., and T. A. Taylor. 1980. *Estimating the costs of landslides in the United States.* U.S. Geological Survey circular 832.

Gori, Paula L., Carolyn Driedger, and Sharon L. Randall. 1999. *Learning to live with geologic and hydrologic hazards.* U.S. Geological Survey Water Resources Investigations Report 99-4182.

Highland, Lynn. 2000. *Landslide Hazards.* U.S. Geological survey fact sheet FS-071-00, May.

Highland, Lynn M. 1993. Slumgullion: Colorado's natural landslide laboratory. *Earthquakes and volcanoes* 24, no. 5: 208.

Highland, Lynn, J. Godt, David Howell, and W. Z. Savage. 1998. *El Niño 1997–1998: Damaging landslides in the San Francisco Bay area.* U.S. Geological Survey fact sheet 089-98 (free).

Highland, Lynn, Ellen Stephenson, Sarah Christian, and William Brown III. 1997. *Debris flow hazards in the United States.* U.S. Geological survey fact sheet FS-176-97.

Holzer, T. L., ed. 1984. Man-induced land subsidence. *Geological Society of America reviews in engineering geology* VI.

Krohn, J. P., and James Slosson. 1976. Landslide potential in the United States. *California geology,* October.

Larsen, Matthew C., Gerald F. Wieczorek, and others. 2001. *Natural hazards on alluvial fans: The Venezuela debris flow and flash flood disaster.* U.S. Geological Survey fact sheet FS 103-01, October.

McPhee, John. 1989. Los Angeles against the mountains. *The control of nature,* ch. 3. New York: Farrar, Straus & Giroux.

Monastersky, Richard. 1999. Against the tide: Venice's long war with rising water. *Science news* 156, no. 4 (24 July): 63.

National Academy of Sciences. 1978. *Landslides: Analysis and control.* Transportation Research Board special report 176. Washington, D.C.: National Academy Press.

Nilsen, Tor H., and others. 1979. *Relative slope stability and land-use planning in the San Francisco Bay region, California.* U.S. Geological Survey professional paper 944. Washington, D.C.: U.S. Government Printing Office.

Pearson, Eugene. 1995. Environmental and engineering geology of central California: Salinian block to Sierra foothills. National Association of Geology Teachers, Far Western Section, Fall Conference, University of the Pacific, Stockton, California.

Reid, Mark E., R. G. LaHusen, and William Ellis. 1999. *Real time monitoring of active landslides.* U.S. Geological Survey fact sheet 091-99.

Saucier, Robert T., and Jesse O. Snowden. 1995. Engineering geology of the New Orleans area. *Geological Society of America annual meeting field trip guidebook,* nos. 6a and 6b.

Schultz, Arthur, and Randall W. Jibson (eds.). 1989. *Landslide processes of the eastern United States and Puerto Rico.* Geological Society of America special paper 236, 102 pp.

State of California, Resources Agency. 1965. *Landslides and subsidence.* Geologic Hazards Conference, Los Angeles.

Tremper, Bruce. 1993. Life and death in snow country. *Earth: The science of our planet,* no. 2.

U.S. Geological Survey. 1982. *Goals and tasks of the landslide part of a ground-failure hazards reduction program.* U.S. Geological Survey circular 880. Washington, D.C.: U.S. Government Printing Office.

Varnes, D. J., and W. Z. Savage (eds.). *The Slumgullion earth flow: A large-scale natural laboratory.* U.S. Geological Survey bulletin 2130.

_____. 1995. *Debris-flow hazards in the San Francisco Bay region.* U.S. Geological Survey fact sheet 112-95 (free).

Varnes, David. 1984. *Landslide hazard zonation: A review of principles and practice.* Paris: UNESCO.

Wieczorek, G. F., B. A. Morgan, and R. H. Campbell. 2000. Debris flow hazards in the Blue Ridge of Central Virginia. *Environmental and engineering geoscience,* February.

Williams, Knox. 2000. Avalanche formation and release. *Rock talk.* Colorado Geological Survey publication no. 4, October.

Wilshire, Howard, Keith A. Howard, C. M. Wentworth, and Helen Gibbons. 1996. *Geologic processes at the land surface.* U.S. Geological Survey bulletin 2149.

ENVIRONMENTAL
Geology⇌Now™

Assess your understanding of this chapter's topics with additional quizzing and comprehensive interactivities at http://earthscience.brookscole.com/pipkingeo4e

as well as current and up-to-date Web links, additional readings, and InfoTrac College Edition exercises.

Fresh-Water Resources
The High Plains Aquifer and Water Sharing

MARK TWAIN, A WRITER WHO COULD TWIST A PHRASE, is credited with saying "whiskey is for drinking, water is for fighting." There is much truth in the saying, as demonstrated in the conflict over water in the early settlement of the western United States. History may record that twenty-first century conflict in the Middle East was not over terrorists, territory, or even oil—but water. Most of the water in this region comes from three rivers: the Jordan, the Tigris-Euphrates, and the Nile. By 2025 the populations of water-scarce countries will be between 2.5 and 3.5 billion people, or about half the world's population in 2003. Syria, Iraq, Israel, Jordan, and Egypt will be no exception to the population increase and they will need more water for their people. You can bet that water-sharing will be high on the agenda of any future peaceful talks between the leaders of these countries.

The same might be true, without the threat of military action, in the water wars over the High Plains aquifer. The High Plains aquifer contains as much water as

Streams will burst forth in the desert, And rivers in the steppe. The burning sands will become a pool, And the thirsty ground, springs of water.

Isaiah 41:18

◆ FIGURE 1 *For decades, ground water from the High Plains aquifer has irrigated our nation's "bread basket." This irrigation system draws water from a well that taps the aquifer. The well's potential yield was estimated at 4,000 liters per minute (about 1,000 gallons per minute), 24 hours a day.*

Lake Huron and underlies 480,000 square kilometers (147,000 mi²) of parts of Kansas, Colorado, New Mexico, Wyoming, South Dakota, Nebraska, Oklahoma, and Texas. Water distributed by the irrigation system in the photo is from a well drilled into the aquifer's water-bearing strata that can produce 4,000 liters per minute (about 1,000 gals/min) 24 hours a day. When the amount of water withdrawn from an aquifer exceeds the amount replenished to it by rainfall and rivers, an overdraft condition exists, a shortfall, and as with shortfalls of all kinds, it will have to be reckoned with later if not sooner. This overdraft condition is called ground-water *mining*, because the resource is being exploited without replenishment, and water levels in the strata have been dropping dramatically (see Figures 8.17 and 8.18). Rather than fight, however, farmers in these states are supporting federal government plans for managing the resource and are withdrawing water from the aquifer conservatively.

If extraterrestrials exist and are able to view earth from afar, they cannot help but be struck by our planet's heavy cloud cover and its expanse of ocean, clear indications that earth is truly "the water planet." Water is indeed what distinguishes our planet from other bodies in our solar system. Three quarters of the earth's surface is covered by water, and even the human body is 65 percent water. Water provides humans with means of transportation, and much human recreation is in water—and *on* it where it exists in solid form as ice and snow. Since it is not equally distributed on the earth's surface, we have droughts, famine, and catastrophic floods. The absence or abundance of water results in such marvelous and diverse features as deserts, rain forests, picturesque canyons, and glaciers. In this chapter we investigate the reasons for this uneven distribution and its consequences. Water is involved in every process of human life; it is truly our most valuable resource.

Water as a Resource

Water is the only common substance that occurs as a solid, a liquid, and a gas over the temperature range found at the earth's surface. Little more than 97 percent of the earth's water is in the oceans, and 2 percent is in ice caps and glaciers; about 0.6 percent is readily available to humans as underground or surface fresh water (◆Figure 8.1). Fortunately, water is a renewable resource, and this is illustrated by the **hydrologic cycle,** the earth's most important natural cycle (◆Figure 8.2). Water that is evaporated from the ocean or land surface goes back into the atmosphere and forms clouds. Precipitation from these clouds that falls over land areas eventually flows in streams back to the sea, and the cycle is repeated. Remarkably, very little water is in the atmosphere at any one time, about 0.001 percent of the total. Because of this, we know that water must recycle continuously in order to supply the precipitation we measure. The average annual rainfall over the surface area of the 48 contiguous states is about 75 centimeters (30 in or 2.5 ft). **Evapotranspiration** is the term for both the direct return

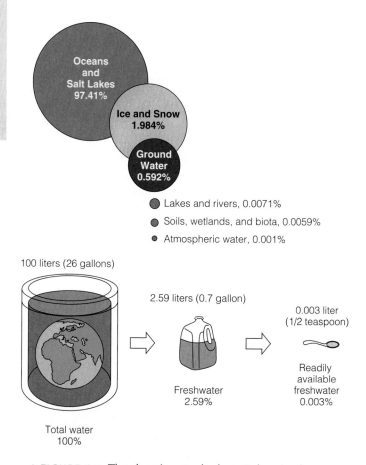

◆ FIGURE 8.1 The planet's water budget. Only a tiny fraction by volume of the world's water supply is fresh water available for human use.

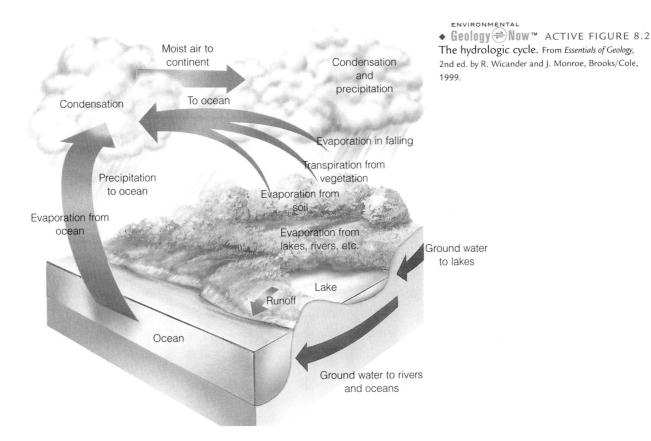

ENVIRONMENTAL
◆ **Geology ⇌ Now**™ ACTIVE FIGURE 8.2
The hydrologic cycle. From *Essentials of Geology*, 2nd ed. by R. Wicander and J. Monroe, Brooks/Cole, 1999.

of surface water to the atmosphere by evaporation and its indirect return through transpiration from the leaves of plants. It accounts for the return of about 55 centimeters (22 in) of the average annual precipitation. River runoff to the sea and infiltration into the ground account for the remaining 20 centimeters, but it varies from 2.5 centimeters per year in the dry West to more than 50 centimeters per year in the East. Even though runoff varies greatly with location, the net water transport from land to sea must over time balance the net transport from sea to land. The sources and users of public water supplies are shown in Figure 8.3. The data do not include water for agriculture or water that is self-supplied.

Fresh Water at the Earth's Surface

Rivers

Rivers provide abundant water to cities, some far from the river's location. Usually rivers, as water suppliers, have to be dammed and the waters metered out downstream to maintain the river and prior users. Planners recognize several categories of river-water use:

- *Instream use*—all uses that occur in the channel itself, such as hydroelectric power generation and navigation.

- *Offstream use*—the diversion of water from a stream to a place of use outside it, such as to your home. The amount of water available for offstream use along any stretch of a river is the total flow *minus* the amount required for instream purposes and for maintaining water quality. (◆Figure 8.4).

- *Consumptive use*—water that evaporates, transpires, or infiltrates and cannot be used again immediately. Forty-four percent of all water that is withdrawn is used consumptively, and agriculture accounts for about 90 percent of that.

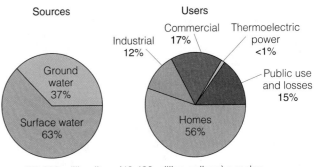

153,000 million liters (40,400 million gallons) per day

◆ FIGURE 8.3 The sources and users of the U.S. public water supply in 1995. This does not include fresh water supplied for agricultural use or water that is self-supplied.

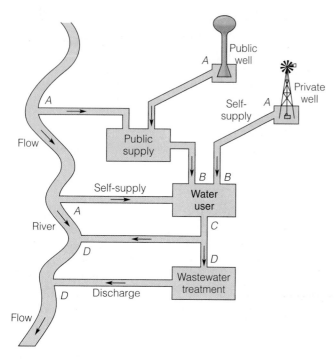

◆ FIGURE 8.4 Water is supplied to users from streams and from public and private wells (*A*). Consumptive use (*C*) is the difference between the amount delivered to users (*B*) and the amount returned to the system (*D*). USGS data

■ *Nonconsumptive use*—water that is returned to streams with or without treatment so that it can be used again downstream. Interestingly, domestic (household) water is used nonconsumptively, because it is returned to the cycle through sewers and storm drains.

In many areas, particularly the western United States, consumption exceeds local water supplies. This necessitates importing water from other areas or mining ground water. For example, about 75 percent of Southern California's water is imported from sources more than 200 miles away. Two aqueducts import water from west and east of the Sierra Nevada, and a third aqueduct brings water from the Colorado River. Irrigation utilizes about 40 percent of the total water withdrawn in the United States, and a startling 80 percent in California. It is not surprising then that California, with the largest population and a huge irrigation system, is the largest U.S. consumer of water, using almost as much as the two next-largest consumers, Texas and Illinois, combined (◆Figure 8.5).

Total world withdrawals of water for offstream purposes amount to about 2,400 cubic kilometers per year (1.7 trillion gal/day, *gpd*), of which 82 percent is for agricultural irrigation. This is expected to increase manyfold to 20,000 cubic kilometers (14.5 trillion gpd) by the year 2050, which is beyond the firm flow of the world's rivers. Availability of

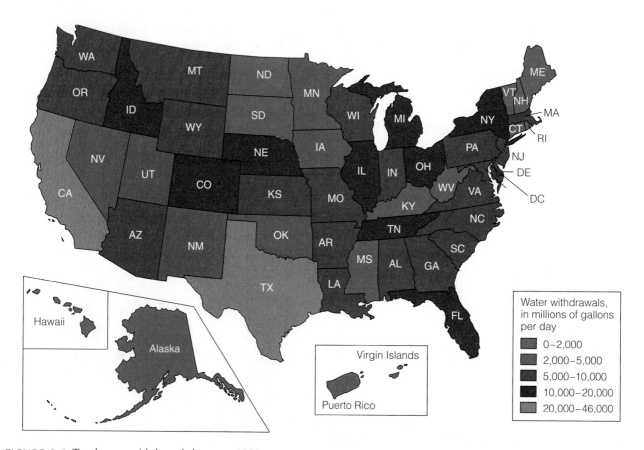

◆ FIGURE 8.5 Total water withdrawals by state, 1995. USGS

water will limit population growth in the twenty-first century, either by statute in developed countries or by natural means, most likely drought and famine, in developing nations. Repetition of the terrible droughts experienced in Africa in the 1970s and 1980s, compounded by overtaxed water supplies, would be disastrous for those countries and their people.

Lakes

Lakes are bodies of water in landlocked basins that are formed by many different geologic processes. The geologic origin of the Great Lakes (except Lake Superior) is glacial, Crater Lake is volcanic, and the Great Salt Lake is a down-faulted basin, to name just a few. Lakes are not important quantitatively in the global water balance, providing less than 0.4 percent of all continental fresh water. However, they are of local importance for agriculture, recreation, and as a source of water. It is significant that about 80 percent of all the water in lakes worldwide is held in less than 40 lakes.

Lakes are important as a source of fresh water, and salt lakes can be useful for transportation and mineral extraction. Whether a lake is fresh, salt, or even dry depends on the balance between precipitation (P) and runoff (R) into the lake, and evaporation (E) and outflow (O) from the lake. If a condition exists where $P + R = E + O$, the result is a fresh-water lake. If $P + R = E$ and there is no outflow, the result is a salt lake with a balance that is similar to the oceans.

However, in the southwestern United States, the condition exists where $P + R < E + O$, producing in time a dry lake bed, also called a *playa*. Dry lake beds, especially the salt flats of the Great Salt Lake, have been used for automobile speed trials because they are among the most level surfaces above sea level. Rogers Dry Lake in California has the distinction of being used as the landing strip for the space shuttle and is the home of Edwards Air Force Base. Playas are also known for a variety of salts and useful clays, products found in their sediments. Some lakes have their source of water cut off by humans, usually for agriculture, occasionally with catastrophic results (⊙ Case Study 8.1 on page 226).

A process that keeps the bottom waters in lakes from becoming stagnant is **overturn.** This occurs to lakes in temperate and cool climates where surface waters are subject to freezing temperatures. The temperature of maximum density of fresh water is 4°C; that is, at this temperature it is heavier per unit volume than water that is warmer or colder. In the summer the lake is stratified and the warmer surface water is separated from the cooler denser bottom waters by the *thermocline,* which acts as a floor for the surface water and a ceiling from the bottom waters. As it gets colder in the fall, the surface water lowers to 4°C and the termocline weakens; that is, the temperature of the lake becomes about equal or *isothermal.* The surface water sinks and the bottom waters that are oxygen-deficient rise to replace them. As ◆Figure 8.6 illustrates, the bottom waters stagnate during the winter but are refreshed again in the spring when the surface-water temperature rises to 4°C and sinks and the bottom waters rise.

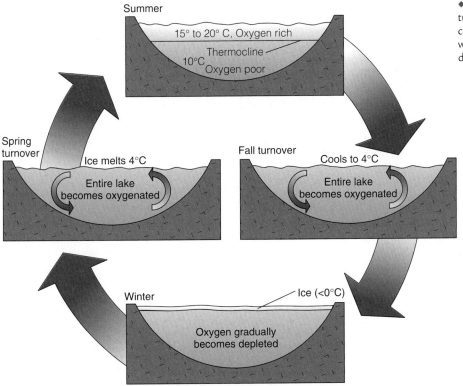

◆ FIGURE 8.6 Fall and spring overturn occurs in lakes in temperate and cold regions. Overturn is a mechanism whereby lake waters are refreshed and do not become stagnant.

The net result is oxygenation of the lake waters and recycling of nutrients, which are transported from the bottom to the surface. This makes for abundant plankton and good fishing.

Another factor in the well-being of lakes and their suitability as a source of water is how well "nourished" they are. Nourishment consists of inorganic compounds (nitrate, phosphate, and silica) that are carried into the lake by streams and produced by decaying organic matter at the bottom. An *oligotrophic* lake is one that is usually deep, has clear water, is low in nutrients, and has a relatively low fish population (◆Figure 8.7). They are a source of good drinking water and are usually visually appealing. Crater Lake is an example. A *eutrophic* lake is "well fed" with nutrients, usually shallow, many plants, a population of so-called trash fish like carp, low oxygen content, and usually lots of algae or other plants at the surface (◆Figure 8.8). In between the two extremes are the *mesotrophic* lakes that have well-balanced nutrient input and output. The result is a good fish population, a variety of shore and bottom plants, and an abundance of plankton (see Figure 8.7).

◆ FIGURE 8.8 A lake in western New York State. It is covered with a growth of slimy algae and bacteria and is not really suitable for swimming, fishing, or as a water supply. Many eutrophic lakes are the result of seepage from cesspools and other human-waste facilities around their perimeter. Eutophication caused by human activities is known as "cultural eutrophication."

Low nutrient input, low plankton and fish population, clear water.

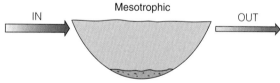

Good nutrient balance, good plant growth, plankton and fish population.

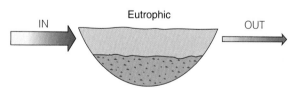

High nutrient input, many plants, low oxygen in bottom waters, low fish population, may have a surface mat of algae.

◆ FIGURE 8.7 Lakes go through a "nutrient" cycle that may be accelerated by waste products from humans living around the lake or by agricultural chemicals.

Fresh Water Underground

Some of the water in the hydrologic cycle infiltrates underground and becomes **ground water,** one of our most valuable natural resources. In fact, most of the earth's liquid fresh water exists beneath the land surface, and its geologic occurrence is the subject of many misconceptions. For example, it is commonly believed that ground water occurs in large lakes or pools beneath the land. The truth is that almost all ground water is in pore spaces and fractures in rocks. Another misconception is that deposits of ground water can be found by people with special skills or powers, people referred to as *water witches* or *dowsers*. They locate underground water using two wires bent into L shapes, forked branches, or other forked devices called *divining rods*. A dowser locates an underground "tube" or lake of water by walking over the ground, holding the divining rod in a prescribed manner until it jerks downward, pointing to the location of ground water, a mineral deposit, or a lost set of keys (◆Figure 8.9). Some dowsers claim they can locate good places to drill water wells using only a map of the land. The divining rod is passed over the map until it jerks downward where presumably a well should be located (X marks the spot). Water dowsers are found throughout the world, particularly in rural areas, and many of them will guarantee finding water. This is one difference between dowsers and ground-water geologists, or **hydrogeologists.** Hydrogeologists will not make this guarantee.

◆ FIGURE 8.9 After Abbé de Villemont, Traite de la Physique Occulte (undated engraving)
© Bettmann/CORBIS

Ground-Water Supply

Ground water provides 40 percent of public water supplies and 30 to 40 percent of all water used exclusive of power generation in the United States. It is by far the cheapest and most efficient source of municipal water, because obtaining it does not require the construction of expensive aqueducts and reservoirs. Thirty-four of the largest 100 cities in the United States depend entirely upon local ground-water supplies; Miami Beach, San Antonio, Memphis, Honolulu, and Tucson are just a few. Ground water provides 80 percent of the water for rural domestic and livestock use and it is the only source of water in many agricultural areas. It is not surprising that some of the largest agricultural states—such as Texas, Nebraska, Idaho, and California—are the largest consumers, accounting for almost half of all the ground water produced.

Ground water is important even to communities that import water from distant areas, because in almost every case, local ground water provides a significant low-cost percentage of their water supply.

Although it is not generally known, ground water maintains streamflows during periods when there is no rain. In arid climates where there are permeable water-saturated rocks beneath the surface, ground water seeping into stream channels may be the only source of water for rivers. The fabled oases of the Sahara and Arabian deserts occur where underground water is close to the surface or where it intersects the ground surface, forming a spring or watering hole.

Location and Distribution

Between the land surface and the depth at which we find ground water is the *zone of aeration,* a zone where voids in soil and rock contain only air and water films. The contact between this zone and the *zone of saturation* below it is the **water table** (◆Figure 8.10). Above the water table is a narrow

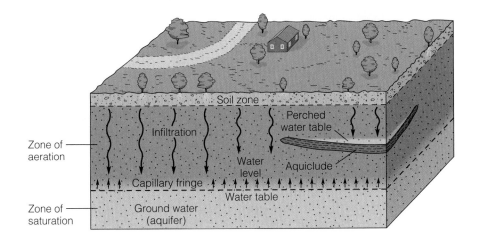

◆ FIGURE 8.10 Schematic cross section of ground-water zones showing a static water table with a capillary fringe. A perched water table exists where a body of clay or shale acts as a barrier to further infiltration.

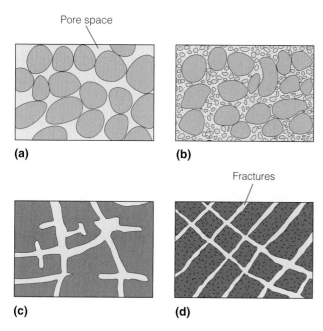

Pore space

(a)

(b)

Fractures

(c)

(d)

◆ FIGURE 8.11 A rock's porosity is dependent upon the size, shape, and arrangement of the material composing the rock. Whereas (a) a well-sorted sedimentary rock has high porosity, (b) a poorly sorted one has low porosity. (c) In soluble rocks such as limestones, porosity can be increased by solution, whereas (d) crystalline metamorphic and igneous rocks are porous only if they are fractured.

moist zone, the *capillary fringe.* In the saturated zone, free (unattached) water fills the openings between grains of clastic sedimentary rock, the fractures or cracks in hard rocks, and the solution channels in limestones and dolomites (◆Figure 8.11).

Water-saturated geologic formations whose porosity and permeability are sufficient to yield significant quantities of water to springs and wells are known as **aquifers.** A rock's **porosity** (root word *pore*) is a function of the volume of openings, or void spaces, in it and is expressed as a percentage of the total volume of the rock being considered. Without such voids, a rock cannot contain water or any other fluid. **Permeability** is the ease with which fluids can flow through an aquifer; it is thus a measure of the connections between pore spaces. **Hydraulic conductivity** is the term hydrologists prefer, but *permeability* will suffice for our purposes, as it has a long history of use and is easy to remember (see ⊙ Case Study 8.2 on page 228). **Aquicludes** are rocks or sediments such as shale and clay that lack permeability and hence will not transmit water. In some places aquicludes occur as *lenses* that trap water and prevent it from percolating down to the water table. Such a condition creates a **perched water table** (see Figure 8.10).

Aquifers are important because they transmit water from *recharge areas*—that is, areas where water is added to the zone of saturation—to springs or wells. In addition, they store vast quantities of drinkable water. In terms of the hydrologic cycle, we can view aquifers both as transmission pipes and as storage tanks for ground water. The Nubian Sandstone in the Sahara Desert, for instance, is estimated to contain 600,000 cubic kilometers (144,000 mi^3) of water, mostly untapped. It is the world's largest known aquifer and is a resource of great potential for North Africa.

Static Water Table

Unconfined aquifers are formations that are exposed to atmospheric pressure changes and that can provide water to wells by draining adjacent saturated rock or soil (◆Figure 8.12). As the water is pumped, a **cone of depression** forms around the well, creating a gradient that causes water to flow toward the well (◆Figure 8.13). A low-permeability aquifer will produce a steep cone of depression and substantial lowering of the water table in the well. The opposite is true for an aquifer in highly permeable rock or soil. This lowering, or the difference between the water-table level and the water level in the pumping well, is known as **drawdown.** Stated differently, low permeability produces a large drawdown for a

◆ FIGURE 8.12 A horizontal (static) water table. The water table will remain static until pumping begins.

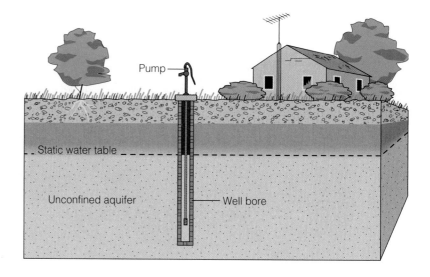

Pump

Static water table

Unconfined aquifer

Well bore

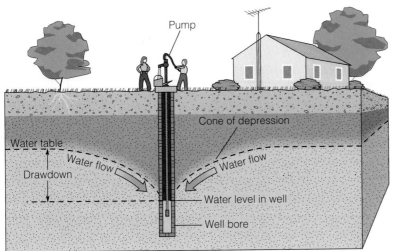

Pump

Cone of depression

Water table

Water flow   Water flow

Drawdown

Water level in well

Well bore

ENVIRONMENTAL
♦ **Geology⇌Now**™ ACTIVE FIGURE 8.13
When pumping begins, the water table is drawn down an amount determined by the pumping rate and the aquifer's permeability. The development of a cone of depression creates a hydraulic gradient, which causes water to flow toward the well and enables continuous water production. The amount of water the well yields per unit of time pumped depends upon the aquifer's hydrologic properties.

given yield and increases the possibility of a well "running dry." A good domestic (single-family) well should yield at least 11 liters (2.6 gal) per minute, which is equivalent to about 16,000 liters (3,700 gal) per day, although some families use as little as 4 liters per minute (1,350 gal/day). As pointed out, a pumping well in permeable materials creates a small drawdown and a gentle gradient in the cone of depression. Some closely spaced wells, however, have overlapping "cones," and excessive pumping in one well lowers the water level in an adjacent one (♦Figure 8.14). Many lawsuits have arisen for just this reason. Water law is discussed briefly in ⊛ Case Study 8.3 on page 229.

The water table rises and falls with consumption and the season of the year. Generally, the water table is lower in late summer and higher during the winter or the wet season. Thus water wells must be drilled deep enough to accommodate seasonal and longer-term, climatic fluctuations of the water table (♦Figure 8.15). If a well "goes dry" during the annual dry season, it definitely needs to be deepened.

Streams that are located above the local water table are called **losing streams**, because they contribute to the underground water supply (♦Figure 8.16a). Beneath such streams there may be a mound of water above the local ground-water table known as a *recharge mound*. A stream that intersects the water table and is fed by both surface water and ground water is known as a **gaining stream** (♦Figure 8.16b). A stream may be both a gaining stream and a losing stream, depending upon the time of year and variations in the elevation of the water table.

Where water tables are high, they may intersect the ground surface and produce a spring. Springs also are subject to water-table fluctuations. They may "turn on and off" depending upon rainfall and the season of the year. ♦Figure 8.17 illustrates the hydrology of three kinds of springs.

> **CONSIDER THIS** *As explained in Case Study 8.3 the legal issues surrounding the withdrawal and use of ground water are complex, and several legal doctrines apply to them. From a philosophical viewpoint, which of the four doctrines seems fairest for multiple users of an aquifer? Please explain your choice.*

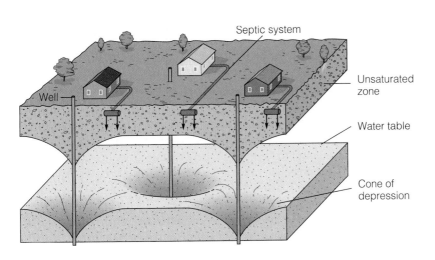

Septic system

Unsaturated zone

Well

Water table

Cone of depression

♦ FIGURE 8.14 Cones of depression overlap in an area of closely spaced wells. Note the contributions to ground water from septic tanks.

◆ FIGURE 8.15 The effect of seasonal water-table fluctuations on producing wells. A shallow well may go dry in the summertime.

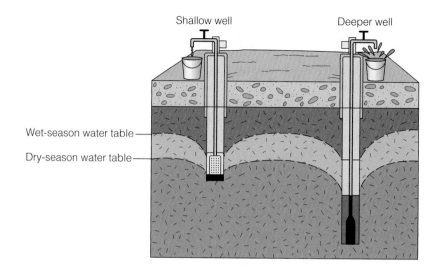

Wet-season water table

Dry-season water table

Pressurized Underground Water

Pressurized ground-water systems cause water to rise above aquifer levels and sometimes even to flow to the ground surface. Pressurized systems occur where water-saturated permeable layers are enclosed between aquicludes, and for this reason, they are called **confined aquifers** (◆Figure 8.18). A confined aquifer acts as a water conduit, very similar to a garden hose that is filled with water and situated such that one end is lower than the other. In both cases a water-pressure

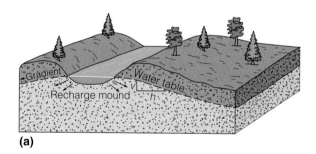

(a)

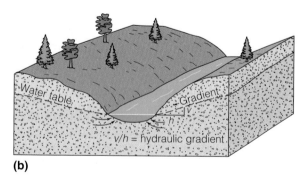

(b)

◆ FIGURE 8.16 Hydraulic gradient (v/h) in (a) a losing stream that contributes water to the water table, and (b) a gaining stream that gains water from the water table.

difference known as a *hydraulic head* is created. Water gushes from the low end until there is no longer a pressure difference. At that point the water level stabilizes, becomes *static*.

Wells that penetrate confined aquifers are called *artesian wells*, after the province of Artois, France, where they were first described. Artesian systems are distinguished by a water-pressure gradient called the **artesian-pressure surface** (◆Figure 8.19). This "surface" is the level to which water will rise in a well at a given point along the aquifer. It is important to note that the slope of the water table of an unconfined aquifer also is a pressure surface. Underground water moves from areas of high hydraulic head to areas of low hydraulic head (pressure). Contour maps of the pressure surface can be very useful, because they indicate the direction of ground-water flow and the slope of the hydraulic gradient (◆Figure 8.20). If an aquifer's hydraulic conductivity can be estimated or measured, the rate and volume of ground-water flow can be determined using Darcy's Law (see Case Study 8.2 on page 228).

Artesian wells differ from water-table wells in that the latter depend upon pumping or topographic differences to create the hydraulic gradient that causes the water to flow. Because the energy required to raise water to the surface is the greatest expense in tapping underground water, artesian systems are much desired and exploited (see ❂ Case Study 8.4 on page 230). Not all artesian wells flow to the ground surface, because the pressure surface lowers as water is extracted; water may rise naturally only part way up the well bore. Nevertheless, a natural rise of water any distance reduces the cost of extracting ground water.

Finally, many bottled water and beverage producers attempt to lead us to believe that deep-well or artesian water is purer than other kinds of water. *Artesian* refers only to an aquifer's hydrogeology and has no water-quality connotation. Do not be misled by advertisements attributing greater purity to artesian or deep-well water. Such waters may or may not have low levels of dissolved contaminants.

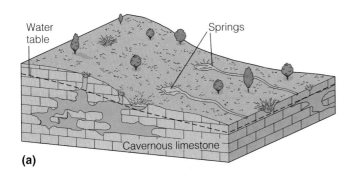

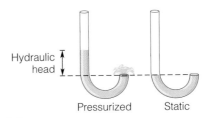

(a)

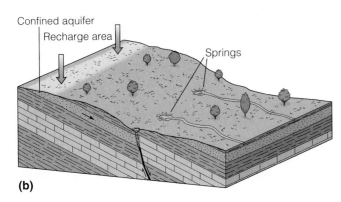

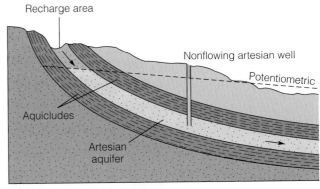

(b)

ENVIRONMENTAL
◆ Geology⇌Now™ ACTIVE FIGURE 8.18 Water under pressure. (a) The hydraulic head in a hose lowers as water flows out the low end. Eventually a static condition prevails. (b) A confined aquifer, commonly called an *artesian* aquifer. The artesian-pressure surface is the level to which water will rise in wells at specific points above the aquifer.

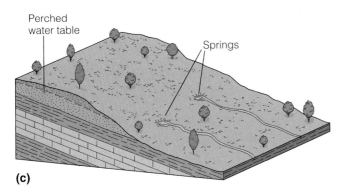

(c)

◆ FIGURE 8.17 Geology and hydrology of three kinds of springs. (a) Springs form where a high water table in cavernous limestone intersects the ground surface. (b) Artesian springs. A fault barrier can cause pressurized water to rise as springs along irregularities in the fault. (c) Springs form where a perched water table intersects the ground surface. Such springs are intermittent, because the small volume of perched water is quickly depleted.

CONSIDER THIS *Some hot springs form geysers that are celebrated for their regularity. Old Faithful in Yellowstone National Park is an example. Cold springs, on the other hand, do not generate attention-grabbing geysers, and their flow is often far from regular. Indeed, many springs operate on a sort of stop-and-go basis. How can we explain this intermittent flow?*

Ground-Water Storage and Management

In order to manage underground water supplies in a ground-water basin intelligently, the hydrologist or planner must have two measurements:

- the quantity of water stored in the basin, and
- the **sustained yield,** the amount of water the aquifers in the basin can yield on a day-to-day basis over a long period of time.

A ground-water basin consists of an aquifer or a number of aquifers that have well-defined geological boundaries and *recharge areas*—places where water seeps into the aquifer (see Figure 8.18b). To determine the amount of usable water in the basin, we rely on the concept of specific yield of water-bearing materials. **Specific yield** is the ratio of the volume of water an aquifer will give up by gravity flow to the total volume of material, expressed as a percentage. For example, a saturated clay with a porosity of 25 percent may have a specific yield of 3 percent; that is, only 3 percent of the saturated porosity will drain into a well. The remaining water in the pore spaces (22 percent) is held by the clay-mineral surfaces

◆ FIGURE 8.19 An artesian well in the Owens Valley, near Bishop, California. Note that artesian pressure raises the water above the ground surface and that the well flow is not regulated in any manner. This is wasteful, as the good well water just flows as a small stream into the Owens River.

and between clay layers against the pull of gravity. This "held" water is its *specific retention*. Similarly, sand with a porosity of 35 percent might have a specific yield of 23 percent and a specific retention of 12 percent. Note that the sum of specific yield and specific retention equals the porosity of the aquifer material.

Thus the total water-storage capacity of an aquifer may be calculated by multiplying the specific yield by the volume of the aquifer determined from water-well data. The process is a bit more complicated for artesian aquifers, but it follows the same procedure. The amount of water available to us in a ground-water basin can be determined fairly accurately if the basic data are available from numerous wells.

As noted, sustained yield is the amount of water that can be withdrawn on a long-term basis without depleting the resource. Sustained yield is more difficult to assess than aquifer storage capacity, because it is affected by precipitation, runoff, and recharge. Nevertheless, it must be at least estimated for intelligent management of the resource.

Ground-Water Mining

As explained in the chapter opening, when the amount of water withdrawn from an aquifer exceeds the aquifer's sustained yield, an overdraft condition called *ground-water mining* exists (◆Figure 8.21). The Ogallala Sandstone makes up about 80 percent of what is collectively known as the High Plains aquifer. The Ogallala's average thickness is 60 meters (about 200 ft), but thicknesses as great as 425 meters (1,400

ft) have been measured. In general, water levels in this important aquifer are dropping. In Texas, overdrafts have exceeded 65 percent, and overdrafts of 95 percent are known in a few places. In the latter case, 20 times more water was extracted from the aquifer than was recharged. This magnitude of overdraft is a serious problem because much of the water in the aquifer was acquired during wetter glacial climates thousands of years ago.

The High Plains aquifer was not massively tapped until the 1950s, and by 1980 the water-saturated zone had decreased an average of 3 meters, and as much as 30 meters under some parts of Texas and Nebraska (◆Figure 8.22). Government subsidies have encouraged growing water-gulping corn, rather than less-water-intensive wheat, sorghum, and cotton.

In the early 1980s water decline rates decreased because of heavy rain and snow, better water management, and new technologies. New *LEPA* (low-energy, precision-application) irrigation nozzles use less water and less energy and decrease evaporation by as much as 98 percent from the amounts lost with spray-type irrigation. Furthermore, treated wastewater is now being recycled on fields, and rainfall has been induced by cloud seeding in attempts to recharge the aquifer in Kansas.

A 1982 Department of Commerce study found that 6 million hectares (15 million acres) were being irrigated with water drawn from 150,000 wells in the Ogallala and other High Plains aquifers. By the year 2020, according to the study report, about a fourth of the aquifer's water will have been mined at the present rate of withdrawal. A 1995 estimate of

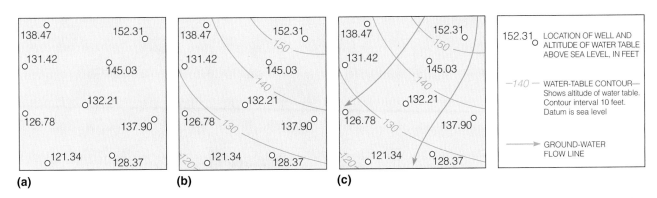

◆ FIGURE 8.20 Using known altitudes of the water table at individual wells (A), contour maps of the water-table surface can be drawn (B), and directions of ground-water flow along the water table can be determined (C) because flow usually is approximately perpendicular to the contours. Elevations are in feet above a datum, usually sea level, as English units are used almost exclusively on pressure-surface contour maps in the United States. USGS circular 1139

the total overdraft of the Ogallala is that it equals one year's flow of the Colorado River. The Ogallala aquifer is certainly the life-blood of High Plains agriculture, and most users of its water are aware that "mining" the resource must be discontinued. Strict management that includes conservative usage, monitoring, and utilization of new technologies is required in attempting to balance its recharge and withdrawals (◆Figure 8.23). As a postscript to the Ogallala story, the aquifer's name is derived from that of the Ogallala Tribe of the Sioux Nation. Led by Chiefs Red Cloud and Crazy Horse, the tribe fought to retain tribal lands in the late 1800s.

Ground-water mining causes shallow wells to go dry, increases the cost of lifting water to the surface as water levels drop, and may eventually cause ground subsidence (see Chapter 7). Currently ground-water pumping is twice the volume of natural recharge in Arizona's Tucson ground-water basin, and it has been necessary to import water from the Central Arizona Project to end the overdraft. One product of nuclear testing in the 1950s and early 1960s was radioactive tritium. This isotope of hydrogen has been used as a tracer of ground water recharged into the Tucson basin on alluvial fans of the adjacent Catalina and Rincon Mountains. The study indicates that a significant amount of the basin's ground water is very young and has infiltrated since the atom-bomb tests. This does not bode well for bringing the water table back to predevelopment levels in the basin.

The Santa Clara Valley of central California is a classic case of regional overdraft. Water levels there declined 46 meters (150 ft) between 1912 and 1959, and downtown San Jose dropped 2.76 meters (9 ft). As a final example, overpumping in Alabama dropped the pressure surface in one aquifer 62 meters (200 ft); it is now below sea level more than 80 miles inland from the sea. The solution to such a problem is *artificial recharge,* or "water spreading," which is accomplished by importing or diverting surface water and ponding it where it can percolate into the aquifer. Such a program was initiated in the Santa Clara Valley in 1959, which, in combination with decreased pumping, has restored the water table to 1912 levels. Raising the water table does not bring the land back to its original level, however, because pore spaces are lost as subsidence occurs.

Ground Water–Saltwater Interaction

Aquifers in coastal areas may discharge fresh water into the ocean, creating bodies of diluted seawater. If the water table is lowered by pumping to or near sea-level elevation, however, saltwater invades the freshwater body, causing the water to become saline *(brackish)* and thus undrinkable. The direction of flow between the two water masses is determined by their density differences. Seawater is 2.5 percent (1/40th) heavier than fresh water. This means that a 102.5-foot-high column of fresh water will exactly balance a 100-foot-high

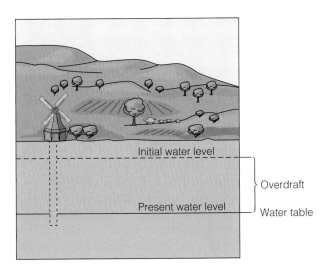

◆ FIGURE 8.21 The effect of ground-water mining on the water table. The drop from the initial level to the present level is the *overdraft,* the amount of water that has been mined.

◆ FIGURE 8.22 Water-level change in the High Plains aquifer, 1980–1995.

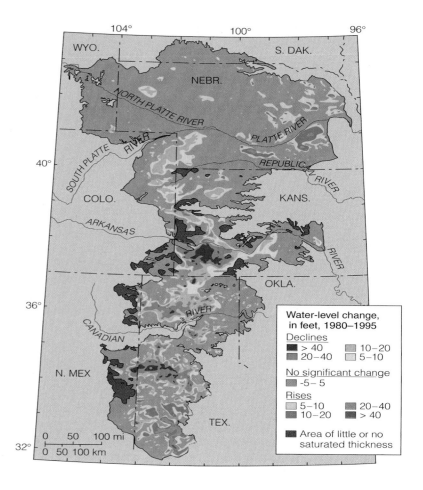

column of seawater (◆Figure 8.24a). Fresh water floats on seawater in the Hawaiian Islands, the Outer Banks of North Carolina, Long Island, and many other coastal areas (◦ Case Study 8.3 on page 229).

This relationship is known as the *Ghyben–Herzberg lens*—named for the two scientists who independently discovered it and for the lenslike shape of the freshwater body (◆Figure 8.24b). In theory, the mass of fresh water extends to a depth below sea level that is 40 times the water-table elevation above sea level. If the water table is 2.5 feet above sea level, the freshwater lens would extend to a depth of 100 feet below sea level. Any reduction of water-table elevation would cause the saltwater to migrate upward into the freshwater lens until a new balance is established. If the water table drops below the level necessary for maintaining a balance with denser seawater, wells will begin to produce brackish water and eventually saltwater. Thus it is important to maintain high water-table levels in coastal zones so that the resource is not damaged by *saltwater encroachment*. This level maintenance is accomplished by good management combined, when necessary, with artificial recharge using local or imported water (water spreading), or with injection of imported water into the aquifer through existing wells.

Ground Water–Surface Water Interaction and Usage Trends

As has been emphasized thus far, surface water and ground water are not separate entities. Nearly all surface-water features—lakes, streams, rivers, and artificial lakes, and wetlands—interact with water underground. Some surface-water bodies gain water and acquire a different chemistry from ground water, and other surface waters contribute to underground water and may even pollute it. Thus, it is incumbent in water planning to have a clear understanding of the linkages between ground water and surface water and of the geology of both.

The greatest volume of fresh water is used for irrigation, and the greatest total volume of water, both fresh and saline, is used for power generation. However, 98 percent of the water used for power generation is returned to streams or other reservoirs; only 2 percent is consumed. How water is used affects the reuse potential of returns flows. For example, irrigation water may be too salty or contaminated by pesticides and insecticides to have any reuse potential unless it is purified naturally or artificially. On the other hand, water used in power generation has great reuse potential, because the principal change is usually just an increase in its temperature.

Since 1950 national water-use statistics have been compiled every five years (see W. B. Solley, For Further Information). There was a steady increase in water usage from 1950 to 1980, the all-time high, as population grew at a steady pace. Between 1980 and 1995, there was a 10 percent decrease in water use, even though the U.S. population grew by 37 million people (◆ Figure 8.25). It was estimated that the United States was saving 144 billion liters (38 billion gal) per day in 1995, enough to fill Lake Erie in a decade. Some of the reasons cited for this decrease were the development and use of more efficient irrigation methods, conservation programs, increased water recycling, and a downturn in farm economies.

Geologic Work by Underground Water

Solution (a chemical reaction discussed in Chapter 6) of limestone and gypsum by underground water leads to formation of caves, caverns, disappearing streams, and prolific ground-water aquifers such as the Edwards aquifer (Case Study 8.2 on page 228). Land area exhibiting these features is called **karst terrane** (German *Karst,* for the Kars limestone plateau of northwest Yugoslavia; ◆ Figure 8.26). Urbanization on karst terrane typically results in problems such as increased flooding, contamination of underground water, and collapse of cavern roofs to form sinkholes.

The first sign of impending disaster in Winter Park, Florida, occurred one evening when Rosa Mae Owens heard a "swish" in her backyard and saw, to her amazement, that a large sycamore tree had simply disappeared. The hole into which it had fallen gradually enlarged, and by noon the next day her home had also fallen into the breach. Six Porsches in a nearby storage yard suffered the same fate, as did a city swimming pool and two residential streets. Deciding to work

◆ **FIGURE 8.23** Should the High Plains aquifer be treated as a bathtub or an egg carton? The egg carton analogy assumes that a user is not subject to water depletion by the action of neighboring irrigators because of the local nature of the cone of depression. The bathtub analogy assumes that the aquifer acts as a common pool and the water level responds as if water were being drawn from a lake or bathtub. The actual situation lies somewhere between these two analogies. USGS circular 1186, p. 47.

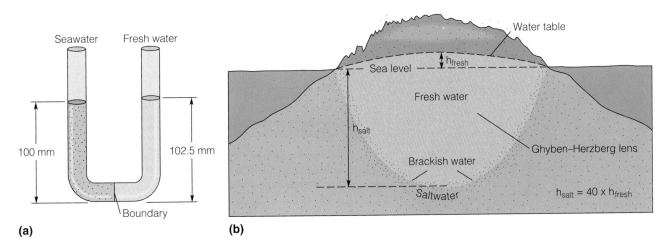

◆ **FIGURE 8.24** (a) Illustration of the density balance between fresh water and seawater. Because seawater is 2.5% heavier than fresh water, 102.5 millimeters of fresh water is required to balance 100 millimeters of salt water. (b) Hypothetical island and Ghyben–Herzberg lens of fresh water floating on seawater. The freshwater lens extends to a depth 40 times the height of the water table above sea level. Not to scale.

◆ FIGURE 8.25 Trends in fresh ground- and surface–water withdrawals and population, 1950–1995. USGS

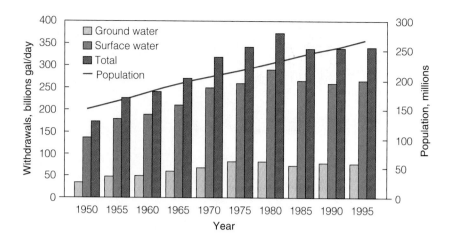

Tens of thousands of sinkholes exist in the United States, and the formation of new ones can be accelerated by human activities. For example, excessive water withdrawal that lowers ground-water levels reduces or removes the buoyant support of shallow caverns' roofs. In fact, the correlation between water-table lowering and sinkhole formation is so strong that "sink-hole seasons" are designated in the Southeast. These "seasons" are declared whenever ground-water levels drop naturally because of decreased rainfall and in the summer, when the demand for water is heavy. Vibrations from construction activity or explosive blasting also can trigger sinkhole collapse. Other mechanisms that have been proposed for sinkhole formation are fluctuating water tables (alternate wetting and drying of cover material reduces its strength) and high water tables (which erode the roofs of underground openings).

with nature, the City of Winter Park stabilized and sealed the sinkhole, and converted it to a beautiful urban lake (◆Figure 8.27). Few people who read of this notorious sinkhole in May 1981 realized the extent of the sinkhole problem in the southeastern United States.

Sinkholes are most commonly found in carbonate ter-ranes (limestone, dolomite, and marble), but they are also known to occur in rock salt and gypsum. They form when carbonic acid (H_2CO_3) dissolves carbonate rock, forming a near-surface cavern. Continued solution leads to collapse of the cavern's roof and the formation of a sinkhole. About 20 percent of the United States and 40 percent of the country east of the Mississippi River is underlain by limestone or dolomite. The states most impacted by surface collapse are Alabama, Florida, Georgia, Tennessee, Missouri, and Pennsylvania.

Although the development of natural sinkholes is not predictable, it is possible to assess a site's potential for collapse resulting from human activities. This requires extensive geo-logic and hydrogeologic investigation prior to site development. Under the best of geologic conditions, good geophysical and subsurface (drilling) data can tell us where collapse is most likely to occur, but they cannot tell us exactly when a collapse may occur. Armed with this information, planners and civil engineers can implement zoning and building restrictions that will minimize the hazards to the public.

CONSIDER THIS *You could become a millionaire if you developed a relatively inexpensive method of locating caverns that may collapse and form sinkholes. Got any ideas? If not, what types of more sophisticated (and expensive) methods might be employed to locate caverns?*

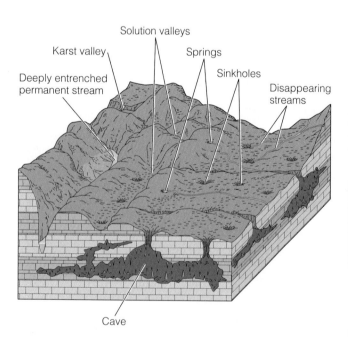

◆ FIGURE 8.26 Surface and subsurface features of karst topography.

ENVIRONMENTAL
Geology ⇌ Now™

Click **Geology Interactive** to work through activities on Tapping the Ground and Potentiometric Surface through Ground Water.

(a)

(b)

◆ FIGURE 8.27 Sinkhole in Winter Park, Florida. (a) Air view that shows its almost perfect symmetry. (b) Plugged and landscaped, the sinkhole area is now an urban park.

Water Quality

Dissolved Substances

Dissolved in natural waters are varying concentrations of all the stable elements known on earth. Some of these elements and compounds, such as arsenic, are poisonous. Others are required for sustaining life and health; common table salt is one and it occurs in almost all natural waters. Years ago the U.S. Public Health Service established **potable** (drinkable) **water** standards for public water supplies based on purity, which is defined as allowable concentrations of particular dissolved substances in natural waters. These concentrations are expressed in weight per volume, which is milligrams per liter (mg/L) in the metric system. The standards can also be expressed in a weight-per-weight system. For example, in the English system, one pound of dissolved solid in 999,999 pounds of water would be one part per million (ppm). The Public Health Service standards allow only minuscule amounts of pollutants in public water supplies. For example, 1 ppm or 1 mg/L is equal to about 1 ounce of a dissolved

solid in 7,500 gallons of water, or 0.0001 percent. The standard for "sweet" water—the preferable quality classification for domestic use—is less than 500 ppm of *total dissolved solids (tds)*, and the standard for "fresh" water is less than 1,000 ppm of total dissolved solids. ✳ Table 8.1 lists some common water-problem symptoms and suggestions for correcting them.

The quality of ground water can be impaired either by a high total amount of various dissolved salts or by a small amount of a specific toxic element. High salt content may be due to solution of minerals in local geological formations, seawater intrusion into coastal aquifers, or the introduction of industrial wastes into ground water. (Organic contaminants, mainly animal wastes, solvents, and pesticides, also degrade water. They are discussed in the next subsection.) ✳ Table 8.2 lists Public Health Service limits for some of the inorganic substances that occur in ground water and impair its safety.

Water with high salt content, particularly bicarbonate and sulfate contents, can have a laxative effect on humans. Dissolved iron or manganese can give coffee or tea an odd

∗TABLE 8.1 Common Water Quality Problems

Problem	Probable Cause	Correction
Mineral scale buildup	water hardness	water softener or hardness inhibitor (such as Calgon)
Rusty or black stains	iron and/or manganese	aeration/filtration; chlorination; remove source of Fe/Mn
Red-brown slime	iron bacteria	chlorination filtration unit; remove source of slime
Rotten-egg smell or taste	hydrogen sulfide and/or sulfate-reducing bacteria; low oxygen content	aeration; chlorination filtration unit; greensand filters
Salty taste	chloride—seawater or other saline water	reduce pumping to raise water table
Gastrointestinal diseases, typhoid fever, dysentery, and diarrhea	coliform bacteria and other pathogens from septic tanks or livestock yards	remove source of pollution; disinfect well; boil water; chlorinate; abandon well and relocate new well away from pollution source
Petroleum smell or film	fuel oil, diesel fuel, or lubricating oil	remove source of pollution

Source: D. Daly, "Groundwater Quality and Pollution" (Geological Survey of Ireland Circular 85-1, 1985), in J. E. Moore and others, *Ground Water: A Primer,* American Geological Institute, 1994.

color, impart a metallic taste, and stain laundry. Concentrations greater than 3 ppm of copper can give the skin a green tinge, a condition that is reversible, and lead is both poisonous and persistent. A *persistent pollutant* is one that builds up in the human system because the body does not metabolize it. The Romans, for example, used lead plumbing pipes, and skeletal materials exhumed from old Roman cemeteries retain high lead concentrations. Lead concentrations in the water supply greater than about 0.1 ppm can build up over a period of years to debilitating concentrations in the human body. Arsenic is also persistent, and the standards for arsenic and lead are the same—less than 0.05 ppm, preferably absent. Nitrates in amounts greater than 10 ppm are known to inhibit the blood's ability to carry oxygen and to cause "blue baby syndrome" (*infantile methemoglobinemia*) during pregnancy. The element with the greatest impact upon plants is boron, which is virtually fatal to citrus and other plants in concentrations as low as 1.0 ppm.

Dissolved calcium (Ca^{2+}) and magnesium (Mg^{2+}) ions cause water to be "hard." In hard water soap does not lather easily, and dirt and soap combine to form scum. Hard water is most prevalent in areas underlain by limestone ($CaCO_3$) and dolomite ($CaMgCO_3$). A water softener simply exchanges sodium (Na^+) for calcium and magnesium ions in the water using a zeolite (a silicate ion exchanger) or mineral sieve. Water with more than 120 ppm of dissolved calcium and magnesium is considered hard (∗ Table 8.3). Because sodium is known to be bad for persons with heart problems (hypertension) and water softeners increase the sodium content of water, those people on low-sodium diets should not drink softened water.

The good news regarding dissolved ions is that fluoride ion (F^-) dissolved in water reduces tooth cavities when it is present in amounts of 1.0–1.5 ppm. The explanation for this is that tooth enamel and bone are composed of the mineral apatite, $Ca_5(PO_4)_3(F, Cl, OH)$, whose structure allows the free substitution of fluoride F^-, chloride Cl^-, or hydroxyl $(OH)^-$ ions in the mineral structure. In the presence of fluoride, the crystals of apatite in tooth enamel are larger and more perfect, which makes them resist decay. Too much fluoride in drinking water, however—concentrations greater than 4 ppm—can cause children's teeth to become mottled with dark spots. This cosmetic blemish, among children in particular, led to the discovery of the beneficial effect of fluorine and to the establishment of concentration standards for it. Fluoride in natural waters is rare, but Colorado Springs, Colorado, is a well-known area with very few children's dentists. Dissolved fluoride is also believed to slow osteoporosis, a bone-degeneration process that accompanies aging.

Ground-Water Pollutants

The term *water pollution* refers to the introduction of chemical, physical, or biological materials into a body of natural water that affect its future use. It is more widespread than formerly thought, as revealed by studies by state and federal agencies and by complaints from users. It is a serious problem because thousands of years may be required to flush contaminated water from an aquifer and replace it with clean water. River water, in contrast, is exchanged in a matter of hours or days. **Residence time,** the length of time a substance remains (*resides*) in a system, is an important concept in water pollution studies. We may view it as the time that passes before the water within an aquifer, lake, river, glacier,

✳TABLE 8.2 Health Effects of Significant Concentrations of Common Inorganic Environmental Pollutants

Inorganic Pollutant	Standard, mg/L	Health Effects
Arsenic	0.05	dermal and nervous-system toxicity effects, paralysis
Barium	1.0	gastrointestinal effects, laxative
Boron	1.0	no effect on humans; damaging to plants and trees, particularly citrus trees
Cadmium	0.01	kidney effects
Copper	1.0	gastrointestinal irritant, liver damage; toxic to many aquatic organisms
Chromium	0.05	liver/kidney effects
Fluoride	4.0	mottled tooth enamel
Lead	0.05	attacks nervous system and kidneys; highly toxic to infants and pregnant women
Mercury	0.002	central-nervous-system disorders, kidney disfunction
Nitrates	10.0	methemoglobinemia ("blue baby syndrome")
Selenium	0.01	gastrointestinal effects
Silver	0.05	skin discoloration (argyria)
Sodium	20–170	hypertension and cardiac difficulties
	70.0	renders irrigation water unusable

✳TABLE 8.3 Water Hardness Scale*

Concentration of Calcium²⁺, mg/L	Classification
0–60	soft
61–120	moderately hard
121–180	hard
>180	very hard

*Hardness is expressed here as milligrams of dissolved Ca^{2+} per liter. Dissolved calcium and magnesium in water combine with soap to form an insoluble precipitate that hampers water's cleansing action.

tling ponds, and chemical-waste dumps (◆Figure 8.28a). In suburban and agricultural areas, septic-tank leakage, fecal matter in runoff and seepage from animal feedlots, and the use of inorganic fertilizers, such as nitrates and phosphates, and of weed and pest-control chemicals all can serve to degrade both surface and subsurface water (◆Figure 8.28b).

Leaking gasoline-storage tanks are a more recent cause of soil and ground-water pollution (◆Figure 8.28c). By federal law, all underground tanks now must be inspected for leakage, and if gasoline is found in the adjacent soil, the tanks must be removed and the soil cleaned. (Although the problem is not amusing, the technology created by the need for hydrocarbon cleanup has come to be lightly referred to as *yank-a-tank*.) If gasoline is found floating on ground water, it must be removed. Both the gasoline and the decontaminated ground water are then usable. Soils can be cleansed of hydrocarbons biologically with bacteria, a process known as **bioremediation,** or by aeration and oxidation of the hydrocarbon films (see Case Study 6.3 on page 170).

✳ Table 8.4 lists some of the detrimental effects of selected organic pollutants. Water standards for common organic pollutants of water bodies are given in ✳ Table 8.5. The very low limits allowed for the pesticides lindane and endrin indicate their toxicities. Some of these toxic chemicals are also carcinogenic (cancer-causing), which makes them doubly dangerous.

Density differences between ground water and particular pollutants often govern the remedial measures that can be used; for example, gasoline and oil float on ground water, whereas salt brines from industry or agriculture sink to lower levels. In some cases it is possible to skim a layer of gasoline from the top of the water table. Gasoline or fuel oil that is floating on the capillary fringe may yield vapors that can rise and be trapped beneath or in the walls of buildings. If these fumes contain BTX (*b*enzene, *t*oluene, and *x*ylene), they are carcinogenic. "Air-stripping" is currently the preferred method of removing volatile organic pollutants such as BTX compounds and trichloroethylene (TCE) and tetrachloroethylene (PCE) cleaning solvents from ground water. (The origin of these substances is usually connected to refinery, industrial,

or ocean is totally replaced and continues on its way through the hydrologic cycle. The average residence time of water in rivers has been found to be a few days; of water in large lakes, several decades; in shallow gravel aquifers, a few days; and in deeper, low-permeability aquifers, perhaps thousands or even hundreds of thousands of years. This knowledge allows us to estimate the approximate time a particular system will take to clean itself naturally and indicates those cases in which a pollutant must be removed from a water body artificially by human means because of a long residence time.

Although most pollution is due to careless disposal of waste at the land surface by humans, some is due to the leaching of toxic materials in shallow excavations or mines. Ground-water pollution occurs in urban environments due to improper disposal of industrial waste and due to leakage from sewer systems, old fuel-storage tanks, wastewater set-

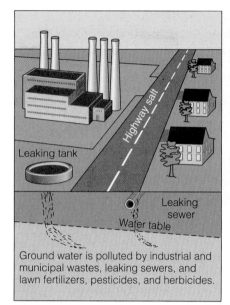

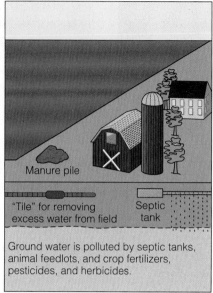

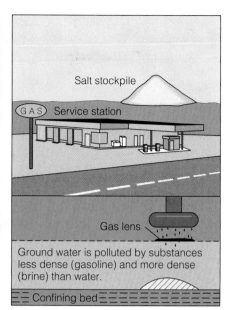

(a) Urban areas

(b) Rural areas

(c) Density effects

◆ FIGURE 8.28 (a) Urban-area and (b) rural-area pollution sources. (c) Some pollutants float and others sink due to differences in density between ground water and the pollutants.

or military activities.) The polluted water is brought to the earth surface, where the contained volatile pollutants (TCE or BTX) are gasified (air-stripped) and wasted to the atmosphere with no adverse effects.

In Canada 10,000 metric tons of PCE are used per year, primarily in the dry-cleaning and metal-cleaning industries. An exhaustive study mandated and sponsored by the Canadian Environmental Protection Act (CEPA) indicates that PCE is entering the environment in significant quantities. Due to its volatility (it evaporates readily), it is found mostly in the atmosphere, where it has a relatively short residence time and does not contribute to global warming. The study concludes that the present atmospheric concentration of PCE is not expected to have adverse effects on Canada's aquatic and terrestrial animals, including humans, but it has the potential of harming trees. Local spills, of course, require cleanup and remediation.

ENVIRONMENTAL
Geology ⇌ Now™

Click **Geology Interactive** to work through an activity on Contamination through Ground Water.

Conservation and Alternative Sources

Many parts of the United States are experiencing drought conditions as this is being written. Excessive ground-water withdrawals have led to water mining in many areas, particularly the Southwest. Remedies to cure water-deficient conditions include desalinization of seawater, recycling wastewater, tugging huge Antarctic icebergs up the coast of South

TABLE 8.4 Health Effects of Common Organic Pollutants		
Organic Substance	**Human Health Effects**	**Environmental Effects**
Dieldrin	convulsions, kidney damage	toxic to aquatic organisms
DDT	convulsions, kidney damage	reproductive failure in animals, eggshell thinning
PCBs	vomiting, abdominal pain, liver damage	eggshell thinning in birds, liver damage in mammals
Benzene	anemia, bone-marrow damage	toxic to some fish and aquatic invertebrates
Phenols	death at high doses	decreased phytoplankton productivity
Dioxin	acute skin rashes, systemic damage, mortality	lethal to aquatic birds and mammals

*Insecticides or compounds used in their manufacture.
Source: U.S. Public Health Service; Environmental Protection Agency.

TABLE 8.5 Water Standards for Common Organic Pollutants

Substance	Maximum Safe Concentration, mg/L
Cyanide	0.05
Lindane	0.004
Endrin	0.0002
2,4-D	0.10
Phenols	0.001

*Insecticides and animal-control compounds.
Source: U.S. Public Health Service; Environmental Protection Agency.

America to drier areas in the Northern Hemisphere, and 35-ton water bags floated down from northwestern United States to Southern California. But in the long run, as population increases, we will be required to conserve water.

Voluntary water conservation has been quite successful, reducing consumption by as much as 15 percent, and even greater reductions have been achieved where rationing is mandatory. Water use is increasing everywhere. The world's 6 billion people are intercepting 54 percent of all fresh water from rivers, lakes, and aquifers. Humans will intercept 70 percent by 2025 as just the result of population growth. If per capita consumption rises at the current rate, humans could be using 90 percent of all available fresh water, leaving 10 percent for the rest of life on earth. We should look for a viable alternative source of water and think "conservation."

CASE STUDY 8.1 Who Shrunk the Aral Sea?

The Aral Sea was once the fourth-largest lake in the world, and second only to the Caspian Sea in the former Soviet Union (◆Figure 1). The level of the lake in 1960 was about 53 meters above sea level and it supported a fish population that employed more than 60,000 persons in fishing and processing. In 2003 the level of the sea was 30 m and it ranked as the world's sixth-largest lake, having lost more than 74 percent of its surface area since 1960. The result of this shrinkage is that the fishing industry is no more (◆Figure 2). Additionally, the salinity of the sea has increased from 10 g/l (about 1%) to over 100 g/l (10%) in some areas, making it sterile to most life. This story leaves a question that begs an answer—What caused this eco-catastrophe?

Agricultural demands of the former Soviet Union deprived this great salt lake of a water inflow sufficient to sustain itself. Two major rivers feed the lake: the Amu Darya (*darya,* river) flowing into the lake from the south, and the Syr Darya, which reaches the sea at its north. The Soviet leaders wanted exportable crops and cotton, a great water gulper, was high on their list. Irrigated agriculture was expanded and the Kara Kum Canal was opened in 1956, which diverted water from the Amu Darya into the deserts of Turkmenistan. Millions of hectares of arid lands came under irrigation. By 1960 the level of the sea began to drop, and where in the past it received 59 km³ of water annually, this amount fell to zero by the early 1980s. As the lake area receded the salinity increased, and by 1977 the fish catch dropped to one-fourth its level before water diversion. By 1990 the area of the Aral was cut in half and its water volume was but a mere one-third its historic amount. The lowering of the sea exposed 30,000 km² of salt- and chemical-filled lakebeds (◆Figure 3). Blowing dust, particularly south of the lake, has resulted in air pollution producing a high incidence of throat cancer, eye and respiratory diseases, anemia, and infant mortality. In addition, the shoreline habitat was lost and delta wetlands desiccated. Animal populations shifted, because conditions favored animals better adapted to drought and high salinity. By any measure the desiccation of the Aral Sea is an ecological disaster. The central Asian republics that surround the sea—Kazakhstan, Turkmenistan, Uzbekistan, and the little-known Karakalpak Republic (within Uzbekistan)—are the poorest of the former Soviet block and have no money to invest in rebuilding irrigation systems. In addition, Afghanistan and Iran, with small areas within the drainage basin, have little interest in the problem.

A sad Uzbek poem goes "You cannot fill the Aral Sea with tears."

◆ FIGURE 1 Location map of Aral Sea Redrawn from Philip Micklin, Western Mich. Univ.

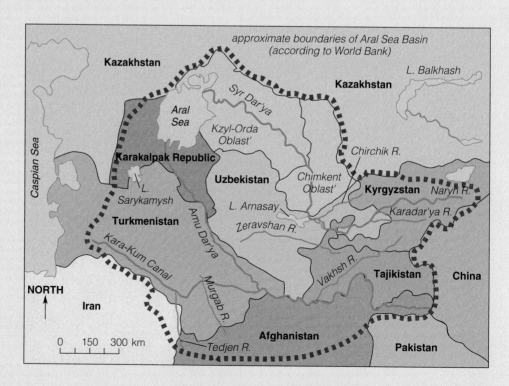

◆ FIGURE 2 Fishing boats grounded with shrinkage of the Aral Sea. This spelled doom to a thriving fishing industry and to an important source of protein for the local inhabitants.

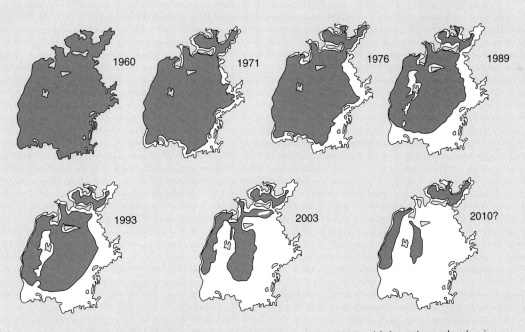

◆ FIGURE 3 The shrinkage in area of the Aral Sea from 1960 to 2003, with its estimated reduction projected to 2010. By 2003, compared to 1960, the sea lost 74 percent of its area and 90 percent of its volume. Redrawn from Philip Micklin, Western Mich. Univ.

CASE STUDY 8.2

Fluid Flow in Porous Rocks

Porosity is the ratio of the volume of pore spaces in a rock or other solid to the material's total volume. Expressed as a percentage, it is

$$P = \frac{V_p}{V_t} \times 100$$

where P = porosity
V_p = pore volume
V_t = total volume

If an aquifer has a porosity of 33 percent, a cubic meter of the aquifer is composed of ⅔ cubic meter of solid material and ⅓ cubic meter of pore space, which may or may not be filled with water. Porosities of 20 to 40 percent are not uncommon in clastic sedimentary rocks.

Permeability is a measure of how fast fluids can move through a porous medium and is thus expressed in a unit of velocity, such as meters per day. Knowing the permeability of an aquifer allows us to determine in advance how much water a well will produce and how far apart wells should be spaced. To determine permeability we use **Darcy's Law,** which states that the rate of flow (Q in cubic meters) is the product of hydraulic conductivity (K), the **hydraulic gradient,** or the slope of the water table (I in meters per meter), and the cross-sectional area (A in square meters). Thus,

$$Q \quad = \quad K \quad \times \quad I \quad \times \quad A$$

rate of flow   hydraulic   hydraulic   cross-sectional
conductivity   gradient   area

Hydraulic conductivity, K, is a measure of the permeability of a rock or a sediment, its water-transmitting characteristics. It is defined as the quantity of water that will flow in a unit of time under a unit hydraulic gradient through a unit of area measured perpendicular to the flow direction. K is expressed in distance of flow with time, usually meters per day.

If we use a unit hydraulic gradient, then Darcy's Law reduces to:

$$Q = K \times A$$

Unit hydraulic gradient is the slope of 1-unit vertical to 1-unit horizontal (I = 1/1 = 45°), and it allows us to compare the water-transmitting ability of aquifers composed of materials with different permeabilities. Flow rates may be less than a millimeter per day for clays, many meters per day for well-sorted sands and clean gravels. In general, ground water moves at a snail's pace (◆Figure 1).

Pumice provides an interesting example. It is a porous but almost impermeable rock. It has porosities as high as 90 percent but few connected pore spaces. Pumice will float for a time because it is light-weight and its low permeability prevents immediate saturation. A cubic meter of pumice may weigh as little as 500 pounds, a fifth of the weight of solid rock of the same composition.

◆ FIGURE 1 The unit hydraulic gradient (I) and unit prism (1 m²) of a sandstone aquifer. When the hydraulic conductivity (K) is known, then the discharge (O) in liters or cubic meters per day per square meter of aquifer can be calculated.

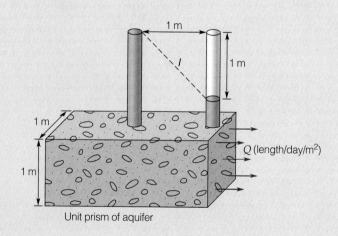

Ground-Water Law

Ownership of ground water is a property right and has traditionally been regulated as such by individual states. Each state applies one or more of four legal doctrines to ground-water rights, none of which considers the geological relationship between surface-water infiltration and ground-water flow. The doctrines are

- *Riparian doctrine* (Latin *ripa,* "river or bank") holds that landowners overlying well-defined underground "streams" have absolute right to that water. They may exercise that right whenever they wish without limitation.

- *Reasonable-use doctrine* restricts a landowner's right to ground water to "reasonable use" on the land above that does not deprive neighboring landowners of their rights to "reasonable use." This doctrine results in a more equal distribution of water when there is a shortage.

- *Prior-appropriation doctrine* holds that the earliest water users have the firmest rights; in other words, "first-come, first-served."

- *Correlative-rights doctrine* holds that landowners own shares to the water beneath their collective property that are proportional to their shares of the overlying land.

In the San Joaquin Valley of California, for example, surface-water rights are governed by the doctrine of prior appropriation (first-come, first-served), whereas ground-water rights are governed by the doctrine of correlative rights (proportional shares). In the late nineteenth century the Sierra Nevada to the east of the valley and the lower ranges to the west maintained high water tables, and they both contributed water to the flow of the San Joaquin River (◆Figure 1). By 1966, however, heavy ground-water withdrawals on the west side of the San Joaquin Valley had reversed the regional ground-water flow, so that the river was losing water underground, rather than gaining. The San Joaquin had become a losing stream. Imagine a courtroom scenario in which hundreds of people with first-come, first-served rights to *surface* water sue thousands of people holding proportional shares rights to *underground* water for causing their surface water to go underground.

As explained throughout this chapter, it is impossible to treat ground water and surface water as separate entities. Surface water contributes to underground water supplies, and ground water contributes to streamflow and springs.

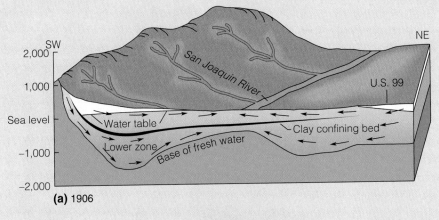

(a) 1906

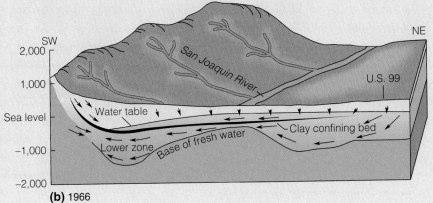

(b) 1966

◆ FIGURE 1 Ground-water flow in the central San Joaquin Valley in (a) 1906 and (b) 1966. Note that the San Joaquin River was a gaining stream in 1906 and a losing stream in 1966. The change reflects intensive development and extraction of underground water. AIPG, *Ground Water: Issues and Answers,* 1985

CASE STUDY 8.4

The Edwards Aquifer, A Texas Bonanza

The Edwards aquifer in south-central Texas is one of the most prolific aquifers in North America (◆Figure 1). It provides water for two million people, including the City of San Antonio. A well drilled for the city in 1991 is reportedly the world's greatest flowing well, yielding 95,000 liters per minute (25,000 gal/min). The aquifer is composed of limestone of Cretaceous age that is a *karstified* flow system; that is, it is honeycombed with solution cavities, caves, and caverns and feeds many springs. About 80 percent of the aquifer's recharge is from losing streams that flow over its exposed outcrops; the rest is from precipitation. Where it dips down below the ground surface toward the south, it is under pressure, and water rises in well bores and sometimes to the ground surface (see Figure 8.19). At depth, the water quality and potability deteriorate to a zone marked on maps by the "bad-water line."

The region served by the aquifer is one of the fastest growing areas in the United States, and water supply is critical to this growth. Recharge areas are environmentally sensitive to pollutants, and the resource is threatened by disregard for this potential problem (◆Figure 2). Threats are from human sewage, highway construction, agricultural chemicals, runoff from paved surfaces, and leaking underground storage tanks, to name a few. Government agencies at all levels are committed to protecting the aquifer. The federal government designated the aquifer a Sole Source Aquifer, which makes it eligible for funding for some protection projects. The state declared the Edwards aquifer the most susceptible to pollution in the state and established conservation districts for overseeing its protection.

Texas Geological Society

◆ FIGURE 2 Midnight Cave in the Edwards aquifer before cleanup, 1993; southern Travis County near Austin, Texas. The trash is household garbage, oil filters, pesticide bottles, partially filled turpentine cans, and automobile parts. Trash had been floated up to higher ledges of the cave during high-water aquifer conditions. Cleanup efforts involved conservation groups, government agencies, and cavers. An estimated 3,000 cubic feet of trash was removed from the cave.

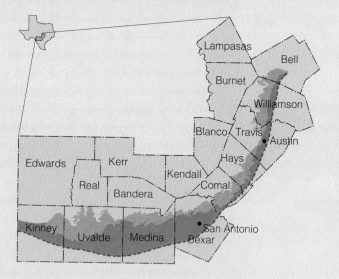

▨ Recharge zone
▨ Extent of Edwards aquifer potable water
----- Bad water line

◆ FIGURE 1 Area of south-central Texas's Edwards aquifer.

CASE STUDY 8.5

Long Island, New York—Saltwater or Fresh Water for the Future?

By far, Long Island's most important source of fresh water is its aquifers, which are estimated to hold 10–20 trillion gallons of recoverable water (◆Figure 1). This water infiltrated underground over centuries and the excess was discharged naturally to the sea. Ground water was exploited as the island was urbanized, causing the water table on the mainland side of the island to drop below sea level. By 1936 saltwater had invaded the freshwater aquifer. Pumping wells were converted to recharge wells using imported water, and by 1965 the water tables had recovered to acceptable elevations above sea level.

To the east in Queens County pumping had increased and recharge had diminished due to the construction of sewers and paving of streets. Treated wastewater was thus being carried to the sea, rather than infiltrating underground as it would with cesspools and septic tanks. With lowered water tables, saltwater invaded the aquifers below Queens County, and the State of New York in 1970 asked the U.S. Geological Survey to conduct ground-water studies.

The basic challenge was to offset ground-water withdrawals of 1.7 million cubic meters per day with equal amounts of surface infiltration or injection. A line of injection wells was proposed for replacing the decreased natural recharge. These wells would inject treated sewage effluent or imported water and build a freshwater barrier against saltwater intrusion (◆Figure 2a). Spreading highly treated sewage effluent and natural runoff into recharge basins could be carried out along with injection into the deeper aquifer. In both the injection-well and the spreading-basin alternatives, the treated sewage is purified to drinking-water standards so that it does not affect local shallow wells. Recharge, like injection, builds a freshwater ridge that keeps saltwater at bay. An unusual but viable method is to allow controlled saltwater intrusion so that a true Ghyben–Herzberg lens of fresh water floating on saltwater develops. This method has the distinct advantage of salvaging much of the fresh water that otherwise would flow through aquifers into the ocean. In addition, this alternative would increase the yield of the aquifer by several hundred million cubic

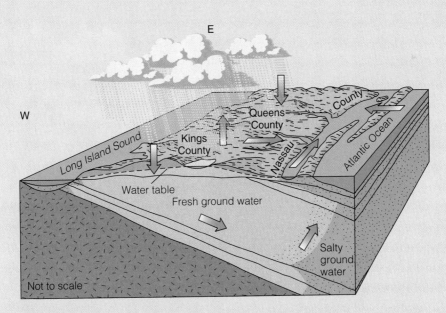

◆ FIGURE 1 Hydrologic relationship between fresh and salty underground water; Long Island, New York. Arrows indicate direction of water movement in the hydrologic cycle. After B. L. Foxworthy, USGS prof. paper 950, 1978

meters per day, although it would decrease the total volume of fresh water in the reservoir.

Each method of balancing outflow and inflow—injection wells, spreading basins, and controlled saltwater intrusion—has advantages and disadvantages. As is the case with many geological problems, applying a combination of methods yields the most fruitful and cost-effective solution. Kings and Queens Counties now rely entirely upon importation for their freshwater needs. Public water in Nassau and Suffolk Counties is entirely underground water, and only a very small amount of seawater intrusion occurs in the southwest corner of Nassau County. Currently, there is a network of recharge basins that collect surface-water runoff and allow it to percolate underground into the shallow aquifer (◆Figure 3). The counties require developers to dedicate land for recharge basins within their housing or commercial projects. This method of recharge has proven effective in maintaining the balance between seawater and fresh water.

◆ FIGURE 2 Method of balancing freshwater outflow and seawater inflow; Long Island, New York. The aquifer is recharged with highly treated wastewater (*T*) using injection wells. This method reverses saltwater intrusion and also improves the outlook for long-term water yield. After B. L. Foxworthy, USGS prof. paper 950, 1978

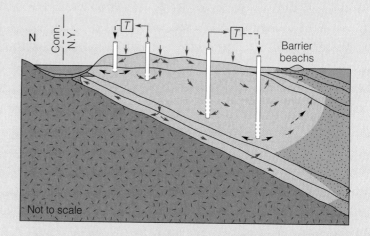

◆ FIGURE 3 A recharge basin for capturing surface-water runoff; Nassau County, New York.

NASSAU COUNTY DEPT. OF PUBLIC WORKS

Ground-Water Wonders

Karst terrane exists in many parts of the United States, but it is particularly prevalent in parts of Florida and Kentucky. Perhaps the most spectacular karst area in the world is along the Li River in the predominantly limestone region of Kwangsi in southern China near Canton (Guangzhou). Solution of limestone has reduced the landscape there but left enormous pinnacles of resistant limestone, called "tower karst," that present an almost extraterrestrial appearance (◆Figure 1). The towers and other karst features draw tourists from all over the world.

In many places, water containing dissolved calcium carbonate percolates downward through soil and rock into caverns, where evaporation and precipitation form needlelike **stalactites**—extending down from the ceiling—and the more irregular **stalagmites**—rising up from the floor (◆Figure 2). These freshwater limestone features in caves are called **dripstone,** because they build up drip by drip. Carlsbad Caverns in New Mexico and Mammoth Cave in Kentucky offer spectacular examples, but literally hundreds of other limestone caves also exhibit these features and are open to the public. **Speleology** is the study of caves, and spelunking, or "caving," is an exciting recreational pastime for many people. Cavers explore and map natural underground openings. Much of what we know about caves and caverns is attributable to these hobbyists' efforts and talents.

Mineralization by underground water is also responsible for transforming organic material into hard mineral water—a process known as *fossilization*. The replacement mineral is typically silica or calcium carbonate, and replacement may take place cell by cell as in the case of petrified wood (◆Figure 3).

◆ FIGURE 1 The Stone Forest, 126 kilometers (80 mi) southeast of Kunming, People's Republic of China, is a high-relief karst landscape formed by the dissolution of carbonate rocks.

◆ FIGURE 2 Dripstone stalactites, stalagmites, and pillars in a limestone cavern.

◆ FIGURE 3 Petrified wood from Arizona. Woody tissue is replaced cell by cell with mineral matter from ground water.

Summary

The Hydrologic Cycle

Only .6 percent of the earth's water is readily available to humans. It is accessible because of the hydrologic cycle: continuous evaporation from the oceans, precipitation on land, runoff back to the oceans, and storage of fresh water in lakes and underground.

Water Use

1. Consumptive use—water that cannot be immediately reused. Where consumptive use exceeds supplies, water must be imported.
2. Irrigation and thermoelectric power plants are the largest consumers. Irrigation water is usually highly degraded, whereas 98 percent of power-plant water is returned to the system, the only change being a higher temperature.
3. Worldwide, about 82 percent of water use is for irrigation. Water availability will limit population growth in the twenty-first century, either by statute in developed countries or by drought and famine in developing nations. The world population projected for 2050 will require water beyond the firm flow of the world's major rivers.

Lakes

Defined

Lakes are bodies of water in landlocked basins formed by many geologic processes.

Budget—depends on precipitation plus inflow equaling outflow plus evaporation. If these do not balance in the long term the result is a salt lake or dry lake known as a playa.

Overturn—in temperate regions overturn keeps lakes fresh.

Nutrients—oligotrophic lakes are relatively sterile and have a small nutrient input. Eutrophic lakes have a high nutrient input and can be choked with plant life. Mesotrophic lakes are in between and have a good balance between inflow and outflow of nutrients.

Rivers and Streams

Dams are built to impound fresh water, for power generation, flood control, and to improve navigation. In the U.S. dams are dimly viewed because of loss of land by inundation and the subsequent relocation of humans and wildlife.

Ground Water

Defined

Loosely defined as all water underground, as distinct from surface water; the water in the hydrologic cycle that infiltrates underground.

Zones

1. The *zone of aeration,* where the voids in rock or sediment are filled with air and water films.
2. The lower *zone of saturation,* where voids are filled with water.

Water Table

The contact between the zone of aeration and the zone of saturation.

Aquifers

Defined

Bodies of porous and permeable rock or sediment that are saturated with water.

1. *Porosity*—the volume of pore spaces in a geologic material, expressed as a percentage of the total volume.
2. *Permeability*—ease with which fluids flow through a geologic material.
3. *Aquiclude*—a geologic material that lacks permeability and inhibits water flow.

Types

1. Unconfined—an aquifer in which the water table is static and that is exposed to atmospheric pressure changes. A slope, or gradient, is required to induce subsurface flowage in a given direction. A well that taps the static water table creates a cone of depression, which provides the necessary gradient for water to flow to the well. Drawdown is the amount of lowering of the water table in a pumping well.
2. Artesian—an aquifer that is confined between aquicludes. The hydraulic head causes water to rise in a well when the aquifer is penetrated. The height to which water will rise at any point along the aquifer is known as the artesian pressure surface.

Ground-Water Management

Defined

Managing the amount of water contained in a ground-water basin so that it will produce water in the future.

1. *Sustained yield*—the amount of water an aquifer will yield on a day-to-day basis.
2. *Specific yield*—the amount of water a body of rock or sediment will yield by gravity alone.
3. *Specific retention*—the amount of water held by a body of rock or sediment against the pull of gravity. Specific yield *plus* specific retention *equals* porosity.

Seawater Encroachment

The invasion of coastal aquifers by saltwater. Management requires maintaining high freshwater tables to balance the denser saltwater.

Ground Water–Surface Water Interaction

Surface water and ground water cannot be treated as separate entities; what affects one impacts the other.

Surface Water Sources

Lakes, rivers, streams, and artificial reservoirs.

U.S. Use Trends

U.S. water use increased from 1950 until 1980. Between 1980 and 1995 it decreased 10 percent, even though population increased by 37 million people. Reasons include more economical application of irrigation water, public conservation programs, recycling, and a downturn in farm economies.

Solution by Ground Water

1. Solution of limestone and gypsum by underground water forms caves, caverns, disappearing streams, sink holes, and prolific ground-water aquifers. When terranes underlain by carbonate rocks exhibit these features, they are said to have *karst* topography.
2. Water and carbon dioxide from carbonic acid, a weak natural acid, which, over long periods of time, will dissolve carbonate rocks and gypsum.
3. Urbanization on karst terrane results in increased flooding, contamination of underground water, and collapse of the ground surface to form sinkholes.
4. Forty percent of the country east of the Mississippi is underlain by carbonates. States most impacted by karst and sinkhole formation are Alabama, Florida, Georgia, Tennessee, Missouri, and Pennsylvania.

Water Quality

Defined

Purity or drinkability (potability) of water as established by the U.S. Public Health Service. Dissolved constituents are measured in milligrams per liter (volume) or parts per million (weight).

Standards

Less than 500 ppm total dissolved solids for "sweet" water, and limits on ions and molecules that impair the quality of water.

Pollution

Defined

Chemical, physical, or biological materials that impair the future use of water and that may be a health hazard.

Residence Time

The average length of time a given substance will stay in a system, such as water in a lake or aquifer. Whereas the residence time of human-generated smog in the atmosphere of large cities is on the order of ten days, pollutants may reside in some aquifers for thousands of years.

Common Pollutants

Hydrocarbons, carcinogenic BTX solvents (benzene, toluene, and xylene), fertilizers, pesticides, livestock fecal matter, and gasoline.

Conservation and Alternative Sources

Conservation means include the use of treated wastewater for irrigation and recharging aquifers, water-saving irrigation devices, water-spreading during wet years, and voluntary water conservation. Desalinization and importation are the only viable alternative sources at present.

Key Terms

aquiclude	evapotranspiration	karst terrane	residence time
aquifer	gaining stream	losing stream	specific yield
artesian-pressure surface	ground water	overturn	speleology
cone of depression	hydraulic conductivity	perched water table	sustained yield
confined aquifer	hydrogeologist	permeability	unconfined aquifer
drawdown	hydrologic cycle	porosity	water table

Study Questions

1. Sketch a geologic cross section that shows the ground-water zones and the static water table.
2. Sketch or define a confined aquifer. Explain the artesian-pressure surface and its importance in production of water from an artesian well.
3. How does topography affect the shape of the static water table?
4. How does one determine the amount of water a well that taps the static water table might produce on a long-term basis?
5. Explain the distinction between gaining and losing streams.
6. Explain the relationship between the amount of drawdown in a water well and the aquifer's permeability.
7. What states are the largest consumers of ground water, and why is their usage so heavy?
8. What is meant by ground-water *mining*? Cite the case history of an aquifer where this has happened.
9. Many beverage manufacturers claim the water in their drinks is purer because it comes from artesian wells or springs. Comment on these claims.

10. What are the Public Health Service standards for total dissolved solids in drinking water?

11. Name three organic compounds sometimes found dissolved in ground water that are hazardous to human health.

For Further Information

Books and Periodicals

Alley, William M., Thomas Reilly, and O. Lehn Franke. 1999. *Sustainability of ground-water resources.* U. S. Geological Survey circular 1186.

American Institute of Professional Geologists. 1985. *Ground water: Issues and answers.* Arvada, Colo.: A.I.P.G.

Baldwin, H. L., and C. L. McGuinness. 1963. *A primer on ground water.* Washington: U.S. Geological Survey.

Cohen, P., O. L. Franke, and B. L. Foxworthy. 1970. *Water for the future of Long Island, New York.* New York Division of Water Resources, Department of Environmental Conservation, in cooperation with the U.S. Geological Survey.

Galloway, Devin, David Jones, and S. E. Ingebritsen. 1999. *Land subsidence in the United States.* U.S. Geological Survey circular 1182.

Hauwerrt, Nico M., and Shawn Vickers. 1994. *Barton Springs/Edwards aquifer: Hydrogeology and ground water quality.* Austin, Texas: Texas Water Development Board, Barton Springs/Edwards Conservation District.

Kidd, Mary A. 1996. *Nutrients in the nation's drinking water: Too much of a good thing?* U.S. Geological Survey circular 1136.

Moore, John E., A. Zaporozed, and James W. Mercer. 1994. *Ground water: A primer.* Environmental awareness series. Alexandria, Va.: American Geological Institute.

Sharp, John M., Jr., and Jay L. Banner. 1997. The Edwards aquifer: A resource in conflict. *G.S.A. Today* (Geological Society of America), August.

Solley, Wayne B., Robert R. Pierce, and Howard A. Perlman. 1998. *Estimated use of water in the United States in 1995.* U.S. Geological Survey circular 1200.

U.S. Geological Survey. 1999. *The quality of our nation's waters— nutrients and pesticides,* circular 1225.

Winter, Thomas C., Harvey, Judson O., Lehn Franke, and William M. Alley. 1998. *Ground water and surface water: A single resource.* U.S. Geological Survey circular 1139.

Zwingle, Erla. 1993. Ogallala aquifer: Well spring of the High Plains. *National geographic* 183, no. 3 (March).

ENVIRONMENTAL
Geology⇌Now™

Assess your understanding of this chapter's topics with additional quizzing and comprehensive interactivities at http://earthscience.brookscole.com/pipkingeo4e

as well as current and up-to-date Web links, additional readings, and InfoTrac College Edition exercises.

CHAPTER 9

Hydrologic Hazards at the Earth's Surface

Rain added to a river that is rank perforce will force it overflow the bank.

William Shakespeare (1564–1616)

The Blue Danube?

DURING THE SUMMER (AUGUST) OF 2002 EASTERN EUROPE and western Russia experienced their worst flooding in a century. The losses have been estimated in excess of $20 billion, which would place it on a par with the Northridge, California, earthquake in terms dollar loss but greater in terms of loss of life. There were 59 deaths in Northridge versus an estimated 105 for the 2002 flood—62 in Russia alone near the Black Sea coast. The following litany is only a sample of the flood's impact:

- Twelve countries were affected: Russia, Ukraine, Romania, Bulgaria, Croatia, Czech Republic, Austria, Slovakia, Switzerland, France, Germany, and Italy.
- Along the Elbe River in Germany a dam broke, threatening a town of 16,500 people.
- More than 400 houses were destroyed and at least 7,000 were damaged in Russia.
- About 220,000 residents living near the Vltava River (German Moldau) in the Czech Republic were displaced.
- More than 20,000 persons were evacuated in Dresden, Germany.
- A 30-meter-long steel bridge in Schwertberg, Germany, was torn loose by the flooded Danube River.
- The Danube River overflowed and appeared brown with suspended sediment.

Probably the worst flood impact, after the human loss, was the toll on the Prague Zoo. Zoos are a habitat that most humans don't think about during a flood. After 40 years of neglect under communist rule, zoo officials had invested heavily in building new pavilions, obtaining new residents, and developing interesting programs. It had become one of the most renowned zoos of the former communist world. At the peak of the flood the lower half of the zoo was completely under water. A 12-year-old seal named Gaston made headlines by swimming 120 kilometers in the river before being recaptured in

The Danube River at Regensburg, Germany, during the great flood of August 2002. Note that the river walk behind the high-water sign is inundated.

B. PIPKIN

B. PIPKIN

(a) **(b)**

Diners enjoy a Sunday lunch in a restaurant below the flood level of the Danube River in Regensburg, Germany. The water is kept at bay by the red inflatable flood-fence around the dining area.

Dresden, Germany. An elephant and hippo were lost, but another hippo, which was believed to have died, was saved after the floodwaters began to recede. Ten big animals were lost, including an aging lion and a bear considered too old to evacuate. The big loss was Pong, a popular gorilla and one of five in the zoo, who is believed to have drowned in the evacuation effort. More than 1,000 animals were saved and things are now back to normal, but many handlers still grieve the loss.

Geologic agents are processes that perform the work of erosion and deposition, and that shape and modify the earth's landscape. Rivers, mass-wasting, glaciers, wind, underground water, and waves are examples. Rivers, however, are the most important geologic agent at work in today's world. Their importance is reflected in the fact that river flooding may well be the most globally pervasive, environmentally diverse, and continually destructive of all natural hazards.

Hydrology is the science that deals with liquid and solid global water, its properties, circulation, and distribution on earth and in the atmosphere. It arose from the very practical problems of human water supply, river flooding of human habitat, and water quality. We mainly address hydrologic hazards in this chapter, after a necessary introduction to the geology of river processes.

River Systems

A stream's ability to erode (Latin *erodere*, "to gnaw away") the land and impact humans is a function of the stream-flow velocity and the amount of water discharged by the stream.

With equal discharge—that is, the same volume of water flowing past a point in a river channel in a given time period—a fast-flowing stream erodes its banks more significantly than does a slow-moving one. In the United States velocity (V) is usually expressed in feet per second (ft/sec), and **discharge** (Q), the amount of water, in cubic feet per second (cfs). A cubic foot contains about 7.4 gallons of water. A third factor, the cross-sectional area of the stream, also is important. A stream channel that is wide and shallow will have a much lower velocity than a stream with the same discharge that is narrow and deep. This is because the wide, shallow channel has greater area of friction between the running water and its banks (◆Figure 9.1). Whatever the mechanics of the stream's flow, with increased velocity and discharge, there are increases in the stream's erosion potential and its sediment-carrying capacity (◆Figure 9.2).

The longitudinal profile, or side view, of a stream channel has a slope known as its **gradient,** which can be expressed as the ratio of its vertical drop to the horizontal distance it travels (in meters per kilometer or feet per mile). The gradient of the Colorado River in the Grand Canyon is about 3.2 meters per kilometer, a moderately steep gradient, whereas in its lower reaches, where it flows into the Gulf of

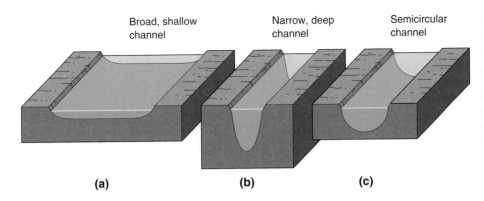

Broad, shallow channel

Narrow, deep channel

Semicircular channel

(a)　　　　**(b)**　　　　**(c)**

◆ FIGURE 9.1 Three stream channels with the same cross-sectional area but different shapes. The semicircular channel, (c), has the least perimeter in contact with the water and thus the least frictional resistance to flow. If all other variables such as roughness and slope of the channel are equal, flow velocity will be greatest in the semicircular channel.

California, its gradient is only a few centimeters per kilometer. A mountain stream may have a gradient of several tens-of-meters per kilometer and an irregular profile with rapids and waterfalls. As the same stream approaches the lowest level to which it can erode, an elevation known as **base level**, its gradient may be almost flat. A stream's ultimate base level is sea level, but it may have intermediate, temporary base levels such as a lake surface or a rock formation that is highly resistant to erosion. A stream whose gradient has adjusted so that an equilibrium exists between its discharge and its sediment supply is known as a graded stream. Longitudinal profiles of graded streams are generally concave-upward from source to end. ◆Figure 9.3 contrasts the profiles of the Arkansas River, which drains the wide east slope of the Rocky Mountains, and the Sacramento River, which drains the much narrower Pacific slope.

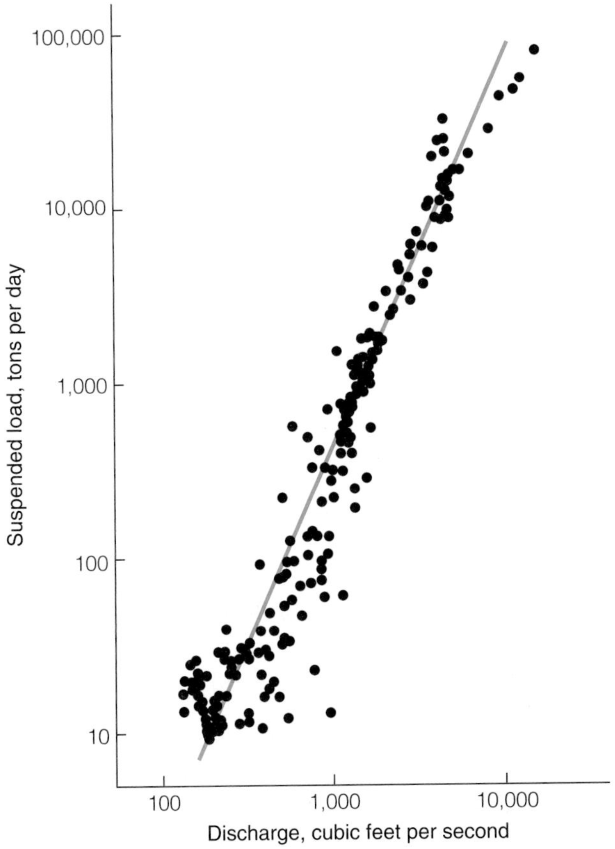

Gradients decrease in the lower reaches of a river, and sedimentation may occur as streams overflow their banks onto adjacent **floodplains.** Decreased water velocity on the floodplain results in the deposition of fine-grained silt and clay adjacent to the main stream channel. This productive soil, commonly known as *bottom land*, is much desired in agriculture (⊘ Case Study 9.1 on page 253).

> **CONSIDER THIS** *We generally associate high sediment loads with high river discharge, but there are other variables that determine the sediment transported by a river. With this in mind, would you expect the suspended sediment load per cubic meter of water to be greater in the Columbia River, which has a high-rainfall drainage basin, or in the Colorado River, which has a low-rainfall basin?*

Stream Features

All sediment deposited by running water is called **alluvium** (Latin, a flooded place or flood deposit). In the arid West, where streams lose velocity as they flow from mountainous terrain onto flat valley floors, we find conspicuous depositional features known as **alluvial fans** (◆Figure 9.4). They are in fact fan-shaped and they build up as the main stream branches, or splays, outward in a series of *distributaries*, each "seeking" the steepest gradient. Alluvial fans can provide beautiful, interesting views of surrounding terrain, but they also can be subject to flash floods and debris flows from intense seasonal storms in nearby mountains.

Deltas form where running water moves into standing water. The Greek historian Herodotus (484–420 B.C.) applied the term to the somewhat triangular shape of the Nile delta in recognition of its resemblance to the Greek letter delta (Δ). Deltas are characterized by low stream gradients,

◆ FIGURE 9.2 The relationship between water discharge (*Q*) and the suspended load for a typical stream. The points represent separate measurements. A tenfold increase in water discharge results in almost a 100-fold increase in sediment load. USGS

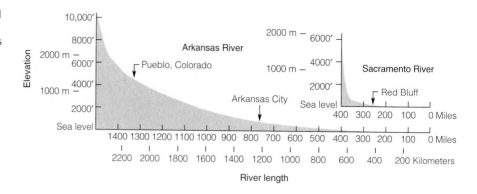

◆ FIGURE 9.3 Measured longitudinal profiles of the Arkansas and Sacramento Rivers. The vertical scale is exaggerated 275 times.

swamps, lakes, and an ever-changing channel system. The "arcuate" (arc-shaped) delta of the Nile and the "bird's foot" delta of the Mississippi also are described by their appearance on maps (◆Figure 9.5). The tremendous sand output of the Nile has created the huge coastal dunes east of Alexandria that formed the background for the famous World War II battle at El Alamein (see the Chapter 12 opener). The Ganges–Brahmaputra delta of Bangladesh consists of tidal flats and many low-lying sand islands offshore (Figure 9.5c). Because deltas are at or near sea level, they are subject to flooding by tropical storms. Coastal flooding is the most deadly of natural disasters, even more deadly than earthquakes or river floods. This type of flooding was particularly

◆ FIGURE 9.4 Alluvial fan just north of Tucson, Arizona. The Catalina Mountains are in the background. At the base of the fan is the Rillito Creek, which has caused flood damage in the past. The city is expanding and growing on the fan, which will cause further problems.

disastrous in Bangladesh in 1970, when more than 500,000 people were drowned by a surge of water onshore from a tropical cyclone.

Streams and rivers with low gradients and floodplains form a number of features that both aggravate and alleviate flood damage. The courses of such rivers follow a series of S-shaped curves called **meanders,** after the ancient name of the winding Menderes River in Turkey (◆Figure 9.6). Straight rivers are the exception, rather than the rule, and it appears that running water of all kinds will meander—ocean currents, rivers on glaciers, and even trickles of water on glass. The key to these regularly spaced curves is the river's ability to erode, transport, and deposit sediments. The meandering process begins when the gradient is reduced to a slope at which a stream stops deepening its valley. As a stream meanders, it erodes the outside curve of each meander and deposits sediment on each inside curve, producing a deposit known as a **point bar.** ◆Figure 9.7 shows the velocity distribution on a stream meander, which explains the erosion–deposition pattern. Meanders also widen stream valleys by lateral cutting of valley side slopes, causing mass wasting and eventually creating an alluvium-filled floodplain adjacent to the river. It is across this surface that the river meanders, sometimes becoming so tightly curved that it shortcuts a loop to form a **cutoff** and **oxbow lake** (◆Figure 9.8).

By the time a river has developed a wide floodplain with meanders, cutoffs, and oxbow lakes, a situation of poor drainage exists that increases the impact of overbank flooding. **Natural levees** build up on both banks of a river when it overflows and deposits sand and silt adjacent to the channel (◆Figure 9.9). Natural levees, or manufactured ones, **artificial levees,** keep flood waters channelized so that the water level in the stream may be above the level of the adjacent floodplain. This is a tenuous situation, because a breach in the levee would cause flooding of the floodplain, which could force the river into a new course. New Orleans on the Mississippi River is an example of a city on a floodplain that is almost perennially below the level of the river that flows through it (see ◎ Case Study 9.2 on page 254). Although levees may prevent flooding of adjacent floodplains, they can cause flooding downstream in areas where there are no levees.

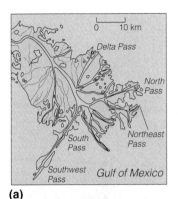

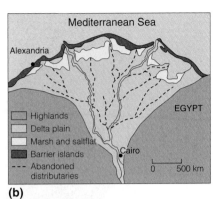

(a)   **(b)**

◆ FIGURE 9.5 River deltas. (a) Birdsfoot delta of the Mississippi River. The main channel and subordinate distributaries are separated by swamps and wetlands. (b) Arc-shaped delta of the Nile River is predominantly wetlands and rice-growing areas. The barrier islands are sand and they contribute to large coastal dunes to the west. (c) The Ganges–Brahmaputra delta of Bangladesh. The small islands rise above the tidal flats. Though subject to catastrophic floods, they are densely populated.

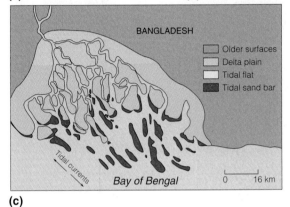

(c)

During the 1993 floods in the Midwest, artificial levees saved some cities in the Mississippi River drainage basin, notably St. Louis, where the flood crested at 49.4 feet and the levee height is 52 feet. Quincy, Illinois, lacking artificial levees, was inundated, but the levees at Hannibal, Missouri, on the opposite side of the river held and spared the town. A levee on the Raccoon River at Des Moines, Iowa, failed, and floodwaters inundated a city water-purification plant. Thousands of residents were without drinking water for 19 days. Some levees were intentionally breached in order to reduce flood-crest elevation downstream. More than 800 of the 1,400 levees in the nine-state disaster area were breached or overtopped. Most of these were simple dirt berms built by local communities, which explains why so many small towns

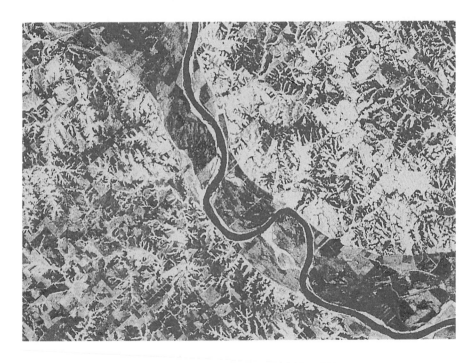

◆ FIGURE 9.6 Meanders on the Missouri River at Glasgow, Missouri. Note that the meander belt swings across the full width of the floodplain and erosion by this method widens the valley. Lighter color represents a change in vegetation.

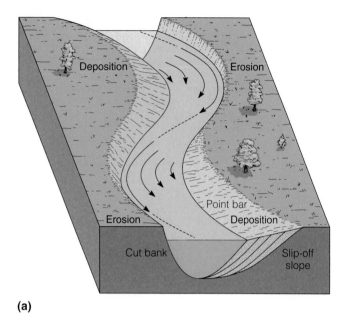

(a)

(b)

ENVIRONMENTAL

◆ Geology⇌Now™ ACTIVE FIGURE 9.7 Erosion and deposition patterns on a meandering stream. (a) Erosion of the cutbank and deposition of a point bar on the gently sloping side of the meander. Arrow length is proportional to stream velocity. (b) Rillito Creek rampaging through Tucson, Arizona, during the flood of 1983. Damage occurred when the cutbank of the meander migrated west (left) into vacant property immediately upstream from the townhouses (center left). This allowed erosion to occur behind a concrete bank that was protecting the townhouse property. Note the prominent point bar that developed (bottom center) as the meander migrated westward.

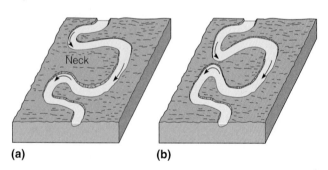

(a)   **(b)**

(c)   **(d)**

ENVIRONMENTAL

◆ Geology⇌Now™ ACTIVE FIGURE 9.8 (a–d) Meander pattern sequence of erosion leading to a cutoff and an oxbow lake. A cutoff forms in (c) and completely isolates the oxbow lake in (d).

were almost completely submerged by the floodwaters. However, more than 30 levees built by the U.S. Army Corps of Engineers also failed during this record flood event.

> **CONSIDER THIS** *Deltas are deposited where a river flows into a body of standing water, such as a lake or an ocean. Why does one of the great rivers of North America, the St. Lawrence River, have no delta, even though it flows into a landlocked estuary?*

When There Is Too Much Water

Excess water, whether in rivers, lakes, or along a coast, leads to flooding. The basic hydrologic unit in fluvial (river) systems is the **drainage basin,** the land area that contributes water to a particular stream or stream system. Individual drainage basins are separated by **drainage divides.** A drainage basin may have only one stream, or it may encompass a large number of streams and all their tributaries. Most of the rain that falls within a particular drainage basin after a dry period infiltrates (seeps) into the uppermost layers of soil and rock; little water runs off the land. As the surface materials become saturated and the rain continues, however, rainfall exceeds infiltration, and runoff begins. Runoff begins

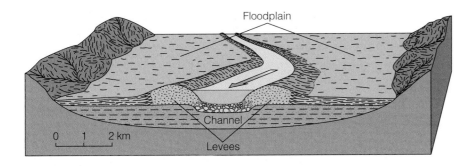

◆ FIGURE 9.9 Relationship of natural levees and floodplain to a hypothetical stream. The coarsest sediment is deposited in the stream channel, where the stream's velocity is greatest, and finer silt and clay are deposited on levees and the floodplain as velocity diminishes during overflows.

as a general sheet flow, then forms tiny rills and gullies, then flows into small tributary streams, and finally is consolidated into a well-defined channel of the main stream in the watershed. Whether or not a flood occurs is determined by several factors: the intensity of the rainfall, the amount of antecedent (prior) rainfall, the amount of snowmelt if any, the topography, and the vegetation.

Upland floods occur in watershed areas of moderate or high topographic relief. They may occur with little or no warning and they are generally of short duration; thus the name *flash floods*. Such floods result in extensive damage due to the sheer energy of the flowing water. Lowland, *riverine floods,* on the other hand, occur when broad floodplains adjacent to the stream channel are inundated. The typical damage is that things get wet or silt-covered. The difference in these types of floods is due to watershed morphology (shape). Mountain streams occupy the bottoms of V-shaped valleys and cover most of the narrow valley floor; there is no adjacent floodplain. Heavy rainfall at higher elevations in the watershed can produce a wall of fast-moving water that descends through the canyon, destroying structures and

endangering lives (◆Figure 9.10). Lowland streams, in contrast, have low gradients, adjacent wide floodplains, and meandering paths.

A **hydrograph** is a graph of a water body's discharge, velocity, stage (height), or some other characteristic over time, and it is the basic tool of hydrologists (◆Figure 9.11). The "synthetic" hydrograph in ◆Figure 9.12 graphically illustrates the basic difference between upland (flash) floods of short duration and the longer-lasting lowland floods. The timing of flood crests on the main stream—that is, whether they arrive from the tributaries simultaneously or in sequence—in large part determines the severity of lowland flooding.

Floods are normal and, to a degree, predictable natural events. Generally, a stream will overflow its banks every two to three years, gently inundating a portion of the adjacent floodplain. Over long periods of time, catastrophic high waters can be expected. Depending upon their severity, these events are referred to as fifty-, one-hundred-, or five-hundred-year floods. The more infrequent the event, the more widespread and catastrophic is the inundation (◆Figure 9.13). The Midwest flood of 1993 that inundated 421 counties in 9 states, with a peak

◆ FIGURE 9.10 A debris flow near Estes Park, Colorado.

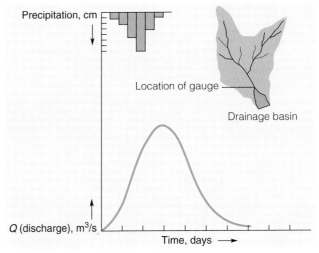

◆ FIGURE 9.11 A hypothetical hydrograph that relates precipitation and discharge to time. Note that peak discharge occurs after the peak rainfall intensity. The drainage basin includes the main stream and all its tributaries.

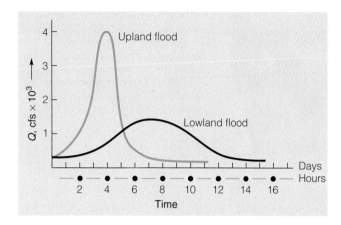

◆ FIGURE 9.12 Hypothetical hydrographs of an upland flash flood and lowland riverine flood.

flow of 720,000 cfs (5.1 million gallons per second) at Boonville, Missouri, is regarded as a 500-year flood. Flood frequency is explained in more detail later in this section.

Floods from all causes are the number-one natural disaster in the world in terms of loss of life. Fifteen disastrous U.S. floods are listed in ✳ Table 9.1. Although by no means a complete listing of severe floods, the table gives an idea of the distribution of riverine floods, flash floods, coastal floods, and dam failures in the United States. Coastal flooding due to hurricanes and typhoons is the greatest single killer. The record-holding river floods are due to dam failures. The legendary 1889 Johnstown flood killed 3,000, and the 1928 St. Francis Dam failure in Southern California killed 450 people.

Flood Measurement

As we have seen, the basic unit in river studies is the drainage basin, which can be defined on any scale. A large river's drainage basin includes the drainage basins of many tributaries, and each tributary's drainage basin includes the drainage basins of all *its* tributaries, and so on. It follows that a large drainage basin has the potential for a larger flood than does a small drainage basin.

Flood studies begin with hydrographs for a particular storm on a river within a given drainage basin. These graphs are generated from data obtained at one or more stream gauging stations, where stage (the water elevation above a previously established datum plane) and velocity are measured and then related to discharge. Rapid runoff in a drainage basin is accompanied by high stream discharge and high water velocity. As discharge increases, the water rises to the *bank-full stage* and then spills onto the adjacent floodplain at *flood stage.* ◆ Figure 9.14 shows such a condition. It is a hydrograph superimposed on the cross section of a hypothetical stream valley. A *stage hydrograph* is most useful for flood planning purposes, because it relates stream discharge to river height, or elevation. One can see at a glance at what point the stream will overflow its banks and at what elevations flood waters will come in contact with specific structures. Note the contrasting hydrographs in ◆ Figure 9.15. Whereas the flash flood in Tucson, Arizona, lasted only about five hours, the North Dakota flood lasted for almost a month. This reflects the climatic and topographic differences between an area with intense, short-duration, monsoonal rains (Tucson), and an area in the north-central United States where a very slow moving low-pressure system stalled and 43 centimeters (16.91 in) of rain fell in a 24-hour period. The Mississippi River Valley flood of 1993 lasted for three months.

Flood Frequency

One of the most useful relationships that can be derived from long-term flood records is that of **flood frequency,** or **recurrence interval;** that is, how often on average a flood of a given magnitude can be expected in a particular location. With this kind of information we are better equipped to design dams, bridge clearances over rivers, sizes of storm drains, and the like.

The information needed for determining recurrence intervals is a long-term record of annual *peak discharge* (the largest flow) for the particular location on the river. The longer the period of record, the more reliable the statistics will be. The peak discharges are then ranked according to their relative magnitudes—with the highest discharge being ranked as 1, the second-largest as 2, and so forth. To serve as an example, selected annual peak discharges on the Rio

◆ FIGURE 9.13 The probable extent of flooding over long time intervals on a hypothetical floodplain. The 2–3-year flood inundates the existing floodplain. The 100-year flood inundates both the existing floodplain and a higher one (Terrace 1), which formed when the river stood at a higher level. The 500-year event inundates an even higher terrace (Terrace 2) and all lower terraces and floodplains.

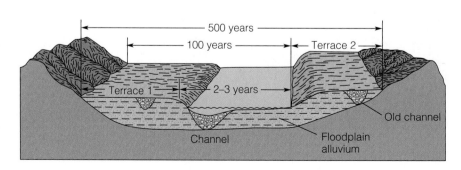

Type	Date	Location	Lives Lost	Estimated Damage $ millions
Regional flood	September 13, 1928	Lake Okeechobee, Florida	1,836	26
	March 1938	Southern California	79	25
	September 21, 1938	New England	600	306
	September–October 1983	Tucson, Arizona	<10	50–100
	Summer 1993	Upper Mississippi River drainage basin	50	12,000
	January–February 1996	Oregon, Washington, Idaho, Montana	1	89*
	October 1998	South-central Texas, San Antonio to Houston, Guadalupe River	29	400
Hurricane	September 8, 1900	Galveston, Texas	6,000	30
	August 17–18, 1969	Mississippi, Louisiana, and Alabama (Hurricane Camille)	256	1,421
	September 1999	North Carolina, Hurricane Floyd and post-hurricane rains	47	5,000+
Flash flood	July 1976	Big Thompson River, Colorado	139	30
Dam failure	May 1889	Johnstown, Pennsylvania	3,000	–
	March 12–13, 1928	California, St. Francis Dam	450	14
	February 1972	Buffalo Creek, West Virginia	125	10
	June 1976	Southeast Idaho, Teton Dam	11	1,000

＊TABLE 9.1 Selected U.S. Floods since May 1889

Source: Various sources, including U.S. Geological Survey.

Grande near Lobatos, Colorado, are listed in ＊ Table 9.2, with their ranks and recurrence intervals. A recurrence interval, *T*, is calculated as

$$T = (N + 1) \div M$$

where *N* is the number of years of record (76 years in our example), and *M* is the rank of the event. The analysis predicts how often floods of a given size can be expected to occur in a watershed. In this example, a flood the size of the 1906 flood ($M = 10$) would be expected to occur on average once every 7.7 years [$(76 + 1) \div 10$]. If 100 years of records are available, the largest flood to occur during that period ($M = 1$) would have a recurrence interval of about 100 years on average. This is the so-called 100-year flood. The 100-year

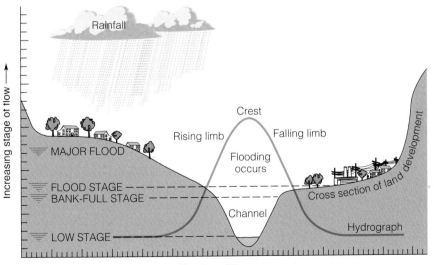

◆ FIGURE 9.14 A stage hydrograph superimposed on a hypothetical stream valley, which illustrates the impact of increasing stage height on the flooded area.

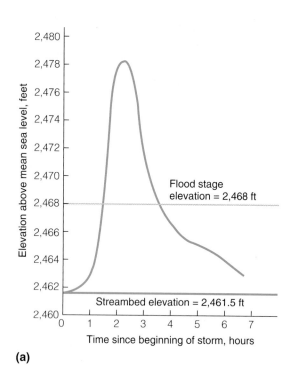

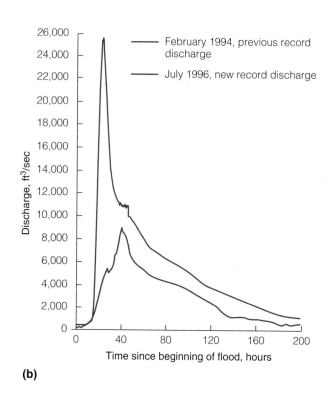

(a) **(b)**

♦ FIGURE 9.15 (a) Stage hydrograph for a 5-hour storm in 1975 on Tanque Verde Creek in Tucson, Arizona. Note that the water rose 10 feet above flood-stage elevation and returned to below flood stage in about 2 hours. All the damage was done in a very short time. (b) Flood discharge of July 1996 on the South Branch Kishwaukee River near Fairdale, Illinois. Of the approximately 120 hours of flooding in 1996, flooding was severe for 3 days.

flood can also be obtained by projecting rainfall or streamflow data beyond the period of record.

Big floods seem to be occurring more often now in the United States, begging the question, How can "the 100-year

*TABLE 9.2	Annual Peak Discharges on the Rio Grande River near Lobatos, Colorado, 1900–1975		
Year	Discharge, m³/sec	Rank (M)	Recurrence Interval (T)
1900	133	23	3.35
1901	103	35	2.20
1902	16	69	1.12
1903	362	2	38.50
1904	22	66	1.17
1905	371	1	77.00
1906	234	10	7.70
1907	249	7	11.00
1908	61	45	1.71
1909	211	13	5.92
1974	22	64	1.20
1975	68	43	1.79

Source: U.S. Geological Survey.

flood" occur so *often*? The term is misleading, because it implies that a flood of the particular size can happen only once in 100 years. "The 100-year flood" is really a statistical statement of the probability that a flood of a given *rank* will occur in any one year; that is, $1/T$. For the 100-year flood, there is a 1 percent chance (1/100) of its occurring in any one year; and for the 50-year flood (1/50), a 2 percent chance. The U.S. Geological Survey says a better term would be "the 1-in-100 *chance* flood." Viewed as a probability phenomenon, it is more understandable that we might get big floods in successive years or every few years. Keep in mind that the longer the rainfall or streamflow record, the better the statistic. Look at the streamflow data collected between 1940 and 1996 on the Chehalis River in Washington State in ♦Figure 9.16. Note that the 1-in-100 chance flood calculated for data from 1940 to 1975 is lower than the same statistical flood based upon data for a period 21 years longer. This illustrates the importance of long-term data collection in assessing the magnitude of the 1-in-100 chance flood. There is, however, no meteorological or statistical prohibition against more than one 100-year flood occurring in a century, or even in one year, for that matter. Recurrence intervals are not exact timetables; they merely indicate the statistical probability of a given-size flood. However, knowing the flood discharge that can be expected in a river system every 25, 50, or 100 years enables us to better plan against the flooding that is certain to occur eventually.

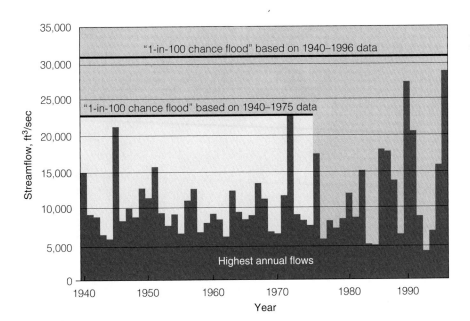

◆ FIGURE 9.16 Streamflow data for the Chehalis River near Doty, Washington. Note the difference in streamflow for the 100-year flood based upon a 35-year data-collection period and one of 56 years in length.

Once we have determined recurrence intervals from the historical record, it is possible to construct flood-frequency graphs like the ones in ◆Figure 9.17. Such graphs allow us to estimate at a glance how often a flood-level or discharge of a particular magnitude should occur. For example, if we wish to build a warehouse with a useful life of 50 years on the floodplain of the Red River at Grand Forks, we see in the figure that the river reaches a reference elevation of about 49 feet every 50 years on average. We would probably want to build the warehouse at or above that elevation. However, urbanization, artificial levees, and other human-generated changes can alter the amount and timing of runoff and cause higher-discharge flood events to become more common. Consequently, the "100-year flood" may become the 70- or 80-year flood.

The current thinking is that a river's response to higher-than-average water stages can change in time. Land-use changes, engineering modifications of the river channel, and long-term climatic changes require that recurrence intervals, flood levels (stages), and floodplain management must be updated periodically. Many researchers believe we should rethink traditional mathematical analysis and include changing climate and climate cycles to yield a more accurate prediction of flood frequency. As an example, let's examine how often we can expect the Red River at Grand Forks, North Dakota, to reach a flood stage of 10 meters above the datum. If we choose the period from 1882 to 1920, when it was wet and cool, we see that there were 12 floods exceeding 10 meters. From 1921 to 1940 it was much drier and there were no floods exceeding 10 meters during that period, but from 1941 to 1980, when it was wetter and warmer, 20 floods were greater than the 10-meter mark. Since 1980 it has been wetter than any preceding period and has produced 12 floods above 10 meters. For this reason it has been proposed that we use analogous climatic periods for the data we plug into our statistical equations (◆Figure 9.18). If we use data from the dry Dust Bowl years of the 1930s and mix it with the current wet-period data, we can drastically underestimate recurrence intervals and flood hazards.

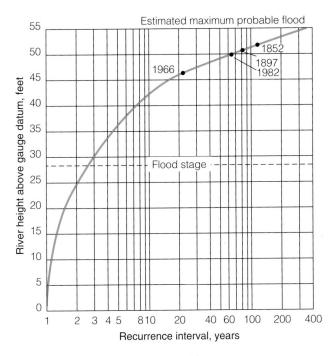

◆ FIGURE 9.17 Flood-frequency curves for two rivers, (a) the Red River at Grand Forks, North Dakota, using water elevation as the standard, and (b) the Rio Grande River near Lobatos, Colorado, using water discharge as the measure.

CONSIDER THIS *Artificial levees proved to benefit some communities at the expense of others during the Great Flood of 1993. What benefits are gained by constructing levees, and what damage may levees cause downstream along a river course?*

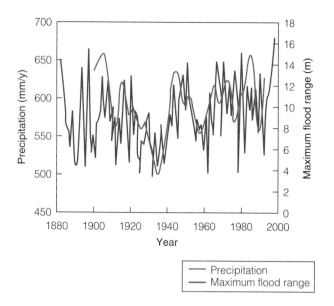

◆ FIGURE 9.18 This plot of long-term flood height at Grand Forks, North Dakota, against smoothed regional precipitation shows correlation between floods and climate cycles.

The Flood of 1993

Forty percent of all the precipitation falling on the contiguous 48 states is collected in the Mississippi River drainage basin. Thus it is not surprising that one of the largest riverine floods attributable to rainfall was the great Mississippi Valley flood of 1993. The flood was the result of a wet spring followed by record-breaking rains in June and July. Above-normal precipitation in the six months preceding the flooding filled reservoirs and produced stream flows that were well above normal. In June and July the cool, dry air of the jet stream dipped abnormally south and collided with moist air from the Gulf of Mexico. This caused heavy rainfall in parts of nine midwestern states. Large streamflows, exceeding those expected only once every 100 years, were recorded at 45 gauging stations in these states. Flooding along the Mississippi and its tributaries inundated 4 million hectares (10 million acres) in the nine states. At least 50 lives were lost, and damage estimates ranged from 10 to 12 billion dollars.

Most of the rain fell in the upper Mississippi drainage basin, and the flood crests migrated progressively downstream from St. Paul, Minnesota (June 26), to Dubuque, Iowa (July 6), to Quincy, Illinois (July 13), and finally to St. Louis, Missouri. Levees were overtopped, and the adjacent floodplains inundated in spite of volunteers' heroic efforts to raise them by sandbagging. The July 19 crest at St. Louis was 46 feet above the datum plane, 3 feet higher than ever before recorded. The next day the river crested at 47.1 feet, setting another record. On August 1, the river surged again to a record 49.4 feet.

At the peak of flooding the Mississippi River valley was referred to as the sixth Great Lake (◆Figure 9.19). Flooding was most severe where artificial levees failed and along stretches of the river where natural levees had not been raised

(a)

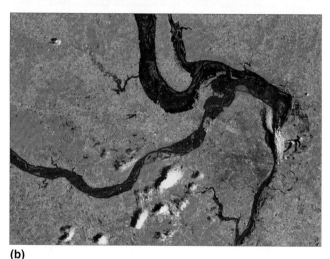

(b)

◆ FIGURE 9.19 Satellite photos of the confluence of the Mississippi, Illinois, and Missouri Rivers during (a) the drought of 1988 and (b) the flood of 1993. The top of the photos is north. The Illinois River flows south to the Mississippi, joining it at Grafton, Illinois. The Missouri River flows eastward across Missouri and empties into the Mississippi just north of St. Louis.

artificially. Rivers flood naturally, and the only way to reduce losses is to limit floodplain development.

Mitigation
Options

Dams: The Solution or the Problem?

In the contiguous 48 states there are more than 75,000 dams more than two meters high, and nearly every major river is controlled by a dam. No one doubts that dams provide such societal benefits as cheap electricity, flood control, recreation, and reduction of fear of drought. However, the negative side is loss of habitats, loss of the natural cycles of flooding and sediment deposition (◎ Case Study 9.3 on page 255), changes in aquatic and riverine biology, and destruction of important spawning grounds for migratory fish such as salmon. The halcyon days of dam construction in the United States are over. Dams are too expensive, they displace huge numbers of people, and the reservoir and downstream environmental impacts are too great (see Case Study 9.3). Although more dams are being built in developing countries, only a few sites in the United States, Canada, and Europe are seriously being considered in the early twenty-first century.

The water emerging from a dam is not the same as the natural river water before damming. Sediment-free outflow scours the riverbed below the dam; because the temperature is different, native fish either are stressed by it, or they die; and new **riparian** communities arise in the downstream area. In addition, some dams fail, and dam-failure floods have accounted for loss of life (see Table 9.1). For all these reasons, dams are being dismantled or "fiddled with" to make them more ecologically acceptable. The Kennebec River in Maine now flows freely, because the 162-year-old Edwards Dam in Augusta was breached in 1999. Other dams scheduled for breaching are the Elwha Dam in Washington (to make way for spawning Pacific salmon) and ten hydroelectric dams on the Deerfield River in Vermont and Massachusetts (to allow passage by Atlantic salmon and other migratory species). This dismantling is occurring because of a little-known 1996 federal law requiring that environmental concerns be addressed before dams can be relicensed. Unfortunately, for the dam haters, many dams sequester polychlorinated biphenyls (PCBs), which are organic pollutants that are toxic to fish and other aquatic and terrestrial life. These dams cannot be removed or "tinkered" with. Lakes are also important to water birds, which further complicates the picture. It will be interesting to see how the call for dam removal plays out.

Artificial Levees: The Solution or the Problem?

Dams, retaining basins, floodwalls, and artificial levees are time-honored means of flood control. All of these are attempts to keep floodwaters within the river channel or to store water for slow release at a later time. Unfortunately, population and property values adjacent to and below flood-control structures increase faster than the land can be protected. That is, floodplain development occurs in response to speculation that flood control has been achieved or is not far off. For example, many artificial levees were built along the Mississippi River to protect farmland. The floodplain was later urbanized, and the existing levees, though adequate to protect farmland, are not adequate for protecting homesites (◆ Figure 9.20).

An artificial levee is built upward upon a natural levee or riverbank so as to increase the amount of water the channel can accommodate and thus prevent the natural periodic flooding on adjacent land. When streamflow is high, the area's would-be floodwater flows downstream to areas that have lower natural levees or no levees, which then experience greater flooding than they would if the artificial levee had not been built upstream. Some of these downstream communities will build their own artificial levees to prevent this

in the future. Then areas without artificial levees downstream from those levees are subject to flooding, so more levees are built there, and so it goes. Artificial levees are thus self-perpetuating, and the Mississippi River today is largely contained within them. Even at nonflood river levels, the water surface is above the adjacent floodplains, so the danger of flooding is ever present.

Riverside land adjacent to the old French Quarter of New Orleans is extremely valuable tourist property. The natural levees here were about 2.1 meters (7 ft) above sea level (*a.s.l.*) in the 1700s, and natural flood stages probably never exceeded 3.0 meters a.s.l. However, confinement of the river upstream by artificial levees raised the flood level to 6.5 meters (21 ft) in modern times, and artificial levees were constructed. Due to the real estate's historic and economic value, it was felt that additional protection should be provided as a safety factor—such as that provided by extending the levees to an elevation of 7.4 meters a.s.l. (24 ft). Because levees require a large land area (to provide a slope of 20 meters horizontal to 1 meter vertical), it was decided to construct a **floodwall** just landward of the existing levee's crest instead. The floodwall has gates that close during flooding events (◆ Figure 9.21). The 13-kilometer-long (8-mi) wall looks innocent enough, but it is built of 2-foot-thick reinforced concrete resting on a continuous wall of wooden piles driven 7.6 meters (25 ft) into the ground. The "sheet" piles in turn rest upon concrete piles that extend down another 15 meters (almost 50 ft). The sheet piles prevent underseepage of water during floods. New Orleans's solution to the flooding problem has served the city well for some time, but eventually the river will win the battle and reclaim its cypress swamps and floodplains, or it will shift its course to another distributary.

Insurance, Floodproofing, and Floodplain Management

The National Flood Insurance Act of 1968 provides federally subsidized insurance protection to property owners in flood-prone areas. The Federal Insurance Administration chose the 100-year-flood level as the **Regulatory Floodplain** on which to establish the limits of government management and set insurance rates. Maps have been developed that delineate Regulatory Floodplains in literally thousands of communities in the United States. The subsidized flood insurance is not available on new homes within the regulatory floodway (◆ Figure 9.22). Pre-1968 houses in the floodway were "grandfathered" (allowed to stay and to be insured under the government program), but they cannot be rebuilt at that site with flood-insurance money if they are more than 50 percent damaged by flooding.

Unfortunately, there are serious problems with the 1968 act. Because the insurance is government-subsidized, owners need not assume the full insurance risk for homes in flood-prone areas. This subsidy serves to encourage *more* building in flood-prone areas, because the land is cheap and the risk is subsidized. Also, as with much well-meaning federal legislation, enforcement is at the local-government level and thus varies considerably from place to place.

◆ FIGURE 9.20 Artificial levees protect land from the river and the sea and sometimes need protection themselves. (a) A levee with a road on the top protects an industrial area on the floodplain northeast of New Orleans. (b) A levee protects homes from the waters of Lake Pontchartrain.

(a)

(b)

Three methods of flood-proofing are (1) raising structures above the 100-year-flood level by artificial filling with soil, (2) building walls and levees to resist floodwaters, and (3) using water-resistant building materials. Good floodplain management limits the uses of floodplains in chronically flood-prone areas. As noted earlier, alluvial fans in southwestern deserts are subject to flash floods and debris flows from intense storms. Nonetheless, they are extensively developed in resort areas of the Southwest such as Palm Springs, California, although they are subject to strict regulation. Fans present a special problem, because the chance of flooding near the fan's apex does not determine the probability or depth of flooding below that point. Flood intensities decrease downfan, from life-threatening channelized debris flows near the apex, to a zone where streams rapidly change channels in a "braided" drainage pattern, to the area of little hazard at the toe of the fan (◆Figure 9.23). The building pads of homes built on alluvial fans near Palm Springs are required to be raised artificially several feet above the natural

◆ FIGURE 9.21 The French Quarter as seen from atop the natural levee on the Mississippi River. Note that the levee slopes downward toward the French Quarter. The wall and gate are part of a floodwall built to hold back floodwaters should the river overtop the levee.

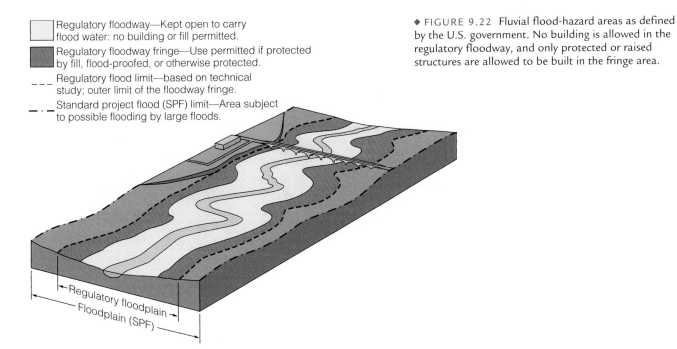

Regulatory floodway—Kept open to carry flood water: no building or fill permitted.

Regulatory floodway fringe—Use permitted if protected by fill, flood-proofed, or otherwise protected.

Regulatory flood limit—based on technical study; outer limit of the floodway fringe.

Standard project flood (SPF) limit—Area subject to possible flooding by large floods.

◆ FIGURE 9.22 Fluvial flood-hazard areas as defined by the U.S. government. No building is allowed in the regulatory floodway, and only protected or raised structures are allowed to be built in the fringe area.

ground so as to locate the structure above possible floodwater levels. When built in this manner, they qualify for federally subsidized flood insurance.

Urban development presents special problems, because it alters the hydrology of a watershed. Because of urbanization floods peak sooner and their total runoff and peak flow are greater. Urbanization represents the replacement of permeable soil with impermeable pavement, and it has been found that when 50 percent of a drainage basin is urbanized, the frequency of overbank flows (floods) is four times what it was before urbanization.

Development has literally poured onto floodplains in the United States. Annual flood damage (in constant dollars) almost tripled between 1950 and the mid 1990s to an average of $3 billion per year. As a result of the great flood of 1993 in the Midwest, the Senate Committee on Environment and

Public Works stopped legislation that would have authorized dozens of new flood-control projects. The new belief is that flood policies must be changed, because traditional structural solutions—levees and dams—contribute to the problem.

Floods are natural and inevitable, as demonstrated in this chapter. A congressional report perhaps summed it up best in noting that "Floods are an act of God. Flood damages result from acts of men."

CONSIDER THIS *You wish to build a house on an alluvial fan in the desert. In order to buy federally subsidized flood insurance you would have to raise the building pad several feet near the apex of the fan, but very little or not at all near the toe. What is it about the hydrology of alluvial fans that justifies these differing requirements?*

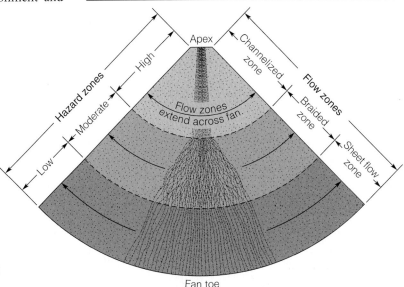

◆ FIGURE 9.23 Flood-hazard zones on an alluvial fan. Hazards exist across the full width of the fan.

◆ FIGURE 9.24 Typical sea-surface temperatures (winter) and surface winds during warm and cool phases of Pacific Decadal Oscillation. These phases are currently thought to last for 20 to 30 years and thus a drought or wetter than normal years could last half a lifetime.

Warm phase

Cool phase

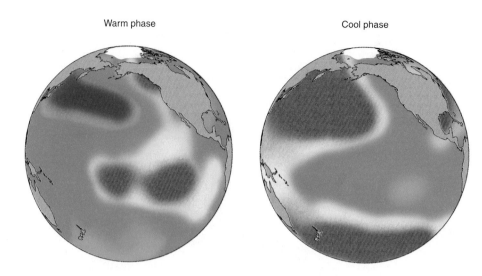

When There Is Too Little Water

Drought is no stranger to the southwestern United States. In 2002, Arizona, Colorado, Utah, Nevada, and Wyoming had one of the driest summers since meteorologists began taking records. The Colorado River ran almost 85 percent below normal, a 100-year low, and the Rio Grande failed to reach the Gulf of Mexico for the first time in 50 years. Of course, water diversion for agriculture and urban use is part of the problem. But what worries climatologists most is a relatively new finding called the Pacific Decadal Oscillation (PDO). It is a long-lived pattern that lasts as long as 20–30 years, not 6–18 months like El Niño. Unlike El Niño (Chapter 10) it is most prevalent in the northern Pacific Ocean and doesn't affect tropical regions (◆Figure 9.24). It greatly impacts the onshore flow of moisture-laden air depending upon the position of the warm and cool phase of the PDO. For instance, during a warm phase Seattle's rain may fall over Alaska. Climatologists have not found a cause for the oscillation, but its recognition is important because it shows that climatic conditions can persist over significantly long time periods. In other words, drought or flood conditions could persist for many years if the PDO turns out to be a reality. More study is needed and, for now, it will become part of the climate-change and global-warming puzzle.

Flood Facts and Flood Planning

The following flood and flood-control information is useful for planning and survival.

- Floods are the most consistently destructive natural hazard in the United States.
- Flooding cannot be controlled completely, but its damaging effects can be mitigated by restricting floodplain development or by government acquisition of floodplain lands.
- Flood-damage costs have been increasing at a rate of 5 percent annually over the past several decades in real dollars.

- Urbanization increases flood peaks on small streams by two to six times.
- Use of the 100-year flood standard for regulation has been effective, but new research indicates that we need to include climate change in our statistical analysis. The design of critical facilities should include elevations beyond the reach of the 500-year flood.
- Do not drive through floodwaters, because a vehicle begins to float when the water reaches its chassis and muddy waters hide washouts and deeper water (◆Figure 9.25).

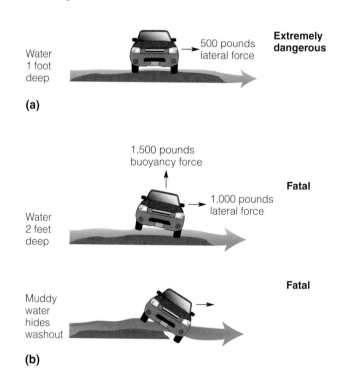

◆ FIGURE 9.25 (a) A vehicle is buoyed up and begins to float when the floodwaters reach about .6 meters (2 ft). This allows lateral forces of the rushing water to push it off the road. (b) Washed-out roadways are obscured by muddy water, allowing vehicles to drop into unexpected deep water. USGS

The Nile River—Three Cubits between Security and Disaster

Flooding is a natural part of the hydrologic cycle, and as farmers, hydrologists, and geologists know, it can be beneficial. The annual flood on the Nile River, the longest river in the world (depending upon how one measures the Amazon), was a source of life for at least 5,000 years. Eighty percent of the water for these floods came from the faraway Ethiopian Highlands, and it replenished the soil moisture and deposited a layer of nutrient-rich silt on the floodplain. The Greek historian and naturalist Herodotus, in 457 B.C., recognized the importance of annual flooding when he spoke of Egypt as "the gift of the Nile." The priests of ancient Egypt used principles of astronomy to calculate the time of the river's annual rise and fall. Records of the height of the flood crests go back to 1750 B.C. (some say to 3600 B.C.), as measured with primitive hydrographs known as Nilometers (◆ Figure 1). Water height was measured in cubits, one cubit being the distance from one's elbow to the tip of the middle finger, approximately 0.5 meter. In Shakespeare's *Antony and Cleopatra*, Mark Antony notes that

> They take the flow o' th' Nile . . . .
> By certain scales . . . they know
> By th' height, the lowness, or the mean, if dearth
> Or foison follow. The higher Nilus swells,
> The more it promises. As it ebbs, the seedsman
> Upon the slime and ooze scatters his grain,
> And shortly comes to harvest.

And, of course, John Lennon of the Beatles supposedly once pointed out that "denial is not a river in Egypt."

The annual flooding cycle has changed since completion of the Aswan High Dam on the Upper Nile in southern Egypt in 1970. The dam's purposes were to increase the extent of arable land, generate electricity, improve navigation, and control the flooding on the Lower Nile, and is accomplishing all of those things. Now that the flooding is controlled, however, crop maintenance depends upon irrigation. Severe environmental consequences have resulted, including depletion of the soil nutrients that are beneficial to farming (necessitating the use of artificial fertilizers) and the proliferation of parasite-bearing snails in irrigation ditches and canals. *Schistosomiasis* is a human disease found in communities where people bathe or work in irrigation canals containing snails serving as temporary hosts to parasitic flatworms (schistosomes). These revolting creatures, commonly called blood flukes, invade humans' skin when they step on the snails while wading in the canals. The flukes live in their human hosts' blood, where the females release large numbers of eggs daily, which cause extensive cell damage. Next to malaria, *schistosomiasis* is the most serious parasitic infection in humans. Its outbreak in Egypt is a result of controls on the river that prevent the generous floods that formerly flushed the snails from canals and ditches each year. Another negative consequence is the loss of a huge shrimp fishery off the Nile delta; the dam traps the nutrients that encouraged the growth of plankton, upon which the shrimp larvae fed. Every dam has a finite useful lifetime that is determined by how fast the lake it impounds fills with sediment. Lake Nasser, behind the Aswan Dam, is rapidly filling with the (precious) sediment that formerly enriched the floodplain of the lower Nile.

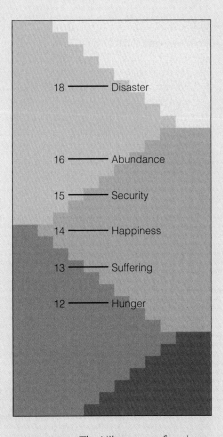

◆ FIGURE 1 The Nilometers of ancient Egypt were primitive hydrographs used to measure the Nile's flood heights in cubits, a cubit being the distance between one's elbow and the tip of the middle finger (usually about 18 in). The environmental interpretations shown here were used during the time of Pliny the Elder (A.D. 23–79) of Mount Vesuvius fame.

CASE STUDY 9.2

Water, Water, Everywhere

New Orleans, the Crescent City, has the flattest, lowest, and youngest geology of all major U.S. cities. The city's "alps," with a maximum elevation of about 5 meters (16 ft) above sea level, are natural levees built by the Mississippi River, and its average elevation is just 0.4 meters (1.3 ft). No surface deposits in the city are older than about 3,000 years. The city was established in 1717 on the natural levees along the river. The levees' sand and silt provided a dry, firm foundation for the original city, which is now called *Vieux Carré* (French "old square"), or "the French Quarter." Farther from the river the land remained undeveloped, because it was mainly water-saturated cypress swamp and marsh formed between distributaries of the Mississippi River's ancient delta. These areas are underlain by as much as 5 meters (16 ft) of peat and organic muck (♦Figure 1). When high-volume water pumps became available about 1900, drainage canals were excavated into the wetlands to the north. Swamp water was pumped upward into Lake Pontchartrain, a cutoff bay of the Gulf of Mexico. About half the present city is drained wetlands lying well below sea and river level. The lowest part of the city, about two meters below sea level, is in the lively, historic French Quarter, noted for its music, restaurants, and other tourist attractions (see Figure 9.21).

Subsidence is a natural process on the Mississippi River delta due to the great volume and the sheer weight of the sediment laid down by the river. The sediment compacts, causing the land surface to subside. The region's natural subsidence rate is estimated to have been about 12 centimeters (4 in) per century for the past 4,400 years. This estimate is based upon C-14 dating of buried peat deposits and does not take into account any rise of sea level during the period.

Urban development in the 1950s added to the subsidence. Compaction of peaty soils in reclaimed cypress swamps coincided with the period's construction of drainage canals and planting of trees, both of which lowered the water table. Peat shrinks when it is dewatered, and also oxidation of organic matter and compaction contribute to subsidence. Most homes constructed on reclaimed swamp and marsh soils in the 1950s were built on raised-floor foundations. These homes are still standing, but they require periodic leveling. Unfortunately, homes built on concrete-slab foundations, a technique that had just been introduced, sank into the muck and became unlivable. Other homes' foundations were constructed on cypress-log piles sunk to a depth of 10 meters or more (at least 30 ft). These houses have remained at their original level, but the ground around them has subsided. This *differential subsidence* requires that fill dirt be imported in order to make the sunken ground meet the house level. It is very common to see carports and garages that have been converted to extra rooms when subsidence has cut off driveway access to them (♦Figure 2). One also sees many houses with an inordinate number of porch steps; the owners have added steps as the ground has sunk. In 1979, Jefferson and Orleans Parishes (counties are called "parishes" in Louisiana) passed ordinances requiring 10- to 15-meter-deep wooden-pile foundations for all houses built upon former marsh and swamp land. This has been beneficial, but differential subsidence continues to damage New Orleans' sewer, water, and natural-gas lines and streets and sidewalks.

♦ FIGURE 1   High-altitude false-color image of New Orleans, a city almost completely surrounded by water. The light spot in the lower center is the city. Note that the Mississippi River meanders through it, Lake Pontchartrain is to the north, and smaller Lake Borgne is to the northeast. Low, swampy delta lands appear brownish to the east and south. The land width in the photo is approximately 200 kilometers (125 mi).

B. PIPKIN

◆ FIGURE 2 Differential subsidence around this pile-supported house has left the carport high and dry. The carport was converted to a family room, and fill was imported to bring the yard surface up to the previous level.

CASE STUDY 9.3

Dams—No Longer in Vogue?

Damming has long been a means for making river water more available to humans. The Three Gorges Dam on the Yangtze River in China is a modern example (◆Figure 1). Begun in 1993, it is 181 meters (594 ft) high and will be completed in 2009 at a cost in excess of $24 billion. Its main functions are power generation, flood control, and improved navigation. In 2003 the first electricity was generated and the first ship lock was operating. The dam has been criticized because of cost, resettlement of displaced persons, environmental impact, and adverse effects on local culture and history. In response, the protagonists say the electrical transmission will eventually pay for the dam, millions of people downstream will be better off, most of the

important historical sites are being moved before they are inundated, shipping will become faster and cheaper, and hydroelectric power is cleaner than other large-scale energy sources. Time will tell whether this ambitious project is a success or a flop.

In the United States new dams are dimly viewed for the same reasons that have been raised at the Three Gorges project. In fact, as this is being written, rivers are being diverted and dams across them removed. There are 7,700 large dams in the country and the total is more than 75,000. Most existing dams serve a useful purpose—for energy, flood control, recreation, and as a water source. However, many environmentalists, geologists, and engineers believe that the era of large dam

building in the United States is over. The main reasons are the loss of land by impounded water and the resulting relocation of wildlife and humans.

The United States has more than 2 million miles of rivers and streams, which includes major waterways and intermittent streams that flow only in wet weather. Rivers and streams cannot be divorced from other sources of fresh water as they connect with lakes and ground water. In fact, it is now common to view ground water and surface water as a single resource.

CONTINUED

◆ FIGURE 1 Three Gorges Dam Project (TGDP) under construction on the Yangtze River in summer of 2002.

CASE STUDY 9.4

Rivers, Gorges, and National Parks

Throughout the world national parks are areas of historical and scientific interest as well as places of great beauty. U.S. National Parks exhibit a variety of terranes and origins; for example, Hawaii Volcanoes and Yellowstone are volcanic, Yosemite is glacial, and the Grand Canyon and Bryce have been shaped by river erosion.

The U.S. government's designation of areas to be protected and preserved began when Abraham Lincoln mandated that the State of California should pro-tect Yosemite Valley. In 1872 Congress enacted legislation that provided funds for the federal government to protect, preserve, and maintain Yellowstone in its natural condition. Thus was the U.S. national park system created, and Yellowstone is considered to be our first national park. Other areas needing protection for ensuring their enjoyment by the public were added to the National Park roster early in the system's development—most notably, Crater Lake National Park in 1902, Glacier National Park in 1910, Hawaii Volcanoes National Park in 1916, and Grand Canyon and Zion in 1919.

Our fifty-fifth national park, Black Canyon National Park in Colorado (◆Figure 1), was the last new park to be established in the twentieth century. The Gunnison River there has cut deeply into imposing Precambrian schist and gneiss laced with veins of granite, and the near-vertical canyon walls are 830 meters (2,700 ft) high. The spectacular Black Canyon gorge is the heart of this park.

B. PIPKIN

◆ FIGURE 1 Black Canyon of the Gunnison River in Colorado. The river has eroded a classic V-shaped gorge in Precambrian schist and gneiss. The area was declared a National Park in 2002.

gallery

Going . . . Going . . . Gone!

To geologists the most famous U.S. dam failure was that of the St. Francis Dam in southern California in 1928. That failure took 450 lives. The dam was part of an aqueduct system bringing water from the east slope of the Sierra Nevada to semiarid Los Angeles and its growing suburbs. The left abutment of the concrete dam was founded in sandstone, and the right abutment in mica schist (◆Figure 1). The two rock formations were in contact along a 2-meter-wide shattered fault zone at the left abutment. The dam was completed in 1926, and more than two years was required to fill the reservoir with 38,000 acre-feet (12.5 billion gallons) of water. On March 12, 1928, just one week after the reservoir had filled, the dam failed. Traditional theory has it that the dam failed along the fault zone in the left abutment, because the rocks there contained gypsum. Water percolating through the fault zone dissolved the gypsum and, eventually, full reservoir pressure blew out the weakened sandstone. Newer analysis, however, suggests that the right abutment was placed into an unrecognized landslide in the schist, and that the slide mass collapsed when the reservoir filled, leading to failure of the right abutment and the dam (◆Figure 1). Whatever the cause, the man who built the dam, William Mulholland, took the failure personally and died a short time later. He is said to be the four hundred fifty-first victim of the flood.

A dam just isn't supposed to mess up a beautiful trout stream, as the one in ◆Figure 2 used to. Whether the hole was due to poor construction or the wrath of an angler or cattle rancher is not known.

(a)

(b)

◆ FIGURE 1 (a) The St. Francis Dam in San Francisquito Canyon, Southern California, early March 1928. The right abutment is in Pelona schist; the left abutment (center of photo) is in the Sespe Formation, which is mainly sandstone. (b) All that remained of the dam on March 13, 1928. Security Pacific National Bank Historical Collection, Los Angeles County Library

◆ FIGURE 2 Pristine trout stream in Colorado and a useless dam that probably should never have been built.

◆ FIGURE 3 Parking is to the right, no parking is to the left; Danube River, 2002.

Summary

River Systems

Drainage Basin

The fundamental geographic unit or tract of land that contributes water to a stream or stream system. Drainage basins are separated by divides.

Discharge

The amount of water flowing in a stream channel depends upon the amount of runoff from the land, which in turn depends upon rainfall or snowmelt, the degree of urbanization, and vegetation.

Erosion

The erosive power of a stream is a function of its velocity—the greater the velocity, the greater the erosion. Stream velocity is determined by discharge, channel shape, and gradient.

Base Level

The lowest level to which a stream or stream system can erode. This is usually sea level, but there are also temporary base levels such as lakes, dams, and waterfalls.

Graded Stream

A stream that has reached a balance of erosion, transporting capacity, and the amount of material supplied to it. This condition is represented by a concave-upward profile of equilibrium.

Stream Features

Alluvium

Sediment deposited by a stream, either in or outside the channel.

Alluvial Fan

A buildup of alluvial sediment at the foot of a mountain stream in an arid or semiarid region.

Delta

Forms where a sediment-laden stream flows into standing water. An example is the delta of the Nile River.

Floodplain

A low area adjacent to a stream that is subject to periodic flooding and sedimentation; the area covered by water during flood stage.

Meanders, Oxbow Lakes, and Cutoffs

Flowing water will assume a series of S-shaped curves known as meanders. The river may cut through the neck of a tight meander loop and form an oxbow lake.

Flooding and Flood Frequency

Floods

Upland floods come on suddenly and move with great energy through narrow valleys. Upland floods are often "flash" floods; the water rises and falls in a matter of a few hours. Riverine floods, on the other hand, inundate broad adjacent floodplains and may take many days, weeks, or even months to complete the flood cycle.

Hydrograph

A graph that plots measured water level (stage) or discharge over a period of time.

Recurrence Interval

The length of time (T) between flood events of a given magnitude. Mathematically, $T = (N + 1) \div M$, where N is the number of years of record and M is the rank of the flood magnitude in comparison with other floods in the record. This enables engineers and planners to anticipate how often floods may occur and to what elevation the water may rise.

Flood Probability

The chance that a flood of a particular magnitude will occur within a given year based on historical flood data for the particular location. Using the calculation for T established for the recurrence interval, probability is calculated as $1/T$. Thus the 100-year flood has a 1 percent chance of occurring in any one year. It is the 1-in-100 *chance* flood.

Mitigation

Structures

Dams, retaining basins, artificial levees, and raising structures on artificial fill are common means of flood protection. They also present problems.

Flood Insurance

The National Flood Insurance Act of 1968 provides insurance in flood-prone areas, provided certain building regulations and restrictions are met. There are many problems with this tax-supported program.

Urban Development

Hydrology

Urbanization causes floods to peak sooner during a storm, results in greater peak runoff and total runoff, and increases the probability of flooding. Floods, including coastal flooding during hurricanes, are the greatest natural hazards facing humankind.

Planning and Survival

Flooding is the most consistently damaging natural disaster in the United States.
Do not drive through flood waters unless you are sure they are shallow.

Key Terms

alluvial fan	discharge	geologic agent	natural levee
alluvium	drainage basin	graded stream	oxbow lake
artificial levee	drainage divide	gradient	point bar
base level (stream)	flood frequency	hydrograph	recurrence interval
cutoff	floodplain	hydrology	Regulatory Floodplain
delta	floodwall	meander	riparian

Study Questions

1. Distinguish between weathering and erosion.

2. How can we prevent floods or at least minimize the property damage caused by flooding?

3. How is the velocity of a stream related to its erosive power, discharge, and cross-sectional area?

4. What has been the impact of urbanization on flood frequency, peak discharge, and sediment production?

5. According to the local newspaper, your town on the Ohio River has just experienced a 50-year flood. What is the probability of a repeat of this event the following year? What is the weakness of such statistical forecasts?

6. How does the hydrograph record of a flash flood differ from that of a lowland riverine flood? Where do we find flash flooding, and what precautions should be observed in such areas?

7. What has the federal government done to alleviate people's suffering due to flooding? Do you believe this is a good thing? Does it promote more development in flood-prone areas? If so, does this development itself have possible negative impacts? Support your answers with examples and rationales.

8. Consider a stream whose headwaters are 1,500 meters (4,900 ft) above sea level and that flows 300 kilometers (180 mi) to the ocean. What is the average gradient of the stream in m/km and in ft/mi?

9. What benefits do river dams impart to humans? What are some negative impacts of dams upon human beings and the environment?

10. What is being done today to reverse some of the negative impacts of dams on the environment? Cite examples.

For Further Information

Books and Periodicals

Baker, Victor. 1997. *Hydrological understanding and societal action.* Forest hydrology symposium presentations, April 10, 1997, Department of Geography, University of British Columbia, Vancouver, B.C.

Carrier, Jim. 1991. The Colorado: A river drained dry. *National geographic*, June.

Collier, Michael, and others. 1996. *Dams and rivers: A primer on the downstream effects of dams.* U.S. Geological Survey circular 1126.

Devine, Robert S. 1995. The trouble with dams. *Atlantic monthly*, August: 64–74.

Dinacola, Karen. 1996. The "100-year flood." U.S. Geological survey fact sheet FS-229-96.

EOS. 1996. *Controlled flood of Colorado River creates stream of data.* Transactions of the American Geophysical Union 77, no. 24 (June 11).

Fischetti, Mark. 2001. Drowning New Orleans. *Scientific American,* October.

Gosnold, William, and others. 2000. Floods and climate. *Geotimes*, May.

Holmes, Robert R., Jr., and Amit Kapadia. 1997. Floods in northern Illinois, July 1996. U.S. Geological Survey fact sheet FS-097-97.

Lucchitta, Ivo, and Luna B. Leopold. 1999. *Floods and sandbars in the Grand Canyon. GSA today* 9, no. 4 (April).

McPhee, John. 1989. *The control of nature.* New York: Farrar, Straus & Giroux.

National geographic. 1993. Water: the power, promise and turmoil of North America's fresh water. Special edition, October.

Perry, Charles A. 2000. Significant floods in the United States during the 20th century. U.S. Geological Survey fact sheet FS-024-00, March.

Rogers, J. David. 1992. Reassessment of the St. Francis dam failure. In Pipkin, B. W., and Richard Proctor, eds., *Engineering geology practice in southern California.* Belmont, Calif.: Star Publishing Co.

Saarinen, T. F., V. R. Baker, R. Durrenberger, and T. Maddock, Jr. 1984. *The Tucson, Arizona, flood of October 1983.* Washington D.C.: National Research Council, National Academy Press.

Teller, R. W., and M. J. Burr. 1998. Floods in north-central and eastern South Dakota, spring 1997. U.S. Geological Survey fact sheet FS-021-98.

U.S. Geological Survey. 1993. Mississippi River flood of 1993. *U.S. Geological Survey yearbook:* 37–40.

ENVIRONMENTAL
Geology⇌Now™

Assess your understanding of this chapter's topics with additional quizzing and comprehensive interactivities at http://earthscience.brookscole.com/pipkingeo4e

as well as current and up-to-date Web links, additional readings, and InfoTrac College Edition exercises.

Coastal Environments and Humans

REEFS IN PERIL

CORAL REEFS ARE UNIQUE AND ARE RIVALED by only a few other ecosystems. Charles Darwin may have been the first to recognize that they are the largest and most important ecosystems in the marine environment. Corals are animals related to jellyfish that secrete a hard skeleton of calcium carbonate (limestone). They grow only in warm water (above 20° C, which is approximately 70° F) within the zone of light. Corals require light because they live in symbiosis with the tiny marine algae *Zooxanthellae*. These plants live within the coral polyp and give the calcareous skeleton its color (◆Figure 1). They provide food and oxygen to the coral, and the coral provides the carbon dioxide *Zooxanthellae* need for photosynthesis. In 1998 an unprecedented bleaching of corals over large areas in the world was noticed, with the most severe bleaching occurring in the Indian Ocean region.

Coral reefs help people in diverse ways. They form barriers to beach erosion, build atoll islands upon which people live, support fisheries, and attract tourism. Unfortunately, they are being destroyed worldwide at a phenomenal rate. Mining reefs for limestone, dumping mine tailings on reefs, fishing by use of explosives or cyanide, land reclamation, and polluted river runoff all take their toll. Seawater temperature increases due to global warming and El Niño events are natural causes of reefs' destruction. Reefs in Indonesia, North America, Hawaii, and the Great Barrier Reef of Australia are just a few at risk. At higher temperature coral polyps expel *Zooxanthellae*, lose their color, and eventually die if the warm water persists. The bleaching turns a beautiful underwater garden into a spiny wasteland (◆Figure 2).

◆ FIGURE 1 *Healthy coral reefs exhibit a variety of colors and diversity. A healthy reef off the Bahama Islands, with the staghorn coral,* Acropora palmata, *growing to the low-tide water level.*

B. PIPKIN

Everyone who hears these words and does not put them into practice is like the foolish man who built his house on sand. The rains came down, the streams rose, the wind blew and beat against the house, and it fell with a great crash.

Matthew 7:26–27

◆ FIGURE 2 *A bleached reef in the Bahamas that has lost most of its symbiotic algae. The limestone of the coral skeleton is clearly visible. If the environmental stress is not removed, the coral polyps will die. The coral here is elkhorn coral,* Acropora cervicornis.

Coastal states in the United States experienced enormous population growth in the late twentieth century. Apparently, many people want to live near a beach. Fifty percent of the U.S. population can now drive to a coastline in less than an hour. In spite of advances in technology and scientific understandings of coastal dynamics, Matthew's warning is largely unheeded. Nearshore environments are constantly being modified by natural and human intervention, usually with negative impacts. For example, Hurricanes Dennis and Floyd dumped huge amounts of rainfall on North Carolina in 1999, displacing thousands of residents; each spring and summer nitrates from agricultural fertilizers wash down the Mississippi River, creating algal blooms and a 7,000-square-mile "dead zone" in the Gulf of Mexico that threatens commercial fisheries; wetlands are being filled or made into marinas; and shoreline modifications continue to silt in harbors and erode beaches.

Wind-driven waves that arise in the open ocean eventually strike shorelines, generating dynamic processes that constantly alter beaches. Beach erosion is a serious, costly problem in most coastal areas. Armed with knowledge of how waves erode beaches and of the fate of eroded materials, we may be able to mitigate many of the human-induced coastal problems.

Wind Waves

Waves in Deep Water and at the Shore

When wind blows over smooth water, surface ripples are created that enlarge with time to form local "chop" (short, steep waves) and, eventually, wind waves. The factors that determine the size of waves at sea and ultimately the size of the surf that strikes the shoreline are

- the wind velocity,
- the length of time the wind blows over the water, and
- the **fetch,** the distance along open water over which the wind blows (◆Figure 10.1).

For a given fetch, wind velocity, and wind duration, wave dimensions will grow until they reach the maximum size for the existing storm conditions. This condition, known as a *fully developed sea,* represents the maximum energy that the waves can absorb for wind of a given velocity. For example, theoretical wave heights of 32 meters (100 ft) are possible for a 50-knot (93 km/h, 58 mph) wind blowing for 3 days over a fetch of 2,700 kilometers (1,600 mi). Wind waves will move out of the area of strong atmospheric disturbance as regularly spaced, long waves known as *swell* and eventually become huge breakers on a distant shoreline.

Important dimensions of wind waves are their length, height, and *period*—the length of time between the passage of equivalent points on the wave. These dimensions are measures of the amount of *energy* the waves contain—that is, their ability to do work (◆Figure 10.2). Although waves with long wavelengths and long periods are the most powerful, both long and short waves are capable of eroding beaches.

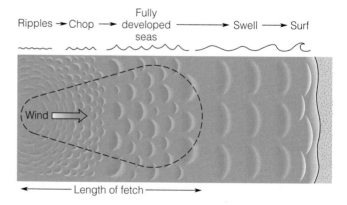

◆ FIGURE 10.1 The evolution of wind-generated sea waves from a rippled water surface to fully developed seas (storm waves). Long waves with very regular spacing and height that outrun the storm area are called *swell.*

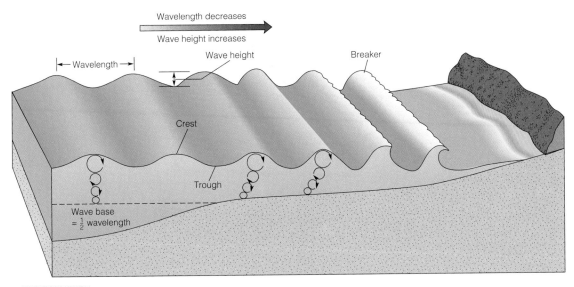

◆ Geology⊜Now™ ACTIVE FIGURE 10.2 Motion of water in a wave and the terminology of wave geometry. Note that the orbital water motion decreases with depth until it is essentially nil at a depth of one-half the wave length (*L*/2), a depth called *wave base*. Energy is transferred from wind to the water to create waves that in turn strike the shoreline and do work.

The actual water motion in a wind-generated wave describes a circular path whose diameter at the sea surface is equal to the wave height. This circular motion diminishes with depth and becomes essentially zero at a depth equal to half the wavelength (*L*/2), the depth referred to as **wave base.** Wave base is the maximum depth to which waves can disturb or erode the sea floor. (This explains why submersibles at depth are not tossed about during storms as surface ships are.) Although the waveform moves across the water surface with considerable velocity, there is no mass forward movement of water. Such waves are called **waves of oscillation.** Floating objects simply bob up and down with a slight back-and-forth motion as wave crests and troughs pass.

As waves approach a shoreline and move into shallow depths, the sea floor interferes with the oscillating water particles. This friction causes wave velocity to decrease. This in turn causes the wave to steepen; that is, the wavelength shortens and the wave height increases, a phenomenon easily seen by surf watchers. Eventually the wave becomes so steep that the crest outruns the base of the wave, and the crest jets forward as a "breaker." Breakers are **waves of translation,** because the water in them physically moves landward in the surf zone. Where the immediate offshore area is steep, we find the classic tube-shaped *plunging breakers* typical of Sunset Beach and Waimea Bay on the north shore of Oahu (◆Figure 10.3). This is because the wave peaks rapidly on a steep sea

◆ FIGURE 10.3 (a) Plunging (tubular) breakers form where there are steep offshore slopes. The front of the wave steepens until the top of the wave "jets" forward and down. This one is at Waimea Bay, Oahu, Hawaii, famous for large waves and surfing tournaments. (b) Spilling breakers form where there are gently sloping offshore areas. The forward face of the wave steepens and the crest spills down the front of the wave.

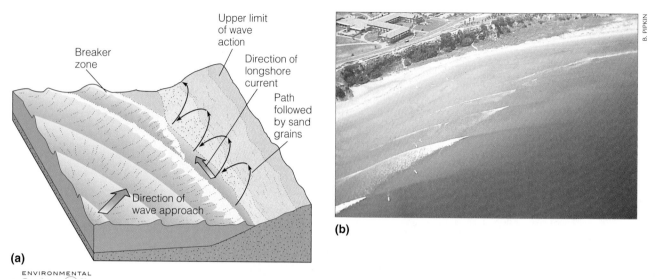

(a)

(b)

B. PIPKIN

ENVIRONMENTAL
◆ Geology⇌Now™ ACTIVE FIGURE 10.4 (a) Refraction, or bending, of wind waves and the generation of a longshore current parallel to the beach. The longshore current transports beach sand, resulting in a dynamic process called *beach drifting*. (b) Wave refraction along a relatively straight shoreline near Santa Barbara, California.

floor. On a flat nearshore slope, on the other hand, waves lose energy by friction, so they build up slowly, and the crest simply spills down the face of the wave. This latter type are called *spilling breakers*. They are typical of wide, flat nearshore bottoms such as that at Waikiki Beach on Oahu (see Figure 10.3). Regardless of the type of wave, *swash*—the water that rushes forward onto the foreshore slope of the beach—returns seaward as *backwash* to become part of the next breaker. A common misconception is the existence of a bottom current called *undertow* that pulls swimmers out to sea. The strong pull seaward one feels in the surf zone is simply backwash water returning to be recycled in the next breaker.

ENVIRONMENTAL
Geology⇌Now™

Click **Geology Interactive** to work through activities on Wave Properties, Wave Progression, and Wind–Wave Relationships through Waves, Tides, and Currents.

Wave Refraction

Where waves approach a shoreline at an angle, they are subject to **wave refraction,** or bending. Waves may be seen to slow first at their shallow-water ends while their deeper-water ends move shoreward at a higher velocity. Thus a fixed point on a wave crest follows a curved path to the shoreline (◆Figure 10.4).

Because waves seldom approach parallel to the shore, some water is discharged along the beach as a weak current within the surf zone called a **longshore current.** The greater the angle at which the wave approaches the shoreline, the greater the volume of water that is discharged into the longshore-current stream. This current is capable of moving swimmers as well as sand grains down the beach. Sand

becomes suspended in the turbulent surf and swash zone and moves in a zigzag pattern down the beach's foreshore slope (see Figure 10.4). Beaches are dynamic; the sand grains present on the foreshore today will not be the same ones we see there tomorrow. Beach-sand transport parallel to the shore in the surf zone is known as **littoral drift,** and its rate is usually expressed in thousands of cubic meters or yards of sand per year.

Another consequence of wave refraction is the straightening of irregular shorelines. This occurs because the concentration of wave energy at points causes shoreline erosion, and the dissipation of wave energy in embayments allows sand deposition (◆Figure 10.5).

Longshore currents are likened to rivers on land, with the beach and the surf zone constituting the two banks. Just as damming a river causes sedimentation to occur upstream and erosion to occur downstream, so does building structures perpendicular to the beach. Such a structure slows or deflects longshore currents, causing sand accretion upstructure and erosion below it (◆Figure 10.6). **Groins** (French *groyne,* "snout") are constructed of sandbags, wood, stone, or concrete, ostensibly to trap sand and create a wider beach. For groins of equal length and design, the severity of the erosion-deposition pattern is a function of the strength of the waves and the angle at which they strike the shoreline. Most groins are short and are intended to preserve or widen a beach in front of a private home or resort. Unfortunately, because this causes neighbors' beaches below the groin to narrow, many of those neighbors build groins to preserve and widen *their* beaches, and so on. Like artificial levees on a river, groins tend to generate more groins, and we generally see fields of them along a given stretch of beach instead of just one (◆Figure 10.7).

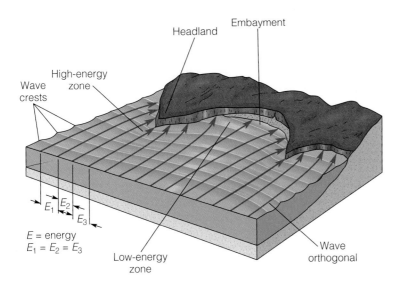

Headland

Embayment

Wave crests

High-energy zone

E_1 E_2
E_3

E = energy
$E_1 = E_2 = E_3$

Low-energy zone

Wave orthogonal

◆ FIGURE 10.5 Wave refraction leads to concentration of energy on points and dissipation in embayments. The arrowed lines, called *orthogonals*, are constructed perpendicular to the wave crest and show the wave path. E_1, E_2, and so on, represent equal parcels of wave energy and, therefore, equal ability to erode. Note that energy is focused toward the points and spread out in the bays.

Just as rivers overflow their banks, longshore currents may overflow through the surf zone, particularly during periods of big surf. Overflow (really *out*flow to the open ocean) occurs because the water level within the surf zone is higher than the water level outside the breakers, making for a hydraulic imbalance. The higher water inshore flows seaward to lower water offshore as a **rip current** through a narrow gap, or *neck*, in the breakers (◆Figure 10.8). Because the location of rip currents, erroneously called rip *tides*, is controlled by bottom topography, they usually occur at the same places along a beach. Rip currents can become a problem to swimmers under big-surf conditions, when velocities in the neck may reach 4–5 knots. Awareness can save your life, so be alert to a gap in the breakers, white water beyond the surf zone, and objects floating seaward—all of which are evidence of a strong rip current flowing out to sea (◆Figure 10.9). Should you find yourself in a rip current, swim parallel to the beach until you are out of the narrow neck region of high velocity.

Impulsively Generated Waves

Tsunamis

Perhaps the most myth-ridden hazard associated with earthquakes are tsunamis (pronounced "soo-nah-meez"). Tsunami is a Japanese word meaning harbor (tsun) and wave

◆ FIGURE 10.6 Accretion and erosion of beach sand at jetties constructed to improve an inlet to a navigable lagoon at Boca Raton, Florida.

◆ FIGURE 10.7 Groin field at Cape May, New Jersey. Note the bread-knife, or sawtooth, appearance that has resulted from sand accretion and erosion above and below the groins.

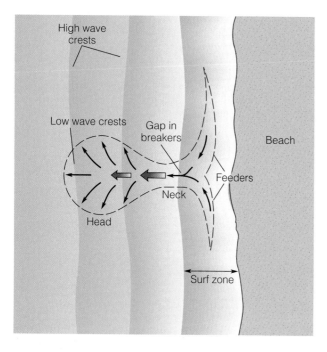

◆ FIGURE 10.8 A rip current is characterized by relatively weak *feeder currents*, a narrow *neck* of strong currents, and a *head* outside the surf zone, where the rip dissipates.

(ami) because harbors are where these waves "resonate" and are most observable after an offshore quake. In Japanese the singular and plural are the same but we will follow the custom of English writers and add the "s" for the plural.

Tsunamis are produced mostly by underwater earthquakes and less commonly by volcanic eruptions, submarine landslides, or meteorite impacts (∗ Table 10.1). These impulses give rise to waves that move out from their source in all directions—analogous to those produced when you drop a rock into a pond. Tsunamis in deep water may have velocities exceeding 800 kilometers per hour (480 mph) and could easily keep pace with a Boeing 747. With wavelengths in excess of 160 kilometers (100 mi) and wave heights of only a meter or two in the open ocean, however, tsunamis have such gentle slopes that they go unnoticed (◆Figure 10.10). As the leading waves move into shallow water they slow to freeway speeds, and the waves behind begin to overtake them—a process called *shoaling*. Thus, the energy contained in many waves is condensed into a smaller volume of water, and the waves increase in height and steepness. Eventually, the wave runs ashore as a huge breaking wave, as a wall of white, foamy water, or as a tidelike flood.

Fortunately, most underwater earthquakes do not generate tsunamis. In almost 100 years (1861–1948), only 124 tsunamis were recorded by more than 15,000 sea-floor earthquakes. The west coast of South America, a seismically active area, experienced 1,100 earthquakes in the same period and only 20 tsunamis. One of the reasons is that many tsunamis have such small amplitudes that they go unnoticed. In addition, earthquake-induced tsunamis require shallow-focus events with surface-wave magnitudes greater than 6.5. The Atlantic Ocean experiences only two percent of the tsunamis, whereas the Pacific Ocean experiences 34 percent of them. To understand the reason for this requires only a basic knowledge of plate tectonics.

Earthquake-induced tsunami may rise to heights of more than 30 meters. The time between crests of the incoming waves is about 20 minutes, and the first crest may not be the largest. The *trough* of the initial wave may arrive first, and many lives have been lost by curious persons who wandered into tidal regions exposed as water withdrew. In Lisbon, Portugal, on All Soul's Day in 1755, many people were in

◆ FIGURE 10.9 Small rip currents. Note the gap in the breakers and the white water seaward of the surf zone.

	Number of Events	Percentage of Events	Number of Deaths	Percentage of Deaths
*TABLE 10.1 Causes of Tsunamis in the Pacific Ocean Region over the Last 2,000 Years				
Cause				
Landslides	65	4.6	14,661	3.2
Earthquakes	1,171	82.3	390,929	84.5
Volcanic	65	4.6	51,643	11.2
Unknown (meteorite?)	121	8.5	5,364	1.2
Total	1,422	100	462,597	100

Source: National Geophysical Data Center and World Data Center A for Solid Earth Geophysics, 1998, and Intergovernmental Oceanographic Commission, 1999, in Bryant, 2001.

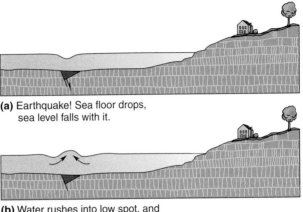

(a) Earthquake! Sea floor drops, sea level falls with it.

(b) Water rushes into low spot, and overcompensates, creating a bulge

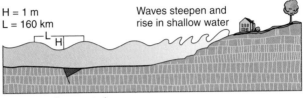

$H = 1$ m
$L = 160$ km

Waves steepen and rise in shallow water

(c) Tsunami generated

◆ FIGURE 10.10 Formation of a tsunami when a portion of the sea floor is down-faulted and the water surface over the fault is lowered. Water rushes in to fill the low spot and over-compensates and creates a bulge. An impulsively generated wave then moves out in all directions with a very high velocity (hundreds of km/h).

church when an earthquake struck. The worshipers ran out-side to escape falling debris and fire, many of them joining a group seeking safety on the waterfront. There was a quiet withdrawal of water followed by a huge wave minutes later. Sixty thousand people died. The length of time of strong shaking has been estimated from historical accounts of how many *Ave Marias* and *Pater Nosters* the churchgoers recited during their fearful experience.

Water withdrawal before the first high wave also con-tributed to fatalities in the 1946 Hawaiian tsunami; many people walked offshore to collect exposed mollusks just before the first wave crest arrived. A good rule of thumb is to head for high ground at once should you witness a sudden recession of water along a beach or coastline.

Tsunamis are generated mostly by large-scale dip-slip faulting in subduction zones that ring the Pacific and Indian Ocean basins. Japanese studies have shown that strike-slip faulting seldom produces tsunamis. In the late twentieth cen-tury earthquake-generated tsunamis around the Pacific Rim damaged Chile (1960), Alaska and the Pacific Northwest (1964), Nicaragua (1992), and New Guinea (1998). The Hawaiian Islands are very tsunami-prone due to multiple tsunami source areas. Between 1850 and 2000 the Islands averaged a damaging tsunami every two years (◆Figure 10.11). A tsunami's damage potential depends upon the earthquake mechanism, the distance the waves travel to the shoreline, the offshore topography, and configuration of the coastline. In 1998 a community on a low, coastal sandbar in Papua, New Guinea, was obliterated by tsunami waves of up to 15 meters (50 ft) high that killed 2,200 people and left few survivors. Experts were puzzled, because the magnitude-7.1 earthquake was considered too small to generate such huge waves. This led to speculation that the earthquake triggered a submarine landslide (see later in this section) that added strength to the tsunami.

In 1992, twenty-six towns along 250 kilometers (150 mi) of Nicaragua's Pacific coast were struck by a 10-meter-high

(33-ft) wave that killed 170 people and left 13,000 homeless. Normally, coastal inhabitants run for high ground when they feel rapid earthquake shaking. Only slight tremors were experienced in this case, however, and people did not suspect danger. Most of the earthquake energy was transmitted in much longer waves (long-period, up to 200 seconds) that were not noticed or were ignored. This kind of quake, known as a "slow" earthquake, was first described by Hiroo Kanamori of the California Institute of Technology. Kanamori believes these quakes may be ten times as strong as the magnitude determined from short-period seismometer readings. Slow earthquakes of local origin are particularly dangerous in coastal areas, because they produce little pre-cursory ground shaking that serves to warn people that a tsunami may be imminent. Scientists believe that up to ten percent of all tsunamis are caused by this kind of earthquake.

The entire Pacific Rim is vulnerable to tsunami inunda-tion. Geologic evidence indicates that within the past 2,000 years several large earthquakes have generated destructive tsunami on the shores of Washington, Oregon, and Northern California. The risk becomes clearer when we recognize that 90 percent of the deaths in the Alaska earthquake of 1964 were tsunami related, and that the geology of the Cascadia subduction zone is very similar to the Alaska coast (◆Figure 10.12). The best defense against tsunami risk is an early

◆ FIGURE 10.11 Tsunami sources and travel paths to the Hawaiian Islands, 1900–1983. USGS data

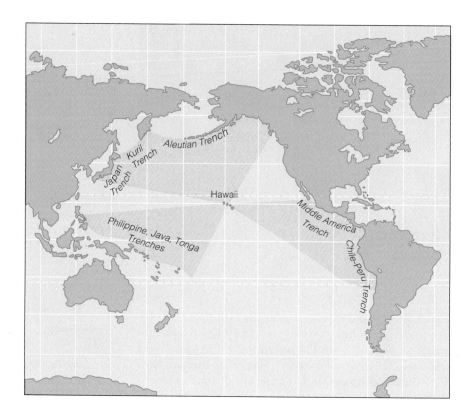

warning system. Useful (life-saving) warning times for the half-million people who live and work in potential tsunami inundation areas of the Pacific Northwest range from a few minutes in the southern part of the Cascadia subduction zone to 20–40 minutes in the northern parts of the zone.

In April 1992, a magnitude-7.1 earthquake generated a small tsunami at Cape Mendocino, California. This event triggered establishment of the National Tsunami Hazard Mitigation Program, involving the five Pacific states and three federal agencies—the U.S. Geological Survey, NOAA, and FEMA. By the late 1990s inundation maps were being prepared for parts of Oregon and Washington (◆Figure 10.13). Most important is the development of a warning system based on real-time tsunami detection. The current Pacific-wide warning system is based on earthquake foci, and about an hour is required to issue a warning. It is therefore useful only for coastal areas more than 750 kilometers (465 mi) from the tsunami source, the approximate distance a tsunami could travel in that amount of time. Regional systems warn in about ten minutes, making them useful within 100–750 kilometers (62–465 mi) of the source, and local systems warn in about five minutes, which is useful <100 kilometers (62 mi) from the source. However, not all earthquakes generate tsunamis, and 15 of the 20 warnings issued between 1946 and 2000 were false alarms. This has led to public distrust of early earthquake-based warning systems (✿ Case Study 10.1 on page 283).

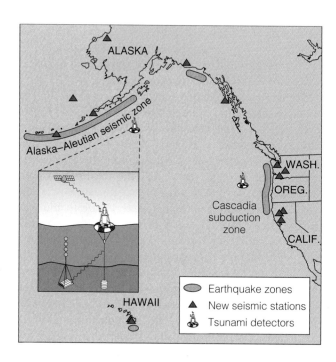

◆ FIGURE 10.12 U.S. earthquake zones capable of generating tsunami are the Alaska–Aleutian seismic zone and the Cascadia subduction zone. The five Pacific states are subject to tsunami. The insert shows a tsunami detector that rests on the sea floor. It transmits acoustic signals to a surface buoy, which relays signals to a NOAA satellite and then to ground-based warning centers. From *EOS* 79, no. 2, June 2, 1998.

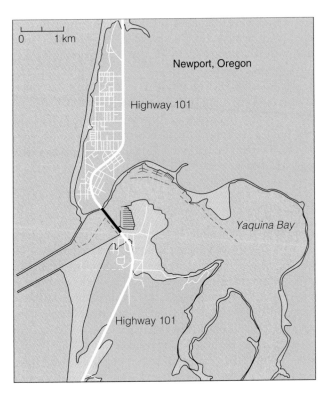

◆ FIGURE 10.13 Potential tsunami inundation areas (green) at Newport, Oregon. From *EOS* 79, no. 2, June 2, 1998

The goal now is to place real-time deep-ocean tsunami detectors on the sea floor that transmit data acoustically to a floating buoy, which relays the information to a satellite that, in turn, transmits it to a ground station (see Figure 10.12). The sensitive instruments can detect tsunami as small as 3 centimeters (1.17 in) high and transmit the data to the ground station in three minutes. For nearby tsunami sources, three minutes from detection to an alert is a life-saving warn-

ing time. Six instruments were in operation by 1998, and 38 instruments will eventually be installed and maintained in the Pacific Ocean. Lest you think that only the Pacific Northwest, Hawaii, and Alaska are at risk, tsunami are purported to have struck Santa Barbara, California, in 1812 and Santa Monica, California, in 1930.

Tsunamis Initiated By Submarine Landslides

Recent research now includes landslides as a significant process in tsunami generation. This new awareness may save lives in population centers around the Pacific margin that are currently under the false impression that tsunamis are not a hazard. San Francisco, Santa Barbara, Los Angeles, and San Diego are some of the cities that could have landslides on the continental slope. Known landslide-generated tsunamis resulted from sea-floor instability after the Great Alaskan earthquake in 1964 and Grand Banks earthquake in 1929. The latter quake is better known for the 12 transatlantic cables severed sequentially by sea-floor failure than for the tsunami that struck the Burin Peninsula about 2.5 hours after the seismic event (◆ Figure 10.14). Numerous landslides on the slope coalesced over a distance of 110 km and eventually became a fast-moving soup of sediment and water known as a *turbidity current*. Communication cables were sequentially broken and their times of failure indicate the submarine flow had to travel between 50 and 70 km/h (30–42 mi/h). Sediment was deposited by the flow over 400 kilometers from the site of the initial slump, making this the largest turbidity current identified either historically or in old sedimentary rocks.

Landslides Generated By Submarine Volcanic Action

The explosion of Krakatoa on August 27, 1883 was one of the largest volcanic eruptions and subsequent tsunamis ever recorded. The volcano collapsed into its own magma chamber and, because it was on an island in the Sunda Strait

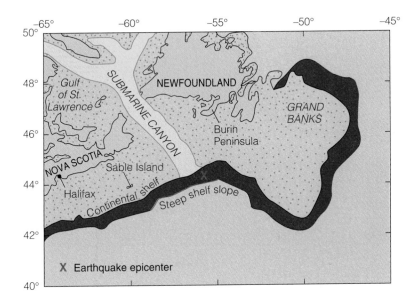

◆ FIGURE 10.14 Epicenter of Grand Banks earthquake and Burin Peninsula.

between Java and Sumatra, the ocean followed into the collapse. The eruption produced many bad results, but the death toll of 36,417 from the tsunami was the worst. The tsunami had a maximum runup of 42 meters and traveled 5 kilometers inland over low-lying areas. The largest wave struck the town of Merak because of the funnel shape of its harbor. Here the 15-meter wave increased to 40 meters due to concentration of energy, and coral blocks weighing over 600 tons were transported onshore. About 5,000 boats were destroyed and, as mentioned, the death toll was staggering.

Meteorite Impacts

Most meteorites come from the asteroid belt and burn up in the atmosphere before reaching earth. However, those that penetrate the atmosphere and land in a body of water are capable of generating tsunamis. It has been proposed that such an impact in the past created a global winter resulting in extinction of the dinosaurs and about 70 percent of all other creatures that lived at that time. It is believed that such a crater, the Chicxulub, has been found. It is 180 kilometers in diameter in the northern Yucatan Peninsula and Gulf of Mexico and its debris marks the Cretaceous–Tertiary boundary, a time of great extinction. Although beyond the purview of this book, it will be interesting to see the results of further research.

Shorelines

Where the sea and land meet we find *shorelines*, some of which are shaped primarily by erosional geologic processes and others by depositional geological processes. Shorelines that are exposed to the open ocean and high-energy wave action are likely to have *erosional* features such as sea cliffs or broad, wave-cut platforms. Shorelines that are protected from strong wave action, by offshore islands or by the way they trend relative to the direction of dominant wave attack, are characterized by *depositional* features such as sand beaches.

Technically, the shoreline is the distinct boundary between land and sea that changes with the tides, whereas the *coast* is the area that extends from the shoreline to the landward limit of features related to marine processes. Thus, a coast may extend inland some distance, encompassing estuaries and bays.

Beaches

In addition to their usefulness as recreational areas, beaches also serve to protect the land from the erosive power of the sea. They are composed of whatever sediment rivers and waves deliver to them (◆Figure 10.15a). For example, in some places in Hawaii, black sand forms from weathered basalt and also from hot lava that disintegrates when it flows into the sea. In contrast, reef-fringed oceanic islands are characterized by beautiful white coral sand and shell beaches (◆Figure 10.15b). Mainland beach materials are largely quartz and feldspar grains derived from the breakdown of granitic rocks. *Shingle* beaches are composed of gravels; they typically occur where the beach is exposed to high-energy surf. Amusing or not, tin cans from a local dump once formed a beach at Fort Bragg, California.

> **CONSIDER THIS** *Your favorite beach disappears during a large winter storm. Will you have to find a new beach for next summer? If not, will the summer beach look any different and, if so, how will it be different? Explain your answer.*

(a)

(b)

◆ FIGURE 10.15 (a) Black sand beach on American Samoa (weathered basalt and explosively fragmented basalt). (b) Coral sand from Cancun, Mexico, in the Caribbean Sea.

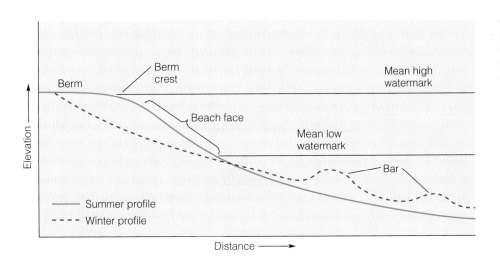

◆ FIGURE 10.16 Winter and summer beach profiles and some beach terminology. *From Essentials of Geology*, 2nd ed., by R. Wicander and J. Monroe, p. 348, Brooks/Cole, 1999

Types of Beaches

Four types of beaches can be recognized, according to the geomorphic setting where they are found:

- coastal-plain, or mainland, beaches,
- pocket beaches,
- barrier islands, or barrier beaches, and
- and spit beaches.

Each of these has its own peculiar stability problems, but all are influenced by seasonal differences in wave action. As a rule, winter beaches are narrow or nonexistent, because high winter-storm waves erode sand from the beach berm and deposit it in offshore sandbars parallel to the beach (◆ Figure 10.16). You may have experienced this while wading in waist-deep water that abruptly shallowed as you walked onto a sandbar. With the onset of long, low summer waves, the sand in the offshore bar gradually moves back onto the beach, and the beach berm widens. Thus, the natural annual beach cycle is from narrower and gravelly in the winter to wider and sandy in the summer. On the East Coast, beach narrowing is related to storm events that may occur at any time of the year.

Barrier islands form the most extensive and tenuous beach system in the United States. They extend southward from Long Island and Coney Island, New York, to Atlantic City, continuing as the Outer Banks of North Carolina (see ◌ Case Study 10.2 on page 284) and then down the coast of Florida to Daytona Beach, Cape Canaveral, Palm Beach, and Miami Beach. The barrier system begins again on the west coast of Florida and continues to the Padre Islands in Texas, interrupted only by the Mississippi River delta. These barrier beaches are transient features; formed by wave action, they are reoriented by wave action and frequently overtopped by storm surges. They are separated from the mainland by shallow lagoons and marshland over which they migrate as large storms overwash the islands, transporting sand from their seaward sides to their landward sides.

Hog Island, Virginia, is a barrier island that was a popular hunting and fishing spot of the rich and famous in the late 1800s. Its town of Broadwater had 50 houses, a school, a cemetery, and a lighthouse. After a 1933 hurricane inundated the island, killing its pine forest, Broadwater was abandoned. Today Broadwater is under water a half-kilometer offshore. Typical shoreward migration rates for an Atlantic-coast barrier island are 2 meters per year, but extraordinary displacements have been reported (Case Study 10.2).

Spits, such as the one at Cape Cod, Massachusetts, are elongated masses of sand attached at one end to the mainland. Spits may be formed by longshore currents when they transport and deposit sand across the mouths of bays or estuaries. Newport Beach, California is a spit deposited across Balboa Bay, an estuary of an abandoned Pleistocene river channel. Coronado Island, a sand spit that protects San Diego Bay, formed from huge volumes of sand delivered from the south by Mexico's Tijuana River (◆ Figure 10.17).

◆ FIGURE 10.17 The sand spit that encloses San Diego Bay is the result of a huge volume of sand transported northward from the Tijuana River in Mexico. Naval Air Station Coronado is at the south entrance to the bay, and rock-supported Point Loma forms the north entrance.

Beach Erosion—The River of Sand Gone Dry

A beach is stable when the sand supplied to it by longshore currents replaces the amount of sand removed by waves. Where the amount of the supplied sand is greater than the available wave energy can remove—typically near the mouth of a river—the beach widens until wind action forms sand dunes. Significant areas of coastal dunes occur in Oregon, on the south shore of Lake Michigan in Indiana, near the Nile River in Egypt, and on the Bay of Biscay on France's west coast, to name but a few (◆Figure 10.18). The more usual case is for wave action and longshore currents to move more sand than is supplied, and here we find eroding beaches or no beaches at all. All 30 of the U.S. coastal states have beach erosion problems of varying magnitudes resulting from both human activities and natural causes (◆Figure 10.19).

Inasmuch as sand comes to beaches by way of rivers, we should look to the land for causes of long-term beach erosion. Water-storage and flood-control dams trap sand that would normally be deposited on beaches. The need for upland flood protection is clear, but it can conflict with other needs, including the need for well-nourished beaches. Urban growth also increases the amount of paving, which seals sediment that might otherwise be eroded into local rivers and ultimately to the ocean.

Significant natural causes of beach erosion are protracted drought and rising sea level. Little runoff occurs during drought conditions, and therefore, little sand is delivered to beaches. This is true even for unmodified watersheds without dams. The impact of drought on beach nourishment is less where the watershed is behind a dam. Dams trap sediment so that even with normal rainfall sand is prevented from reaching the beach.

A rise in sea level could have catastrophic results, as indicated by the map of coastline changes in the eastern United States over the past 20,000 years (◆Figure 10.20). On a shorter time scale, it has been estimated that as the earth warms over the next century, sea level will rise 1 to 3 meters (up to 10 ft) due to thermal expansion and melting of glacial ice. On flat coastal plains where slopes are as gradual as 0.2–0.4 meters per kilometer, a sea-level rise of 2.54 centimeters (1 in) would cause the shoreline to retreat between 62 and 125 meters (200–400 ft). If the coast is subsiding, this shoreline migration landward would be magnified. On the other hand, tectonic uplift along parts of the Alaska and California coasts is faster than sea-level rise, which makes for a net shoreline retreat seaward. Careful tidal gauge measurements at New York City, where the shoreline is tectonically stable, indicated a sea-level rise of 25 centimeters (almost 10 in) between 1900 and 1970. If that trend continues, plan to take a boat to the Statue of Liberty's big toe and to say goodbye to Long Island, Miami Beach, and much of Texas's Galveston County by the middle of the next century.

Local Erosion

Long-term beach erosion is aggravated locally by poorly conceived engineering works, such as *breakwaters* built to create quiet water for safe yacht anchorage, and long **jetties** built to prevent sedimentation in harbor or river inlets. Such structures cut off the littoral flow, causing accretion of sand in their immediate vicinity and erosion on their downcurrent

(a)

(b)

◆ FIGURE 10.18 Indiana Dunes National Lakeshore. (a) Glacially deposited sands were reworked by wave action and then by wind to form these impressive coastal dunes on the south shore of Lake Michigan. (b) Someone's beach home was buried when sand was blown to the lee side of the 25-meter-high (80-ft) coastal dunes. This is an example of dune migration (see Chapter 12).

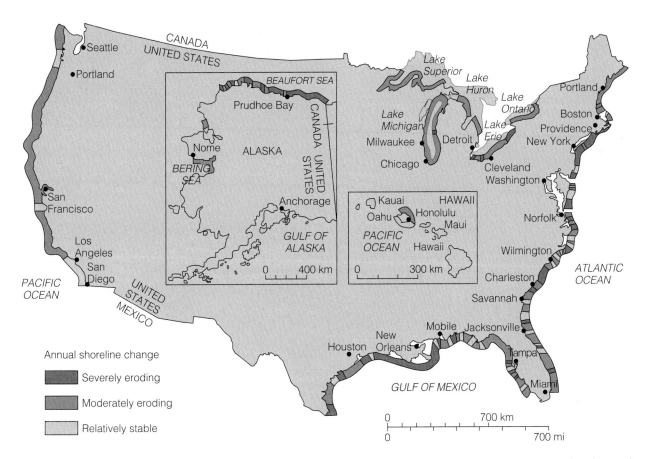

◆ FIGURE 10.19 Present coastal erosion in the United States. Note that most of the shorelines are eroding. Uncolored coastal areas on the map represent shorelines for which data are not available. *After R. Dolan and others, 1985, "Coastal Erosion and Accretion," USGS National Atlas*

side. Rates of erosion and accretion are calculated in cubic yards of sand per year, and it is a rule of thumb that a cubic yard of sand equals a square foot of beach. Thus, the littoral drift rates given in ✳ Table 10.2 can be roughly translated into gains and losses of beach area. However, you can bet that where beach sand has accreted and a beach has widened, it has occurred at the expense of some beach downcurrent.

Ocean City, Maryland, is at the south end of Fenwick Island, one of the chain of barrier islands extending down the U.S. east coast from New York to Florida. Long jetties were constructed there in the 1930s to maintain the inlet between Fenwick and Assateague Island to the south. As a result, the beach has widened at Ocean City—because the north jetty blocked the southward longshore transport of sand—and the Assateague Island beach has been displaced 500 meters (1,650 ft) shoreward due to sand starvation (◆Figure 10.21). Even though the Ocean City beach had widened, a strong storm in March 1962 inundated all but the highest dune areas, causing $7.5 million worth of damage.

In the 1920s Santa Barbara's civic leaders decided to mold their community into the "French Riviera" of California. In spite of unfavorable reports by the Corps of Engineers, city officials authorized construction of a long, L-shaped breakwater for a yacht anchorage in 1929 (◆Figure 10.22). The predominant drift direction here is southeasterly, and as the beach west of the breakwater grew wider, near-record erosion occurred on the beaches east of the structure. Eventually, the southeast-moving littoral drift traveled around the breakwater, forming a sandbar in the harbor entrance. The bar built up, diminishing the size of the harbor and creating a navigational hazard that remains to this day. When the Santa Barbara city fathers asked Will Rogers, the noted American humorist, what he thought of the harbor in the 1930s, he is alleged to have replied, "It would grow great corn if you could irrigate it." The drift rate here is about 250,000 cubic yards per year, which necessitates dredging sand from the harbor and pumping it to the beach about a mile downdrift. Summerland Beach, about 16 kilometers (10 mi) downdrift, retreated 75 meters (250 ft) in the decade following construction of the breakwater.

As a postscript, the City of Santa Monica to the south built a breakwater parallel to the shoreline in hopes of avoiding the kinds of problems experienced in Santa Barbara. The parallel breakwater created a *wave-energy shadow* on its

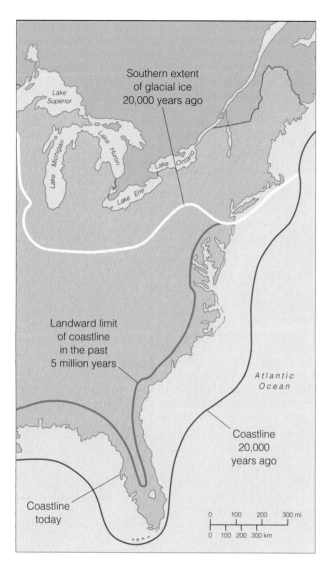

◆ FIGURE 10.20 Position of the coastline in the eastern United States in late Pleistocene time (20,000 years ago), Miocene time (5 million years ago), and today. USGS data

Location	Drift Rate, yds³/yr	Direction
West Coast (all California)		
Newport Beach	300,000	south
Port Hueneme	290,000	south
Santa Barbara	250,000	south
Santa Monica	160,000	south
Anaheim Bay	90,000	south
East Coast		
Fire Island Inlet, N.Y.	350,000	west
Rockaway Beach, N.Y.	260,000	west
Sandy Hook, N.J.	250,000	north
Ocean City, N.J.	230,000	south
Palm Beach, Fla.	130,000	south
Great Lakes		
Waukegan, Ill.	43,000	south
Racine Co., Wis.	23,000	south

✳TABLE 10.2 Rates of Littoral Drift at Selected U.S. Locations

Sources: U.S. Army Corps of Engineers and other sources.

Sea Cliffs

Most people enjoy ocean views, particularly from atop a 100-foot-high sea cliff. Unfortunately, such cliffs are eroded both by wave action at the toe and by mass-wasting processes on the face and the top. An **active sea cliff** is one whose erosion is dominated by wave action. The result is a steep cliff face and little rock debris at the base. The erosion of an **inactive sea cliff** is dominated by running water and mass-wasting processes. Its slopes are much gentler, and angular debris accumulates at the toe. The nature of the debris at the foot of a sea cliff can aid in determining whether it is active or inactive. Wave erosion produces rounded cobbles, whereas mass-wasting debris (talus) is angular. Rates of cliff erosion vary from a few centimeters to tens of centimeters per year, depending on the geology, climate, tectonic setting, and relative sea-level changes.

It is standard practice to consider 50 years as the lifetime of a house or other structure. Therefore, a structure should be set back from the top of a cliff at least 50 times the annual rate of cliff retreat. In all too many cases, the rates are not known or people choose to ignore them, with inevitable results (◆Figure 10.24). Orrin Pilkey of Duke University (see For Further Information) has developed a checklist for people to use when they are considering buying property on cliffs and bluffs. Pilkey's list includes such items as looking

landward side, causing sand to accumulate behind the structure but no serious downdrift erosion (◆Figure 10.23). However, dredging is still required to keep the sand from building out to the breakwater and filling in the harbor.

Communities along the shores of the Great Lakes have erosion problems, also. Early in this century, protective structures were built on the Lake Michigan shoreline. Many of these have now deteriorated, particularly during the record high lake-level of 1984 (582.9 feet above mean sea level). Lake levels began dropping around 1960, and there was little concern at the time about shoreline erosion. Since 1964 (record low of 576.6 feet above mean sea level), levels have been rising, and there is increasing public pressure to put more protective structures on the shoreline.

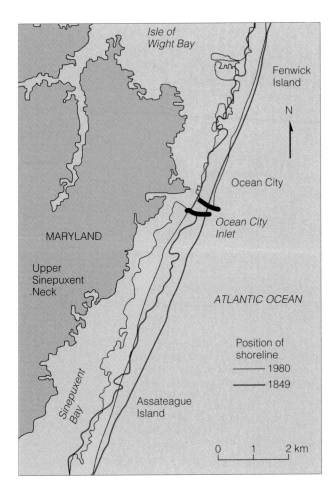

◆ FIGURE 10.21 Jetties built at Ocean City inlet in the 1930s cut off the supply of sand from the north. The north end of Assateague Island is now 500 meters shoreward of the south end of Fenwick Island. J. S. Williams and others, *Coasts in Crisis*, USGS circular 1075

◆ FIGURE 10.22 The Santa Barbara, California, breakwater in 1988. Sand accumulation in the harbor is clearly visible, as is the dredge used to pump sand to the beaches downdrift.

◆ FIGURE 10.23 Beach at Santa Monica, California. Note the sand accumulation in the "wave shadow" of the breakwater. The breakwater is the line of whitewater parallel to the shoreline.

for evidence of failure at the toe of the cliff, determining rock hardness, and learning the history of cliff recession in the area from local land records and photographs.

CONSIDER THIS *Several states have outlawed hard protective structures between beaches and the land. Why did they do this? Should other states, including those on the Great Lakes, enact similar legislation?*

Mitigation of Beach Erosion

Most coastal engineering works are built to protect the land from the sea. This battle against the sea is older than the tale of the Danish king Canute who commanded his men to whip the ocean in order to keep back the rising tide. It is perhaps presumptuous to think we can build works that will stop the rise of sea level or withstand enormous storm surges, but that does not keep us from trying. Although the sea will ultimately win the battle, we should attempt to build protective works that conform to nature as much as possible and that may outwit her for awhile.

Rock revetments and concrete seawalls are built landward of the shoreline and parallel to the beach not to protect the beach but, rather, to protect the land behind them. Commonly, these "hard" structures reflect wave energy downward, removing sand and undermining the structure. There is a question among coastal engineers and geologists that goes something like "Do you want a revetment, or do you want a beach?" Because waves reflecting off hard shoreline structures remove the beach material in front of them, the best-designed rock revetments have sufficient permeability to absorb some of the wave energy. Another problem is

(a)

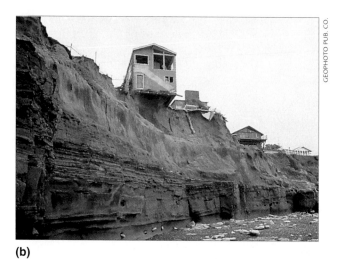

(b)

◆ FIGURE 10.24 (a) Rapid sea-cliff retreat near Big Lagoon, Humboldt County, California. These cliffs near the Oregon border are composed of sandy and gravelly uplifted marine sediments. As shown by the photograph, they are subject to direct wave attack. (b) Drastic undercutting by wave action has left this house in San Diego County, California, in a precarious position. The near-vertical cliff and the rounded cobbles at the base indicate that this is an active sea cliff.

that wave action erodes around the ends of revetments. When that happens, they must be extended to prevent damage to structures located there.

Groins are constructed to trap sand for the purpose of widening beaches. They interrupt sand flow on their upcurrent sides and deprive downcurrent beaches of sand. Once sand has built out to the end of a groin, it flows around the tip of it and begins filling in the eroded area downcurrent. Frequently another groin is built to widen *that* beach, and the cycle is repeated. The result is a beach with groins every few hundred feet and a shoreline with a scalloped or bread-knife appearance. The northern New Jersey shore and Miami Beach are textbook examples of multiple-groin shorelines. Permeable groins are often constructed today. They allow some sand to filter through them for nourishing downcurrent beaches.

Jetties are the real culprits of beach erosion, because they are very long and they usually occur on both sides of a harbor entrance or where a river flows into the sea. They have caused entire beaches and whole communities to disappear. The jetties at Ocean City, Maryland, have already been cited.

Breakwaters intended to create quiet water for yacht sanctuary can create unintended environmental problems. We have already seen the results of breakwater construction at Santa Barbara and Santa Monica in California. Venice Beach, California, a community with a reputation for the unusual, built a parallel breakwater. It was constructed so close to shore that the sand deposited in the wave "shadow" behind the breakwater has connected the breakwater to the land. This sand feature is called a **tombolo** (◆Figure 10.25).

At present, the designs of most long structures to be built perpendicular to the shoreline incorporate permanent *sand-bypassing works*. These features allow for the pumping of sand from the upcurrent side of the structure to nourish the beaches downcurrent. In addition, downcurrent beaches can be artificially nourished by pumping in sand from offshore areas or by transporting it to them from land sources. Unfortunately, artificially nourished beaches are generally short-lived. One reason is that the imported sand is often finer than the native beach sand and is thus easily removed by the prevailing waves. The other reason is that artificial beaches usually have a steeper shoreface and are thus more subject to direct wave attack than is a gently sloping, natural shore, which dissipates wave energy. Artificial beach nourish-

◆ FIGURE 10.25 This breakwater at Venice, California, was built so close to the beach that sand deposited behind it has connected the breakwater to the beach, forming a tombolo.

ment is expensive, and it is most commonly done where a government project caused the erosion problem.

Because theory is lacking, thorough shoreline-impact studies using models should be performed for all large projects. Such a model is an exact replica of the project at a scale of about 1:100 (◆Figure 10.26).

Estuaries

Estuaries (Latin *aestuarium,* "tidal") are semienclosed basins in which river water (fresh water) mixes with tidally influenced seawater. The river water flowing into the estuary "floats" on the denser seawater, which is constantly moving landward and seaward with the tide. The mixing of fresh water and saltwater generally occurs below the water surface, and the resulting mixture is dilute, or brackish, water. There are almost 900 estuaries along U.S. coasts, covering 68,000 square kilometers (27,000 mi^2). Based upon their geology, four major types of estuaries are recognized:

1. *Coastal-plain estuaries* form where the land subsides or sea level rises and submerges a river valley. Chesapeake Bay is an example of this type, as are Delaware Bay, the Hudson River, the Mississippi River, the Thames River in England, and the Seine River in France.

2. *Fiords* are drowned U-shaped valleys formed by glaciers (see Chapter 11). They are common in Scandinavia, Greenland, Alaska, British Columbia, and New Zealand.

3. *Fault-block estuaries* (grabens; see Chapter 3) form on tectonically active coasts such as the southern part of San Francisco Bay.

4. *Bar-formed estuaries* occur as lagoons between barrier islands and the mainland. Examples are Pamlico Sound

within the Outer Banks of North Carolina and Laguna Madre in Texas.

Chesapeake Bay (◆Figure 10.27), a coastal-plain estuary, has the added distinction of having been impacted 35 million years ago by a body from the far reaches of space. A **bolide** (meteor or comet) plopped into the coastal plain between Richmond and Norfolk, Virginia, leaving a crater 1.3 kilometers (4,250 ft) deep and 85 kilometers (53 mi) in diameter—deeper than the Grand Canyon and twice the size of Rhode Island. It is three times larger than any other U.S. crater and is the sixth-largest crater known on the planet today.

Estuaries and the marsh wetlands on their shores are habitats for flourishing biological communities. The waters of estuaries have high nutrient contents that stimulate phytoplankton (plant) growth, which in turn supports a rich and diverse fauna. Estuaries are the habitat of a host of shellfish—including blue crab and common shrimp—and the breeding place of many fish species. The tidal marshes on their shores support a variety of waterfowl. It is estimated

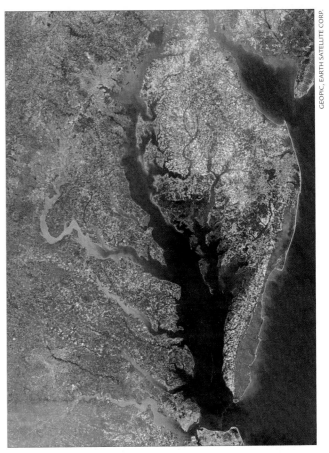

◆ FIGURE 10.27 Chesapeake Bay, a coastal-plain estuary that formed as sea level rose and drowned the valleys of the Susquehanna, Potomac, and James Rivers and many smaller watersheds. The bay is 311 kilometers long and up to 40 kilometers wide (193 mi x 25 mi).

◆ FIGURE 10.26 Scale model of King Harbor Marina, Redondo Beach, California. The model was built after the big storm of 1988 to aid in assessing potential damage from waves of varying height and source direction. Data thus obtained will be used to make future design improvements.

that 95 to 98 percent of the important commercial fish species are spawned in estuaries.

Urbanization around estuaries has placed conflicting demands on their waters. For example, the fishing and shellfish industry that the Hudson River estuary once supported has drastically declined since 1960. The Hudson is now the emergency water supply for New York City and also the receptacle for the sewage of 16 million people. Because of the Clean Water and Ocean Dumping Acts of 1972, no raw sewage or sludge (solids) may be discharged into the river, but discharges of treated sewage have resulted in elevated bacteria levels, the presence of heavy metals and organic compounds in bottom sediments, and a reduced dissolved-oxygen content in the water column. The aesthetic amenities and resources of this estuary have been severely degraded by human activities.

CONSIDER THIS *"Why sweat a swamp?" is what some developers say about the fuss over preserving estuaries and marshlands. Is this attitude justified, or are these features valuable to the environment?*

Hurricanes and Coastal Flooding

Andrew, Mitch, Camille, Opal, Gilbert, and *Floyd* are innocent-sounding names, but onshore surges of water and intense rainfall produced by hurricanes with such names are the greatest natural threat to humans. Hurricanes (called *typhoons* in the North Pacific Ocean and *cyclones* in the South Pacific and Indian Oceans) get their energy from the equatorial oceans in the summer and early fall months when seawater is warmest, usually above 25°C (about 80°F). Hurricanes begin as tropical depressions when air that has been heated by the sun and oceans rises, creating reduced atmospheric pressure, clouds, and rain. As atmospheric pressure drops, the warm air moves toward the center of low pressure, the eye of the impending hurricane. When the rising air reaches higher elevation it releases heat, causing it to rise even faster and creating even lower atmospheric pressure and greater wind velocities. Wind blowing toward the center of low pressure is given a counterclockwise rotation by the Coriolis force in the Northern Hemisphere (◆Figure 10.28).

Thus, what begins as a localized tropical disturbance can grow to be 800 kilometers (500 mi) across, with wind velocities far in excess of 120 kilometers per hour (74 mph), the lower velocity limit for hurricane designation. Hurricanes travel as coherent storms with velocities of 10 to 35 knots (12–40 mph), but their exact course and where they will strike land is not completely predictable. If we divide a storm system in the North Atlantic into four equal quadrants, we find the strongest winds in the northeast quadrant (designated as quadrant I), because the cyclonic winds and the storm system are moving in the same direction there (◆Figure 10.29). When a storm makes a landfall, the greatest wind damage usually occurs in this sector.

A "hill" of water called a **storm surge** is pushed ahead of the hurricane that may cause sea level to rise many meters above the normal high-tide level. The reduction in atmospheric pressure associated with a hurricane also causes sea level to rise—like fluid rising by suction in a soda straw. Average sea-level atmospheric pressure is 1,013 millibars (or 29.92 inches of mercury), and the rise in sea level is about 1 centimeter for every millibar of pressure drop. Hurricane Camille in 1969, the strongest storm to hit the U.S. mainland, had a near-record low barometric pressure of 905 millibars (26.61 inches of mercury). Hurricane surge and winds devastated the entire Gulf Coast from Florida to Texas.

Twenty-five residents of the Richelieu Apartments in Pass Christian, Mississippi, on the state's south coast were enjoying a hurricane party when the police chief came to alert them of the danger. They laughed when he requested the names of their next of kin because they did not appear to be willing to take his advice and evacuate. The storm surge at Pass Christian reached 7 meters (23 ft) above mean sea level, and only two people survived the party. One of them was rescued from a floating sofa 8 kilometers (5 mi) from the apartment complex.

Camille killed 130 people and three thousand homes simply disappeared. Yet real estate was resold, and new homes were built on the same sites where homes had been removed by the storm. Camille was rated as a category-5 hurricane, the highest category on the Saffir–Simpson Scale (✳Table 10.3). Only two category-5 hurricanes were recorded in the United States in the twentieth century.

Some hurricanes veer eastward and follow a track up the U.S. East Coast, causing flooding and beach erosion from the Carolinas to Maine. In 1989 Hurricane Hugo struck the coast just north of Charleston, South Carolina, with sustained winds of more than 215 kilometers per hour and a storm surge of 6 meters (20 ft; ◆Figure 10.30). Hurricane Floyd was the 1999 storm in North Carolina that wouldn't go away. Although not a strong hurricane, it created flooding of biblical proportions in the eastern part of the state, aided by Hurricane Dennis that had dropped 25 centimeters (10 in) of rain two weeks earlier. Thousands of square miles were underwater for weeks, and the flood area became a fetid swamp with the bodies of 10,000 hogs and 2.5 million chickens and turkeys. Hurricane Floyd dropped 55 centimeters (20 in) of rain, leading a resident of Rocky Mount, one of the cities most impacted by the storm, to say, "The last time it rained like this, Noah was alive."

In 1970 a tropical cyclone in East Pakistan (now Bangladesh) resulted in perhaps the deadliest coastal flooding of all times with a death toll of more than 300,000 people. Dissatisfaction with the central government of Pakistan's response to the tragedy led to the formation of Bangladesh, a new country separated from Pakistan. Another tragedy in 1991 that killed 60,000 led to the formation of a tropical cyclone forecasting center. Experts at the center recommended construction of raised concrete bunkers above storm-surge level for refuge. A fierce cyclone

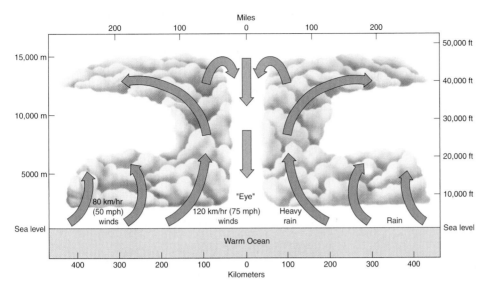

(a)

(b)

ENVIRONMENTAL
◆ **Geology⇌Now**™ ACTIVE FIGURE 10.28 Circulation patterns within a hurricane, showing inflow of air in the spiraling arms of the cyclonic system, rising air in the towering circular wall cloud, and outflow in the upper atmosphere. Subsidence of air in the storm's center produces the distinctive calm, cloudless "eye" of the hurricane. (b) Satellite photograph of Hurricane Gilbert practically filling the Gulf of Mexico, September 15, 1988. Gilbert's rotation and central eye are clearly visible.

in 1998 left thousands homeless but killed only 200 people. One of the problems in third-world countries is overpopulation that forces people to settle on land that is subject to life-threatening natural hazards. Bangladesh is about the size of Wisconsin with a population of 117 million, about half the population of the United States. Even in areas where people were warned of the 1991 storm, many were reluctant to leave their few possessions to looters.

Lest we think such tragedies are limited to third-world countries, we should look to Galveston, Texas, in September 1900. A hurricane traveled across the Gulf of Mexico and struck the barrier island upon which Galveston is built. The hurricane struck with such force that it smashed 3,600 wooden structures and drowned over 6,000 persons, the largest toll from a natural disaster in American history. Nellie Carey, who lived through the disaster, wrote in a letter: "Thousands of dead in the streets—the gulf and bay strewn with dead bodies . . . . Not a drop of water—food scarce. The dead are not identified at all—they throw them on drays and take them to barges, where they are loaded like cord wood, and taken out to sea to be cast into the waves . . . ." The letter was reprinted in its entirety in a book published shortly after the disaster. Today, Galveston is one of the leading cities to deal with evacuation in case of hurricane.

El Niño and the Coastal Zone

El Niño is a perturbation, or disruption, of the normal cold-water conditions that exist along the west coasts of North and South America. During El Niño years high-pressure cells replace the normally low atmospheric pressure of the western Pacific Ocean. Because the pressure difference that drives the trade winds between the eastern Pacific and the western Pacific lessens, the trade winds stop, or may even reverse

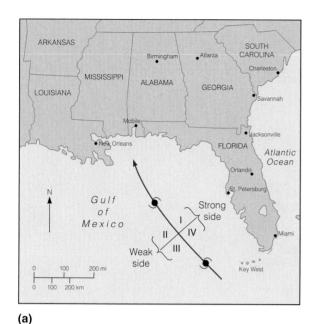
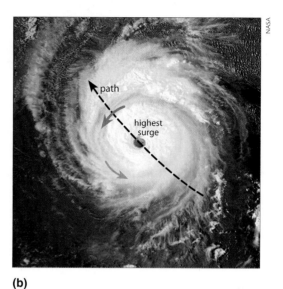

(a)   **(b)**

◆ **FIGURE 10.29** Northern Hemisphere hurricane quadrants. The highest winds occur on "the strong side." (b) Hurricane Andrew as it approaches the Louisiana coast on August 25, 1992. The storm surge is greatest in the northeast quadrant of the disturbance. This is true for all cyclones in the Northern Hemisphere.

direction. A bulge of warm surface water (as much as a meter high) called a *Kelvin wave* surges eastward toward the coast of the Americas and displaces the cooler water there (◆Figure 10.31). First noticed by Peruvian fishermen around Christmastime in 1892, this invasion of warm water was dubbed the *Corriente del Niño* ("Current of the Christ Child"). Today, the atmospheric pressure reversal in the western Pacific region, called the *southern oscillation,* is known to be the trigger for El Niño events. We now call this connection and the ensuing warm-water events the **El Niño–Southern Oscillation,** usually shortened to its easily articulated acronym, **ENSO.**

*TABLE 10.3 Saffir–Simpson Hurricane Scale				
	Wind Velocity			
Category	*km/h*	*mph*	**Damage**	**Examples**
1	119–153	74–95	minimal	Agnes, 1972
				Juan, 1985
2	154–178	96–110	moderate	David, 1979
				Bob, 1991
3	179–210	111–130	extensive	Betsy, 1965
				Frederic, 1979
4	211–250	131–155	extreme	Hugo, 1989
				Andrew, 1992
				Opal, 1995
5	>250	>155	catastrophic	Labor Day, 1935
				Camille, 1969

Source: Coastal Weather Research Center, University of South Alabama, Mobile.

Significant El Niño events of 1972–1973 and 1976–1977 and the super El Niño of 1982–1983 established it as a global phenomena. An almost continuous El Niño condition persisted from 1991 to 1994 (◆Figure 10.32), and then another super El Niño occurred in 1997–1998. During El Niño years, because of atmospheric convection shifts, there have been drought in Africa, Australia, and Indonesia; flooding in California, Ecuador, Peru, and Tahiti; and impressive storm waves and beach erosion along both the east and west coasts of the United States. More bad news was the unprecedented number of landslides and debris flows in California, Oregon, and Washington during the 1997–1998 event. According to Jim O'Brien at the Center for Ocean–Atmosphere Prediction Studies (COAPS) at Florida State University, there is also good news. El Niño:

- decreases Atlantic hurricane activity;
- reduces forest fires in Arizona, Texas, and New Mexico;
- reduces tornadoes;
- produces better citrus and vegetable crops; and
- improves the fishing off California because of warm water.

Interestingly, Hawaiian hurricanes occur only during El Niño years.

El Niño is a good example of the earth functioning as a system—that is, the atmosphere, hydrosphere, and lithosphere interacting to create natural events. El Niño's "sister," La Niña, appears to be a media concoction. La Niña is a cold-water condition, a condition that most experts call the normal condition for the west coast of the Americas. Even though it has been linked to drought and increased hurricane activity by the media, a direct scientific connection has yet to be demonstrated.

(a) **(b)**

◆ FIGURE 10.30 (a) Homes at beautiful Folley Beach, North Carolina, before Hurricane Hugo. Note the rock revetment facing the beach to prevent sand erosion and damage to structures. (b) The area had five fewer houses after Hurricane Hugo, but the palm trees and the rock revetment remained. Three houses in the background were damaged by battering-ram action of the demolished first-row structures.

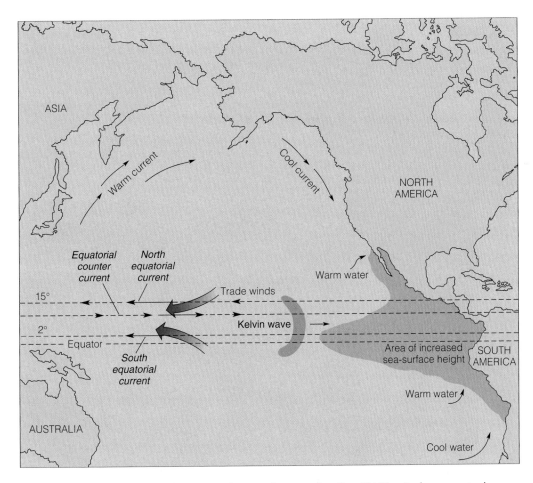

◆ FIGURE 10.31 Oceanography, meteorology, and geography of an El Niño. At the onset, trade winds maintain warm equatorial currents toward the east. As El Niño begins, the trade winds weaken, and the equatorial countercurrent strengthens. This creates Kelvin waves, surges of warm water moving eastward. When a Kelvin wave reaches the Americas, it divides northward and southward, traveling along the west coasts.

◆ FIGURE 10.32 The numbers on the left side of the graph are an ENSO index that incorporates these measured parameters: air temperature, atmospheric pressure, cloudiness, and wind speed and direction. El Niño events are shown in red (+); La Niña events, in blue (−). From *Meteorology Today*, 6th ed., by C. D. Ahrens, Brooks/Cole, 2000

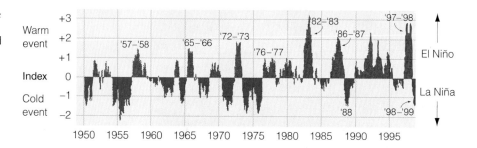

ENVIRONMENTAL
Geology⇌Now™

Click **Geology Now** to work through an activity on El Niño through Weather and Climate.

Hurricane Predictors

We know that hurricanes are part of the tropical severe weather pattern and that they will occur seasonally every year. According to meteorologist William Gray at Colorado State University, certain predictors can foretell the severity of the hurricane season. The predictors Gray uses require a massive data-collection effort, some of which are:

- Tropospheric winds up to an altitude of 12,000 meters (40,000 ft). Strong winds in the troposphere work against hurricane circulation by "shearing off" the tops of the vertical buildups that are required for storms to reach hurricane strength.

- West African climate. Dry years promote westerly (from the west) winds at 12,000 meters above the tropical Atlantic. This promotes wind shear and reduction of hurricane-strength buildups. During wet years winds blow from the east at 12,000 meters, greatly reducing wind shear and thus promoting hurricane-strength storms.

- El Niño. The warming of the equatorial Pacific Ocean promotes westerly winds that blow into the Caribbean and Atlantic basins, where trade winds blow at lower levels from the east. This promotes wind shear that knocks down hurricane buildups. Between 1991 and 1994 repeated El Niño resulted in few hurricanes. In 1995 and 1999 El Niños gave way to a cooling of the equatorial Pacific Ocean (La Niña), and hurricane activity greatly increased.

- Atmospheric pressure. Low atmospheric pressure in the Atlantic Ocean and Caribbean Sea indicates areas of warm water, which fuels hurricanes, promotes convergence of air masses, and stimulates vertical buildups of moisture-laden clouds (◆Figure 10.33).

The most difficult part of hurricane forecasting is predicting where and when a hurricane will make landfall. Unfortunately, this is the information that coastal-state residents need most of all. A few computer forecasts have been successful, but there is still much research to be done before landfall prediction is perfected.

◆ FIGURE 10.33 A series of 19 tropical storms, of which 11 were hurricanes and 5 were major hurricanes, occurred during the 1995 season in the Atlantic Ocean. Seen marching across the equatorial Atlantic on August 24, 1995 is Hurricane Luis (right center of the photo). Luis was preceded by Felix and followed by Hurricanes Marilyn, Opal, and Roxanne.

Distant Tsunamis—The Silent Threat

Ample warning had been given the residents of Hawaii as the Chilean tsunami raced across the Pacific Ocean at jetliner speeds. A warning had been issued about 6:45 P.M. on May 22, 1960 that large waves were expected to reach Hilo on the Island of Hawaii about midnight. Coastal warning sirens had wailed at 8:30 P.M., and continued for half an hour.

At midnight a wave arrived, but it was only a few feet high. Hundreds of people had stayed home, and those that had left assumed the danger was over and returned home. The highest wave struck at 1:04 A.M. Sixty-one people died, and another 282 suffered severe injuries (◆Figure 1). A restaurant memorialized the event with a waterline on its window (◆Figure 2).

The city of Hilo, with its bay facing northeast toward the Aleutian trench, is a coastal section with high expected wave heights. This is due to its orientation and the bay's funnel shape, which concentrates tsunami energy. An engineering study in the aftermath of the 1960 tsunami resulted in some recommendations for mitigating tsunami damage in that city. Parking meters in the runup area were bent flat in the direction of debris-laden wave travel (◆Figure 3), with the direction varying throughout the affected area. It was recommended that new structures be situated with their narrowest dimension aligned in the direction indicated by nearby bent parking meters. Several waterfront structures had open fronts that allowed water to pass through them. Waterfront buildings are now designed with open space on ground level, so that damage from water impact can be avoided.

◆ FIGURE 2 A line painted on the window of a rebuilt restaurant commemorates the height to which the tsunami waters rose.

◆ FIGURE 3 Parking meters bent by the 1960 tsunami at Hilo, Hawaii. The meters resemble arrows aligned with the wave runup.

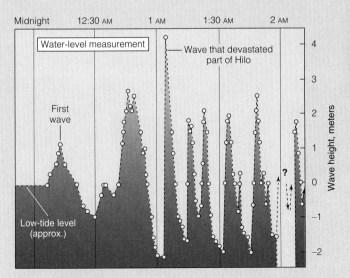

◆ FIGURE 1 Water-level measurements beneath the Wailuku River bridge during the first hours of the tsunami of May 23, 1960.

A Moving Experience

Cape Hatteras Lighthouse, with its black and white candy-cane paint job, is the tallest brick lighthouse in the world. Built in 1870 with state-of-the-art lenses, the 63-meter (208-ft) structure is so distinctive that it is on the National Register of Historic Places. Besides saving lives, the lighthouse has hosted millions of visitors. It is one of the most enduring landmarks in North Carolina and the nation. This part of the Outer Banks, designated the Cape Hatteras National Seashore by Congress, is a protected environment (◆Figure 1).

Fixed structures built on barrier islands are inevitably at the mercy of beach erosion and, ultimately, direct wave action. This is because all U.S. barrier islands have been "migrating" shoreward due to the rise in sea level over the past few thousand years. When it was built in 1870 the lighthouse was 463 meters (1,500 ft) from the ocean. By 1919 the sea had come to within 100 meters (310 ft) of its base and, by 1935, to within 30 meters (100 ft). When it comes to a battle of man against the sea, the sea always wins. Illustrating the futility of shore-protection efforts during times of rising sea level, the National Park Service and the Corps of Engineers spent millions of dollars attempting to save the beloved lighthouse—constructing groins to trap sand, replenishing beaches and artificially depositing a barrier-sand-dune system in front of the structure, installing offshore wave-energy dissipaters, and finally, constructing hard structures such as seawalls that reflect wave energy. By the 1980s the sea was lapping at the base of the lighthouse, and direct measures were required. The National Park Service decided to move the lighthouse in 1990.

Moving structures of great weight has become easier with the development of modern hydraulic systems. The massive Cape Hatteras Lighthouse was moved by using horizontally mounted "pusher" jacks that pushed it along a track system in five-foot

◆ FIGURE 1 North Carolina's Outer Banks are a chain of islands composed of sand that form a barrier between lagoons and the ocean. Cape Hatteras, near the center of the photo, juts the farthest out into the Atlantic.

increments. The move took just 22 days, 19 of which were "pushing" days. A remarkable engineering feat, indeed (◆Figure 2).

The powerful light that had been out for nine months was relit on Saturday, November 13, 1999 at ceremonies on the original lighthouse site in Buxton, North Carolina.

◆ FIGURE 2 Ready . . .

Rolling . . .

Home safe

Reminders of Nature's Force

The following photographs are good examples of the earth system at work—that is, the interaction of the solid earth, the atmosphere, and the oceans. ◆Figure 1 is a sort of defiant parting shot that Hurricane Hugo left as a reminder of the weakness of human efforts to hold back the sea. Hurricane Camille also left her reminder of nature's forces (◆Figure 2). The State of New York has proved to be a slow learner as elevated stilt houses, formerly at sand level, have made their appearance because of big surf or hurricanes (◆Figure 3).

◆ FIGURE 1 A revetment's half-ton to one-ton stone blocks were lifted and hurled through the walls and windows of a living space they were intended to protect; Hurricane Hugo, 1989.

◆ FIGURE 2 Delta House Movers appeared to have added boat-moving services after Hurricane Camille in 1969; Gulfport, Mississippi.

◆ FIGURE 3 "Watch that first step!" would be a good warning to anyone exiting these stilt houses that once rested on the beach. Strong waves reduced the beach level almost four meters during one storm; Westhampton, New York.

Summary

Coastal Ocean

Importance

1. Impact on local climate, weather, and shoreline structures
2. Estuaries
3. Beaches, an important natural resource

Waves

Cause

Wind blowing over the water surface creates ripples, then stormy irregular seas, and finally regularly spaced swell that outruns the source area.

Factors of Size

Wave size depends upon wind velocity, the length of time wind blows over the water, and fetch. A fully developed sea has the highest waves attainable for the three variables.

Motion

Water parcels move in circular paths to a depth of half the wavelength called *wave base*. This wind causes movement of the waveform known as *waves of oscillation*. Water physically moves shoreward when waves break in the surf zone. These breaking waves are called *waves of translation*.

Types of Breakers

Steep offshore slopes develop plunging breakers, and gentle slopes produce spilling breakers.

Wave Refraction

Waves generally approach the shore at some prevailing angle. This results in longshore currents that move sand along the beach. The transported sand is known as *littoral drift*. Groins act as dams to littoral drift, causing sand to be deposited upcurrent and eroded downcurrent. Wave refraction concentrates wave energy on points and dissipates it in bays, thus tending to straighten an irregular shoreline.

Impulsively Generated Waves

Also known as tsunamis (Japanese for harbor wave), they are mostly generated by seismic activity but also are generated by submarine volcanic action, landslides, or meteor impacts. They have long wave lengths (hundreds of kilometers) and low amplitudes (about a meter) at sea, but transform to an impressive breaker at the shoreline. Causes and death tolls over the last 2,000 years are given, as well as regions most susceptible to tsunami.

Beaches

Defined

Narrow strips of shore that are washed by the waves or tides, usually covered by sand or pebbles.

Types

1. Beaches backed by coastal plains
2. Pocket beaches along cliffed shorelines

3. Barrier islands (barrier beaches)
4. Spits

Beach Erosion

Defined

Less sand is supplied by littoral currents than is removed by wave action.

Natural Causes

1. Drought
2. Rising sea level

Human Causes

1. Dams that impound sediment
2. Groins, jetties, and breakwaters that impound sediment
3. Hard structures such as seawalls and revetments

Mitigation

Design of groins, breakwaters, and jetties can incorporate features that allow sand to pass through or around them. Seawalls, revetments, and artificial nourishment are only temporary solutions. Model studies of existing structures and shorelines help us to better understand the problem and thus to effect remedies.

Sea Cliffs

Defined

Cliffs formed by wave action.

Types

Active and inactive.

Erosion

Annual rates between a centimeter and tens of centimeters.

Mitigation

Structures' optimal (safest) setback from the edge of a cliff can be determined when the erosion rate is known.

Estuaries

Defined

Basins open to the sea in which fresh water mixes with seawater.

Types

1. Coastal-plain estuaries, drowned river valleys (Chesapeake Bay, site of an ancient bolide impact)
2. Fiords, drowned glacial valleys (Alaska)
3. Fault-block estuaries (south San Francisco Bay)
4. Bar-formed estuaries landward of barrier beaches (Laguna Madre, Texas)

Importance

Estuaries and their surrounding wetlands are areas of high biological productivity for marine life and waterfowl.

Human Impact

Used for dumping sewage, commercial fishing, transportation, and recreation. Structures are built on filled marshland.

Coastal Flooding

Hurricanes

Storms generated at sea characterized by counterclockwise winds greater than 120 kilometers/hour (74 mph). They travel across the sea at velocities of 10–35 knots (12–40 mph).

Flood Danger

A storm surge is a "hill" of water due to high winds and low atmospheric pressure. It may be many meters above normal high-tide levels. Storm surges on low-lying deltas such as the Ganges River delta in Bangladesh are the greatest natural hazard to human life.

Case Histories

Hurricanes Camille (1969), Hugo (1988), Andrew (1992), and Floyd (1999) are described.

El Niño (ENSO)

Associated with floods, landslides, and accelerated beach erosion. Also associated with decreased hurricane activity in the Atlantic, fewer tornadoes in the Midwest and fewer forest fires in the Southwest, and better fruit and vegetable crops. Hawaii experiences hurricanes only in ENSO years.

Hurricane Prediction

Utilizes data on tropospheric winds, West African climate, El Niño, sea-surface temperature, and atmospheric pressure in the Caribbean Sea. Heat is the fuel that drives hurricanes.

Key Terms

active sea cliff	estuary	littoral drift	tombolo
barrier island	fetch	longshore current	wave base
bolide	groin	rip current	wave of oscillation
El Niño–Southern Oscillation (ENSO)	inactive sea cliff	spit	wave of translation
	jetty	storm surge	wave refraction

Study Questions

1. Why do submarines and submersibles not experience severe storms at sea?
2. What shoreline structures impede littoral drift and cause beach erosion? How may these structures' impact be mitigated to minimize damage?
3. Where and when do hurricanes originate, and how do they obtain their energy? Distinguish a hurricane from a typhoon and a cyclone.
4. Name two types of breaker shapes and explain what causes one or the other to occur along a given shoreline.
5. How does the width of a beach vary naturally over the year? Where is the sand stored that is removed from the beach in winter and returned to the foreshore in summer?
6. What is a storm (hurricane) surge, and why is it such a danger in low-lying coastal communities?
7. How do longshore currents develop, and how do they impact beaches and *swimmers* in the surf zone?
8. How can hazardous currents in the surf zone be recognized?
9. Why are there fewer hurricanes during El Niño years than during "La Niña" times?
10. Why do warm ocean temperatures and low atmospheric pressure stimulate hurricane development?

For Further Information

Books and Periodicals

Bascom, Willard. 1964. *Waves and beaches.* Garden City, N.Y.: Doubleday.

Beardsley, Tim. 2000. Dissecting a hurricane. *Scientific American,* 282, no. 3, March.

Bryant, Edward. 2001. *Tsunami—the underrated hazard.* Cambridge, U.K.: Cambridge University Press.

Dean, Cornelia. 1999. *Against the tide: The battle for America's beaches.* New York: Columbia University Press.

Field, M. E., Susan Cochran, and Kevin R. Evans. 2002. *U.S. coral reefs—imperiled national treasures.* U.S. Geological Survey fact sheet 025-02, 4 pp.

Griggs, G., and L. Savoy, eds. 1985. *Living with the California coast.* Durham, N.C.: Duke University Press.

Kaufman, W., and Orrin Pilkey. 1979. *The beaches are moving.* Garden City, N.Y.: Anchor Press/Doubleday.

National Oceanographic and Atmospheric Administration (NOAA). 1996. *Hurricanes—a preparedness guide.* Washington D.C.: U.S. Government Printing Office.

Scientific American. The oceans. 1998. *Scientific American Quarterly.* Fall.

Sheets, Bob, and Jack Williams. 2001. *Hurricane watch.* New York: Vintage Books.

Shinn, E. A. 1993. Geology and human activity in the Florida Keys. *Public issues in energy and marine geology* (U.S. Geological Survey), October.

Williams, J. S., K. Dodd, and Kathleen Gohn. 1990. *Coasts in crisis.* U.S. Geological Survey circular 1075.

Wolanski, Eric, Robert Richmond, Laurence McCook, and Hugh Swanson. 2003. Mud, marine snow and coral reefs. *American Scientist* 91, pp. 44–51, January.

ENVIRONMENTAL
Geology⇌Now™

Assess your understanding of this chapter's topics with additional quizzing and comprehensive interactivities at http://earthscience.brookscole.com/pipkingeo4e

as well as current and up-to-date Web links, additional readings, and InfoTrac College Edition exercises.

Glaciation and Long-Term Climate Change

PERHAPS THE MOST BENIGN CLIMATE FOR MODERN HUMANS began in the fifth century, about the time of the fall of the Roman Empire, and lasted until the beginning of the fourteenth century. During this period, a relatively storm-free Atlantic Ocean was warmer, and Europe experienced mild temperatures. About 1300, however, the climate began to deteriorate, and glaciers in Europe and elsewhere advanced, in the period now known as the Little Ice Age. Paintings and written records from that time reveal images of European glaciers whose waxings and wanings affected human activity.

Glacier activity during the Little Ice Age is known. In Norway, advances between 1700 and 1750 displaced people and overrode and destroyed several farms. Glaciers in Iceland increased greatly in size during the 1500s, and by 1700 a glacier had overridden farms at Breitha and Fjall. By the early 1900s, the ice had shrunk back, and today, once productive farmlands are barren glacial deposits. Paintings and photographs of the Unterer Grindelwald Glacier near Interlaken, Switzerland (see ◆Figure 1a), dating as early as 1600, record that the glacier has retreated and advanced several times over about a third of a mile and has lost more

The ice was here, the ice was there, The ice was all around; It cracked and growled and roared and howled, Like noises in a swound!

Samuel T. Coleridge, "The Rime of the Ancient Mariner" (1798)

OFFENTLICHE KUNSTAMMLUNG, KUPFERSTICHKABINET, BASEL

(a)

D. D. TRENT

(b)

◆ FIGURE 1 *(a) The* Unterer Grindelwald, *painted by Samuel Birmann (1793–1847) in 1826. The Grindelwald Valley has two famous glaciers, the Upper and the Lower (German* Unterer*) Grindelwald. (b) The Unterer Grindelwald Glacier today. The village of Grindelwald has expanded into the area formerly covered by the glacier's toe.*

than a mile of its length since 1860. It began a short-lived advance again in 1977 but has been retreating since then.* As the figures testify, land covered with ice during the Little Ice Age is now dotted with houses.

*Michael Hambrey and Jürg Alean, *Glaciers* (New York: Cambridge University Press, 1992); Bjørn G. Anderson and Harold W. Borns, Jr., *The Ice Age World*, (Oslo: Scandinavian University Press, 1994) and Thorleifur Einarsson, *Geology of Iceland* (Reykjavík; Mál og menning, 1994).

Glaciers advance and retreat in response to cyclic global climate changes and consequently may provide evidence of past climates. In order to better understand global climate changes over time, we will first examine glaciers and glaciation. Important in understanding climates are the interrelationships of the atmospheric, oceanic, and biologic systems with plate tectonics and the orbital elements of the earth–sun–moon system.

Glaciation

About 97 percent of the earth's water is in the oceans, and three fourths of the remainder (2.25%) is in **glaciers.** A tenth of the earth's land area is covered by glaciers at present, the same amount that is cultivated globally. During the Pleistocene Epoch, which radiometric dating places as beginning 1.6 million years ago and lasting until about 10,000 years ago, as much as 30 percent of the earth was covered by glaciers (◆Figure 11.1). Today's glaciers store enormous amounts of fresh water, more than exists in all of the lakes, ponds, reservoirs, rivers, and streams of North America. Farmers in the Midwest grow corn and soybeans in soils of glacial origin, Bunker Hill in Boston is a glacial feature, and the numerous smooth, polished, and grooved bedrock outcroppings in New York City's Central Park were eroded by a glacier that flowed out of Canada. Glaciers in the Himalayas, Norway, Switzerland, Alaska, Washington State, Alberta, and British Columbia exert a vital effect on regional water supplies during dry seasons. They do this by supplying a natural base flow of meltwater to rivers that helps to balance the annual and seasonal variations in precipitation. Meltwater from glaciers serves as the principal source of domestic water in Switzerland. The Arapahoe Glacier is an important water source for Boulder, Colorado, so much so that Boulder residents take pride in telling visitors that their water comes from a melting glacier. (Actually, the glacier is only one part of a drainage basin of several tributaries that together supply the city's water.)

The steady water supply provided by melting glaciers during long periods of summer drought is widely used as a water resource for generating hydroelectric power in many temperate regions of the world. Utilization of glacial meltwater is highly developed in Norway and Switzerland, for example. Switzerland's hydropower stations, built mainly in the 1950s and

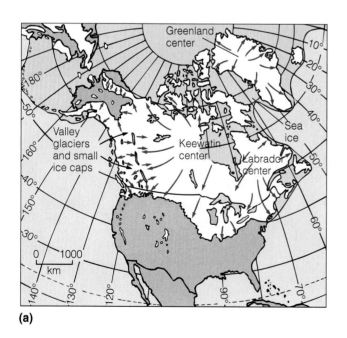

(a)

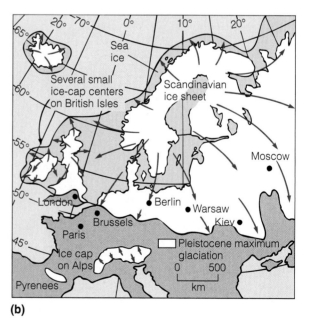

(b)

◆ FIGURE 11.1 Major centers of ice accumulation and the maximum extent of Pleistocene glaciation in (a) North America and (b) Europe.

1960s, use water from melting snow and glacial ice. Dams store the meltwater in the summer, when electrical consumption is low in that country, and in winter, when demand is highest, half of the country's electrical production is generated by water released from the reservoirs.

Origin and Distribution of Glaciers

Glaciers form on land where for a number of years more snow falls in the winter than melts in the summer. Such climatic conditions exist at high latitudes and at high altitudes. The pressure increases that occur as the thickness of the snowpack increases cause the familiar six-sided snowflakes to change into a coarse, granular snow called **firn** ("corn snow" to skiers). Continued accumulation increases the pressure and causes much of the trapped air to be expelled from the firn. Recrystallization into larger crystals occurs, and eventually the dense crystals of "blue" glacial ice form.

Once the ice attains a thickness of about 30 meters (100 ft), it begins to deform as a viscous fluid, flowing due to its own weight. It moves downslope if in mountains, or radially outward if on a relatively flat surface. This kind of behavior of an apparently solid material, described as *plastic flow*, takes place in the lower part of the glacier called the **zone of flowage** (◆Figure 11.2). The upper, surficial layer of ice behaves in a more familiar manner as a brittle solid, often breaking into a jagged chaotic surface of *crevasses*—ominous deep cracks—and *seracs*—towering prominences of unstable ice. It is the deformation by plastic flow that distinguishes glaciers from the perennial snow fields that persist at higher elevations in many mountain ranges. Glaciers are more than just masses of frozen water. Although they are largely ice, they also contain large quantities of meltwater and vast amounts of

rock debris acquired from the underlying bedrock or mountains where they originated and across which they have moved on their journey downslope. There are several types of glaciers, a classification of which appears in ✳ Table 11.1.

Glacier Budget

When the rate of glacial advance equals the rate of wasting, the front (terminus) of the glacier remains stationary, and the glacier is said to be in equilibrium. Glaciers **ablate,** or waste away, by melting or, if they terminate in the ocean, by calving. **Calving** is the breaking off of a block of ice from the front of a glacier that produces an iceberg. It is important to realize that glacier ice always flows toward its terminal or lateral margins, irrespective of whether the glacier is advancing (when the net gain of snow and ice in the accumulation zone exceeds the net loss in the ablation zone), is in equilibrium (when the accumulation zone covers about two-thirds of the glacier's total length), or is retreating (when the net loss in the ablation zone exceeds the net gain in the accumulation zone). Glaciers move at varying speeds. Where the slopes are gentle, they may creep along at rates of a few centimeters to a few meters per day; where slopes are steep, they may move 8–10 meters (26–32 ft) a day (◆Figure 11.3). Some glaciers, under exceptional conditions, may episodically accelerate with speeds up to 100 meters (328 ft) a day, a condition known as *surging* (see ◉ Case Study 11.1 on page 317).

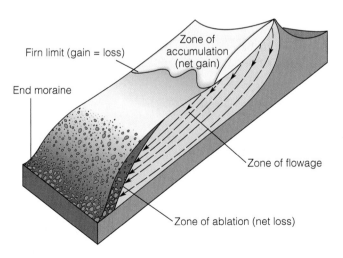

◆ FIGURE 11.2 Typically, about two-thirds of a glacier lies at elevations above the firn limit in the zone of accumulation, where more snow falls in winter than melts in summer. Below the firn limit is the zone of ablation, where glacial ice and the entrained rock debris are exposed by melting in the summer. The ridge of rock debris formed at the edge of the melting ice is an end moraine.

✳TABLE 11.1	Classification of Glaciers	
Type	**Characteristic Location**	**Example**
Valley (alpine)	mountain valleys	Vaughan Lewis Glacier, Alaska
Ice-field	extensive mountain glaciers that bury many adjoining valleys, leaving only the highest peaks and ridges rising above the ice	Juneau Ice Field, Alaska
Ice-cap	relatively smooth land surfaces in high latitudes or mountain summits where ice flows radially outward	Vatnajökull, Iceland
Piedmont	at the foot of mountain ranges where several valley glaciers emerge, flow laterally, and coalesce	Malaspina Glacier, Alaska
Continental (ice-sheet)	vast land areas	Greenland and Antarctica

WILFRIED HAEBERLI

◆ FIGURE 11.3 Scientists use a jet of hot water to drill a borehole in a glacier. A stake will be placed in the hole that will enable scientists to measure the glacier's movement and summer ablation.

Probably the best-documented and most dramatic retreat of a glacier is found at Glacier Bay National Park, Alaska. When George Vancouver arrived there in 1794, he found Icy Strait, at its entrance, choked with ice and Glacier Bay only a slight dent in the ice-cliffed shoreline. By 1879, when John Muir visited the area, the glacier, by then named the "Grand Pacific," had retreated nearly 80 kilometers (50 mi) up the bay. The Grand Pacific Glacier had retreated another 24 kilometers (15 mi) by 1916, and today a 104-kilometer (65-mi)-long fiord occupies the area that just 200 years ago held a valley glacier that was as much as 1,220 meters (4,000 ft) thick (◆Figure 11.4).

> **CONSIDER THIS** *If an Alaskan glacier's ice accumulates at an average rate of 50 centimeters a year and the glacier is 500 meters thick, how many years will be recorded in a complete core drilled to the base of the glacier?*

Glacial Features

Glaciers carry all manner of rock debris that is deposited directly by melting ice as unsorted rock debris, or **glacial till.** **Moraines** are landforms composed of till and named for their site of deposition. **End moraines** (also called *terminal moraines*) are deposited at the ends of the melting glaciers. **Recessional moraines** are series of nested end moraines that record the stepwise retreat, or meltback, at the end of an ice age (◆Figure 11.5).

After melting, continental and ice-cap glaciers leave a subdued, rounded topography with a definite "grain" that indicates the direction of glacial movement. They may also leave behind numerous **kettle lakes,** water-filled, bowl-shaped depressions without surface drainage (◆Figure 11.6). For example, there are about a dozen kettle lakes in northern Indiana's Valparaiso moraine. These lakes are believed to have formed when large blocks of stagnant ice that were wholly or partly buried in the deposits left behind by retreating glaciers melted (◆Figures 11.6 and 11.7). End moraines now form the hilly areas extending across the Dakotas, Minnesota, Wisconsin, northern Indiana, Ohio, and all the way to the eastern tip of Long Island, New York. Some Chicago suburbs have names that reflect their location on high areas of hilly moraines: Park Ridge, Palos Hills, and Arlington Heights, for example. In addition, poor drainage or deranged drainage characterizes many areas that have undergone continental glaciation.

Following a period of glaciation, a **fiord** is formed when the seaward end of a coastal U-shaped valley, eroded by a valley glacier, is drowned by an arm of the sea (◆Figure 11.8).

> **CONSIDER THIS** *How can a "retreating" glacier still be flowing downhill during a time of global warming?*

The Effects of Glaciation

Although the great ice sheets of the Pleistocene Epoch seem remote in time and modern glaciers seem remote in distance, both have impacts on modern society that are far from trivial.

Distribution of Soils

Farming and forestry generally depend upon the distribution and quality of Pleistocene deposits. Some of the best agricultural soils in the world lie on sediments deposited in former glacial lakes and on glacio-eolian (loess) deposits (see Chapter 6). As Pleistocene glaciers expanded and moved over western Europe and North America, they scraped off soils from the underlying land and redeposited them as moraines of glacial till where the ice melted. Today some till-covered areas are excellent forestland and excellent-to-poor farmland (◆Figure 11.9). Locally, in some locations such as south of Buffalo, New York, where the till is clay-rich, the clay may become unstable in places like road cuts along highways, and flow following significant wetting.

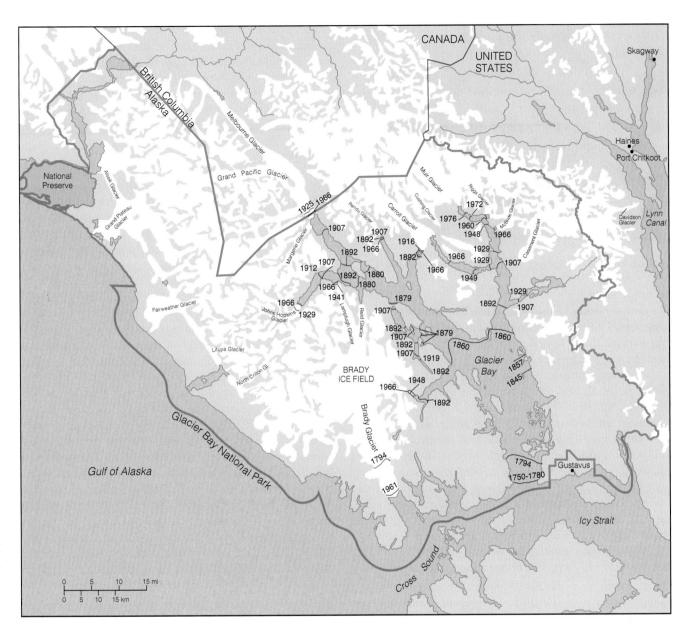

◆ FIGURE 11.4 Map of Glacier Bay National Park, Alaska, showing the terminal positions of the retreating Grand Pacific and other glaciers from about 1750 to the present. NPS data

Ground-Water Resources

Glacial outwash is sediment that was deposited by meltwater rivers originating at the melting edge of a glacier. The great Pleistocene ice sheet that extended out of Canada into the northern Mississippi River valley released copious amounts of sediment-laden meltwater, resulting in vast sheets of outwash sediments. Some of these glacial-outwash deposits serve as important ground-water aquifers in the end-moraine regions of the Midwest. Moraines associated with the outwash are poor aquifers due to the low porosity and permeability of the clays they contain. Thus geophysical exploration, test-drilling, and careful mapping are necessary in order to delineate the aquifers of porous and permeable sands within the buried stream channels. Glacial outwash can serve as an important source of sand and gravel (◆Figure 11.10).

Sea-Level Changes

The transfer from the oceans to the continents of the millions of cubic miles of water that became the enormous Pleistocene continental ice sheets caused sea level to lower about 100 meters (330 ft). This exposed extensive areas of continental shelves. Consider that a drop of this amount today would

◆ FIGURE 11.5 Recessional moraines; Coteau des Prairies, near Aberdeen, South Dakota.

increase the area of Florida real estate by more than 30 percent, would leave New Orleans high and dry, and would cause the Panama Canal to drain. Conversely, if all the polar ice were to melt, sea level would rise on the order of 40 meters (130 ft), and London, New York, Tokyo, and Los Angeles would be mostly under water. As sea level dropped during the Ice Ages, the rivers in coastal regions began downcutting across the exposed continental shelves as they adjusted to the new conditions. When sea level rose to its present level, about 5,000 years ago, the mouths of the coastal rivers were drowned, forming what are now harbors and estuaries.

The formation of land bridges due to the lowering of sea level during the Ice Ages was of major importance in establishing the current biogeography of the earth. Among the land bridges at this time was the Bering bridge. It had existed during much of Cenozoic time, and is especially significant because the two-way traffic across the bridge accounts for horses, mam-

moths, and other large mammals crossing between North America and Asia. About 20,000 to 30,000 years ago, during the lowered sea level of the last major Ice Age, humans used the bridge to cross from Asia into North America. Rising sea level about 13,000 years ago drowned the bridge, and the 90-kilometer (54-mi) stretch of ocean between Alaska and Siberia now blocks temperate-climate organisms from migrating.

Isostatic Rebound

The Ice Age continental glaciers and ice caps, with thicknesses of two miles or more, were so massive that the crust beneath them was depressed by their weight. In the time since the ice retreated at the end of the last ice age, 8,000 to 11,000 years ago, the land regions exposed by the retreating ice sheets in Scandinavia, Canada, and Great Britain have risen due to the geologically rapid removal of the great load of ice. Uplift of the crust due to unweighting is caused by **isostatic rebound**, the slow transfer by flowage of mantle rock to accommodate uplift of the crust in one area that causes subsidence in another area. Evidence of isostatic rebound in deglaciated areas is seen in the raised beaches around Hudson Bay, the Baltic Sea, and the Great Lakes (◆Figure 11.11). Careful surveying in Europe shows that as Scandinavia and Great Britain have been rising, a corresponding subsidence of Western Europe has occurred that is especially pronounced in the Netherlands. The uplift and subsidence in these regions remain active. Above the modern shorelines of the Gulf of Bothnia in Finland and Sweden are raised beaches that show a maximum uplift of 275 meters (900 ft). It is estimated that an additional 213 meters (700 ft) of rise will occur before equilibrium is reached. The rise is so rapid in some places in Scandinavia that docks used by ships are literally rising out of the sea. Uplift in Oslo

◆ FIGURE 11.6 Kettle lakes at Kewaskum, Wisconsin.

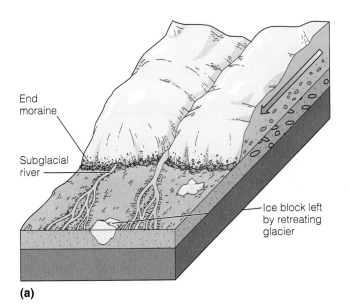

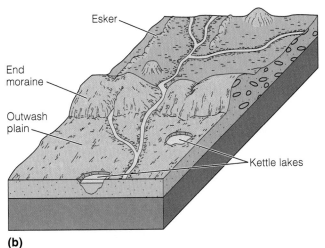

(a)

(b)

◆ FIGURE 11.7 Formation of end moraines, kettle lakes, and eskers (a) during glaciation and (b) after glaciation.

◆ FIGURE 11.9 Rocky soil developed on glacial till; Adirondack Mountains, New York.

Fjord, Norway, is about 6 millimeters (¼ in) per year, which adds another 60 centimeters (2 ft) of elevation per century. Sea-level mooring rings the Vikings used for tying up their dragon boats a thousand years ago are now 6 meters (20 ft) above sea level. Not even the tallest Viking would be able to moor his boat to those rings today. Similar uplifting is recorded along the northern shores of Hudson Bay, where the average uplift is about a meter (3 ft) per century (◆Figure 11.12).

Uplifting in the Great Lakes region is slow but continuing, and eventually it may cause problems for the residents of Chicago and elsewhere in Illinois. The rebound threatens to cause a drainage change that will force more of Lake Michigan to drain southward into the Mississippi River system.

◆ FIGURE 11.8 A fiord, Gastineau Channel near Juneau, Alaska.

◆ FIGURE 11.10 Part of a vast plain, underlain by Pleistocene glacial outwash near Elgin, Illinois. Outwash sediments, deposited along the margins of the melting continental ice sheets, are spread for thousands of square kilometers across the north-central United States. The ponds in the photograph are the excavations remaining from sand and gravel mining operations.

◆ FIGURE 11.11 Isostatically raised beach ridges at Richmond Gulf, Hudson Bay, Quebec, Canada. The highest ridge in the photo lies about 300 meters (984 ft) above sea level and is about 8,000 years old.

The rapid unloading of the crust at Glacier Bay, Alaska, in the last 200 years has resulted in measurable rebound in the lower parts of the bay. Small islands are slowly emerging, and existing islands are rising at a rate of about four centimeters (1½ in) per year.

Human Transportation Routes

Beginning in 1980, the Columbia Glacier on Prince William Sound, Alaska, began a catastrophic retreat by calving thousands of icebergs each year, some of them as large as houses. The glacier had been in equilibrium at much the same position for at least 200 years—it was first observed in the late eighteenth century—when it began to show evidence of retreat in the late 1970s. Fortunately, the moraine shoal at its terminus is sufficiently shallow to trap the larger icebergs and prevent them from drifting into Prince William Sound, where they would pose a major threat to the supertankers passing through these waters to and from the oil-loading terminal at Valdez. Because the terminus of the glacier has

◆ FIGURE 11.12 Locations of maximum postglacial (after glaciation) isostatic uplift in the Hudson Bay region and of land features resulting from the retreat of the late Pleistocene ice sheet about 10,000 years ago. Proglacial lakes (temporary, ice-marginal lakes) overflowed at different times through various spillways. One of the arrows east of the Finger Lakes represents the spillway pictured in Figure 11.17. The spillway draining Lake Agassiz in South Dakota is pictured in Figure 11.18. Windblown loess was deposited adjacent to major glacial meltwater systems. Waters from Lake Michigan drained southwestward into the Mississippi River at the time, which may occur again as isostatic uplift continues. After Dott and Batten, *Evolution on the Earth,* 3rd ed., p. 514, McGraw-Hill, 1981

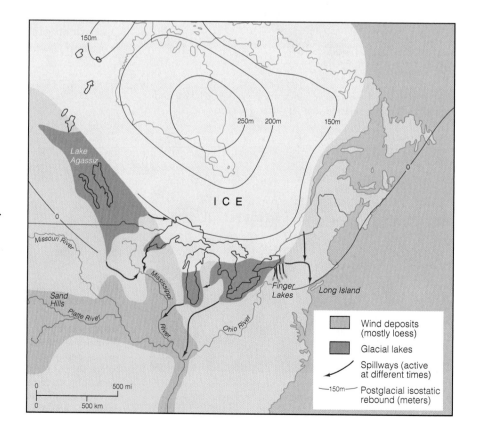

retreated some 15 kilometers (9 mi) since 1982, it appears that a new fiord is forming.

Juneau, Alaska, is the only state capital in the United States without highway access, even though it is only about 160 miles (by air) west of the nearest highway in Canada. A highway connecting Juneau with Canada would be desirable, but it is not to be, because the advancing Taku and Hole-in-Wall Glaciers threaten to close off the only possible route along the west shoreline of Taku Inlet (◆ Figure 11.13).

Whereas glaciers impede transportation in some areas, in other areas they have served to facilitate it. In wet boggy regions in the early days of New England, the crests of **eskers** provided good travel routes. Eskers are long, winding, steep-sided ridges of stratified sand and gravel deposited by subglacial or englacial streams that flowed in ice tunnels in or beneath a retreating gla-

cier. Hence, especially in Maine, the eskers were called *horsebacks* (Figure 11.7b and ◆ Figure 11.14).

Pleistocene Lakes

The cooler climates of the Pleistocene brought increased precipitation, less evaporation, and the runoff of glacial meltwater. Lakes dotted the basins of North America. Accumulations of water, mostly in valleys between the fault-block mountain ranges in the Great Basin, formed hundreds of lakes far from continental glaciers (◆ Figure 11.15). Today many of these valleys contain aquifers that were probably charged by waters of the Ice Age lakes. California had dozens of lakes, now mostly dried up; among them were Owens Lake, Lake Russell (the ancestor of Mono Lake), and China Lake. Lake Manley

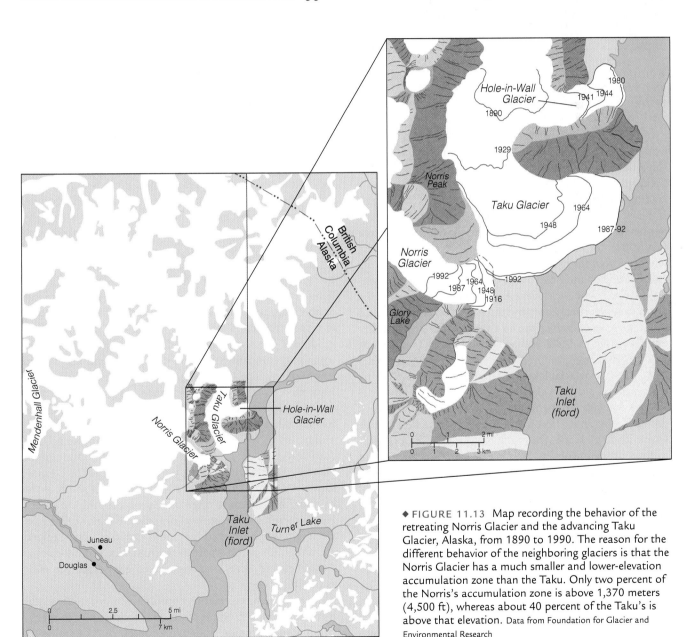

◆ FIGURE 11.13 Map recording the behavior of the retreating Norris Glacier and the advancing Taku Glacier, Alaska, from 1890 to 1990. The reason for the different behavior of the neighboring glaciers is that the Norris Glacier has a much smaller and lower-elevation accumulation zone than the Taku. Only two percent of the Norris's accumulation zone is above 1,370 meters (4,500 ft), whereas about 40 percent of the Taku's is above that elevation. Data from Foundation for Glacier and Environmental Research

ROLF A. LARSSON

◆ **FIGURE 11.14** Esker with a road on its crest; Malingarna area, southern Sweden.

covered the floor of what is now Death Valley National Park, being named after the Manley party that suffered great hardships there in 1849, and Rogers Dry Lake, near Lancaster, California, is the location of Edwards Air Force Base and a landing site for the space shuttle. The Great Salt Lake in Utah is all that remains of the once-vast Pleistocene Lake Bonneville (◆Figure 11.16). The Bonneville Salt Flats, an area that is ideal for setting automobile speed records, formed by evaporation of the ancient Pleistocene lake. Further east, the Great Lakes and the Finger Lakes of New York State are of glacial origin. In the Pleistocene Epoch, when drainage to the north from the Great Lakes and the Finger Lakes was blocked by ice, these lakes drained through various spillways at different times (Figure 11.12 and ◆Figure 11.17). Major disruptions in drainage produced large temporary ice-marginal lakes, called **proglacial lakes,** which were dammed partly by moraines and partly by glacial ice.

Ice Age lake sediments are the basis of the rich soils of the northern United States and southern Canada. The rich wheat lands of North Dakota and Manitoba, for example, formed as sediments in Pleistocene Lake Agassiz (Figure 11.12). Centered in present-day Manitoba, the lake was the largest known North American lake; its area of more than 518,000 square kilometers (200,000 mi²) was more than twice that of the five present-day Great Lakes combined. (It was named after Louis Agassiz, the nineteenth-century scientist who fostered the theory of the Ice Ages.) The lake formed when the great Canadian ice sheet blocked several rivers flowing northward toward Hudson Bay. For a time, a large spillway channel carried outflow southward from the lake, forming what is now the 200-foot-deep Browns Valley on the border between South Dakota and Minnesota (◆Figure

11.18). The many remnants of Lake Agassiz include Lake Winnipeg and Lake of the Woods. Northwest of this area are Great Bear Lake, Great Slave Lake, and Lake Athabaska, descendants of other large ice-marginal lakes in the present-day Northwest Territories, Alberta, and Saskatchewan.

> **CONSIDER THIS** *If there had been no large-scale Pleistocene continental glaciation in North America, what differences might we find in today's midwestern states?* (Hint: *Carefully examine Figures 11.1a and 11.2*).

Glaciation and Climate Change

In the past century it has come to be recognized worldwide that Pleistocene—commonly referred to as *the Ice Age*—glacial deposits of different ages exist. The oldest deposits, weathered and gullied from erosion, have been found beneath younger ones, which are fresher-appearing, less weathered, and less eroded. By applying the Laws of Superposition and Cross-cutting Relationships to these observations, it is clear that multiple stages of glaciation occurred during the Pleistocene Epoch. It is now generally accepted that there have been at least four major glacial stages in North America and Eurasia. Independently, oceanographers studying fossil shells contained in cores of sea-floor sediments from three oceans have determined that the oceans experienced twenty or more periods of deep cooling with intervening warm periods during the Pleistocene. Because evidence of glaciations on land is easily destroyed by later ice advances, it is assumed that the continental record is probably incomplete.

Global temperatures during the Pleistocene Ice Ages averaged only about 4° to 10° C (7° to 18° F) cooler than the temperatures at present. The lowering of the average Ice Age temperature was not as significant as the range of seasonal extremes; winters were much colder, but summers were moderate. Consequently, climate zones shifted toward the equator, accompanied by similar shifts of plant and animal communities. The Rocky Mountains, the Sierra Nevada, the Alps, the Caucasus, and the Pyrenees were topped with ice fields similar to the modern ice field in the mountains near Juneau, Alaska. Small mountain glaciers existed in high mountains of the midlatitudes much closer to the equator than any similar modern glaciers. Even the summit of Hawaii's Mauna Loa volcano, only 19° north of the equator, shows evidence of glaciation. It also was cooler south of the equator, but because the landmasses in the Southern Hemisphere's middle latitudes are small or narrow, extensive glaciation was limited to the South Island of New Zealand and the southernmost part of the Andes in South America. Antarctica and Greenland were ice-covered much the same as they are today.

The recognized relationships between cyclic global climate change and multiple stages of glaciation lead to speculation about common underlying causes. In order to better

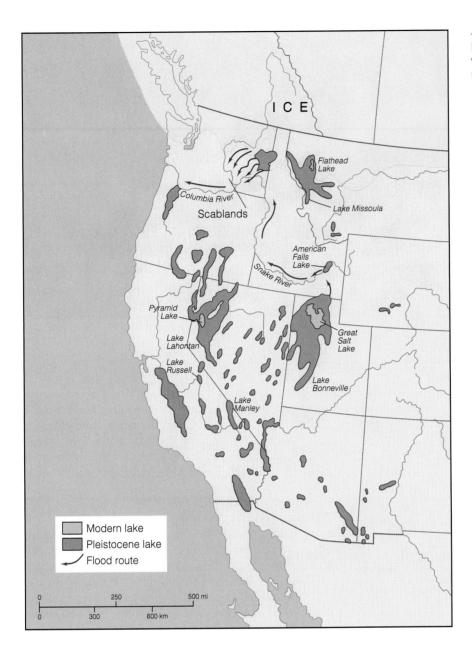

◆ FIGURE 11.15 Pleistocene pluvial lakes in the western United States. After Weiss and Newman "The Channeled Scablands of Eastern Washington," USGS pamphlet, 1973

understand global climate changes over time and why the earth experiences periodic ice ages, we next survey the factors that control climate.

Climate

Factors That Control Climate

Climate is the long-term average of day-to-day weather. Its definition also includes the annual extremes of weather elements within a specific region. Understanding climate requires recognizing that it is determined by the interactions of the geospheres that comprise the earth system—that is, the atmosphere, hydrosphere, biosphere, and lithosphere,

coupled with the orbital elements of the earth–sun–moon system (see Chapter 2). The **climate controls,** the relatively constant factors that dictate a region's climate, are

1. the intensity of the sun, which heats the earth and the atmosphere,
2. the distribution of land and water masses,
3. winds and atmospheric circulation,
4. circulation of the oceans, and
5. mountain barriers.

Some of these controls are treated in more detail in Chapter 12.

Heating of the Earth and Atmosphere

The earth's surface and atmosphere is warmed by radiant energy from the sun. Because part of the sun's radiation is

◆ FIGURE 11.16 Terraces on the flanks of the Wasatch Mountains that mark former shorelines of Pleistocene Lake Bonneville, at Brigham City, Utah. See Figure 11.15.

reflected back to space by clouds and the atmosphere, about 51 percent of the solar energy that could reach the earth actually does. This energy received from the sun is mainly visible light and near-infrared radiation, particular wavelengths of solar radiation that do not "see" the atmosphere through which they pass on their way to the earth, where they heat the ground and oceans. The intensity of the sun's heat at a particular place on the earth's surface varies with the angle at which the sun's rays hit the earth (◆Figure 11.19). This incoming solar radiation is sometimes referred to as *short-wave radiation*. Because all warm objects radiate heat, the warmed surface of the earth radiates energy back to the atmosphere, but at longer wavelengths known as *infrared*, or *long-wave, radiation*. The atmosphere is warmed by this long-wave radiation due mainly to the absorption and reradiation of heat by atmospheric CO_2, water vapor, and other trace gases. Consequently, the atmosphere acts as a blanket that absorbs most of the long-wave radiation and keeps the earth warm. This process, called the **greenhouse effect** by analogy to the solar-heating method that warms greenhouses, serves to keep the earth's surface temperature from dropping excessively during winter and night.

Winds and Atmospheric Circulation

Wind is movement of air from a region of high pressure to a region of low pressure due to unequal heating of the earth's surface. Because the sun's rays strike the earth more directly at the equator than at other places, more radiation (heat) per unit area is received there (◆Figure 11.20). Solar heating at the equator warms the air, and it rises, creating a zone of low atmospheric pressure known as the **intertropical convergence zone (ITCZ),** and the **doldrums.**

In polar regions, the sun rays strike the earth at a low angle, resulting in less-intense heating per unit area. There the colder, drier, denser air creates a zone of high pressure known as the **polar high.** The cold, dry, polar air sinks and moves laterally into areas of lower pressure at lower latitudes. Upper-level air flow from the equatorial regions moves toward the poles. When it reaches approximately 30 degrees north and south of the equator, it sinks back to the earth's surface as dense, dry air to form high-pressure belts known as the **subtropical highs,** or the **horse latitudes.** Winds, then, are driven by differences in air pressure caused by unequal heating, but as they move they do not follow straight paths. Instead, they are deflected by the Coriolis effect.

The **Coriolis effect** is caused by the rotation of the earth. A simple analogy of the Coriolis effect is to consider two people playing catch on a merry-go-round (◆Figure 11.21). If the merry-go-round is stationary, the ball will follow a

◆ FIGURE 11.17 Typical subdued terrain resulting from continental glaciation at Lake George, New York. The lake is dammed by an end moraine. The break in the ridge on the right marks the channel of the ancient river that carried meltwater southward from the Pleistocene glacier. See Figure 11.12.

◆ FIGURE 11.18 Big Stone Lake now occupies part of Lake Agassiz's spillway channel. It is in Browns Valley on the boundary between western Minnesota and northeastern South Dakota. See Figure 11.12.

straight path. An apparent paradox occurs when the merry-go-round is turning, however. As seen by stationary observers not on the merry-go-round, the thrown ball still travels in a straight path, but to the people on the turning merry-go-round, the ball appears to follow a curved path. The reason for this paradox is that the merry-go-round and its passengers are turning below the ball as it travels from one person to the other. To the people on the merry-go-round, the path of the ball appears to be deflected as though some force is acting on it. The curve will be to the right if the merry-go-round is turning counterclockwise, the same direction the earth rotates if observed from above the North Pole. The deflection of freely moving objects also occurs on the earth. Rifle bullets, ballistic missiles, ocean currents, and winds that travel long distances all appear to follow curved paths because of the earth's rotation. The path is deflected to the right in the Northern Hemisphere and to the left in the Southern Hemisphere.

The rising warm air in the equatorial belt expands and cools as it rises. Cooling reduces the air's ability to hold moisture, resulting in the formation of clouds and heavy precipitation. For this reason, the intertropical convergence zone (ITCZ) is one of the rainiest parts of the world. We generally do not find deserts there. Air also rises at latitudes of approximately 60° north and south, where cold polar surface air and mild-temperature surface air converge but tend not to mix. At this boundary, called the **polar front,** surface air rises, forming a zone of low pressure called the **subpolar low.** It is along the polar front that rising air causes storms to develop. In the subtropics, in contrast, cool, dry air sinks, forming high-pressure regions. As the air sinks, it is compressed, which causes its temperature to increase. This warming enables the air to absorb water and precludes the formation of clouds. Thus the climate belts of subtropical highs, from about 30° to 35° north and south of the equator, the so-called horse latitudes, experience warm, dry winds with generally clear skies and sunny days. It is here that we find most of the world's hot deserts. (Deserts are the topic of Chapter 12.)

CONSIDER THIS *The old expression "caught in the doldrums" is sometimes used to describe a person who lacks motivation and is listless. What is the relationship of this expression to the traditional term for the ITCZ?*

Oceanic Circulation

Unequal heating of the earth's surface by the sun moves the atmosphere, and the resulting winds drive the surface currents of the oceans. Each major ocean basin is distinguished by currents that rotate clockwise in the Northern

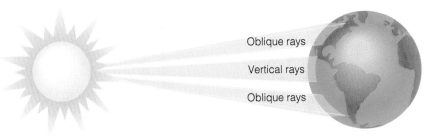

Oblique rays

Vertical rays

Oblique rays

◆ FIGURE 11.19 The intensity of the sun's heat depends upon the angle at which the sun's rays hit the earth. The intensity and heating is greatest where the rays hit vertically.

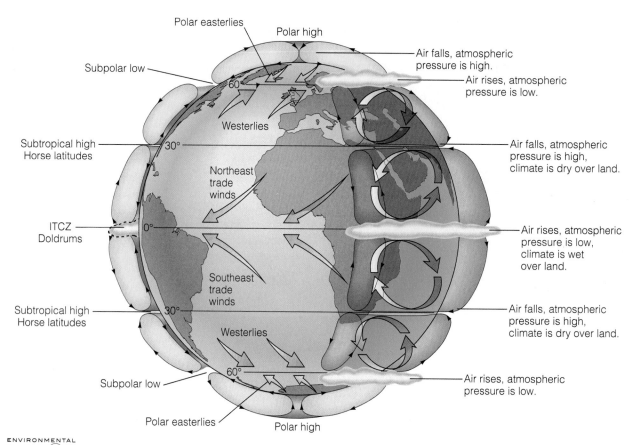

ENVIRONMENTAL
◆ Geology⇌Now™ ACTIVE FIGURE 11.20 In the general scheme of global atmospheric circulation, heat transfer occurs in steps generating three cells between each pole and the equator. This results in persistent global air-circulation patterns that generate atmospheric pressure zones and surface wind belts. Descending air causes zones of high pressure (the polar highs and the subtropical highs), and rising air produces zones of low pressure (the ITCZ and the subpolar lows). From B. Murphy and D. Nance, *Earth Sciences Today*, p. 331, Brooks/Cole, 1999

Hemisphere and counterclockwise in the Southern Hemisphere in accordance with the Coriolis effect (◆Figure 11.22). In both hemispheres, these currents carry heat from the equator toward the poles along the west sides of the basins and, in turn, carry cooled water from polar latitudes back toward equatorial regions along the east sides of the basins. This transfer of heat occurs in all oceans. For example, the California Current cools the west coast of North America, imparting mild summer temperatures to San Diego. In contrast, the warm Gulf Stream on the east coast

imparts hot, humid summers to Savannah, Georgia, at the same latitude.

A major oceanic process affecting coasts in the "trade-wind" belts north and south of the ITCZ (Figure 11.20) is **upwelling.** Upwelling occurs when winds blowing from the land to the sea *(offshore winds)* sweep surface water away from the shore, creating a condition of hydraulic instability. The displaced nearshore water is replaced by nutrient-rich, cold waters from depths in excess of 200 meters (650 ft; Figure 12.2, near the beginning of the next chapter, illustrates

ENVIRONMENTAL
◆ Geology⇌Now™ ACTIVE FIGURE 11.21 The Coriolis effect in a game of catch. (a) The path of the ball is straight when the game is played on a stationary platform, and an observer perceives it as straight. (b) On a rotating platform the ball still follows a straight path, but to an observer on the platform, the path appears to be curved. This is the Coriolis effect.

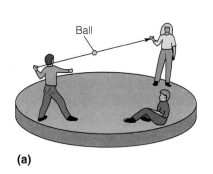

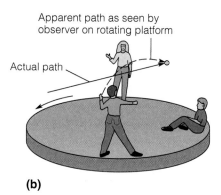

(a)   (b)

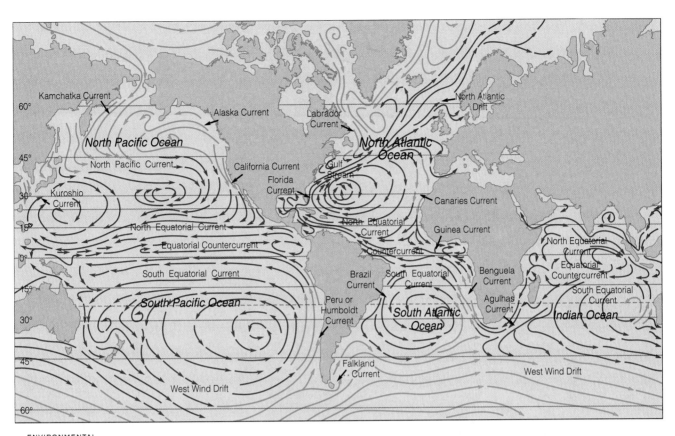

ENVIRONMENTAL
◆ **Geology⇌Now**™ ACTIVE FIGURE 11.22 Major wind-driven oceanic surface currents. Warm currents are shown in red; cool currents in blue. Currents rotate clockwise in the Northern Hemisphere and counterclockwise in the Southern Hemisphere.

this). During upwelling, coastal waters nearshore are colder than offshore water, which further moderates local weather. Conditions for coastal upwelling are most commonly found along the western sides of continents in the trade-wind belts, as, for example, along the coasts of Peru and California. Upwelling brings dissolved phosphates and nitrates to the surface, which leads to the proliferation of marine plankton, which in turn promotes large fish populations and gives rise to commercial fisheries. However, if the trade winds break down or weaken and upwelling ceases, major climatic upsets such as El Niño may occur, causing floods and severe storms in some areas (see Chapter 10).

ENVIRONMENTAL
Geology⇌Now™

Click **Geology Interactive** to work through an activity on World Currents and Dissect a Current through Waves, Tides, and Currents.

Causes of Climate Change

Variations in the Earth's Orbit

Once scholars recognized that cyclic global climate changes were responsible for the multiple stages of glaciation, they began to seek answers to questions about underlying causes. An early answer was provided in the late nineteenth century by Scottish geologist John Croll. Serbian astronomer Milutan Milankovitch expanded upon Croll's work with elegant mathematics in 1941. Both of these scientists recognized that the cyclic pattern of the Pleistocene ice ages could result from periodic minor changes in the earth's rotation and revolution around the sun, with the cold periods of ice advance occurring when three variations in the earth's rotational and orbital elements were synchronous. The three variations are

1. the **obliquity,** or tilt, of the earth's axis;
2. the **eccentricity,** or shape, of the earth's orbit; and
3. the **precession,** or wobble, of the earth's axis.

See ◆ Figure 11.23. The earth's *obliquity* relative to the sun has been found to vary between 21.2° and 24.5° over a 41,000-year cycle. This slight change causes variations in the temperature range between summer and winter. When the earth is at the minimum tilt, sunlight hits the polar regions at a higher angle, and the global seasonal temperature variation decreases. Every 100,000 years, the orbit of the earth changes from nearly circular to more *eccentric*. Seasonal variation in temperature is greater when the orbit is more eccentric. The axis of the earth, which now points to Polaris, the North Star, *precesses* like a spinning top when its axis is disturbed. This precession causes the axis to trace out a cone in space, taking 23,000 years to trace out a complete cone. Consequently, the

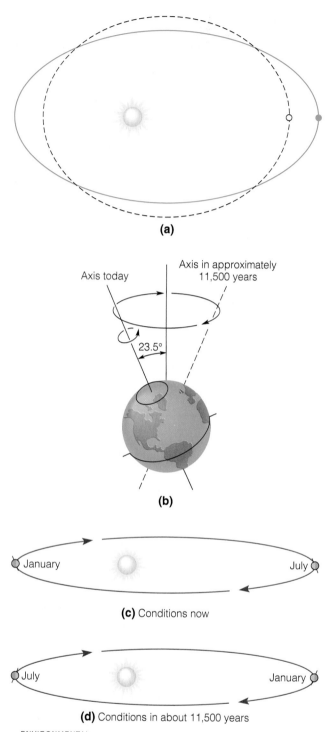

(a)

(b)

(c) Conditions now

(d) Conditions in about 11,500 years

ENVIRONMENTAL
◆ **Geology⇌Now**™ ACTIVE FIGURE 11.23 Geometry of variation in the earth's orbital eccentricity, axial precession, and axial obliquity. (a) The orbital eccentricity varies from nearly circular *(dashed line)* to elliptical *(solid line)* to nearly circular and back again over a period of about 100,000 years. (b) The earth's axis slowly precesses, like a spinning top tracing out a cone in space, taking about 23,000 years for a complete cycle. (c) Conditions today. The earth is closest to the sun in January while the Northern Hemisphere is experiencing winter. (d) Conditions in about 11,500 years. The earth will be closest to the sun in July while the Northern Hemisphere is experiencing summer.

Northern and Southern Hemispheres "trade" their respective winter and summer seasons every 11,500 years.

Each of these periodic factors alone could cause the limited global cooling that might lead to a "little ice age." More important, however, is the interaction of the three cycles. When their cooling effects coincide, there is a sufficient change in the global heat budget to trigger glacial–interglacial episodes, though not enough to explain fully the 4° to 10° C cooling required for the major glacial cycles of the Pleistocene Epoch. In other words, the rotational and orbital factors adequately explain the timing of the glacial–interglacial cycles, but not the amount necessary to cause the *changes in global temperature.* There must be other factors.

Plate Tectonics and Variation in Atmospheric Carbon Dioxide

For glaciers to exist, the continents must be in the middle to higher latitudes, and high-standing mountains are "desirable." Plate tectonics can account for both the positioning of continents at "appropriate" latitudes and the uplifting of vast continental areas. Such repositioning of continents and higher elevations of land areas alters the circulation patterns of both the atmosphere and oceans, which leads to long-term climate change. In addition, the higher the continental land masses, the greater is the rock surface area exposed to chemical weathering by atmospheric gases. As the rocks weather, they combine with water and atmospheric carbon dioxide (CO_2), which reduces the amount of the gas in the atmosphere. Further, input of CO_2 from volcanic activity is reduced when, at the end of a tectonic cycle, the volcanic fires at oceanic ridges cool down and the ridges become less active. Because CO_2 is the major greenhouse gas, these reductions promote global cooling. (In the earlier phases of the cycle, the ridges are active, which increases the CO_2 input to the atmosphere, and sea level is high, which reduces weathering. These two factors reinforce each other to increase atmospheric CO_2.) The relationship of CO_2 concentrations in the atmosphere, oceans, and biosphere is enormously complicated, and several feedback mechanisms exist that are not understood (⊙ Case Study 11.2 on page 318). For example, diatoms, the microscopic oceanic plants that can be thought of as the "grass of the sea," may be part of a feedback loop in the biosphere that affects climate. Diatoms may step up productivity when there is excess atmospheric CO_2, thus drawing down the CO_2 content and cooling the climate.

Variations in Solar Radiation

Measurements by sophisticated instruments carried by satellites indicate that the output of solar radiation may vary considerably more than was once thought. Furthermore, it is known that the sun's energy output changes slightly with sunspot activity. *Sunspots*—enormous magnetic storms on the sun's surface that mark cooler regions surrounded by very bright areas radiating great bursts of energy—occur in cycles, with the maximum sunspot size and number occur-

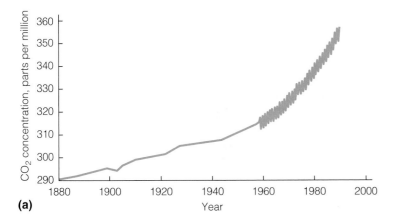

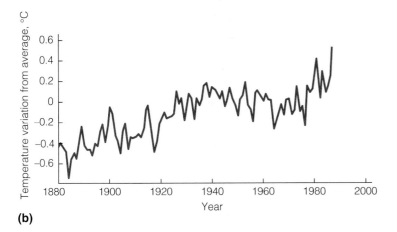

◆ FIGURE 11.24 Atmospheric CO_2 and temperature changes since 1880. (a) Changes in concentration of CO_2 obtained from ice cores (smooth curve) and measured annual oscillations at Mauna Loa, Hawaii. (b) Average annual surface-air temperature at Mauna Loa, plotted as a deviation from the average, denoted as zero (0). After Nansen and Lebedeff, Global Trends of Surface-Air Temperatures," *Journal of Geophysical Research* 29, no. D11 (Nov. 20, 1987), fig. 15, p. 13370.

ring about every 11 years. During such times the sun produces about 0.1 percent more energy than it does at times of sunspot minimum. The 11-year cycle appears to collapse from time to time, as it did in the seventeenth century (at a time known as the *Maunder minimum*), when there were few if any sunspots. This "minimum" occurred during the coldest period of the Little Ice Age, when the average global temperature was about 0.5° C (1° F) lower than the long-term average.

Correlated with the 11-year sunspot cycle is a 22-year cycle, the time required for completion of the sun's *magnetic cycle*—and an interval that closely corresponds to the 20-year cycle of drought on the Great Plains. The sun's general magnetic field increased 2.3 times between 1901 and 2000, and because the general magnetic field varies in time with the 11-year sunspot cycle, the varying magnetic field is also related to the sun's brightness. Furthermore, the impact of the sun's magnetic field on the earth appears to be related to cloudiness. (The field influences the amount of cosmic rays that enter the earth's atmosphere, which in turn, may affect the formation of electrically charged particles, on which water condenses to form cloud droplets.)

Moreover, solar brightness also affects the stratospheric ozone layer. The 0.1 percent increase in solar energy pro-

duced during the 11-year maximum is ultraviolet radiation. Ultraviolet rays' wavelengths are absorbed by stratospheric ozone, the level of which also varies with the sunspot cycle. Because the time of maximum ozone content is also the time of maximum sunspot activity, additional ozone at such times absorbs the additional ultraviolet radiation.

If increased solar radiation were currently involved in global warming, it would be warming the upper and lower atmosphere as well as the earth's surface. This is not the case. The upper atmosphere is cooling, while the lower atmosphere and the earth surface are warming. Eventually it may be found that fluctuations in solar radiation account for some long-term climate changes, but much more research is needed. Correlation is not necessarily causation.

Greenhouse Gases

Scientists have been measuring and recording the level of carbon dioxide (CO_2) in the atmosphere since 1880. Their measurements over the years show that the concentrations of CO_2 and other greenhouse gases, as well as average global temperatures, have risen steadily (◆Figure 11.24). This increase in atmospheric CO_2 is unprecedented in human history, and it is believed to be largely due to humankind's dependence upon fossil fuels. Other "greenhouse gases" include:

- methane (CH_4), which is expelled as flatulence from the world's cattle population, from rice paddies, and of all things, from termites in tropical rainforests;
- nitrogen oxide (NO_x) from fertilizers and automobile exhausts; and
- **chlorofluorocarbons (CFCs)**, largely from air conditioners and spray-can propellents.

These substances combine to increase global warming.

In addition to their being a greenhouse gas, CFCs play another role in the atmosphere; they break down the upper-atmospheric (the stratospheric) ozone layer 10 to 48 kilometers (6–30 mi) above the earth surface. Ozone, a form of oxygen, serves as a vital shield that protects the earth's organisms from the harmful solar ultraviolet radiation that bombards the earth. The chlorine in one molecule of CFC has the potential to destroy as many as 100,000 ozone molecules before it becomes inert. Any thinning of the ozone shield leads to global change.

Using data from satellites and airborne sensors, scientists discovered in the late 1970s and early 1980s that atmospheric ozone levels were decreasing at an alarming rate. By the spring of 1987, an **"ozone hole"** the size of the United States and Mexico combined had developed over Antarctica. The hole has continued to enlarge, appearing to document the continuing effects of human activity on earth. Ozone thinning is now recognized over the North Pole and the middle latitudes as well. Ozone loss in the stratosphere causes warming on the earth's surface as well as dangerous cooling in the stratosphere.

A thinning of the atmospheric ozone layer threatens humans with increased risks of skin cancer and cataracts and of weakening the immune system. The increased ultraviolet radiation reaching the earth also threatens to damage crops and phytoplankton, the microscopic plants that are the basis of ocean food chains. Some medical authorities reported that a person's chance of developing skin cancer by the 1990s was about one in 75, up from one in 1,500 about 40 years earlier. Humans, of course, can protect themselves from ultraviolet radiation, but other animals cannot. For example, ultraviolet-radiation damage to DNA has been shown to account for severe losses of fertilized eggs in at least two species of Cascade Mountain frogs.

Also contributing to the depletion of the ozone layer is the burning of aviation fuel at altitudes above 9 kilometers (5.5 mi). Nitrogen oxides (NO_x) and water-vapor molecules released at altitudes between 9 and 13 kilometers (5.5 and 8 mi)—the usual cruising altitude for commercial jetliners—stay in the atmosphere about 100 times longer than those released near ground level. Although it seems contradictory, NO_x emissions can either increase or decrease ozone concentrations in the atmosphere. In the lower atmosphere, where we live, NO_x, together with hydrocarbons, forms ozone (a major component of smog) in the presence of ultraviolet light. In the upper atmosphere, however, NO_x breaks down

ozone. CFC climate modeling, which does not take NO_x emissions from air traffic into account, has underestimated the amount of ozone depletion. According to researchers, the amount of ozone loss over the Northern Hemisphere is about twice that calculated using only CFC climate modeling. Regular (subsonic) air traffic is believed to be responsible for a 10 percent increase in the nitric acid concentration at 21 kilometers (13 mi) above sea level, which may be the most critical altitude for ozone holes. Clearly, the chemistry of the atmosphere is complex, and it is the subject of continuing scientific research.

In the 1990s, a few radio talk-show hosts and some government policymakers expressed doubts about the danger of CFCs and the depletion of the stratospheric ozone layer. They argued that chlorine from CFCs is trivial in comparison to the great quantities of chlorine released to the atmosphere from natural sources such as volcanic eruptions, which have been contributing chlorine to the atmosphere for many hundreds of millions of years. In spite of this, the ozone layer still exists, so, they argued, the concern about CFCs is overstated. Their argument was spurious, however, because natural chlorine is chemically reactive in the lower atmosphere and soluble in water; consequently, it gets rained out before it can reach the stratosphere. CFCs, in contrast, are absolutely inert and thus insoluble in the lower atmosphere. This allows them to rise through the lower atmosphere and into the stratosphere, where ultraviolet rays cause the release of chlorine. It was recognition of their chemical inactivity in the lower atmosphere—they react with *nothing*—that made them so appealing as coolants and propellants.*

Dust and Aerosols

Studies of glacial ice cores not only verify the Croll–Milankovitch theory of cyclic ice ages (Figure 11.23), but they also reveal that the atmosphere was dusty during glacial stages. Winds apparently lifted fine dust from glacial deposits and arid regions, causing the skies to become hazy. This haziness would have scattered incoming solar radiation back into space, causing further cooling of the earth.

Volcanic activity has been suggested as a contributor to global cooling, both because of the dust ejected by eruptions and due to volcanic releases of sulfur dioxide (SO_2), which is converted to sulfate-aerosol droplets in the atmosphere. The sulfate aerosols, acting like tiny mirrors, cool the atmosphere by reflecting sunlight away from the earth. An illustration of the role played by volcanic eruptions in global climate is pro-

---

*U.S. production of CFCs ended in December 1995. Now compounds that lack chlorine or that decompose in the lower atmosphere—the hydrofluorocarbons (HFCs) and the hydro*chloro*fluorocarbons (HCFCs)—are being used. In the rest of the world production of CFCs ended on January 1, 2000. The changeover requires that both refrigeration units and manufacturing facilities be modified to use the new refrigerants. Older refrigerators and air conditioners presently using CFCs are to be recharged only with recycled CFCs, but eventually they will be discarded or will be retrofitted to use the newer refrigerants.

vided by the 1815 eruption of Tambora in Indonesia. After that event, with the spread of the eruption's aerosols and dust through the stratosphere, the amount of solar radiation reaching the earth was so reduced that the ensuing global cooling caused widespread crop failure. The starvation that followed caused the deaths of more than 90,000 people. Because the global cooling resulted in rain, late snows, and failed crops in Europe and in New England, 1816 became known as the "year without a summer."* Similar reductions in average global temperatures due to stratospheric sulfate aerosols and dust following volcanic eruptions in modern times are recognized, such as those after the 1982 eruptions of El Chichón in Mexico and (for two years following) the 1991 eruption of Mount Pinatubo in the Philippines (see Chapter 5).

Scientists once thought that all sulfate aerosols made clouds more reflective, thus helping to *cool* the earth. But that is true only of aerosols resulting from volcanic emissions, as they are blasted high into the stratosphere. In contrast, it is now recognized that sulfate aerosols released by coal-burning power plants and industry and by motor vehicles reside in the lower atmosphere, where they form droplets that make clouds dirtier and less reflective, thereby *increasing* global warming.

Change in Ocean Currents

Vagaries in ocean currents, too, seem to play a role in plunging the earth into glacial and interglacial climates. A giant, salty "conveyor-belt" current snakes through the world's oceans (◆Figure 11.25). Originating in the North Atlantic by northward-flowing warm, salty (and thus dense) water because of evaporation, the belt is chilled as it moves northward, which further increases its density. It sinks into the depths flowing southward out of the Atlantic and into the other oceans, much of it rising in the Pacific, where it offsets the evaporation losses there. This conveyor is the main agent of oceanic heat transfer, with one part of it a deep, cold, salty (and thus dense) current (shown in green in Figure 11.25), and the other part of it a warm, less salty, surface current (shown in yellow). The northward-flowing surface water averages 8° C warmer than the cold, deep water flowing south.

When the conveyor is operating, the climate of the North Atlantic is mild, like that of today. It is vulnerable to abrupt disruptions and shutdowns, however, which can cause drastic shifts in global climate (see ◎ Case Study 11.3 on page 320) and force the North Atlantic into bitter, ice-age cold.

---

*This cold summer made the *Old Farmer's Almanac* famous. When the printer was setting type for the 1816 issue, he found July's weather forecast to be incomplete. He decided to predict snow for the Fourth of July. Astoundingly, on that date, it *did* snow in New England, where a large proportion of Americans lived at the time, thereby establishing a solid reputation for that almanac's reliability.

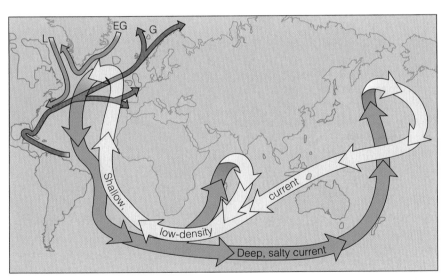

ENVIRONMENTAL
◆ **Geology ⊜ Now™** ACTIVE FIGURE 11.25 The giant "conveyor belt" current that snakes through the world's oceans. Two surface currents of major importance in determining the modern climates in Greenland, eastern North America, and western Europe are shown. The cold Labrador (*L*) and East Greenland (*EG*) currents (both shown in blue) cool the adjacent coasts. The Gulf Stream (*G*, red) transfers heat northward, influencing Arctic air currents and warming the coasts of Iceland, Greenland, and western Europe. This current was deflected southward toward North Africa during the coldest part of the last maximum glaciation. The importance of these currents, in combination with the atmospheric wind system, is illustrated by comparing the modern equitable temperature of northern Scandinavia, which is essentially unglaciated, with that of Greenland, which, at the same latitude, is covered by an enormous ice sheet. Data from W. S. Broecker, "Chaotic Climate," *Scientific American* 273, no. 5 (Nov. 1995).

When excessively large volumes of cold fresh water from melting icebergs and river runoff enter the North Atlantic, the conveyor shuts down and winter temperatures drop by five or more degrees.

Glacier Evidence of Climate Change

Glacial ice provides evidence of cyclic changes in atmospheric chemistry, temperature, and amount of atmospheric dust that correlate with periods of cooling and warming during the Pleistocene Epoch. As snow accumulates and is packed down into ice, it carries information with it that gets locked into glaciers. The relative abundance of two slightly different oxygen atoms—the oxygen-18 to oxygen-16 isotope ratio, symbolized $^{18}O/^{16}O$—in ice molecules indicates what the temperature was when the snow fell from which the ice formed. During a glacial period, when sea water evaporates from a cold ocean and precipitates on land to form glaciers, water containing the lighter isotope, ^{16}O, is more readily evaporated than is water containing the heavier ^{18}O. Thus, the oceans come to contain more of the heavy isotope, and glaciers to contain more of the light isotope. Furthermore, each year's snowfall forms an easily recognized line in glacial ice. When core samples of glaciers are examined, these lines act like tree rings, allowing scientists to determine the years, and the $^{18}O/^{16}O$ ratios allow each year's temperature to be estimated. Together, these data allow a generalized overview and time scale of global climate change.

The bubbles of ancient atmosphere trapped in glacial ice obtained from 3,348-meter (10,981-ft)-deep cores drilled at Vostok Station in Antarctica and the Greenland Ice Sheet Project 2 (GISP2) provide a 420,000-year record of variations in the atmosphere's temperature and composition (◆ Figure 11.26). The ice-core record and correlative deep-sea-sediment cores show that ice ages are marked by climatic jiggles—intricate cycling within the larger cycles of glacial advance and retreat due to orbital forcing by the Croll–Milankovitch effect. Peculiarities in the lower part of the GISP2 ice core, presumably representing atmospheric and climatic conditions existing about 120,000 years ago, imply abrupt reversals every 2,000 or so years between glacial climate and warm interglacial climate, with the change occurring in as little as three years. Another study, the Greenland Ice-Core Project (GRIP), drilled about 11 kilometers (7 mi) away from GISP2, provided similar information. The climate interpretations of the upper part of the Vostok core and the two Greenland cores agree, but the wild changes portrayed by the Greenland cores are not seen in the Vostok core. The analysis is far from complete, and there is some evidence that the GISP2 and GRIP cores have been disturbed by intermittent melting and by layering distortions from nearby bedrock. Thus, it is believed that the Vostok record is the only undisturbed ice core that records full glacial–interglacial cycles—that is, the conditions during four major interglacial periods and the preceding glacial periods (◆ Figure 11.27). The amounts of

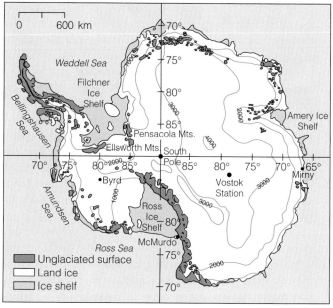

(a)

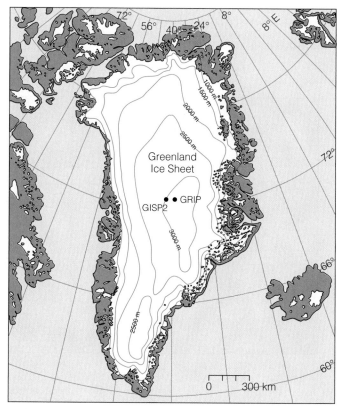

(b)

◆ FIGURE 11.26 The earth's two continental glaciers. (a) The largest glacier complex covers practically all of Antarctica, a continent that is more than half again as large as the 48 states. The ice sheet's thickness averages about 2,160 meters (7,085 ft), with a maximum of 5,000 meters (16,500 ft). (b) Greenland's ice sheet has a maximum thickness of 3,350 meters (10,988 ft). Ice thicknesses are contoured in meters. Vostok, GRIP, and GISP2 are the sites of core drilling projects discussed in the text.

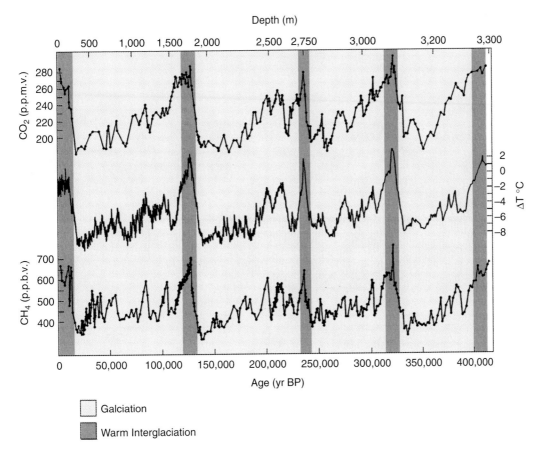

◆ FIGURE 11.27 Temperature variations for the last 420,000 years relative to the present, and CO_2 and CH_3 atmospheric concentrations recorded in the Vostok ice core, Antarctica. Core depths on upper axis with corresponding time scale on lower axis (p.p.m.v. = parts per million by volume; p.p.b.v. = parts per billion by volume; ΔT °C= Antarctic air temperature relative to present). *Note:* Four complete Ice Age cycles are recorded and the atmospheric concentrations of CO_2 and CH_3 correlate well with Antarctic air temperature throughout the record. (Source: J. R. Petit, and others, 1999, Climate and atmospheric history of the past 420,000 years from the Vostock ice core, Antarctica. *Nature* 399, pp. 429–36. Used by permission.)

methane (CH_4) and CO_2, the principal greenhouse gases, in all three cores are lower in the layers accumulated during glacial stages and higher in the layers representing warm interglacial stages. The synchroneity of increasing and decreasing amounts of greenhouse gases with increasing and decreasing average temperature suggests that these changes are related. This correlation seems compelling, but a cause-and-effect relationship is still unknown: does the change in atmospheric gas content initiate a glacial stage, or does the global cooling of a glacial stage cause a change in the gas content? That greenhouse gases play a significant role in the glacial–interglacial cycle seems obvious, but it is still uncertain precisely what causes their percentages to change. On a human time scale, the ice reveals that since preindustrial times, the level of CO_2 has increased by about 40 percent and that CH_4 has increased more than 100 percent.

ENVIRONMENTAL
Geology ⇌ Now™

Click **Geology Interactive** to work through an activity on Ozone Holes through Weather and Climate.

Today's Global Warming

I'll take ten-to-one-odds against anyone who says global warming is not well underway. And I'll take all the money they care to offer.

Jerry Mahlman, Director, NOAA Geophysical
Fluid Dynamics Laboratory, Princeton University

Evidence of Global Warming

Increasing Temperatures of the Earth's Surface and Lower Atmosphere

Temperature records from land areas and oceanic shipping lanes reveal that the average earth temperature increased about 1.1° F during the twentieth century, with half of that increase after 1970 (◆Figure 11.28). The greatest amount of warming occurred at night over land in the higher latitudes. Some areas of Canada, Alaska, and Eurasia warmed as much as 5.5° C (10° F) between 1965 and 2000. These differences may seem small, but such changes in the past have had major

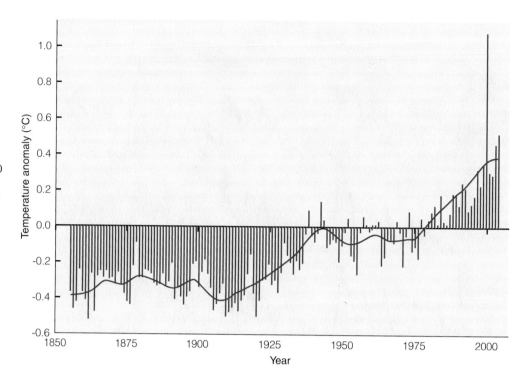

◆ FIGURE 11.28 Time-series record of the combined global marine and land average surface temperature from 1856 to 2002. The 1990s were the warmest decade in the series. The warmest two years of the series are 1998 and 2002, with the former the warmest at 1.16°C (2°F) above the 1961–1990 average, the warmest year in 341 years. Data are from 400 proxy climate series (ice cores, corals, tree rings) historical and instrumental records. Courtesy of Jean Palutikof, Climatic Research Unit, University of East Anglia

consequences. As an example, the earth has warmed by only 3° to 5° C (5°–9° F) since the depths of the last ice age some 200,000 years ago.

The earth's present average temperature is the warmest it has been since 1400, which is as far back as we can go with certainty by direct and indirect evidence. Some authorities believe this may be the warmest period in 100,000 years. The warmest years of the last six centuries were 1998 and 2002 (Figure 11.28) with 1998 the warmest on record, even after correcting for the warming influence due to the active El Niño at the time.

Changes in Modern Glaciers

Modern glaciers provide relevant information because glaciers respond to climate changes by either advancing or retreating. Their behavior serves as an indicator of climate change and also as a "filter," smoothing out the record of seasonal and annual variations in temperature and precipitation. Because of their small size and ice volume, mountain (alpine and ice cap) glaciers in the mid- and low latitudes are remarkably sensitive indicators of climate change: they respond quickly even to small perturbations in climatic elements. A worldwide survey of 160,000 mountain glaciers and ice caps reveals that the volume of the world's glacier ice is declining, and the rate of loss is increasing. For instance, in Glacier National Park in Montana's Northern Rocky Mountains, glacial fluctuations have been studied for a century, with reports dating back to 1914. By 2003, all but 35 of the park's 150 ice fields have melted since 1850, with the cold slivers that remain estimated to disappear about 2050 at the current rate of shrinking. Similar findings in mid-latitudes are reported from Switzerland, where alpine glaciers have lost as much as half their mass since 1850; the Caucasus Mountains of Russia, where glacial ice has

decreased by about 50 percent during the twentieth century; and in New Zealand, where, on average, the 127 glaciers in the Southern Alps have become 38 percent shorter and 25 percent smaller in area than when first studied in the twentieth century. At low latitudes, the largest glacier on Africa's Mount Kenya has lost 92 percent of its mass in the last few decades.

At higher latitudes, Alaskan glaciers are in dramatic retreat, having thinned twice as fast in the last five years as during the preceding years. Using a laser altimeter attached to an airplane, a USGS research team has flown over 67 glaciers in Alaskan mountains, in order to compare their present surface elevations with those that were mapped in the 1950s. Since the 1950s, most glaciers show several hundred feet of thinning at low elevations, and about 18 meters (60 ft) of thinning at higher elevations. Furthermore, the glaciers were thinning twice as fast between 1997 and 2002 as they did from the 1950s to the mid-1990s. This equates to a rise in sea level of 0.1 millimeter (0.004 in) per year from the mid-1950s to the mid-1990s, making Alaska the largest contributor to rising sea level of any ice-bound region on earth.

But mountain glaciers contain only six percent of the world's ice. Of more concern are the enormous continental glaciers of Greenland and Antarctica, which together contain about 90 percent of the world's fresh water. It is estimated that if only 10 percent of the water in these enormous ice sheets were to melt and be added to the ocean, sea level would rise by over six meters (20 feet) and cause coastal devastation worldwide. Based on satellite data, the largest glacier of the West Antarctic Ice Sheet is losing mass at about four gigatons (giga = 1 billion) a year, equivalent to a sea-level rise of 0.01 millimeters (0.0004 in). These glaciers are responding to the higher global temperature between 1977 and 2002, which was 0.4° C (0.7° F) above the 1940–1976 mean.

Thawing in Boreal and Subarctic Regions

Climate changes in Alaska and the Bering Sea region have already had socioeconomic impacts on Alaskans, most of them negative. For example:

1. With less sea ice in the Bering Sea, weather events such as storm surges have increased in frequency and severity, causing increased coastal erosion, inundation, and threats to structures.

2. Subsistence lifestyles of the native peoples in the region have been adversely affected by such things as changes in sea ice conditions. Hunting on the ice, for example, is more dangerous.

3. Slope instability, landslides, and erosion are increasingly common in thawing permafrost terrain. This threatens roads and bridges and causes flooding.

4. The warmer climate has increased forest fire frequency and insect outbreaks, reducing timber yields.

5. The speed of permafrost thawing has caused major changes in the forest, bog, grassland, and wetland ecosystems that have affected land use.

6. Over the past 30 years, the number of days that oil company personnel can travel on Alaska's frozen North Slope to search for oil prospects has shrunk in half—from 200 to 100. The tundra, which must be completely frozen before running vehicles on it, has been freezing later and thawing earlier.

Rising Sea Level

Sea level was about 120 meters (393 ft) lower than the current level during the last glacial maximum, about 20,000 years ago, and its rate of rise has varied. In the nineteenth century it rose about 12 centimeters, and during the twentieth century the rise appears to have been about 25 centimeters (10 in). The accelerated rate is due to some combination of the thermal expansion of the oceans and the melting of mountain and Antarctic glaciers. (Sea level rises about 24 centimeters with each 1° C increase in temperature, with a time lag of about 20 years.)

The effect on sea level by the melting of Antarctic ice remains uncertain. Globally, sea level is rising about two millimeters (0.06 in) per year, but that amount could be explained by the increased volume of sea water due to thermal expansion. Melting of mid-latitude glaciers may exert a greater influence on sea level because their meltwater is added directly to the ocean, whereas so much polar ice is floating shelf ice that already displaces its own weight. A factor that may moderate a global rise of sea level is that a warmer atmosphere will hold more water vapor and cause more snow to fall in polar regions.

Biological Response

Observations in North America and Europe reveal that wildlife in various habitats are responding to warming climates. ✿ Case Study 11.4 on page 321 provides summaries of some research studies.

Thinning of the Polar Ice Cap

Scientists believe that the Arctic Ocean ice cap is displaying a substantial climatic signal. Comparing ice thicknesses measured by submarines between 1958 and 1976 with measurements made in the 1990s revealed that, on average, the ice cap had thinned by 1.3 meters (4.3 ft) and that it had lost 40 percent of its volume in that time. Such a major change in the Arctic could have a significant impact on Northern Hemisphere weather.

Arctic research was stepped up in the 1990s—a major role in the research was the U.S. Navy's Scientific Ice Expeditions between 1993 and 1997. Their efforts revealed a 1.1° C (2° F) temperature increase in a large part of the Arctic Ocean and recognized that at least part of the increase was due to incursions of warmer water from the Atlantic Ocean. Since 1998 it appears that the torrid pace of thinning of Arctic Ocean ice has slowed. Computer models forecast that Arctic ice will continue to shrink by more than half its 1955 volume by mid-twenty-first century, enough that the thinning and shrinkage will play havoc with Arctic life (see Case Study 11.4) and open a Northwest Passage in summers that may prove useful as a commercial shipping route.

It is uncertain what has caused the decrease in ice—whether it is due to human activities, natural global warming, a natural Arctic climatic fluctuation, or some combination of these factors. A key factor in explaining the retreat of Arctic Ocean ice is the influence of a decades-long atmospheric cycle called the Arctic Oscillation (AO). The AO varies from a cold, still air phase to a warm and windy phase. Normally, the phases shift every ten years or so, but the Arctic has been locked into the warm, windy phase since the 1980s. Some scientists believe that the locked-in warmer mode is due to atmospheric pollutants such as CO_2 enhancing the greenhouse effect and that it forces the decades-long atmospheric swing toward the AO warm, windy phase on which the decadal swings are superposed.

CONSIDER THIS *With what you have learned about the increasing concentration of greenhouse gases in the atmosphere, what changes might you expect over the next 50 years where you live if global temperature should rise by 2° C?*

Causes of Today's Global Warming

For every complex problem there is always a simple answer, and it's usually wrong.

H. L. Mencken (1880–1956), American writer

Of the five factors most likely responsible for the large-scale temperature changes presently being experienced, three vary naturally:

- stratospheric volcanic sulfate aerosols,
- climate variability, and
- solar radiation

and two, which have changed decisively, are due to human activity:

- greenhouse gases—mainly CO_2, but also CH_4, NO_x, and CFCs—and
- anthropogenic sulfate aerosols.

A decisive climatic change due to a change in one or more of these factors is called a *forcing*. Some scientists believe that solar forcing may have caused warming in the early 1900s, although natural climate variability and an **anthropogenic** factor (a human factor) cannot be ruled out. But warming from about 1950 to the end of the century, irrespective of any natural forcing, volcanic influence, or increasing solar activity, is most likely attributable to human activity. It was such an analysis that led the United Nations' Intergovernmental Panel on Climate Change (IPCC) in 1995 to recognize that, according to the best estimates from climate models, the climate changes due to human activity add up to just about the amount of expected change. The atmospheric patterns of temperatures match well the increased concentrations of greenhouse gases, sulfate aerosols (mainly from burning fossil fuels), and the reduction in stratospheric ozone from CFCs. In light of this combination of factors, the panel concluded that "the balance of evidence suggests that there is a discernible human influence on global climate."

Weather's Puzzling Signals

In August 1995 a record-breaking dry spell in the Northeast brought drought warnings to New York City, Westchester County, and the Catskills. The following winter brought record-breaking snowfall to the entire region. Puzzling weather anomalies have become almost commonplace. Since the mid 1980s these anomalies have included wide swings in the weather from cold winters (except for a peculiarly warm one in 1995) in New England to warm winters in Alaska, a six-year drought in California, a bitter cold snap for the 1994 winter Olympics in Norway yet a unique mild Norwegian winter in 2003, record-breaking flooding on the Mississippi River valley in 1993, and summer heat waves the same year and in 2003 in Europe. Worldwide, the 1990s was the warmest decade on record in spite of cooling due to dust and aerosols emitted by the 1991 eruption of Mount Pinatubo.

Are such weather extremes likely to continue, and do they indicate global warming or are they just coincidence? Scientifically, it is hard to tell, because climate is fickle. This episode of warming could be part of the natural changes that scientists have traced over the millennia as the earth has cooled, then warmed, then cooled again (Case Study 11.3). Weather patterns change naturally in the short term, and climate changes naturally in the long term. Thus, any human-induced contributions to climate change are superimposed on natural changes.

Clouds: A Wild Card in Global Warming

Clouds perform an important role in global temperature control, but scientists are unsure of their exact role. For instance, when the earth is warmer, there is more water vapor in the atmosphere, and consequently there are more clouds. Recognizing this relationship leads to the question, will clouds be the same as we know them today if the earth becomes warmer? Will they be at the same elevation, and will the different types of clouds exist in the same proportion as they do today?

To illustrate some of the uncertainty about the effect of cloud types, consider the global-warming feedback differences between low-altitude clouds and high clouds. Wispy, high cirrus clouds composed of tiny ice crystals are very effective at trapping infrared radiation that is emitted from the earth's surface and thus act to warm the atmosphere; they give positive feedback. On the other hand, the heat-trapping property of low clouds is very poor (negative feedback). In addition, low clouds reflect the sun's radiation back to space, which cools the earth; they give negative feedback in two ways.

The size of the water droplets in clouds is also decisive. Whereas clouds formed of large numbers of small droplets reflect more of the incoming solar radiation away from the earth (giving negative feedback), clouds with larger, fewer droplets give positive feedback, amplifying global warming.

Because of such uncertainties, many scientists have turned away from the debate about whether human-induced global warming is occurring and are now attempting to predict exactly what we can expect to happen in the next century.

Living with Uncertainty

Atmospheric scientists are still uncertain about why climate changes, and they lack techniques for distinguishing between long-range climate trends and short-term weather events. "The temptation . . . is often to attribute any of these temporal and sometimes local variations to a wider and more pervasive change in climate," note several atmospheric scientists with the National Oceanic and Atmospheric Administration (NOAA). A respected senior climate scientist has shown that U.S. weather has become more extreme since 1980 and has estimated with 80 percent certainty that the extremes are due to human-induced greenhouse warming. In the case of greenhouse warming, built-in delays in the earth's climate system compound the potential dangers of waiting for 100 percent certainty; decades are required to warm or cool the atmosphere and oceans. If the world's policymakers wish to head off the potential for greater climate changes, they must act before the major effects of greenhouse-gas pollution are apparent. Regrettably, global warming has become political, and communication between scientists and policymakers is sometimes difficult (◆Figure 11.29). If it *is* occurring, it presents a dilemma. If it is due to human activity, we must take measures now to slow or halt it, and enormous up-front costs will be required for mediating the impact. But the determination of certainty must be based on the uncertainty of the past record, and any actions taken may prove to be expensive mistakes if humans are not the primary cause for the apparent climate change.

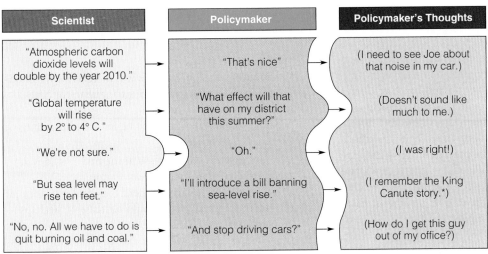

◆ FIGURE 11.29 A hypothetical scientist briefs a hypothetical policymaker on the dangers of the greenhouse effect. *After R. Byerly, Jr., "The Policy Dynamics of Global Change," EarthQuest, Spring 1989*

*Canute was a king of Denmark, England, and Norway, 1016–1035, who supposedly commanded the oceanic tide to stop rising.

A few climatologists have been skeptical about global warming, and some of them have doubted that it is occurring. A few feel that warming is indeed happening, but believe that the apocalyptic projections for the future are overblown. Nevertheless, most atmospheric scientists, basing their judgment on a variety of observed changes and studies, express growing confidence that human-induced global warming is occurring. Moreover, computer modeling of the world's climate at NOAA's laboratory throughout the 1990s successfully matched forecasts with actual conditions.

Implications for the Future

There can be no genuine security if the planet is ravaged by climate change.

Tony Blair, Prime Minister of the United Kingdom

Getting a group of scientists to agree on nearly anything, such as forecasting the long-term effects of global warming, is (according to a group of scientists) like trying to herd a bunch of cats. Scientists are especially argumentative when they are thrown a data curveball, as they were in 1999. The "curveball" was the finding that the rate of greenhouse-gas buildup in the atmosphere had decreased by about 25 percent since 1980. The amount of CO_2 from burning fossil fuels had increased only 0.4 percent in 1998 (despite the emissions from electric power plants increasing by 3.2 percent that year), down from a one percent per year increase on the previous few decades. This does not mean that greenhouse warming has stopped; only that it slowed, thereby giving humankind more time to cope with what many scientists think is inevitable global climate change. The decreasing rate, if it continues, means that the greenhouse-gas buildup will not double by 2050, as many climate forecasters had anticipated; the doubling time has merely been postponed until

early in the twenty-second century. Make no mistake, there has been no reduction of greenhouse *emissions,* only in the atmospheric buildup. The reduction in build-up is caused by some unknown process, with the "missing" gases being sequestered in some mysterious manner in oceans, forests, lakes, or wetlands (see Case Study 11.2).

In May 2001, President George W. Bush asked the National Academy of Sciences to form a committee to examine what is known and not known about global warming. The Committee on the Science of Climate Change so formed included 11 eminent climate scientists from various U.S. institutions, and their conclusions can be summarized simply: "(1) Earth has become warmer during the past several decades; (2) the warming is likely due to human activities, mainly atmospheric changes caused by the burning of fossil fuels (coal, oil, gas, and wood); (3) Earth will continue to warm, but we are not sure how fast temperatures will rise or how particular regions on Earth will be affected by climate change."* The scientists also concluded that it is unlikely that the present warming is caused by natural factors. The scientific data collected indicate that the present warming "is related to an enhancement of the greenhouse effect . . ." and they recognize that the average global temperatures could increase by 1.5° to 5° C (2.7° to 9° F) by 2100.

Most of the scientific community supports global efforts to cut back emissions of greenhouse gases, such as the **Kyoto Protocol,** which set standards for reductions in CO_2. Under conditions of the 1997 Kyoto Protocol, the United States would need to reduce emissions to seven percent below 1990 levels between 2008 and 2012. The Kyoto Protocol has been ratified by some industrialized nations since 1997, but in 2001 the U.S. government, citing economic concerns, withdrew from the agreement. The United States and some other countries favored giving nations credit for CO_2 reductions by

*The complete report is available at http://books.nap.edu/books/0309075742/html/

considering the amount of CO_2 that is naturally absorbed by forests, plants, and soil. However, those nations that lacked large areas of farmland and forests objected to this idea. Other issues addressed by scientists that remain unresolved are rates of consumption, energy efficiency, and population growth.

A wide range of projections exist on how much warmer it may get, how storm and drought patterns may respond, and what the effects will be on ecosystems, agriculture, and health. Nevertheless, many climate experts say that despite uncertainties, some effects can reasonably be predicted. Let us examine a scenario of anticipated effects of global warming developed by the Forum on Global Modeling, assuming a continuing anthropogenic (human-caused) increase in greenhouse gases based on a one percent per year increase in CO_2.

Very Probable:

- Globally, the mean surface temperature will increase by the mid twenty-first century by about 0.5° to 2° C (0.9° to 3.6° F). Such warming will be due to the concentrations of greenhouse gases alone, assuming that no significant actions are taken to reduce these gases. If the CO_2 concentrations in the atmosphere should double, there is a range of estimates for the warming that will result. The best estimate is 1.5° to 4.5° C (2.7° to 8° F), and the most probable estimate is 2.5° C (4.5° F).

- Globally, mean precipitation will increase, but how it will be distributed is uncertain. A warming of the earth's surface temperatures will increase global mean precipitation because of the well-established relationship between evaporation rates and surface temperature.

- Northern Hemisphere sea ice will be reduced. The projected effects and timing of changes in Southern Hemisphere sea ice are uncertain.

- Arctic land areas will experience wintertime warming. Paleoclimate studies and computer modeling offer evidence for polar warming and reduction of snow cover on land areas. The magnitude of the warming is uncertain because of the complexity in understanding all of the variables.

- Globally, sea level will rise at an increasing rate, although the rate of rise may not be significantly greater than the last few decades of the twentieth century. The most reasonable estimate for the rate of sea level rise is 5–40 centimeters (2–16 in) by 2050, as compared to a rise of 5–12 centimeters (2–4.7 in) if the rates of the twentieth century continue. Incorporated in the modeling is a component due to seawater expansion, which is closely dependent on the amount of atmospheric warming, and an estimate of water derived from the retreat of mountain glaciers. The volume of water contributed by the melting of the polar ice sheets is less certain. If global temperatures increase by 3° C by the year 2050, as some climatologists believe, it is estimated that meltwater from mountain glaciers alone will raise sea level by another 20 centimeters (8 in).

Probable:

- Precipitation at high latitudes will increase, with potential feedback effects related to the influence of additional fresh water to the oceans causing changes in oceanic circulation. There will be increased precipitation on the polar ice caps.

- Summer dryness will increase in the mid latitudes of the Northern Hemisphere. Evaporation increases strongly with increasing temperatures. Soil moisture in the Northern Hemisphere could decrease by perhaps 40 percent if greenhouse gases continue to increase at predicted rates. This would make "dry" farming impossible or nearly so in much of North America and Europe.

- Episodic, explosive volcanic eruptions will cause short-term relative cooling of a few tenths of a degree lasting up to a few years.

Uncertain:

- Details of the climate changes are uncertain.

- Biosphere–climate feedbacks are expected, but whether these feedbacks will modify or amplify climate change is uncertain.

- Changes in climate variability will occur.

- Regional-scale climate changes will differ from the global averages. Unfortunately, there is limited capability to estimate how various regions will respond to climate change.

- Tropical-storm intensity may change. An increase in intensity is likely but uncertain because of the potential changes in poleward heat flow and other conditions.

Overall, warming will stress human activities because it will demand rapid adjustment of economic patterns and changes in much of society's infrastructure. There is also a growing indication that the potential effect on human health is no less serious.

In their worst-case scenarios, some global climate models point to the possibility of tens of millions more cases of infectious diseases, as mosquitoes and other pests expand their ranges, and of hundreds of thousands of additional deaths each year due to an increasing incidence of heat waves. Epidemiologists say the most deadly punch of global warming will be delivered to countries of the developing world. Their models predict an increase in such diseases as sleeping sickness, malaria, schistosomiasis, and yellow fever. These diseases already afflict more than 600 million people each year, killing more than two million. Some researchers estimate that global warming would increase the populations of insects and snails that transmit these diseases because cold-blooded insects and invertebrates respond to subtle changes in temperature.

Changes in rainfall distribution and amounts in some areas would affect river levels and discharge, perhaps causing more flooding. Many high-latitude regions underlain by permafrost (see Chapter 6) would eventually thaw, allowing more bacterial action, and decomposition of organic matter would release methane. The new source of methane, an even more efficient greenhouse gas at absorbing long-wave radiation than CO_2, would add positive feedback to the global warming regime.

What Individuals Can Do

What can we do about global warming? We can do nothing about any of its natural causes, but there are things we can do individually that may help to slow the pace of human-generated global warming that would hardly affect our life styles. More than 80 percent of the world's energy and most of the greenhouse-gas emissions comes from burning fossil fuels—oil, coal, and natural gas—and currently we are only getting close to cost-effective alternatives (see Chapter 14). But much can still be done. In the United States there are three main things individuals can do.

1. Buy products that reduce energy consumption.
2. Use energy more efficiently.
3. Actively recycle.

✱ Table 11.2 provides some more specific suggestions. The consequences of our choices and life styles are catching up with us, and even small steps will help.

What Industry and Governments Can Do

I find myself increasingly persuaded that a climate effect may be occurring.

Mark Moody-Stuart, CEO, Royal Dutch/Shell Oil Company

There is mounting evidence that the concentration of carbon dioxide in the atmosphere is rising and the temperature of the earth's surface is increasing.

John Browne, CEO, British Petroleum

In debates about global warming, environmentalists and industrialists have commonly found themselves hollering at each other through clouds of scientific and economic jargon, respectively. This situation began to change in 1998, when several major corporations acknowledged that excess greenhouse gases could well be the cause of global climate change. Among the companies that have made the cross-over, warning that "climate change is serious business for all of us," are Boeing, British Petroleum (BP), 3M, United Technologies, Lockheed Martin, and Maytag. Royal Dutch/Shell, for example, announced that it was investing more than half a billion dollars on renewable-energy technology over five years. BP

✱TABLE 11.2	Reducing Global Warming Begins at Home	
Action to Take	**Estimated Reduction of CO_2, pounds per year**	
1. Insulate your home, install water-saver shower heads, clean and tune furnace.	2,480	
2. Buy a fuel-efficient automobile with a rating of at least 32 mpg to replace your most-used car.	5,600	
3. Do not drive your car two days a week.	1,590	
4. Recycle all home waste metal, glass, newsprint, packaging, and cardboard.	850	
5. Install solar heating to help provide hot water.	720	
6. Replace washing machine with low-water-use, low-energy model.	420	
7. Buy food and other products with recyclable or reusable packaging.	230	
8. Replace refrigerator with a high-efficiency model.	220	
9. Use a push lawn mower instead of a power mower.	80	
10. Plant two trees.	20	

Source: Oregon Energy Office

made a new commitment to investment in solar-energy research, and both of these companies backed out of the Global Climate Coalition, an industry lobbying group that opposes the 1997 Kyoto treaty.

Ford Motor Company, whose critics point out that it sells more sport utility vehicles (SUVs) and trucks—both major greenhouse-gas emitters—than its competitors, reported that its new minivans and SUVs would qualify as low-emission vehicles. General Motors is experimenting with hybrid electric power trains that will reduce fuel usage and are currently putting hybrids on the road in mass transit bus applications, are working on hybrids for personal vehicles, and are investing in a long-term project of developing fuel-cell vehicles. Currently three hybrid personal vehicles are available in the United States that are rated either as Super Ultra Low Emission Vehicles (the cleanest emissions rating after zero emission vehicles) or Ultra Low Emission Vehicles. The Toyota Prius hybrid is rated at 48 miles per gallon and emits about half the CO_2 as the average passenger vehicle on the road today.

Some researchers believe that most of the proposed global emission-reduction plans will do relatively little to slow greenhouse-gas emissions, because they consist mainly of modest, voluntary policies. Furthermore, the industrialized countries' CO_2 emissions may not be the major problem. The United Nations' economic and science advisors on global climate change predict that by 2025 the developing countries and those that are moving away from centrally planned economies (including the former Soviet-bloc countries) will be responsible for nearly 70 percent of all energy-related CO_2 emissions.

A U.S. trade group representing manufacturers, utilities, mining companies, coal and oil producers, and railroads insisted that reducing industrial emissions by 20 percent or more below the 1990 levels, as proposed at the 1995 Intergovernmental Panel on Climate Change (IPCC) conference, would be disastrous for the U.S. economy. They estimated that cuts of this magnitude could cost an average of 600,000 jobs a year for several years.

Offering hope for mitigating CO_2 buildup is a chemical process that combines the gas with materials such as calcium oxide and magnesium oxide to form stable minerals. The process is viewed as a short-term insurance policy in case political pressure resulting from the threat of global warming forces industry to limit CO_2 emissions. The monetary costs of capturing CO_2 on an industrial scale would be very high. For example, it is estimated that removing the gas from coal-fired electric-power generating plants would double or triple the cost of electricity. Oceanographers have been cautiously exploring the feasibility of dumping finely ground iron particles into the ocean to stimulate the growth of phytoplankton, which, like all plants, metabolize CO_2 by photosynthesis. From initial experiments it is estimated that spreading a half-ton of iron across 100 square kilometers (39 mi^2) of a tropical ocean would stimulate enough plant growth to absorb some 350,000 kilograms (771,800 pounds) of CO_2 from seawater. Performed on a much larger scale, the iron fertilization of sea water could absorb billions of tons of CO_2, offsetting perhaps a third of global CO_2 emissions. The environmental side effects of such large-scale "iron fertilization" are difficult to foresee, but the enhanced bloom of plants could enrich the entire oceanic ecosystem.

In the short term—that is, on the scale of human lifetimes—it appears that there will be global warming. From a long-term, geological perspective, however, the warming is seen as a brief event and a nonrepeatable one—nonrepeatable because the earth's reserves of fossil fuels will be consumed within the next few hundred years. After reserves are consumed, atmospheric conditions should eventually return to normal. True, human-induced greenhouse warming may last a thousand years, but ultimately the variations in the earth's orbital geometry will change, and the earth will be thrust into yet another ice age.

CASE STUDY 11.1

When a Glacier Surges It's a Moving Experience

An uncommon but spectacular type of glacier movement is **surging.** A surge begins when ice in the upper part of the glacier reaches some critical thickness. Subsequently the ice surface breaks up into thousands of crevasses, which rapidly spread downglacier, and then the ice movement accelerates from perhaps only a few centimeters a day to as much as 100 meters (328 ft) a day. If the surge reaches the glacier front, the glacier advances dramatically. Surging may cause a glacier to advance several kilometers within a few weeks and be accompanied by a spectacular release of meltwater that causes great damage downvalley. Alaska's Bering Glacier—the earth's largest surging glacier—began surging in May 1993. After a minor retreat for seven months, it resumed surging in April 1995 and was still surging in March 1996. The surge advanced with maximum rates of 7–10 meters a day during the winter of 1993–94, pushing the glacier terminus forward by about 12 kilometers (7 mi) and producing a large number of icebergs, several of them more than a half-kilometer (1,640 ft) long. The U.S. Geological Survey and several university research teams are monitoring the surge using aerial and field observations, paying special attention to iceberg production. As with the icebergs produced by the Columbia Glacier in the 1980s, any icebergs entering the Pacific Ocean could pose a threat in the oil-tanker lanes leading to and from nearby Prince William Sound.

Surges are generally not considered climatically induced events, and it is important to understand them in order to predict surges that could endanger installations or people. Such is the case of another Alaskan glacier, the Black Rapids Glacier, which is being monitored because a surge by it could threaten a major highway and the Alaska Pipeline. That the concern is justified was demonstrated by its surge in 1936–37. The owners of a lodge on the Richardson Highway correctly perceived that the glacier's 3-kilometer (1.8-mi)-wide front was moving toward them and notified authorities. The glacier was found to be advancing at rates of up to 65 meters (213 ft) a day. Fortunately, the surge halted a short distance from the highway; had it continued, it would have destroyed the lodge, dammed a major river, and cut through the highway.

Surges of the Plomo Glacier in the Argentinean Andes have led to the formation of ice-dammed lakes. When one of these lakes bursts through the ice dam, the resulting flood of water can create destruction downvalley. Several deaths were caused by the Plomo surge of 1934, and 13 kilometers (8 mi) of the Mendoza–Santiago railway and seven bridges were washed out. Similar events in the Karakoram Mountains on the border of Kashmir and China have caused extensive damage and loss of life.

Why glaciers surge is still uncertain, but research on the 1982–83 surge of Alaska's Variegated Glacier provided information on surging behavior. Located at the northern end of the Alaskan panhandle, the Variegated Glacier (◆ Figure 1) flows off a shoulder of the St. Elias Range. It is 20 kilometers (12 mi) long and is normally a well-behaved tongue of ice, flowing at a rate of about 0.1–0.2 meters

JÜRG ALEAN

◆ **FIGURE 1** During the surge of the Variegated Glacier, the glacier's accelerated motion was measured using surveying equipment. The umbrella was used to protect the heat-sensitive electronic equipment from the heat of the sun.

(0.33–0.67 ft) a day. For some time before the onset of the surge, studies revealed considerable thickening in the upper reaches of the glacier and thinning in the lower part. The consequent steepening of the glacier increased the stress at the base of the ice in the upper area, which gradually closed the subglacial channels that carried away meltwater. Squeezed out of the channels, the water was forced to spread laterally as a layer across the entire glacier bed. This buoyed up the glacier slightly, which reduced the friction and allowed it to slide faster. Rapid movement began in the upper area, and as the faster-moving upper ice pressed on the slower-moving downglacier ice, stresses began transferring to the lower, thinner part of the glacier, and the surge spread downglacier as a wave. The surge's velocity tripled in its upper reaches and doubled in the middle. After the stresses were relieved, the subglacial channels began to reopen, the subglacial reservoir of water emptied in a concluding flood, and the glacier dropped down on its bed and stopped sliding.

CONTINUED

The Bering Glacier, the world's largest surging glacier, near Yakutat, Alaska, experienced a 17-month surge in 1993–1994, and resumed surging again in 1995. Part of Bering's terminus advanced about 9 kilometers (5.5 mi) in the earlier surge, and advanced about 750 meters (2,325 ft) between May 19 and June 1 in the 1995 event.

Near the Bering Glacier is the Hubbard Glacier, the largest calving glacier in North America, 25 percent larger than Rhode Island. It is often cited as an example of a surging glacier because of its spectacular advance across the entrance of Russell Fiord in 1986 and in June 2002, in both instances temporarily turning the fiord into a lake. But the evidence for the Hubbard having surged is considered tenuous. Because the Hubbard's accumulation area (see text Figure 11.2) is 95 percent of the entire glacier area and, like the seven other growing calving glaciers in Alaska, it is an advancing glacier, it is far from being in equilibrium with climate.

Sources: Michael Hambrey and Jürg Alean, *Glaciers* (New York: Cambridge University Press, 1992); Barclay Kamb and others, "Glacier Surge Mechanism," *Science* 227 (February 1, 1985), pp. 469–479; P. J. Fleisher and others, 1995, The surging advance of Bering Glacier, *Journal of Glaciology* 41: n. 137, pp. 207–213; and Bruce F. Molina, 1995, Bering Glacier resumes its surge, *Earth in Space* 8, no. 3, 6: American Geophysical Union.

CASE STUDY 11.2

The Carbon Cycle

Complex interactions of the earth's geospheres—the pedosphere, the atmosphere, the hydrosphere, the biosphere, and the lithosphere—produce climate. The interactions are difficult to analyze and are not well understood. By extending understandings about today's climate into the remote geological past, however, geologists have been able to shed some light on the workings of the climate system. They have reconstructed ancient climates by studying sedimentary layers and deducing their depositional environments, by studying fossil animals and plants, and by carrying out isotope studies that yield information on past temperatures of the earth's oceans and surface. Such *paleostudies* have laid important groundwork for understanding the earth's carbon cycle.

Presently, the earth's reserves of fossil fuels are being burned at a prodigious rate, and the large amount of carbon dioxide (CO_2) being released to the atmosphere by the burning appears to be impacting our planet and its inhabitants. Carbon, the essential element that distinguishes living and once-living matter from other matter, is involved in the changing atmospheric composition that may be causing long-term climate change. Carbon exists in various forms in each of the geospheres:

- The biosphere is the system of all living and some decaying dead organisms, each cell of which contains carbon.

- The hydrosphere contains dissolved carbon dioxide.

- The atmosphere and pedosphere contain CO_2 gas.

- Carbon is stored in the lithosphere in deeply buried organic materials, in petroleum and coal, and as calcium carbonate in limestone and marble.

The transfer of carbon among these geospheres is termed the carbon cycle (◆ Figure 1). The carbon cycle is not just the cycling of CO_2; it incorporates all of the earth's carbon compounds, and there are many of them.

The biosphere is the key in the carbon cycle. Plants continuously remove CO_2 from the atmosphere and utilize it, by photosynthesis, to produce their food (sugar), much of which, in turn, is consumed by animals. Plant and animal waste materials contain carbon, which ultimately is carried to the oceans by surface runoff. Dead organisms decay by combining with atmospheric oxygen to form more CO_2, which is then cycled back to the atmosphere. So fast is the cycling of carbon through the biosphere that a complete recycling of atmospheric CO_2 occurs about every four years.

In the ocean, plankton, shellfish, and other marine life use CO_2 and calcium to form shells and skeletons. These materials eventually sink to the sea floor, where their carbon becomes sequestered in calcareous sediment. By the motion of the crustal plates, ocean-floor sediments eventually are cycled into the mantle by subduction. Heat drives the CO_2 out of the sediment, and CO_2 dissolved in magma eventually rises to the surface through volcanism. When a volcano erupts, large amounts of CO_2 are returned to the atmosphere to restart the cycle.

Not all dead organic matter in the biosphere decays to CO_2. Some is sequestered in sedimentary rock, forming reservoirs of carbon in petroleum and coal, the so-called fossil fuels. Some of this material naturally reenters the atmosphere when uplift and erosion expose the rock and its trapped carbon compounds (◆ Figure 2).

Atmospheric CO_2 is the smallest of all the carbon reservoirs, accounting for only about 0.03 percent of the atmosphere. Because of the atmosphere's relatively small carbon content, it is very sensitive to changes in the larger carbon reservoirs with which it exchanges. Of special concern is the recent acceleration of human-induced CO_2 production by the clearing of forests and the burning of fossil fuels. Both of these release CO_2 to the atmosphere from the biosphere and the lithosphere much faster than would occur naturally. The rate at which natural processes remove CO_2 from the atmosphere has not changed, however, and that is why there is a buildup of CO_2 in the atmosphere. Accordingly, this anthropogenic atmospheric-CO_2 imbalance to the greenhouse effect is the focus of the concern about global warming.

Despite the availability of high-speed computers and our ability to do sophisticated computer modeling, our knowledge about the carbon cycle is remarkably scant. Scientists have yet to account for about 25 percent of the sequestered carbon. Apparently, it is not in the ocean, but somewhere in the terrestrial biosphere—that is, in land-living plants and animals. Solving the puzzle of the "missing" carbon might be a big step toward solving the bigger global-climate-change problem. Such an effort will require the collaboration of paleoclimatologists, ecologists, ecosystem modelers, geochemists, and others. Such coordinated problem-solving is the key to understanding the carbon cycle and global change.

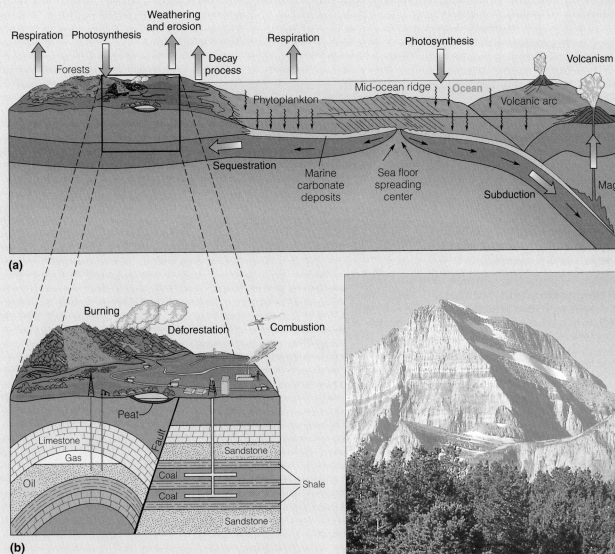

(a)

(b)

◆ FIGURE 1 The carbon cycle. (a) Carbon enters the atmosphere naturally by biological respiration, volcanism, and chemical weathering. Photosynthesis transfers it from the atmosphere to the biosphere. (b) Humans release carbon to the atmosphere when they burn fossil fuels and forests.

◆ FIGURE 2 These mountains in the Montana Rockies are composed largely of ancient carbonates that were formed as sea-floor sediments about 1.5 billion years ago. Such rock bodies constitute reservoirs of carbon that have been isolated from the carbon cycle. Combined uplift, erosion, and weathering processes are now slowly releasing the long-sequestered carbon back into the carbon cycle.

Is There a New Beat to the Rhythm of the Ice Ages?

If you think today's weather seems weird, try to imagine weather conditions during the last ice age. Oceanographers' studies of sea-floor sediments have revealed that massive fleets of icebergs have moved across the North Atlantic Ocean every 2,000 to 3,000 years as the climate has flip-flopped back and forth between warm interglacial and glacial periods. The melting icebergs have chilled the North Atlantic and left trails of ground-up rock debris in layers on the ocean floor. After two or three such cycles there would be a larger flood of icebergs, which left a layer with a different composition. These greater floods, occurring every 7,000 to 12,000 years, have become known as *Heinrich events* (named for the German oceanographer who discovered them), and the smaller, more frequent events are informally called *flickers*.

It is now believed that the Heinrich events, recognized by layers of carbonate-rock grains in the sea-floor rock debris, resulted when the giant Keewatin ice sheet (text Figure 11.1a) surged into Canada's Hudson Strait, causing enormous num-

bers of bergs to calve and drift into the North Atlantic. The more numerous flickers, in contrast, left layers of reddish, iron-coated sand on the ocean floor that has been traced to the present-day St. Lawrence Valley, which also was ice-covered at the time. The significance of the Heinrich and flicker events in the oceanic record is their implication that in the late stages of the Pleistocene ice ages, the Northern Hemisphere experienced numerous rapid climate changes unlike anything seen in historic times (◆ Figure 1).

Meanwhile in Greenland, glaciologists' studies of oxygen isotope ratios in ice cores have revealed wild flip-flops in climate from bitter cold to warm, between 115,000 and 140,000 years ago (an interval known as the *Eemian*), with the same timing as the Heinrich and flicker events recorded in the sea-floor rocks. What does all this mean? One hypothesis is that the St. Lawrence ice sheet accumulated and surged several times as the Keewatin ice sheet was growing, and that when the Keewatin ice sheet surged, the ice over the St. Lawrence did

also. The two ice sheets, each acting independently and having different characteristics, must have surged as a consequence of climate warming. Variable currents in the North Atlantic—perhaps related to abrupt shutdowns in the oceanic conveyor system (see text Figure 11.25), brief ones of a century or less—may cause temperature swings every 2,000 to 3,000 years in a cycle that may run all the time, even today.

The debate continues on the explanations for the cycles of sediment deposition, the oceanic conveyor system, rapid swings in climate, and the discrepancies in the Greenland and Antarctic ice-core data. Conflicting data from different sources may drive the scientific community to develop more complicated, yet more realistic, views of the earth's climate and why it changes. With the earth expected to be bulging with people in 50 to 100 years, we should spare no effort in deciphering the seemingly chaotic behavior of the global climate system.

◆ FIGURE 1 The continuous GRIP temperature record (based on oxygen-18/oxygen-16 ratios) from the Greenland ice core. This record shows that during the Eemian interglacial interval (roughly 115,000–142,000 years ago), generally warm conditions suddenly gave way to drastic cold on several occasions. Compare this record with that of the Vostok core from Antarctica (Figure 11.27). Data from GRIP, "Climate Instability During the Last Interglacial Period Recorded in the GRIP Ice Core," *Nature* 364, no. 6434 (July 15, 1993), pp. 203–207

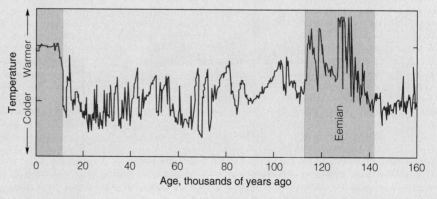

CASE STUDY 11.4

Early Effects of Global Warming on Plants and Animals

With physical evidence of global warming continuing to accumulate, are there any indications of global warming in biologic systems? Yes. Late-twentieth century studies in Europe and North America show that some plants and animals are indeed responding to global warming.

In Britain, a 20-year study of populations of 101 bird species found that, coincidentally with the recent warming trend, most species had extended their northern range by an average of 19 kilometers (30 mi). Another study of the nesting habits of 20 species of British birds found that the species were laying eggs earlier in the spring. Similar results were found in Michigan, where some species were laying eggs as many as 21 days earlier than in the past.

Comparing data collected between 1992 and 1996 at more than 150 sites with historical records of the Edith's checkerspot butterfly *(Euphydryas editha)*, a wide-ranging species of western North America, showed that population extinctions were four times higher at the southern edge of its range (in Mexico) than at the northern edge (in Canada). Furthermore, the rate of extinctions at lower elevations was more than double the rate at elevations above 2,440 meters (8,000 ft). Thus, the butterfly's range was found to be shifting north-

ward and upward. A European study of the geographic ranges of 35 species of nonmigratory butterflies had similar findings. Two thirds of the species' habitats had shifted northward by 35 to 240 kilometers (21–149 mi) coincidentally with the warming of Europe and only 3 percent had shifted south.

In the mountain forests of Monteverde, Costa Rica, 20 of 50 species of frogs and toads disappeared following population "crashes" in 1987. These crashes probably resulted from ecosystem changes that altered the communities of reptiles, amphibians, and birds due to recent warming. The changes are associated with patterns of dry-season mist frequency, which are related to changes in sea-surface temperatures in the equatorial Pacific since the mid 1970s. The implication is that atmospheric warming has caused the base of the mountain-top cloud bank to raise in elevation and has negatively impacted the mountain forest ecosystem.

Temperature-limited environments in mountains, subarctic, and arctic regions also reveal evidence of warming. In the Swiss and Austrian Alps in the early 1990s, a study compared the plant life on mountain summits exceeding 3,000 meters (9,849 ft) to records collected 70 to 90 years earlier, when the annual temperature averaged 0.7° C (1.26° F) cooler. Altitudinal vegetation

belts were found to have shifted to higher elevations by as much as 4 meters (13 ft) per decade, although many belts' shifts were less than 1 meter (3 ft) per decade. A study in Alaska found that **boreal** forests (forests in mountainous and northern regions of North America in which the mean temperature of the hottest season does not exceed 18° C (64.4° F) expanded northward at a rate of 97 kilometers (60 mi) per 1° C (2° F) of temperature increase, and that the length of the growing season increased 20 percent in the last few decades.

Less ice and a longer ice-free season in the Arctic spells trouble for some of its inhabitants. Masses of diatoms, the silica-encased "grass" of the sea, hang from the undersides of floating ice and drop to the sea floor where they are the base of a food chain that extends up to bottom-feeding whales and walruses. Fewer diatoms at the base of the food web means less food for animals higher up the food chain, and research in 1998 shows walruses already to be leaner. On the other hand, disappearance of summer ice could boost the growth of phytoplankton, the floating algae that work up the food web to feed Arctic cod, which in turn become cuisine desired by both seals and humans. For polar bears, however, less Arctic ice spells trouble. Polar bears spend spring to midsummer on ice floes hunting for seals to eat (Figure 1) in

DAN GURAVICH, NATIONAL AUDUBON SOCIETY COLLECTION/PHOTO RESEARCHERS

◆ FIGURE 1 Walking on thin ice. Between 1958–1976 and the 1990s the Arctic Sea ice lost about 40 percent of its volume, thinning by 1.3 meters (4.3 ft) on average. The consequences have already created hard times for polar bears.

CONTINUED

order to build energy reserves to get them through the leaner season after the ice has melted and they must move onto land. Once on land, they fast, often for months. The warmer temperatures of the 1990s made for early spring melting and later fall freezing, and seal dinners were harder to come by. The polar bears suffered, their weight dropping by 10 kilograms (22 lbs) for each week that the ice broke up earlier than normal. Furthermore, cub survival is contingent on a well-fed nursing mother and an adequate food supply after weaning. The fate of the world's largest carnivore is uncertain should global warming continue.

Sources: Georg Grabherr and others, 1994, Climate effects on mountain plants, *Nature* 369: 448; Richard A. Kerr, 2002, A warmer Arctic means change for all, *Science* 297: 1490–91; J. Alan Pounds and others, 1999, Biological response to climate change on a tropical mountain, *Nature* 398: 611–15; Bruce F. Molina, 1998, Climate change workshop results begin to appear, *GSA Today* 8, September: 10–11; Camille Parmesan, 1996, Climate species' range, *Nature* 382: 765–6; Camille Parmesan, and others, 1999, Poleward shifts in geographical ranges of butterfly species associated with regional warming, *Nature* 399: 579–83; and Chris D. Thomas and Jack J. Lennon, 1999, Birds extend their ranges northwards, *Nature* 399: 213.

Wonders of Glaciation

Some of the world's most-visited scenic areas are popular because of the spectacular landscapes that remain as superlative legacies of the work of Pleistocene glaciers. Figure 1 to 5 are a collage of photographs exemplifying the beauties of glaciers and glaciated landscapes.

◆ FIGURE 3 A glacial "erratic" that was transported by the Pleistocene valley glacier that occupied Lee Vining Canyon, California. This erratic boulder's location is more than 25 kilometers (15 mi) from the nearest outcropping of the particular bedrock type that could have yielded it.

◆ FIGURE 1 The Vaughan Lewis Glacier, a valley glacier in the Coast Mountains near Juneau, Alaska.

◆ FIGURE 4 A cirque, an armchair-shaped depression at the head of a glaciated U-shaped valley, viewed from the summit of 4,313-meter (14,150-ft)-high Mount Sneffels, Colorado.

◆ FIGURE 2 Glaciated landscape including a U-shaped glacial valley, two lakes behind end-moraine dams, three lakes in glacially eroded basins, a small alpine glacier, and a glacial-carved knife-edge ridge. The Arrigetch Peaks area, Gates of the Arctic National Park, Alaska.

◆ FIGURE 5 A glacially grooved and polished rock outcrop in New York City's Central Park.

Summary

Glaciers

Description

Large masses of ice that form on land where, for a number of years, more snow falls in winter than melts in summer, and which deform and flow due to their own weight because of the force of gravity.

Location

High latitudes and high altitudes.

Classification

Major types are continental (ice-sheet), ice-field, ice-cap, valley (alpine), and piedmont.

Some Important Glacial Features

End moraine—ridge formed at the melting end of a glacier composed of glacially transported rocky debris.
Kettle lakes—water-filled, bowl-shaped depressions formed as glaciers retreated.
U-shaped canyon—remains after a valley glacier melts.
Fiord—U-shaped canyon drowned by the sea.

Effects of Pleistocene Glaciation

1. Some soil conditions, including loess deposits transported by Ice Age winds and some areas of glacial till.
2. Local ground-water conditions.
3. Shore-line configuration due to sea-level changes and isostatic rebound.
4. Human transportation routes.

Climatic Controls

Wind

Movement of air due to unequal heating of the earth's surface. Caused by:

1. Variations in the heating of the earth at different latitudes because they receive varying amounts of solar heat energy due to the differing angles of the sun's rays.
2. Due to the earth's rotation, the Coriolis effect deflects winds to the right in the Northern Hemisphere and to the left in the Southern Hemisphere.

Air Pressure, Wind, and Climate Belts

Polar highs—polar regions of high pressure; subsiding air produces variable winds and calms.
Westerlies—zones of generally consistent winds lying between 35° and 60° north and south latitudes.
Subtropical highs—zones of high pressure and subsiding air masses lying between 30° and 35° north and south latitudes; have variable winds and calms and, commonly, clear and sunny skies. Often called the *horse latitudes*, these zones are the regions in which most of the earth's deserts are located.

Trade wind—belts of generally consistent easterly winds lying between 5° and 30° north and south latitudes.
Equatorial low and intertropical convergence zone—zone of variable winds and calms lying between latitudes 5° north and 5° south. Sometimes called *the doldrums.*

Heating of the Earth

Solar Radiation

Warming the earth is by ultraviolet radiation from sun.

Reradiated Energy

Infrared energy from earth that warms the atmosphere.

Greenhouse Effect

The warming of the atmosphere by an increase in the amount of certain gases in the atmosphere, primarily CO_2 and CH_4, which increase the retention of heat that has been reradiated by the earth.

Greenhouse Gases

Trace amounts of H_2O, CO_2, CH_4, and CFCs that trap heat in the atmosphere.

Oceanic Circulation

Description

Each ocean has a characteristic current pattern, the general pattern being clockwise in the Northern Hemisphere and counterclockwise in the Southern Hemisphere. The net effect is the movement of cool water from polar regions toward the equator.

Importance

Currents moderate coastal climate and weather, and coastal upwelling brings cool water and nutrients that are important for biologic productivity.

Ice-Age Climate

Description

Temperatures 4° to 10° C cooler than at present.

Causes

A number of interrelated factors contribute to climate changes in ways that are not clearly understood. They include:

1. Obliquity of earth's axis
2. Eccentricity of earth's orbit
3. Precession of earth's axis
4. Variations in content of atmospheric gases
5. Changes in "greenhouse gases"
6. Changes in landmass positions due to plate tectonics
7. Tectonic changes in the elevation of continents

8. Volume and temperature of the oceans
9. Changes in ocean currents
10. Changes in solar radiation
11. Dust and aerosols
12. Oceanic conveyor belt

Some Model-Based Predictions of Climatic Change

1. Globally, average surface temperatures will increase.
2. Globally, average precipitation will increase, but its distribution is uncertain.

3. Northern Hemisphere ice will decrease; Southern Hemisphere ice may increase.
4. Arctic land areas will experience wintertime warming.
5. Sea level will rise at an increasing rate with drowning of low-level coastal plains.
6. Decreasing soil moisture in the Northern Hemisphere may make farming nearly impossible in much of North America and Europe.
7. Tropical-storm (hurricane) intensity and frequency may increase.
8. Plants and animal ranges will expand in some cases, shrink or disappear altogether in others.

Key Terms

ablate	esker	isostatic rebound	recessional moraine
aerosols	firn	kettle lake	subpolar low
anthropogenic	fiord	Kyoto Protocol	subtropical highs (horse
boreal	glacial outwash	moraine	latitudes)
calving	glacier	obliquity	surging, glacial
chlorofluorocarbons (CFCs)	global warming	ozone hole	till, glacial
climate controls	greenhouse effect	polar front	upwelling
Coriolis effect	greenhouse gases	polar high	wind
eccentricity	intertropical convergence zone	precession	zone of flowage
end moraine	(ITCZ), "doldrums"	proglacial lake	

Study Questions

1. What makes the wind blow? Why do winds in the Northern Hemisphere deflect to the right, and winds in the Southern Hemisphere to the left?
2. Explain the generalization "Glaciers are found at high latitudes and at high altitudes."
3. Glacial ice begins to flow and deform plastically when it reaches a thickness of about 30 meters (100 ft). What is the driving force that causes the ice to flow?
4. Describe the behavior of a glacier in terms of its *budget;* that is, the relationship between accumulation and ablation. How does a glacier behave when it has a balanced budget; a negative budget; a positive budget? What kind of a budget does a surging glacier have?
5. Areas far beyond the limits of the ice sheets were affected by Pleistocene continental glaciation. What are some of the effects? (*Hint:* Consider coastlines, today's arid regions, and modern agricultural regions.)

6. Describe isostatic rebound. What effect has the melting of the great ice sheets had on the crustal regions that were covered by thick masses of ice until 8,000 to 11,000 years ago? What are the effects of this rebound on society?
7. Describe the greenhouse effect. What causes greenhouse warming?
8. What can be done to modify (or even control) greenhouse warming?
9. What are some potential effects of greenhouse warming on human life?
10. What causes ice ages? Will there be more ice ages? Explain.
11. Using the "rule of 70" for determining the approximate number of years for doubling (70 *divided by* the rate of increase *equals* years for doubling), how long would it take for the amount of atmospheric CO_2 to double if its rate of increase is one percent per year; if the rate is 0.75 percent per year?

For Further Information

Books and Periodicals

Ahrens, C. Donald. 2000. Possible causes of climatic change (pp. 510–525). *Meteorology today.* Pacific Grove, Calif.: Brooks/Cole.

Alley, Richard. 2001. *The Two Mile time machine: Ice cores, abrupt climate change, and our future.* Princeton University Press.

Andersen, Bjørn G., and Harold W. Borns. 1994. *The Ice Age world.* Oslo: Scandinavian University Press.

Broecker, Wallace S. 1995. Chaotic climate. *Scientific American* 273, no. 5:62–68.

Cicerone, Ralph. 2003. What we know about global warming. *The Americana annual, 2002. Yearbook of the Encyclopedia Americana:* 230–239. Grolier.

Greenland Ice Core Project (GRIP). 1993. Climate instability during the last interglacial period recorded in the GRIP ice core. *Nature*, 364, no. 6434 (July 15):203–207.

Gutzler, David S. 2000. Evaluating global warming: a post-1990s perspective. *GSA Today* 10, no. 10:2–7.

Hambrey, Michael, and Jürg Alean. 1992. *Glaciers.* New York: Cambridge University Press.

Kerr, Richard A. 2002. A warmer arctic means a change for all. *Science* 297:1490–1492.

Kirk, Ruth. 1983. Of time and ice. *Glacier Bay: Official National Park handbook.* Washington, D.C.: Division of Publications, National Park Service, U.S. Department of the Interior.

MacCracken, Michael, and Tom Karl. 1997. Is the climate changing? Indeed it is. *Bob Ryan's 1997 almanac & guide for the weatherwise.* Washington, D.C.: WRC-TV.

MacKenzie, Fred T. 1998. *Our changing planet: An introduction to earth system science and global environmental change.* Upper Saddle River, N.J.: Prentice Hall.

Monroe, James S., and Reed Wicander. 1997. *Physical geology: Exploring the earth,* 3d ed. Belmont, Calif.: Wadsworth.

Petite, J. R., and others. 1999. Climate and atmospheric history of the past 420,000 years from the Vostok ice core, Antarctica. *Nature* 399:429–434.

Singer, S. Fred. 1999. *Hot talk cold science; global warming's unfinished debate,* revised 2d ed. Oakland, Calif.: The Independent Institute.

Tett, Simon F. B., and others. 1999. Causes of twentieth-century temperature change near the earth's surface, *Nature* 399:569–572.

Turco, Richard P. 1997. The climate machine (pp. 320–364) and greenhouse warming (pp. 365–406). *Earth under siege: From air pollution to global change.* New York: Oxford University Press.

ENVIRONMENTAL
Geology⇌Now™

Assess your understanding of this chapter's topics with additional quizzing and comprehensive interactivities at http://earthscience.brookscole.com/pipkingeo4e as well as current and up-to-date Web links, additional readings, and InfoTrac College Edition exercises.

Arid Lands and Desertification

*Deflation Basins, Winds, and the
North Africa Campaign of World War II*

◆ FIGURE 1 Northwest Egypt near El Alamein during the North Africa campaign, 1942. A British soldier is watching an approaching dust storm generated by *gibleh* winds.

THE QATTARA DEPRESSION IN NORTHWESTERN EGYPT is an immense, hot wasteland of salt marshes and drifting sand that covers an area of about 18,000 square kilometers (7,200 mi^2) at an elevation nearly 135 meters (443 ft) below sea level (◆Figures 1 and 2). Its origin is partly due to deflation by wind. Wind scour has lowered the land surface to the capillary fringe at the top of the water table. This has led to evaporation of the saline water and crystallization of the salt as a sparkling white crust on the depression floor.

Deserts are natural laboratories in which to study the interactions of wind and sometimes water on the arid surfaces of planets.

A. S. Walker, U.S.G.S. geologist

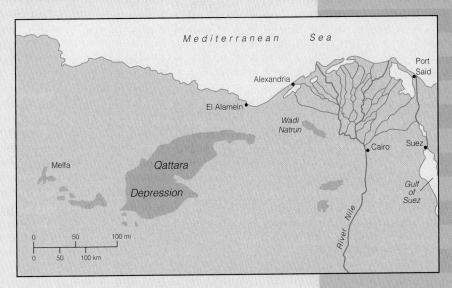

◆ FIGURE 2 Locations of the Qattara Depression and El Alamein. The decisive battle between the forces of Montgomery and Rommel was fought north of the Qattara Depression.

This impressive and forbidding quagmire played an important role during World War II in a decisive Allied victory in 1942. The impassable trough served to block German Field Marshal Erwin Rommel's Afrika Korps' attempt to turn the flank of the British Eighth Army led by General Bernard Montgomery. The geography forced Rommel to attack the Eighth Army's heavily fortified positions on El Alamein, a 64-kilometer (40-mi)-wide strip of passable land between the Mediterranean on the north and the hills to the south that mark the edge of the impassable Qattara. Rommel outran his lines of supply, sustained heavy losses, and was forced into retreat. This pushed him out of Egypt, and eventually westward across Libya.

Throughout the two-year North Africa campaign, the Allied and Axis forces had a common enemy: the *gibleh,* the searing hot winds that blow out of the Sahara. These winds raised temperatures as much as 19° C (35° F) in a few hours and sent sand grains swirling across hundreds of square kilometers. The onslaught of a gibleh dictated the actions of men and machines, preempting military tactics and strategies. Airplanes had to be grounded, and battlefield action stopped for days, as the sun was blotted out and visibility dropped to zero. Soldiers wore tight-fitting goggles to protect their eyes from the clouds of stinging sand and gasped for air through gas masks or makeshift cloth masks. Winds of hurricane strength that tore up camps and turned over vehicles were common. Swirling clouds of sand generated static electricity that rendered useless the magnetic compasses needed for navigating across the desert. Storm-generated static electricity during one gibleh detonated and destroyed an ammunition dump.

Deserts

The term *desert* means different things to different people. For some, a desert is a dry barren region of sand dunes; for others, it is a desolate region of bare, rocky outcrops and cactus; for still others, it is a treasure trove of scenic wonders, unique plants, and animals. Regardless of how they are perceived, deserts comprise about one-third of the earth's surface and are characterized by meager rainfall, scanty vegetation, a distinct landscape, and a limited population of people and animals. In deserts, the effects of running water, tectonic forces, and wind are clearly apparent. Because these factors combine in different ways in different places, the appearance of various desert landscapes differ as well. Deserts may be hot or cold. They may be mountainous and rocky, or vast, flat sand sheets, and some are covered with gravel. But all deserts are dry. Although deserts appear harsh and rugged, they are enormously fragile, and the misuse of arid lands is a serious problem in many parts of the world.

Kinds of Deserts

A **desert** (Latin *desertus,* "deserted, barren") is defined as a region with mean annual precipitation of less than 25 centimeters (10 in), with a potential to evaporate more water than falls as precipitation, and with so little vegetation that it is incapable of supporting abundant life. Notice that high temperatures are not necessary for a region to be called a desert; only low precipitation, a dry climate, and limited biological productivity. On this basis we can identify four kinds of desert regions: polar, subtropical, mid-latitude, and coastal (✳ Table 12.1).

- **Polar deserts** are marked by perpetual snow cover, low precipitation, and intense cold. The most desertlike polar areas are the ice-free dry valleys of Antarctica. The Antarctic continent and the interior of Greenland are true deserts. They are very dry, even though most of the fresh water there is in the form of ice.
- **Subtropical deserts** (also called *trade-wind deserts*) are the earth's largest dry expanses. They lie in the regions of subsiding high-pressure air masses in the western and central portions of continents. They receive almost no precipitation and have large daily temperature variation. The Saharan, Arabian, Kalahari, and Australian deserts are of this type.
- **Mid-latitude deserts,** at higher latitudes than the subtropical deserts, exist primarily because they are positioned deep within the interior of conti-

Type	Characteristic Location	Examples*
Polar	region of cold, dry, descending air with little precipitation	ice-free, dry valleys of Antarctica
Subtropical	a belt of dry, descending air at 20°–30° north or south latitude	Sahara, Arabian, Kalahari, Australia's Great Sandy and Simpson
Mid-latitude	deep within continental interior remote from the influence of an ocean	Gobi, Takla Makan, Turkestan
Coastal	coastal area in middle latitudes where a cold, upwelling oceanic current chills the shore (Figure 12.2)	Atacama, Namib
Rain-shadow	leeward of a mountain barrier that traps moist ocean air (Figure 12.1)	Mojave, Great Basin

★TABLE 12.1 Classification of Deserts

*Figure 12.4 shows the locations of these deserts.

nents, separated from the tempering influence of oceans by distance or a topographic barrier. The Gobi Desert of China, characterized by scant rainfall and high temperatures, is an example. Marginal to many middle-latitude deserts are **semiarid grasslands,** extensive treeless grassland regions. Drier than prairies, they are especially sensitive to human intervention. In North America and Asia the mid-latitude and subtropical arid regions merge to form nearly unbroken regions of moisture deficiency. Also in the mid-latitudes are mountain ranges that act as barriers to the passage of moisture-laden winds from the ocean, a situation leading to the formation of **rain-shadow deserts.** As water-laden maritime air is drawn across a continent, it is

forced to rise over a mountain barrier. This causes the air to cool and to lose its moisture as rain or snow on the windward side of the mountains. By the time the air descends on the mountains' leeward side, it has lost most of its moisture, and there we find desert conditions (◆Figure 12.1). The Mojave Desert and the Great Basin, leeward of the high Sierra Nevada, are examples of rain-shadow deserts. Death Valley, the lowest and driest place in North America, is in the rain shadow of the Sierra Nevada. The Sierra Nevada receives heavy precipitation, and Mount Whitney, its highest point and the highest point in the conterminous 48 states, is a scant 100 kilometers (60 mi) from arid Death Valley.

■ **Coastal deserts** lie on the coastal side of a large land mass and are tempered by cold, **upwelling** oceanic currents (◆Figure 12.2). Their seaward margins are perpetually enshrouded in gray coastal fog, but inland, where the air temperatures are higher, the fog evaporates in the warmer dry air. The Atacama Desert in Chile, which may be the driest area on earth, is a rain-shadow desert as well as a coastal desert. The Andes on the east block the flow of easterly winds, and air from the Pacific Ocean on the west is chilled and stabilized by the Humboldt Current, the ocean's most prominent cold-water current.

Collectively, mid-latitude deserts and semiarid grasslands cover about 30 percent of the earth's land area and constitute the most widespread of the earth's geographic realms.

CONSIDER THIS *Why are the major deserts of the world located at low latitudes?*

Desertification

Deluges of blowing sand, such as the one in the North African Desert described in the chapter opener, are common in many arid regions. In the past these winds have been rare

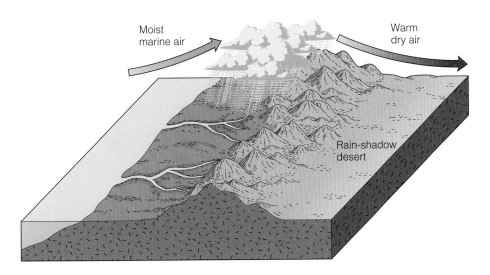

ENVIRONMENTAL
◆ **Geology⊜Now**™ ACTIVE FIGURE 12.1 The origin of rain-shadow deserts in middle- and high-latitude regions. Moist marine air moving landward is forced upward on the windward side of mountains, where cooling causes clouds to form and precipitation to fall. The drier air descending on the leeward side warms, producing cloudless skies and a rain-shadow desert.

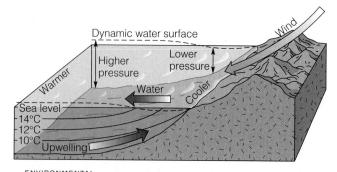

ENVIRONMENTAL
◆ **Geology⇌Now™** ACTIVE FIGURE 12.2 The mechanics of wind-driven upwelling of cool water from depth along a coastline. Upwelling delivers nutrients such as phosphates and nitrates to surface waters, where they promote plant plankton blooms and more marine life.

in semiarid grasslands marginal to middle-latitude deserts, but this is changing. We now see a new kind of desert forming where such winds are more common. These new deserts, which may be called *deserts of infertility,* are devouring more than 20,000 square miles of the earth's arid grasslands each year—an area about the size of the state of West Virginia—and putting another 70,000 square miles at risk. In fact, so much dust from some of these arid lands in Africa blows across the Atlantic Ocean during the summer that air quality in Florida and some other East Coast states is seriously degraded by it. If the new air-quality standards proposed by the Environmental Protection Agency are adopted, some of these states may be found to be in violation of them.

Deserts of infertility are not controlled as much by climate as by human misuse. The expansion of deserts into for-merly productive lands because of excessive human use is called **desertification,** and it is occurring at alarming rates in countries that can least afford to lose productive land. Understanding the causes of this class of desert is important, because much of humankind's future in many parts of the world depends on proper and careful use of arid and semiarid lands. In nearly 70 countries worldwide, deserts are expanding onto grazing lands and agriculturally productive lands (◆Figures 12.3 and 12.4). The result is that millions of people are dying of starvation or having to migrate, in many cases to areas where more hunger and malnutrition await them.

Globally, overgrazing of sparse vegetation (◆Figure 12.5) and trampling of the soil by livestock are the major causes of desertification. Other practices that lead to desertification include:

- clearing land of trees and brush for fuel without reforesting,
- surface mining without reclaiming the land,
- the rapid growth of large desert cities such as Phoenix and Las Vegas in the southwestern United States and other examples in countries such as Egypt and Saudi Arabia;
- depletion of ground water for desert irrigation and urban growth,
- replacement of native vegetation in semiarid regions with cultivated crops, and
- soil salinization due to evaporation of irrigation water (see Case Study 6.1).

⊙ Case Study 12.1 on (page 335) tells of the desertification in Africa's Sahel region.

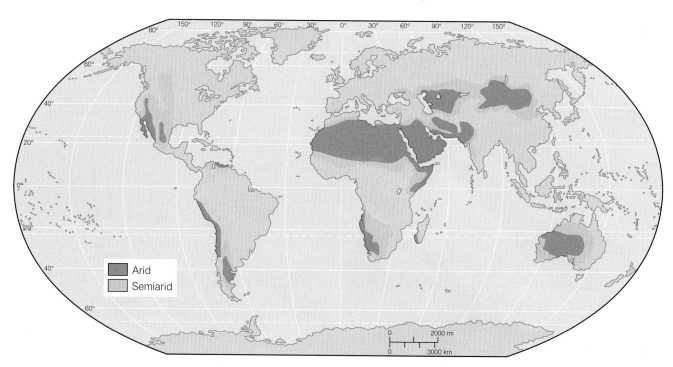

◆ FIGURE 12.3 Distribution of the earth's arid and semiarid regions.

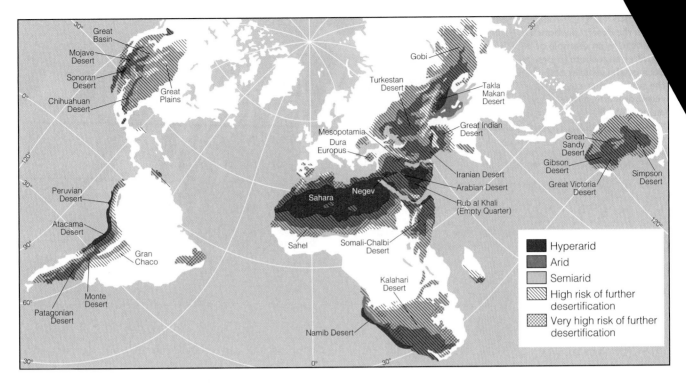

◆ FIGURE 12.4 Arid and semiarid regions of the world currently experiencing desertification.

For years the question has been asked, does climate change cause desertification, or does desertification cause climate change? There has been no clear answer to this question until studies in the 1990s by the University of Arizona's Remote Sensing Center on the border between the United States and Mexico revealed a clear *signal:* higher temperatures were recorded on the Mexican side of the border, where the earth's resources were being exploited more heavily, than on the U.S. side.

Desertification is a complex and subtle process of deterioration that may often be reversible, but disturbed desert and marginal grassland ecosystems are slow to recover. This is because those ecosystems are characterized by slow plant growth, little bacterial activity in the soil for recycling nutrients, little species diversity, and little water. Already sparse vegetation that has been destroyed by overgrazing or off-highway vehicle use (◆Figure 12.6) may require decades, perhaps centuries, to be restored to its natural state. The best program for reversing the trend is to halt exploitation immediately and to adopt measures such as planting trees, developing and using water resources properly, establishing protective ecological screens—grass belts or strips of trees—as windbreaks that will help reestablish natural ecosystems, building sand fences, and covering dunes with boulders or oil or paving areas to interrupt the force of wind near the face of the dunes to prevent sand from moving. In areas marginal to China's Gobi and Tengger Deserts (see ◎ Case Study 12.2 on page 338), efforts are being made to reclaim misused marginal lands that have become desert. At a research station along the margin of the Gobi Desert, Chinese scientists, in partnership with others from Japan and Israel, experiment with techniques for halting desertification. Their research is directed at finding drought- and saline-tolerant grasses, trees, and shrubs, and in utilizing drip irrigation for raising such crops as apples, onions, corn, squash, and watermelons in hope of recovering lands lost to desert dust to once again become productive farmland.

◆ FIGURE 12.5 Desertification in progress: intensely overgrazed rangeland (left) and lightly grazed (right).

d vehicle dam-
Canyon in
ow human
...ification.
...pact and alter
...use reductions in
permeability, which
...moisture content and thus
...local biological systems. Desert
...ns have eroded the tracks down the
slope, transforming them into gullies,
which increases runoff and sediment
yield.

HOWARD G. WILSHIRE, USGS

A good example of natural recovery is seen in Kuwait. It is an unexpected consequence of the 1991 Gulf War. Thousands of unexploded land mines and bombs are scattered across the desert in the western part of the country. These have discouraged off-road driving by sport hunters of the region's animals. The native vegetation is reestablishing, and the desert now resembles a U.S. prairie. The number of birds in the area is much greater than it was before the war.

> **CONSIDER THIS** *In the marginal lands in China and the Sahel that are threatened by desertification, how would you go about convincing local farmers that they should invest money and effort in planting windbreaks and installing drip irrigation?*

Wind as a Geologic Agent in Deserts

Wind erodes by deflation and abrasion. **Deflation** occurs when things are blown away. This produces deflation basins where loose, fine-grained materials are removed, sometimes down to the water table (◆Figure 12.7). **Abrasion** occurs when mineral grains are blown against each other and into other objects. Abrasion by sand can remove paint from vehicles, frost windows, and produce unusual natural features. (See the Gallery).

Once fine-grained rock material is eroded it moves, commonly forming migrating dunes—small hills of migrating, wind-blown sand—and drifting sand, which are common in arid regions, can adversely affect desert highways, railroad lines and even communities (◆Figure 12.8), but they are not limited to arid regions. Strong sea breezes in humid coastal regions coupled with human activities also can cause dune migration (see Figures 10.18b and ◆12.9). The problems of migrating

sand have been mitigated by building fences, planting windbreaks of wind- and drought-tolerant trees such tamarisk (◆Figure 12.10), and oiling or paving the areas of migrating sand. Dune stabilization in humid coastal areas of Europe, the eastern United States, and the California coast has been accomplished by planting such wind-tolerant plants as beach grass (*Elymus*) and ice plant "Hottentot fig" (*Carpobrotus*).

Arid Regions, Winds, and Human Health

Each year as much as two billion tons of dust is estimated to be lifted into earth's atmosphere. Most is stirred up by storms, the most important of which are appropriately

D. D. TRENT

◆ FIGURE 12.7 The Devil's Cornfield, a deflation basin that has been lowered to the capillary fringe at the top of the water table; Death Valley National Park, California. The rugged plants growing here—arrow weed, which resemble shocks of corn—are able to tolerate heat and saline soil, and their roots can withstand the effects of exposure to episodic sandblasting.

◆ FIGURE 12.8 Migrating sand dunes are encroaching on Mauritania's capital city, Novakchott, on northwest Africa's Atlantic coast.

named dust storms. A single large dust storm originating in the arid lands of North Africa may drop more than 200 tons of sediment in the Northern Amazon Basin of South America (◆Figure 12.11). North African dust storms routinely affect the air quality in the Middle East and Europe; a fine layer of African dust on snow in Western Europe is not uncommon. Carried along with the dust are pollutants such as pesticides, herbicides, and microorganisms—viruses, bacteria, and fungi. Microbes detected in Caribbean air during African dust events include about 25 percent that are plant pathogens and about 10 percent that are opportunistic human pathogens, organisms that may infect humans. Evidence suggests that fallout has direct consequences on the health of coral communities and may be responsible for high rates of asthma in the Caribbean—a 17-fold increase in

asthma cases in Barbados since 1973 corresponds with the period of increased African dust associated with the drought in the Sahel (see Case Study 12.1). Furthermore, African dust events are connected to the meningococcal meningitis pathogen as outbreaks of meningitis in sub-Saharan Africa that often follow Saharan dust storms.

Arid lands in Asia are also a significant source of airborne sediments to the Northern Hemisphere during the spring. According to the U.S. National Aeronautics and Space Administration, dust from a March 2001 "Yellow Dragon" sand storm in China (see Case Study 12.2 on page 338) traveled across the Pacific Ocean causing hazy skies in California, and a dust event in April 2002 sullied the skies over Colorado. Some 4,000 tons of sediment per hour may fall in the Arctic during a large dust storm in Asia, and fallout of herbicides and pesticides into the Arctic is common from storms originating in China's semiarid farmlands. These

◆ FIGURE 12.9 Coastal dune field near Pismo Beach, California. In the 1960s these dunes began migrating leeward onto agricultural lands due to the destruction of stabilizing plant cover by uncontrolled off-road vehicle use in the dune field.

◆ FIGURE 12.10 Windbreaks of salt-tolerant *tamarisk* (salt cedar) keep drifting sand from burying railroad tracks; Palm Springs area, California.

◆ FIGURE 12.11 Giant dust storm in February 2001, originating in northwestern Africa and blowing out into the Canary Islands. In addition to dust, such storms appear to carry toxic chemicals, viruses, and other infectious microbes that are now believed to pose a significant hazard to the Caribbean and other regions of the Americas. The image was taken by the Sea-viewing Wide Field-of-view Sensor (SeaWiFS) on the OrbView-2 satellite, which has recorded such storms every year since its launch in 1997.

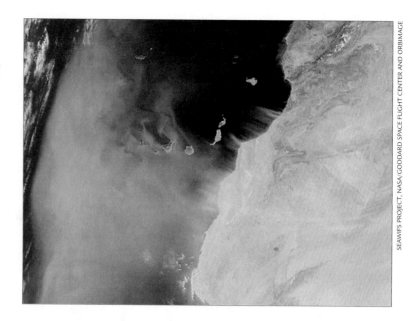

chemicals are found in animal tissues and human breast milk among people living in the Arctic.

Recognition that airborne sediments from one continent can be carried over great distances and affect air quality thousands of miles from their source raises several important but unanswered questions that demonstrate the depth of our ignorance: Can pathogens transported in airborne dust cause outbreaks of disease? Can harmful chemicals transported halfway around the world impact the well-being of ecosystems and human health? What kinds of microorganisms can survive long-range transport in the atmosphere, and can they successfully compete with the indigenous microbial community at a new location?

Sahel's Seventeen-Year Drought: Evidence of Climate Change, Misuse of the Land, or Western Pollution?

The **Sahel** is a belt of semiarid lands along the southern edge of North Africa's Sahara Desert that stretches from the Atlantic Ocean on the west to the Red Sea on the east, and is 300 to 1,500 kilometers (180–940 mi) wide in which a large population resides (◆Figure 1). The region typically has dry winters and wet summers, when the intertropical convergence zone **(ITCZ)** moves into the region, bringing rain. Rainfall amounts vary from year to year, averaging 10 to 20 centimeters (4–8 in), but some years it may be as much as 50 centimeters (20 in) in the southern part where the Sahel blends into the belt of humid savannah grasslands. Desertification in the Sahel begins with the man-induced process of slashing and burning the natural brushland and forest in order to clear land for agricultural zones. In Niger alone, about 250,000 hectares (965 sq mi) are lost each year to desertification (◆Figure 2). After harvesting millet or other crops, the farmers cut the stalks, leaving their fields exposed to strong winds until the next planting. The winds carry away the topsoil, uproot

seedlings and seeds, and suffocate small plants elsewhere where the soil accumulates.

In 1968 the summer rains began decreasing, exacerbating the man-induced vegetation changes, and initiating a 17-year period of severe drought conditions (◆Figure 3) that captured the world's attention with grim images of skeletal mothers and starving children with bloated bellies. Before the rains returned, over 250,000 people and 12 million livestock died in one of the world's most devastating famines. Many more would have perished had it not been for the massive influx of outside aid.

Under normal climatic conditions, such semiarid grasslands on the fringes of true deserts are parts of a delicately balanced ecosystem that is adapted to natural climate variations. The vegetation indigenous to these marginal lands can survive the rigors of a few dry years. Cultivated crops, such as millet and sorghum, however, lack drought tolerance, and thus, crop failure is common during drought. This leaves the soil bare

and susceptible to erosion by wind and rain. In 1972, the Sahel received almost no rain. In 1973, at the climax of the drought, total rainfall was about half the long-term annual average. By then, the Sahara Desert had migrated as much as 70 kilometers (100 mi) southward into the Sahel, and 50 to 70 percent of the region's livestock had died.

In 1985 the rains returned, and by the late 1990s much of the Sahel had become astonishingly green again and was pushing back the desert. The most spectacular growth has been in northern Burkina Faso, so much so that the region is now a "carbon sink," a savannah grassland that soaks up carbon dioxide from the atmosphere. Yields of sorghum and millet in one province of Burkina have increased by 70 percent and former residents driven out by the drought are now returning. Nevertheless, much of the Sahel is still threatened by desertification (◆Figure 4).

It is still unclear just why the drought occurred. The wetter years of the mid-twentieth century were due to the favorable

◆ **FIGURE 1** North Africa's semiarid Sahel is bordered by grasslands on the south and by the Sahara Desert on the north. Adapted from C. D. Ahrens, *Meteorology Today,* 6th ed., Brooks/Cole, 1999

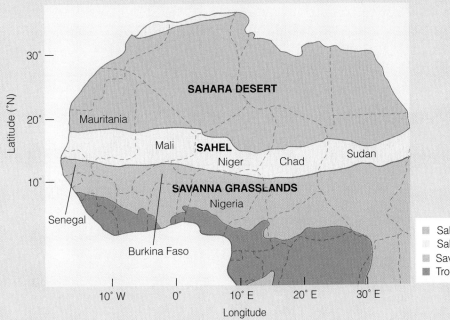

CONTINUED

◆ FIGURE 2 A sharp line marks the boundary between natural pasture and a dune encroaching from the Sahara in the west African country of Mali; a graphic example of desertification. The Sahara Desert is creeping southward into the semi-arid Sahel region, and the Sahel, in turn, is expanding into the grasslands to the south of it.

positioning of the ITCZ, which brought intense summer rains; then the dryer years came with less intense rains. It might be that an initial decrease in rainfall and the corresponding reduction in plant cover (due to the introduction of cultivated crops) modified the land surface sufficiently to reduce convective activity and create a positive-feedback mechanism that increased atmospheric drying even more. (Positive and negative feedback are explained in Chapter 11).

In 2002, research by a group of scientists from Canada and Australia suggested that the drought was triggered by **aerosols,** tiny particles of sulfur dioxide emitted by coal-fired electric power plants and factories thousands of miles away in Europe, Asia, and North America. Curiously, the aerosols didn't even have to travel to Africa to wreak havoc with Africa's weather conditions. Instead, they could alter the physics of cloud formation thousands of miles away and reduce African rainfall as much as 50 percent. The process of such aerosol pollution may explain the summer 2002 drought experienced in much of the midwestern United States.

However, the researchers are careful to state that the cause of Sahel's drought is likely due to a combination of atmospheric aerosols and natural climate variability; El

Niño and overgrazing may also be factors. Interestingly, the return of the Sahel's rains in the 1990s correlated with the enactment of emissions laws in the same period by the industrialized West that reduced atmospheric aerosol pollution. The correlation may be coincidence, but many scientists think not. Fortunately, the deep-rooted pessimism of the 1970s about Sahel desertification has eased somewhat, but widespread decimation of the native plants and

cultivated crops still exists. In some agricultural zones the grain yields are still insufficient and many villagers talk of still being hungry. In Sudan, sand dunes have formed from eroded soil originating inside agricultural zones where the native plants have been removed. Thus, the threat to many farming villages is not from the Sahara but from desertification within agricultural zones.

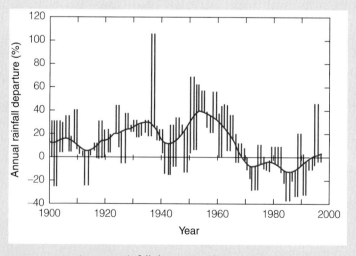

◆ FIGURE 3 Average rainfall departures, by percentage, from the 1961–1990 average for the African Sahel, 10°N to 20°N latitude. The smoothed curve plots the ten-year overlapping means. Mike Hume, Climatic Research Unit, Univ. of East Anglia, England

(a)

(b)

◆ FIGURE 4 (a) A farm in Niger, on the southern edge of the Sahel, after one year's participation in a desert reclamation program. (b) A farmer and his farm in Niger after seven years of participation in a desert reclamation program. The young trees are perennials that will eventually provide a market crop and form a wind break to stabilize the soil in this region of nearly persistent drought. Millet stalks remain following the farm's annual harvest.

CASE STUDY 12.2

Is Desertification Swallowing China?

In the People's Republic of China, Longbaoshan, a farming area about 65 kilometers (40 mi) northwest of Beijing, is at the leading edge of China's war against desertification. The advancing desert is driven by a combination of drought, poor land management, overpopulation, and overgrazing. In the 1990s, China was hit with more than 20 sand storms a year, up from eight in the 1960s. Between 1994 and 2000, desert lands in China grew by nearly 53,000 square kilometers (20,300 square miles)—an area more than the combined area of New Hampshire and Vermont. In the country's driest regions, such as Inner Mongolia (◆Figure 1), drifting sands smother grasslands in scenes reminiscent of the Dust Bowl Days of the 1930s in the midwestern United States (see Chapter 6). China, a quarter of which is arid land, has 20 percent of the world's population but only 2 percent of the world's arable land. China's current desertification causes financial losses of $6.7 billion a year, affects the lives of 400 million people, and threatens farmers and herders. China will experience an unprecedented ecological and human disaster if the marching sands are not halted.

In March 2002, the worst sandstorm in a decade, known as a Yellow Dragon, turned Beijing's sky yellow, blinded its residents by cutting visibility to less than 90 meters (300 ft), and dumped 30,000 tons of sand on the city. Yellow Dragon sandstorms sometimes merge with rain clouds to inundate the city with mud.

The government is trying to halt the moving sands by planting green buffer zones in critical regions. A project intended to protect Beijing, and prepare the region for the 2008 summer Olympic Games, requires reclaiming a large area recently lost to desertification. In Xuanhua County, about 150 kilometers (90 mi) northwest of Beijing, 250,000 soldiers are trying to stop an adjacent desert in a 10-year project of planting pine trees and poplars around the Yanghe Reservoir. Officials are also encouraging farmers to plant trees in dry fields where they once grew crops, paying the peasants with grain and money if 80 percent of the trees survive the first year. Raising trees in this barren mountainous moonscape is arduous because water must be hauled, in some cases up to several kilometers, by oxcart or backpack.

Another strategy for halting drifting sands has been successful in stabilizing windblown sand that had regularly buried the tracks of a main railroad line in China's Tengger Desert on the edge of Inner Mongolia. Inexpensive, low-tech straw mats placed along the railroad right-of-way stabilize the dunes and give desert-adapted plants the time to become established (◆Figure 2).

Water, which is in short supply in China, is the most important weapon against desertification. The demands of rising living standards and heavy industry have created water shortages and caused rivers to dry. The lower Huang He (Yellow River), China's "mother" river, ran dry for 225 days in 1997, and for most of the year the river failed to reach the Yellow Sea. A similar disappearance of once raging rivers is repeated elsewhere in the country. For example, in Beijing is a twelfth-century stone bridge made famous by Marco Polo, the thirteenth-century Venetian explorer. It stretches 300 meters (900 ft) across what was once the Yongding River (◆Figure 3). As recently as the 1970s people fished in the waters and swam in the river beneath the bridge, but today it is nothing but gravel, dust, and weeds.

◆ FIGURE 1 China and Mongolia index map

GEORGE STEINMETZ, NATIONAL GEOGRAPHIC

◆ FIGURE 2 Straw barrier mats being placed along the railroad line in the Tengger Desert region of China's Inner Mongolia stabilize the desert's drifting sands and give desert plants time to become established.

FRANK LANGFITT, SUNSOURCE/ THE BALTIMORE SUN

◆ FIGURE 3 Southwest of Beijing an abandoned paddleboat sits on the dry bed over which the Yongding River once flowed. The twelfth-century bridge in the background was publicized by Marco Polo after his visit to China in the thirteenth century.

Desert Oddities

Arid regions, too, have unique and spectacular characteristics that are distinct from other climatic realms.

◆ FIGURE 1 Mushroom rock; Death Valley National Park, California. Rigorous wind erosion in the abrasion zone just above the ground surface can produce unusual landforms such as this.

◆ FIGURE 2 Remnants of a 1920s plank road, an early attempt to maintain a road in a region of drifting sand; Algodones dune field between El Centro, California, and Yuma, Arizona.

◆ FIGURE 3 Playa lake scraper; Racetrack Playa, Death Valley National Park, California. The boulder's track is preserved on the mud-cracked surface. The boulder, 25 cm (11 in) at its widest, moved or was moved during a winter wet period due to just the right combination of rain-soaked mud and gusty high wind.

◆ FIGURE 4 Devil's Golf Course, Death Valley National Park, California, the lowest elevation in the Western Hemisphere. The salt-encrusted surface has formed at the capillary fringe on the top of the underlying water table.

Summary

Deserts

Description

Regions where annual precipitation averages less than 25 centimeters (10 in) and that are so lacking in vegetation as to be incapable of supporting abundant life.

Causes

1. High-pressure belts of subsiding, warming air that absorbs water and precludes cloud formation.
2. Isolation from moist maritime air masses by position in deep continental interior.
3. Windward mountain barrier that blocks passage of maritime air.

Classification

May be classed into four categories determined by climate belts and a fifth category due to human misuse.

Types

1. *Polar deserts,* regions of perpetual cold and low precipitation.
2. *Mid-latitude deserts* are found within the interior of continents in the middle latitudes, remote in distance from the influence of oceans.
3. *Subtropical deserts,* the earth's largest realm of arid regions, lie in and on the equatorial side of subtropical zones of subsiding high-pressure air masses in the western and central portions of continents.
4. *Coastal deserts* occur on the coastal side of a land or mountain barrier in subtropical latitudes. Because they are bordered by ocean, they are cool, humid, and often foggy.
5. *Deserts of infertility,* generally variants of mid-latitude deserts, are not so much due to climate as to human misuse.

Desertification

The expansion of deserts into formerly productive lands because of excessive human use or misuse.

Wind as a Geologic Agent

Erosion

1. *Deflation*—earth materials are lifted up and blown away.
2. *Abrasion*—mineral grains are blown against each other and into other objects.

Control of Migrating Sand

Accomplished with sand fences, paving, windbreaks, and stabilizing plants.

Key Terms

abrasion
coastal desert
deflation
desert

desertification
ITCZ
mid-latitude desert
polar desert

rain-shadow desert
semiarid grassland
subtropical desert

Sahel
upwelling

Study Questions

1. Explain why the world's largest deserts are located on or near the horse latitudes.
2. What factors other than drought are responsible for the desertification that has resulted in starvation in Africa and elsewhere? (*Hint:* Consider carrying capacity, as discussed in Chapter 1.)
3. Explain why ground water can be found only a few centimeters to a meter (3 ft) or so below the salt-encrusted surface in some of the driest places in the world—for example, the Devil's Golf Course, at about 75 meters (250 ft) below sea level in Death Valley, and the Qattara Depression, at about 135 meters (443 ft) below sea level in the Sahara Desert.
4. List measures that can be taken to halt sand-dune migration.
5. Explain the interrelations between the biologic, atmospheric, and solid-earth systems in arid regions. You may use words, pictures, or both.
6. List agricultural practices in arid and semiarid regions of the United States (such as California's Coachella Valley and western Nebraska) that increase the likelihood of desertification.
7. List measures that can be taken to halt desertification.

For Further Information

Books and Periodicals

Ahrens, C. Donald. 2000. Global climate (pp. 470–502). *Meteorology today,* 5th ed. Pacific Grove, Calif.: Brooks/Cole.

Cooke, R. A., A. Warren, and A. Goudie. 1993. *Desert geomorphology.* London: UCL Press.

Ellis, W. S. 1987. Africa's Sahel: The stricken land. *National geographic* 172, no. 2:140–179.

Griffin, D. W., C. A. Kellogg, V. H. Garrison, and E. A. Shinn. 2002. The global transport of dust. *American Scientist* 90, no. 3, p. 228–235.

Grainger, Alan. 1990. *The threatening desert.* Earthscan Publishers, Ltd.

Monroe, James S., and Reed Wicander. 1998. The work of wind and deserts (pp. 511–535). *Physical geology,* 3d ed. Belmont, Calif.: Wadsworth.

Sheridan, D. 1981. *Desertification of the United States.* Washington, D.C.: Council on Environmental Quality.

Scott, Ralph C. 1992. Climates of the world, (pp. 181–220). *Physical geography,* 2d ed. St. Paul, Minn.: West Publishing Co.

Sletto, B. 1997. Desert in disguise. *Earth* 6, no. 1:42–49.

Walker, A. S. 1996. *Deserts: Geology and resources.* Denver: U.S. Geological Survey.

Webster, D., and George Steinmentz. 2002. China's unknown Gobi. *National Geographic* 201, no. 1, p. 48–75.

ENVIRONMENTAL
Geology⇌Now™

Assess your understanding of this chapter's topics with additional quizzing and comprehensive interactivities at http://earthscience.brookscole.com/pipkingeo4e

as well as current and up-to-date Web links, additional readings, and InfoTrac College Edition exercises.

Mineral Resources and Society

Nevada's New Bonanza

WE ARE CURRENTLY IN THE MIDST OF THE BIGGEST gold-mining boom in U.S. history. The bonanza is the Carlin Trend, a 50 by 5 kilometer (80 by 8 mi) line of enormous low-grade gold deposits in north-central Nevada. In 2002 the Trend poured its 50 millionth ounce of gold since its discovery in 1962, making it the third location in the world to achieve that milestone.

About 20 major mines and several smaller ones, most of them open-pit cyanide heap-leach operations, were mining the Trend in the early twenty-first century. The Trend is a valuable asset to Nevada's economy, producing nearly 4 million ounces of gold per year, contributing $1.8 billion to Nevada's economy. It is estimated that there are another 50 million ounces of gold reserves and resources on the Carlin Trend.

The Gold Quarry Mine, the largest of Newmont Gold Company's several open-pit mines, is representative of the large mines on the Carlin Trend (◆Figure 1). Newmont began operations on the Trend in 1965, and the combined 2001 production of its three open-pit mines exceeded 1.5 million ounces. The cost of

Our entire society rests upon—and is dependent upon—our water, our land, our forests, and our minerals. How we use these resources influences our health, security, economy, and well-being.

John F. Kennedy,
February 23, 1961

D. D. TRENT

◆ FIGURE 1 *Newmont's Gold Quarry Mine, Carlin, Nevada. Shown are the open pit and mill and concentrator works in the distance.*

recovering gold from Newmont's mines runs about $240 per ounce. An additional cost is reclamation bond money, about $1 for each ounce of gold produced that is set aside to meet the eventual costs of reclamation.

After mining ends, which is estimated to be in 2030, the pits will be closed off with berms and fences and the waste-rock dumps will be encapsulated with a 4-foot cover of topsoil and planted with native plants in accordance with the mine's operating permits. A network of drainage ditches is designed to collect and divert surface flow from the waste piles. It should be noted that the protective berms will only be a few feet high, and the fences must be maintained forever for reasons of public safety because the berms could create attractive jumps for ATV riders.

Some Nevadans warn that the true cost of mining the Trend is being externalized on the Humboldt River drainage basin. Approximately 18 mines in the region are working below the water table, which requires massive pumping of ground water. Between 1997 and 2000, the five largest dewatering mines pumped nearly 800,000 acre-feet just to keep the mine pits dry. Such a withdrawal is believed to be creating a water deficit in the basin. Some predictions suggest that when mining ends it would require all of the flow of the Humboldt River at Winnemucca for 32 years to fill the anticipated 5,000,000 acre-foot deficit. It seems highly unlikely that the river will go dry for decades, but it illustrates the concern of some that recharging the deficit may deplete rancher's wells and many of the normal ground-water discharges including springs.

At the end of mining, the remaining leach heaps, some covering hundreds of acres, become waste products. Abandoned heaps contain water, which may drain out for many years, and rainwater may seep out forever, even in Nevada's arid climate. Because the heaps contain tons of cyanide, mercury, salt, and selenium, drainage water at times may be hazardous. Such hazardous waste could threaten water quality over much of northern Nevada.

Every material used in modern industrial society is either grown in or derived from the earth's natural mineral resources. The mineral resources are usually classified as either metallic or nonmetallic, as shown in ＊ Table 13.1. Production and distribution of the thousands of manufactured products and the food we eat is dependent upon the utilization of metallic and nonmetallic mineral resources. The per capita U.S. consumption of mineral resources used directly or indirectly in providing shelter, transportation, energy, and clothing is enormous (◆Figure 13.1). The availability and cost of mineral and rock products influences our nation's standard of living, gross national product, and position in the world.

With the value of nonfuel minerals produced in the United States running $40 billion per year and the value of materials processed from minerals in 1998 estimated at $415 billion (about 5 percent of the gross national product), it is amazing that the general public has relatively little knowledge of where these minerals naturally occur, the methods by which they are mined and processed, and the extent to which we depend on them. It is important that the public recognize that exploitable mineral resources occur only in particular places, having formed there due to unique geologic conditions, and that all deposits are exhaustible. Furthermore, development and exploitation of mineral resources is capital-intensive, requiring substantial long-term investment, and has environmental effects.

Understanding the origins, economics, and methods of mining and processing of mineral resources enables an individual to understand many of the difficult resource-related issues that currently face all levels of government. Increased land-use competition among the interests of mining, housing, wilderness preservation, agriculture, logging, recreation, wetlands preser-

*TABLE 13.1 Classification of Mineral Resources	
Metallic Mineral Resources	**Nonmetallic Mineral Resources**
Abundant Metals	***Minerals for Industrial and Agricultural Use***
iron, aluminum, manganese, magnesium, titanium	phosphates, nitrates, carbonates, sodium chloride, fluorite, sulfur, borax
Scarce Metals	***Construction Materials***
copper, lead, zinc, tin, gold, silver, platinum-group metals, molybdenum, uranium, mercury, tungsten, bismuth, chromium, nickel, cobalt, columbium	sand, gravel, clay, gypsum, building stone, shale, and limestone (for cement)
	Ceramics and Abrasives
	feldspar, quartz, clay, corundum, garnet, pumice, diamond

Source: James R. Craig, David L. Vaughan, and Brian J. Skinner, *Resources of the Earth* (Englewood Cliffs, N.J.: Prentice Hall, 1988).

vation, and industrial development is forcing our government officials to make decisions that affect mineral-resource development and management and, thus, all of our lives.

Mineral Abundances and Distribution

Economic Mineral Concentration

The elements forming the minerals that make up the earth's crust exist in many forms in a great variety of rocks. The higher the abundance, the richer the deposit and the more

economically feasible it is to extract and process the desired element. Locally rich concentrations of minerals are called **mineral deposits.** If they are sufficiently enriched, they may be **mineral reserves,** deposits of earth materials from which useful commodities are economically and legally recoverable with existing technology. Reserves of metallic minerals are called **ores,** and the minerals comprising the deposits are referred to as **ore minerals.**

A particular ore deposit may be described in terms of its **concentration factor,** its enrichment expressed as the ratio of the element's abundance in the deposit to its average continental-crustal abundance. * Table 13.2 illustrates that the minimum concentration factors for profitable mining

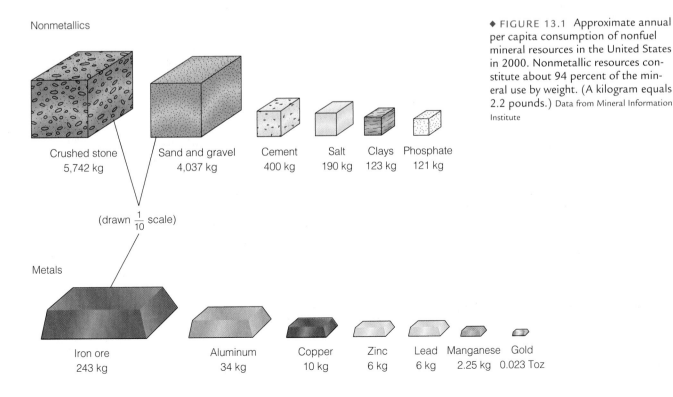

◆ FIGURE 13.1 Approximate annual per capita consumption of nonfuel mineral resources in the United States in 2000. Nonmetallic resources constitute about 94 percent of the mineral use by weight. (A kilogram equals 2.2 pounds.) Data from Mineral Information Institute

⋆TABLE 13.2 Concentration Factors for Profitable Mining of Selected Metals

	Percentage Abundance		Concentration Factor
Metal	*Average in Earth's Crust*	*In Ore Deposit*	
Aluminum	8	35	4
Copper	0.0063	0.4–0.8	80–160
Gold	0.0000004	0.001	2,500
Iron	5	20–69	4–14
Lead	0.0015	4	2,500
Mercury	0.00001	0.1	10,000
Silicon	28.2	46.7	2
Titanium	0.57	32–60	56–105

Source: U.S. Geological Survey professional paper 820, 1973.

vary widely for eight metallic elements. Note in the table that aluminum, one of the most common elements in the earth's crust, has an average crustal abundance of 8 percent. A "good" aluminum-ore deposit is defined at this time as one that contains about 35 percent aluminum; thus, the present concentration factor for profitable mining of aluminum ore is about 4. In contrast, deposits of some rare elements such as uranium, lead, gold, and mercury must have concentration factors in the thousands in order to be considered mineral *resources.*

Some elements are extremely common. In fact, 97 percent of the earth's crust is comprised of only eight elements: oxygen, silicon, aluminum, iron, sodium, calcium, potassium, and magnesium. For most elements, the **average crustal abundance,** the amount of a particular element present in the continental crust, is only a fraction of a percent (Table 13.2). Copper, for example, is rather rare; its average crustal abundance is only 0.0063 percent. In some localities, however, natural geochemical processes have concentrated it in mineral deposits that are 2 to 4 percent copper. A deposit with a high concentration of a desired element is called a **high-grade deposit.** A deposit in which the mineral content is minimal but still exploitable is called a **low-grade deposit.** For either grade of deposit, localities where desirable elements are concentrated sufficiently for economic extraction are relatively few.

Factors That Change Reserves

Reserves are not static, because they are defined by the current economics and technology as well as by the amount of a mineral that exists. Reserves fluctuate due to several factors: changing demand, discoveries of new deposits, and changing technology. In a free economy, a mineral deposit will not be developed at an economic loss, and prices will rise with demand. Consequently, low-grade deposits that are marginal or submarginal in today's economic climate may, if demand and prices rise, eventually become profitable to exploit. Advancements in technology also may increase reserves by lowering the cost of development or processing.

For example, the exploitable reserves of gold increased dramatically in the United States in the late 1960s due to a combination of changes in government policies and advances in technology. In 1968, when the price of gold was $35 per troy ounce, the Treasury Department suspended gold purchases to back the dollar and began allowing the metal to be traded on the open market. (A troy ounce equals 31.103 grams, whereas the avoirdupois ounce of U.S. daily life equals 28.330 grams.) Then in the 1970s the federal government removed restrictions prohibiting private ownership of gold bullion. In 1980 the gold price rose to more than $800 an ounce, and by 2003 it had settled at about $330 an ounce. Thus in a period of about 35 years the price of gold increased by 800 percent, meaning that many submarginal mining claims became economically profitable mines in that time. Furthermore, gold exploitation became more cost-efficient in the 1980s with the development of the new technology of cyanide heap-leaching, by which disseminated gold is dissolved from low-grade deposits and then recovered from the solution. A combination of changing economics and the new heap-leach technology initiated a new gold rush in the West beginning in the 1980s (see the chapter opener).

Other factors may reduce reserves, because extracting the earth's riches almost always requires the trade-off of aesthetic and environmental consequences (⋆ Table 13.3 provides some examples). For example, large low-grade mineral deposits can be mined only by open-pit methods. Not only do these methods devastate the landscape, their excavations also may reach below the water table, which can lead to ground-water contamination, and they produce enormous amounts of waste rock and tailings that can pollute surface-

TABLE 13.3	Some Negative Environmental Consequences of Mineral Extraction and Processing Operations	
Location	**Type of Operation**	**Consequences**
Questa, New Mexico	molybdenum mining	Drainage from waste rock from an abandoned molybdenum mine in upper Rio Grande Valley has contaminated watersheds, destroyed wildlife habitat, killed fish, polluted irrigation ditches and water wells, affected agricultural production, and caused illness to area residents.
Baia Mare, Romania	gold mining	A spill of about 1,000,000 m³ (3.5 million cu ft) in January 2000 of cyanide-laced tailings operations ravaged a 250-mi stretch of the river, killing all aquatic life including thousands of fish.
North-central Montana	gold mining	The Zortman–Landusky open-pit, large-scale cyanide heap-leach gold mine suffered cyanide leaks and spills, causing surface and ground-water contamination, and bird and wildlife fatalities. When faced with cleanup costs, the mining company declared bankruptcy in 1998, forfeiting its cleanup bond, which is estimated to be $8.5 million short of cleanup costs.
Summitville, San Juan Mountains, Colorado	gold mining	Leaking leach pad, contaminated with cyanide and toxic metals, poisoned 27 km (17 mi) of the Alamosa River, on which the region's agriculture depends. When cleanup costs exceeded the posted bond, the Canadian mining company and its parent declared bankruptcy. Cleanup will cost U.S. taxpayers $170–250 million.
Guyana, South America	gold mining	Cyanide spill from the Omai Mine—backed by the same investors as Colorado's Summitville Mine—released 860 million gallons of cyanide-laced tailings into one of Guyana's largest rivers. The spill killed fish, causing a panic in Guyana's seafood markets.

water runoff. **Smelters,** large industrial plants that process ore concentrates and extract the desired elements, produce more air pollution, in the form of flue dust, than any other single industrial activity. Smelters also produce slag, a solid residue that can contain thousands of times the natural levels of lead, zinc, arsenic, copper, and cadmium.

The real and perceived negative environmental impacts that serve to rule out exploitation and development of known deposits also reduce reserves. In recent years there have been several widely publicized instances of wildlife and wilderness values being judged more important than exploitation of mineral or petroleum deposits. Examples include:

- the extraordinary agreement reached between President Clinton and Crown Butte Mines in 1996 that halted proposed mining activity on the doorstep of Yellowstone National Park;
- the two-year moratorium on hard-rock mining along a 100-mile-long region of Montana's Rocky Mountain Front, one of the nation's most important wildlife areas; and
- the decision by the Office of Surface Mining and Reclamation not to allow surface mining of coal in the watershed of Fern Lake in Kentucky, because it would be incompatible with Cumberland National Historic Park's land-use plan, would destroy historic and natural systems, and would damage the region's water quality.

We can expect that controversies over land use will be increasingly common in the future as more and more lower-grade deposits will need to be exploited when higher-grade deposits become exhausted.

Political factors may be related to the economics of mineral resources. The United States imports substantial amounts of mineral resources it needs from other countries. (This is discussed in detail later in the chapter.) Some geopolitical authorities estimate that about a quarter of the roughly 50 wars and armed conflicts in 2001 had a strong "resource dimension." Exploitation of diamonds by Angolan rebels in 1992–2001 resulted in an estimated income of about $4 billion, and emeralds and lapis lazuli from the mid-1990s to 2001 netted Afghan groups (Taliban and the Northern Alliance) $90–100 million per year. Furthermore, resource availability may affect diplomatic policy. In spite of the general trade sanctions the United States enacted in the 1980s against South Africa to protest its apartheid policy, ten minerals the United States imports from South Africa were exempted because of their importance to the United States.

Distribution of Mineral Resources

The global distribution of mineral resources has no relationship to the locations of political boundaries or technological capability. Some mineral deposits, like those of iron, lead, zinc, and copper, originate by such a variety of igneous, sedimentary, and metamorphic processes that they are widely distributed, though in varying abundance. On the other hand, bauxite, essentially the only ore mineral of aluminum (one of the most common elements in the earth's crust), is concentrated by only one geochemical process:

deep chemical weathering in a humid tropical climate. For this reason, economically exploitable deposits of bauxite and other such minerals are unequally distributed and restricted to a few, highly localized geologic and geographic sites. Such localization of mineral deposits means that desired mineral resources do not occur in all countries. When this unequal supply pattern results in the scarcity of particular mineral resources that are crucial to a nation's economy, those minerals become critical or even strategic in the event of a national emergency. What constitutes "critical" minerals changes from time to time, mainly due to technological advances in mining or processing technologies.

Throughout history, nations' domestic mineral-resource bases and their needs for minerals unavailable within their borders have dictated international relations. Empires and kingdoms have risen and fallen due to the availability or scarcity of critical mineral resources. Two classic examples are those of the Greek and Roman empires. Silver mined by the Greeks near Athens financed the building of their naval fleet that defeated the Persians at the Bay of Salamis in 480 B.C. A century later, Greek gold supported the ambitious conquests of Alexander the Great. The Romans helped to finance their vast empire by mining tin in Britain, mercury in Spain, and copper in Cyprus. In the twentieth century, the efforts of Japan and Germany to secure mineral and fuel resources lacking within their borders were major factors leading to World War II, and the Soviet Union's annexation of the Karelian highlands of Finland following World War II provided the Soviets with new reserves of copper and nickel. More recently, Morocco acquired valuable phosphate deposits when it occupied part of the Spanish Sahara in 1975.

Origins of Mineral Deposits

Specific geological processes must occur in order to concentrate minerals or native elements in a mineral deposit. A simple classification of mineral deposits is based on their concentrating processes (∗ Table 13.4). The geologic processes responsible for many mineral deposits may be understood in terms of plate tectonics, as illustrated in ◆ Figure 13.2, with mineral concentrations occurring in distinct tectonic regimes. For example, the great porphyry cop-

∗TABLE 13.4 Genetic Classification of Mineral Deposits		
Genetic Type	**Example of Mineral(s) Formed**	**Location**
Igneous Processes		
Pegmatite	tourmaline	Oxford County, Maine; San Diego County, California
	beryl	Minas Gerais, Brazil
Crystal settling	chromite-magnetite-platinum	Bushveld complex, South Africa
Disseminated	copper-molybdenum porphyries	western North and South America
	diamonds	kimberlite pipes in South Africa, Canada, and Australia
Hydrothermal	gold-quartz veins	Mother Lode belt, California
	lead-zinc deposits in carbonate rocks	Mississippi River valley; Leadville, Colorado
Volcanogenic	massive copper sulfide deposits	Wrangell Mountains, Alaska
Sedimentary Processes		
Chemical precipitates in shallow marine basins	banded-iron formation	Lake Superior region
Marine evaporites	potassium salts	Carlsbad, New Mexico
Nonmarine evaporites	carbonates and borax minerals	Searles Lake, Trona, California
Deep-ocean precipitates	iron-manganese nodules	ocean floor, worldwide
Placer deposits	gold	Yukon Territory, Canada
Weathering Processes		
Lateritic weathering	bauxite	Weipa, Australia; Jamaica
Secondary enrichment	copper minerals	Miami, Arizona
Metamorphic Processes		
Contact-metamorphic	tungsten and molybdenum minerals	eastern Sierra Nevada, California
Regional metamorphic	asbestos	Quebec, Canada

◆ FIGURE 13.2 Relationships between metallic ore deposits and tectonic processes.

	Mid-ocean ridge	Oceanic crust	Subduction zone	Volcanic arc basin	Volcanic island arc	Marginal basin	Granitic intrusions (batholiths)
Metals	Copper and zinc as metallic oxides and sulfides	Manganese, cobalt, nickel	Chromium	Copper, lead, zinc	Copper, molybdenum, gold, silver, lead, mercury, tin	Copper, zinc, gold, chromium (petroleum)	Copper, tungsten, tin, iron (Kiruna type), gold, silver, molybdenum
Deposit type	Volcanogenic massive sulfide	Sea-floor nodules	Chromite in serpentine rock (high-pressure, low-temp., metamorphic)	Stratabound	Porphyry copper; hydrothermal veins	Volcanogenic massive sulfide; evaporites	Contact-metamorphic; hydrothermal veins

per deposits of the Western Hemisphere were formed by igneous processes at convergent plate margins at the sites of former volcanic island arcs. On the other hand, deposits of tungsten, tin, some iron ore, gold, silver, and molybdenum originated by hydrothermal activity or by contact metamorphism accompanying the emplacement of granitic plutons (batholiths) within continental lithospheric plates near convergent boundaries.

Igneous Processes

Intrusive Deposits

Pegmatites are small, tabular-shaped, very coarse grained, intrusive igneous bodies that may be important sources of mica, quartz, feldspar, beryllium, lithium, and gemstones. Reserves of pegmatites in the Western Hemisphere occur in the Black Hills of South Dakota; at Dunton, Maine; at Minas Gerais, Brazil; and in the tin-spodumene deposits of North Carolina (◆Figures 13.3 and 13.4).

Crystal settling within a cooling magma chamber of ultramafic composition appears to be responsible for forming layers according to density, with the densest and earliest formed on the bottom and the less-dense and later-to-crystallize above or toward the top (◆Figure 13.5), although alternate ideas for the layering have been proposed. Examples are the layered deposits of chromite at Red Mountain, Alaska, and the Bushveld Complex in the Republic of South Africa, the largest single igneous mass on earth (roughly the size of Maine). Chromium, nickel, vanadium, and platinum are obtained from this source. Some layered nickel-copper-cobalt deposits also are important sources of platinum-group elements.

◆ FIGURE 13.3 Granitic pegmatite, a very coarse grained igneous rock. The grains in this sample have lengths up to 3 centimeters.

◆ FIGURE 13.4 Tourmaline crystals from Dunton, Maine. Tourmaline is commonly found as an accessory mineral in granitic pegmatites.

◆ FIGURE 13.5 Layered mineral deposit; Red Mountain, Alaska. The black layers are chromite ($FeCr_2O_4$); the light layers are granitic rocks.

Disseminated Deposits

Disseminated (scattered) **deposits** occur when a multitude of mineralized veinlets develop near or at the top of a large igneous intrusion (see Chapter 2). Probably the most common of these are the **porphyry copper** deposits (◆Figure 13.6), in which the ore minerals are widely distributed throughout a large volume of granitic rocks that have been emplaced in regions of current or past plate convergence. The world's largest open-pit mine, at Bingham Canyon, Utah, has extracted more than $6 billion worth of copper and associated minerals from a porphyry copper body (◆Figure 13.7).

Hydrothermal Deposits

Hydrothermal (hot water) **deposits** originate from hot, mineral-rich fluids that are squeezed from cooling magma bodies during crystallization. Solidifying crystals force the liquids into cracks, fissures, and pores in both the magmatic rocks and the adjacent, "country" rocks (Figure 13.6). Hydrothermal fluids also may be of metamorphic origin, or they may form from subsurface waters that are heated when they circulate near a cooling magma. Regardless of their origin, the heated waters may dissolve, concentrate, and remove valuable minerals and redeposit them elsewhere. When the fluids cool, minerals crystallize and create hydrothermal **vein deposits,** which miners often refer to as **lodes.** The specific minerals formed will vary with the composition and temperature of the particular hydrothermal fluids, but they commonly include such suites (associations) of elements as gold-quartz and lead-zinc-silver.

Volcanogenic Deposits

When volcanic activity vents fluids to the surface, sometimes associated with ocean-floor hot-spring **black smoker** activity, **volcanogenic deposits** are formed (◆Figure 13.8). These deposits are so-named because they occur in marine sedimentary rocks that are associated with basalt flows or other volcanic rocks, and the ore bodies they produce are called **massive sulfides.** The rich copper deposits of the island of Cyprus, which have supplied all or part of the world's needs of copper for more than 3,000 years, are of this type. (The word *copper* is derived from Latin *Cyprium,* "Cyprian metal.") Cyprus's copper sulfide ore formed millions of years

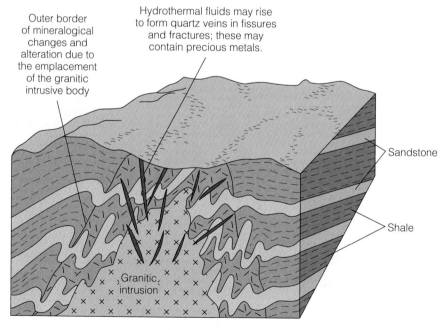

◆ FIGURE 13.6 A disseminated porphyry copper deposit. After "Alaska's Oil/Gas and Minerals," *Alaska Geographic* 9, no. 4, 1982

Disseminated deposits of copper form when mineralized solutions invade permeable zones and small cracks.

◆ FIGURE 13.7 Kennecott Corporation's Bingham Canyon mine south of Salt Lake City, Utah, is the world's largest open-pit mine and largest human-made excavation. By 2002, the pit was more than 800 meters (½ mi) deep and nearly 4 kilometers (2½ mi) wide.

ago adjacent to hydrothermal vents near a sea-floor spreading center. Warping of the copper-rich sea floor due to convergence of the European and African plates brought the deposits to the surface when the island formed.

Sedimentary Processes

Surficial Precipitation

Very large, rich mineral deposits may result from evaporation and direct **precipitation** (the process of a solid forming from a solution) of salts in ocean water, usually in shallow marine basins. Minerals formed this way are called **evaporites.** Evaporites may be grouped into two types: marine evaporites, which are primarily salts of sodium and potassium, gypsum, anhydrite, and bedded phosphates; and nonmarine evaporites, which are mainly calcium and sodium carbonate, nitrate, sulfate, and borate compounds. Other important sedimentary mineral deposits are the **banded-iron formation** ores (◆Figure 13.9). These deposits formed in Precambrian time, more than a billion years ago, when the earth's atmosphere lacked free oxygen. Without free oxygen, the iron that dissolved in surface water could be carried in

solution by rivers from the continents to the oceans, where it precipitated with silica to form immense deposits of red chert and iron ore. Banded-iron deposits are found in the Great Lakes region, northwestern Australia, Brazil, and elsewhere, and they are enormous; they provide two hundred

◆ FIGURE 13.8 A tubular "black smoker" hydrothermal vent in the Pacific Ocean. The "smoke" is hot water saturated with dissolved metallic-sulfide minerals, which precipitate into black particles upon contact with the cold seawater.

◆ FIGURE 13.9 Outcrop of banded-iron formation; northern Michigan.

STEW MONROE

years or more of reserves even without substantial conservation measures.

Deep-Ocean Precipitation

Manganese nodules (◆Figure 13.10) are formed by precipitation on the deep ocean floor. The nodules are mixtures of manganese and iron oxides and hydroxides, with small amounts of cobalt, copper, nickel, and zinc, that grow in onionlike concentric layers by direct precipitation from ocean waters. Any commercial recovery of this resource appears to be many decades away because of technological, economic, international political, and environmental limitations, but eventually it may be necessary to exploit the deposits.

Placer Deposits

Mineral deposits concentrated by moving water are called **placer deposits** (Spanish *placer,* "reef"; pronounced

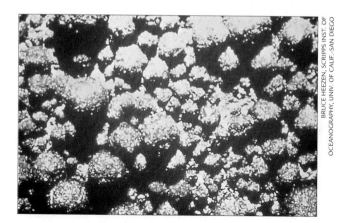

BRUCE HEEZEN, SCRIPPS INST. OF OCEANOGRAPHY, UNIV. OF CALIF.–SAN DIEGO

◆ FIGURE 13.10 Manganese nodules, an example of deep-ocean precipitation. The nodules consist mainly of manganese and iron, with lesser amounts of nickel, copper, cobalt and zinc.

"plass-er"). Dense, erosion-resistant minerals such as gold, platinum, diamonds, and tin are readily concentrated in placers by the washing action of moving water. The less-dense grains of sand and clay are carried away, leaving gold or other heavy minerals concentrated at the bottom of the stream channel. Such deposits formed by the action of rivers are referred to as **alluvial placers;** ancient river deposits that are now elevated as stream terraces above the modern channels are called **bench placers** (◆Figures 13.11 and 13.12 and ✿ Case Study 13.1 on page 373). Considerable uranium is mined from Precambrian placer deposits in southern Africa. The deposits formed under the unique conditions existing on the earth's surface before free oxygen was present in the atmosphere. Some placer deposits occur on beaches, the classic example being the gold beach placers at Nome, Alaska.

Sand and gravel concentrated by rivers into alluvial deposits are important sources of aggregate for concrete. River deposits formed by **glacial outwash**—that is, resulting from the runoff of glacial meltwater—are another type of placer deposit.

CONSIDER THIS *How can a prospector use placer mineral deposits as a guide to finding ore deposits in bedrock?*

Weathering Processes

The deep chemical weathering of rock in hot, humid, tropical climates promotes mineral enrichment, because the solution and removal of more soluble materials leaves a residual soil of less-soluble minerals. Because iron and aluminum are relatively insoluble under these conditions, they tend to remain behind in *laterite,* a highly weathered, red subsoil or material that is rich in oxides of iron and aluminum and

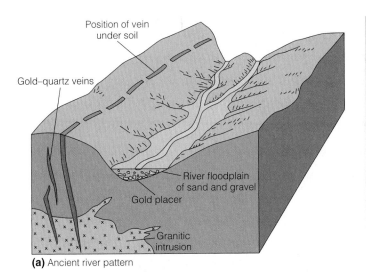

Position of vein under soil

Gold–quartz veins

River floodplain of sand and gravel

Gold placer

Granitic intrusion

(a) Ancient river pattern

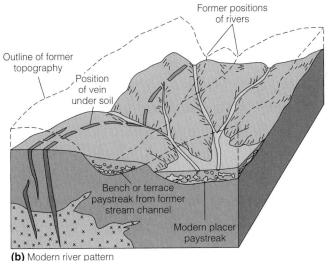

Former positions of rivers

Outline of former topography

Position of vein under soil

Bench or terrace paystreak from former stream channel

Modern placer paystreak

(b) Modern river pattern

◆ FIGURE 13.11 The origin of gold placer deposits. (a) An ancient landscape with an eroding gold-quartz vein shedding small amounts of gold and other mineral grains that eventually become stream sediments. The gold particles, being heavier, settle to the bottom of the sediments in the channel. (b) The same region in modern time. Streams have eroded and changed the landscape and now follow new courses. The original placer deposits of the ancient stream channels are now elevated above the modern river valley as "bench" placers. After "Alaska's Oil/Gas and Minerals," *Alaska Geographic* 9, no. 4, 1982

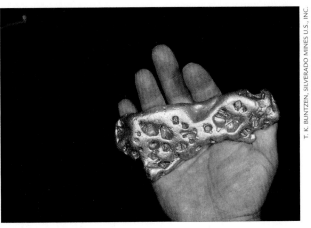

◆ FIGURE 13.12 Gold nugget weighing 41.3 troy ounces, mined from a placer deposit in the Brooks Range, Alaska.

less than one percent copper, but secondary enrichment has improved the grade to more than three percent.

Metamorphic Processes

The high temperatures, high pressures, and ion-rich fluids that accompany the emplacement of intrusive igneous rocks produce a distinct metamorphic halo around the intrusive body. The result, in concert with the accompanying hydrothermal mineralization, is known as a **contact-metamorphic deposit.** If, for example, granite intrudes limestone, a diverse and colorful group of contact-metamorphic minerals may be produced,

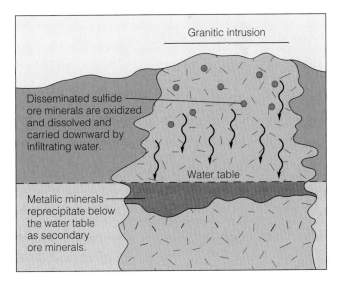

Granitic intrusion

Disseminated sulfide ore minerals are oxidized and dissolved and carried downward by infiltrating water.

Water table

Metallic minerals reprecipitate below the water table as secondary ore minerals.

◆ FIGURE 13.13 The development of ores by secondary enrichment. Descending ground water oxidizes and dissolves soluble sulfide minerals, carrying them downward and leaving a brightly colored residue of limonite and hematite. Below the water table new metallic minerals are precipitated as secondary ore minerals.

lacking in silicates. When the iron content of the parent rock is low or absent, however, this lateritic weathering produces rich deposits of **bauxite,** the principal ore mineral of aluminum (see Chapter 6, Figure 6.16).

Ground water moving downward through a disseminated sulfide deposit may dissolve the dispersed metals from above the water table to produce an enriched deposit below the water table by **secondary enrichment** (◆Figure 13.13). At Miami, Arizona, for example, the primary disseminated copper-ore body is of marginal to submarginal grade, containing

◆ FIGURE 13.14 The relationship of limestone and a granitic intrusion in the origin of contact-metamorphic mineral deposits. After "Alaska's Oil/Gas and Minerals," *Alaska Geographic* 9, no. 4, 1982

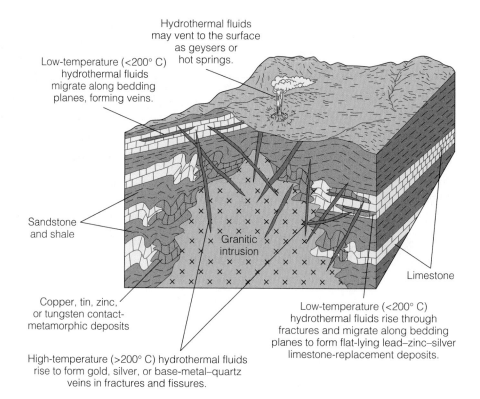

Hydrothermal fluids may vent to the surface as geysers or hot springs.

Low-temperature (<200° C) hydrothermal fluids migrate along bedding planes, forming veins.

Sandstone and shale

Granitic intrusion

Limestone

Copper, tin, zinc, or tungsten contact-metamorphic deposits

Low-temperature (<200° C) hydrothermal fluids rise through fractures and migrate along bedding planes to form flat-lying lead–zinc–silver limestone-replacement deposits.

High-temperature (>200° C) hydrothermal fluids rise to form gold, silver, or base-metal–quartz veins in fractures and fissures.

such as a tungsten-molybdenum deposit (◆Figure 13.14). Asbestos and talc originate by **regional metamorphism,** metamorphism that affects an entire region.

Mineral Reserves

Metallic Mineral Reserves

The metallic minerals are often grouped on the basis of their relative crustal abundance.

The Abundant Metals

The geochemically abundant metals—aluminum, iron, magnesium, manganese, and titanium—have abundances in excess of 0.1 percent by weight of the average continental crust. Economically valuable ore bodies of the abundant metals—such as iron and aluminum, for example—need only comparatively small concentration factors for profitable mining (Table 13.2) and are recovered mainly from ore minerals that are oxides and hydroxides. Even though they are abundant, these metals require large amounts of energy for production. Because of this, it is not surprising that the world's industrialized nations are the greatest consumers of these metals.

The Scarce Metals

The scarce metals are those whose crustal abundances are less than 0.1 percent by weight of average continental crust. Most of these metals are concentrated in sulfide deposits, but the rarest occur more commonly, or even solely, in rare rock types. Included within this category are copper, lead, zinc, and nickel—which are widely known and used, but nevertheless scarce—along with the more obvious gold and platinum-group minerals. U.S. deposits of chromium, manganese, nickel, tin, and platinum-group metals are scarce or of submarginal grade. Important, less-well-known scarce metals include cobalt, columbium, tantalum, and cadmium, which are important in space-age ferroalloys; antimony, which is used as a pigment and as a fire retardant; and gallium, which is essential in the manufacture of solid-state electronic components. The United States has all of these critical metals in the National Defense Stockpile, a supply of about 100 critical mineral materials valued at about $4 billion that could supply the nation's needs for three years in a national emergency.

Nonmetallic Mineral Reserves

The yearly per capita consumption of nonfuel mineral resources in North America exceeds 10 tons, 94 percent of which is nonmetallic, with the major bulk being construction materials (Figure 13.1). The nonmetallic minerals are subdivided for convenience into three groups based on their use in industry, agriculture, and construction. The number and diversity of nonmetallic mineral resources is so great that it is impossible to treat the entire subject in a book of this sort. Thus, this subsection discusses only those non-metallic resources generally considered most important.

Industrial Minerals

Industrial minerals include those that contain specific elements or compounds that are used in the chemical industry—sulfur and halite, for example—and those that have important physical properties—such as the materials used in ceramics manufacture and the abrasives, such as diamond and corundum. Sulfur, one of the most widely used and most important industrial chemicals, is obtained as a by-product of petroleum refining (◆Figure 13.15) and from the tops of salt domes in the central Gulf Coast states, where it is associated with anhydrite ($CaSO_4$) or gypsum ($CaSO_4 \cdot 2H_2O$). Salt domes originate from evaporite beds that are compressed by overlying sedimentary rock layers, causing the salt to be mobilized into a dome, which rises, piercing the overlying rocks and concentrating sulfur, gypsum, and anhydrite at the top. Superheated water is pumped into the sulfur-bearing caprock to melt the sulfur, which is then piped to a processing plant. More than 80 percent of domestic U.S. sulfur is used to make sulfuric acid, the most important use for which is the manufacture of phosphate fertilizers. Salt domes are also important sources of nearly pure halite ($NaCl$), common salt, and many domes have oil fields on their flanks (see Chapter 14).

In addition to being mined from salt domes, halite is mined from flat-lying evaporite beds in New Mexico, Kansas, and Michigan. The most important use of halite is for manufacturing sodium hydroxide and chlorine gas; about 15 percent of halite production is used for de-icing highways, and significant amounts are used in food products and in the manufacture of steel and aluminum.

Diamond is by far the most important industrial abrasive, although corundum, emery (an impure variety of corundum), and garnet also are used. Traditionally, natural diamonds have been important both as gemstones and as industrial abrasives, but artificial diamonds manufactured from graphite now hold about 70 percent of the market for diamond-grit abrasives. The major source of natural diamonds is kimberlite (named for Kimberley, South Africa), a rare and unique ultramafic rock that forms in pipelike bodies. Not only are kimberlite pipes rare, but most of them lack diamonds. The important diamond-bearing kimberlites are in South Africa, Ghana, the Republic of Zaire, Botswana, Australia, Belarus (formerly Belorussia of the USSR), Yakutia (the largest republic of the Russian Federation), and in Canada's Northwest Territories—where North America's first diamond mine began operations in 1998.

The metamorphic minerals corundum and emery, natural abrasives for which U.S. reserves are scarce or nonexistent, have largely been replaced by synthetic abrasives such as carbides, manufactured alumina, and nitrides. The United States still imports limited amounts of corundum from South Africa.

Agricultural Minerals

With the world's human population doubling every 40 years and the global population passing the 6 billion mark in September 1999, it seems obvious that food production and the corresponding need for fertilizers and agricultural chemicals also will continue to expand. Therefore, nitrate, potassium, and phosphate compounds will continue to be in great demand. Whereas nearly all agricultural nitrate is derived from the atmosphere, phosphate and potassium come only from the crust of the earth. Phosphate reserves occur in many parts of the world, but the major sources are in the United States, Morocco, Turkmenistan, and Buryat (the latter two are republics of the former Soviet Union). The primary sources of phosphate in the United States are marine

◆ FIGURE 13.15 Sulfur awaiting shipment to Asia at Vancouver, British Columbia. The sulfur originated as a contaminant in natural gas produced from gas fields in Alberta. Processing plants have separated the sulfur from the natural gas in order to produce a gas whose combustion products meet environmental air-quality standards. The sulfur is a useful by-product of the process.

sedimentary rocks in North Carolina and Florida (see Figure 13.28, later in the chapter), and there are other valuable deposits in Idaho, Montana, Wyoming, and Utah. The main U.S. supply of potassium comes from widespread nonmarine evaporite beds beneath New Mexico, Oklahoma, Kansas, and Texas, with the richest beds being in New Mexico. Canada also has large reserves of potassium salts.

Construction Materials

About two-thirds of the value of industrial minerals mined in the United States in 2001 were construction materials, with aggregates (crushed stone, sand, and gravel) valued at $13.5 billion and cement (made from limestone and shale) valued at $6.5 billion. Eighty percent of the aggregate was used in road building. Other major construction materials are clay, for bricks and tile, and gypsum, the primary component of plaster and wallboard. It is the mining of aggregate that is probably the most familiar to urban dwellers, because quarries customarily are sited near cities, the major market for the product, in order to minimize transportation costs. The quantity of aggregate needed to build a 1,500-square-foot house is impressive: 67 cubic yards are required, each cubic yard consisting of one ton of rock and gravel and 0.7 ton of sand. This amounts to 114 tons of rock, sand, and gravel per dwelling, including the garage, sidewalks, curbings, and gutters. Thus, it is no wonder that aggregate has the greatest commercial value of all mineral products mined in most states, ranking second only in those states that produce natural gas and petroleum. The annual per capita production of sand, gravel, and crushed stone in the United States amounts to about six tons, the total value of which exceeded $13 billion in 1998. The principal sources of aggregate are open-pit quarries in modern and ancient floodplains, river channels, and alluvial fans. In areas of former glaciation, aggregate is mined from glacial outwash and other deposits of sand and gravel that remained after the retreat of the great Pleistocene ice sheets in the northern United States, Canada, and Northern Europe (see Chapter 11).

> **CONSIDER THIS** *How does the nation's need for mineral materials affect U.S. foreign policy?*

Mineral Resources for the Future

Just as the world's fossil-fuel reserves are limited (Chapter 14), so are its mineral reserves. The earth has been well-explored, and fairly reliable estimates exist on the reserves of most important mineral resources. By dividing the known reserves of a resource by its projected rate of use, it is possible to calculate the number of years remaining until the resource is depleted. It was determined in the 1970s that reserves of many important elements would be exhausted by some time early in the twenty-first century. Although more recent studies have found that the situation is not that serious, a few experts still view the exhaustion of some mineral deposits in the near future as likely.

U.S. consumption and production data for selected metallic and nonmetallic mineral resources appear in ✳ Tables 13.5 and 13.6, respectively. Notice in Table 13.5 that we consume greater amounts of many metallic mineral commodities than we produce. In the cases of chromium, cobalt, manganese, nickel, tin, and platinum-group metals, all or

✳TABLE 13.5	U.S. Consumption and Production of Selected Metallic Minerals, 1995		
Mineral	**Consumption, thousands of metric tons**	**Primary Production, thousands of metric tons***	**Primary Production as Percentage of Consumption****
Aluminum	7,300	3,300	45.2
Chromium	387	0	0
Cobalt	7.2	0	0
Copper	2,800	1,840	65.7
Iron ore	69,300	57,000	82.3
Lead	1,500	365	24.3
Manganese	695	0	0
Nickel	131	0	0
Tin	46	negligible	0
Zinc	1,220	540	44.3
Platinum-group metals, metric tons	127	8.3	16.3

*Primary production is production from new ore.
**Primary production divided by consumption and rounded to the appropriate number of significant figures.
Source: *Mineral Commodity Summaries*, U.S. Bureau of Mines, 1995.

∗TABLE 13.6 U.S. Consumption and Production of Selected Nonmetallic Minerals, 1995

Mineral	Consumption, thousands of metric tons	Primary Production, thousands of metric tons*	Primary Production as Percentage of Consumption**
Clay	37,900	42,300	112
Gypsum	26,000	17,300	66.5
Potash	5,390	1,425	26.4
Salt	47,400	39,500	83.3
Sulfur	13,450	11,300	84
Sand and gravel	922,000	922,000	100
Crushed stone	1,198,000	1,195,000	95.7

*Primary production is production from new ore.
**Primary production divided by consumption and rounded to the appropriate number of significant figures.
Source: *Mineral Commodity Summaries*, U.S. Bureau of Mines, 1995.

virtually all of those materials must be imported. The extent of U.S. import reliance for these and other mineral resources is shown in ◆Figure 13.16. Thus, the United States is far from self-sustaining. By the turn of the twenty-first century, the United States imported an estimated $60 billion worth of raw and processed mineral materials annually. If nothing else, the data constitute a clear case against isolationism if we hope to sustain our present lifestyle.

A brighter forecast is shown by the data in Table 13.6. Note in the table that the United States was self-sustaining in its consumption of domestic nonmetallic and rock resources as of 1995 and that it was producing a surplus of one mineral

commodity. There are some apparent disagreements between these tables and Figure 13.16; the low need for imported lead reported in Figure 13.16, for example, does not seem to agree with the value shown in Table 13.5. The discrepancy results from extensive recycling of lead in the United States. Further, the high domestic production of aluminum reported in Table 13.5 results from a high reliance on imported bauxite (reported in Figure 13.16), because domestic sources of bauxite are insignificant.

As resources diminish in the coming decades, we can expect shortages of some minerals that presently are common. In many cases technological advances will develop

Mineral Resources	Important Sources, 1996–1997	Percent
Graphite	China, Mexico, Canada	100
Manganese	South Africa, Mexico, Australia, Gabon	100
Strontium (celestite)	Mexico, Germany	100
Bauxite and alumina	Australia, Guinea, Jamaica, Brazil	100
Platinum-group metals	South Africa, U.K., Russia, Germany	93
Potash	Canada, Russia, Belarus, Germany	80
Tantalum	Japan, Australia, Thailand, China	80
Tin	Brazil, China, Indonesia, Bolivia, Peru	79
Cobalt	Zambia, Canada, Norway, Finland	75
Tungsten	China, Russia	70
Chromium	South Africa, Russia, Turkey, Zimbabwe, Kazakhstan	63
Zinc	Canada, Mexico, Kazakhstan	60
Nickel	Canada, Norway, Australia, Russia	43
Antimony	China, Mexico, South Africa, Belgium	41
Magnesium compounds	China, Canada, Austria, Australia	35
Diamond (industrial)	Ireland, China, Ukraine	27
Lead	Canada, Mexico, Peru, Australia	18
Iron and steel	European Union (EU), Japan, Canada, Mexico	14
Iron ore	Canada, Brazil, Venezuela, Australia	11
Cadmium	Canada, Australia, Belgium	6

◆ FIGURE 13.16 U.S. net import figures for selected nonfuel mineral materials during 2002. (L. McCarten, D. E. Morse, S. F. Sibley, and P. A. Plunkert, U.S. Geological Survey. 2003. The United States 2002 Annual Review. *Mining Engineering* 55, no. 5, p. 23.)

substitute materials that will suffice, but in other cases, we will have to do without. Some minerals will become sufficiently scarce that deposits now classed as "submarginal" will be upgraded to "low-grade" and will be economically exploitable due to market forces.

The need for mineral resources may require exploration and exploitation in more remote regions and more hostile environments. Coal has been mined for more than a half-century well north of the Arctic Circle on Norway's island of Spitsbergen (Svalbard), for example, and sources of fuels and minerals are currently being sought and exploited in the harsh climates of Canada's Northwest Territories and Arctic Islands (◆ Figure 13.17). Exploration by drilling cores has been done at a lead-zinc-silver prospect in western North Greenland. The Red Dog zinc-lead-silver mine, North America's largest zinc producer, on the lands of the Inupiat Eskimos north of the Arctic Circle in Alaska went into full production in 1991. Even environmentally sensitive Antarctica, which is known to have extensive deposits of coal and iron, may be the target of future exploitation as more is learned about that continent's mineral wealth. Deeper parts of the continents and the sea floor may also be targeted for exploration and exploitation as geophysical innovations and technological advances open up these presently inaccessible regions.

In the future we will experience more recycling of critical metals, such as has been done with gold for thousands of years. The gold in the crown on your tooth, for example, may have been in jewelry during the age of Pericles in Ancient Greece. Such metals as lead, aluminum, iron, and copper have been recycled for years. In many cities the large-scale recycling of metals, especially lead and aluminum, as well as

of paper, glass, and plastics, is being conducted successfully as part of municipal trash collection (see Chapter 15). Significant amounts of several metals consumed in the United States are now being obtained from recycled scrap. In 2000 these amounts were as follows:

Metal	Recycled (from old and new scrap (metric tons)	Amount Obtained by Recycling
aluminum	3,450,000	36%
copper	1,310,000	32%
iron and steel	74,000,000	55%
lead	1,120,000	63%
tin	15,100	29%
titanium	18,500	50%
zinc	436,000	27%

The recycling rates in 2000 for steel-containing materials were 95 percent for automobiles that had reached the end of their useful life, 84 percent for appliances, and 58.4 percent for steel cans; 62 percent of all aluminum beverage cans were recycled. Collectively, the value of recycled metal and mineral scrap amounted to about 30 percent of the nation's mineral production.

Mining and Its Environmental Impacts

. . . when the ores are washed, the water which has been used poisons the brooks and streams, and either destroys the fish or drives them away.

Georgius Agricola, *De Re Metallica*, 1550

The General Mining Law of 1872 was important legislation that helped establish the mining industry as a fundamental element in the U.S. economy. At the same time, the law was the ultimate in laissez-faire regulation, as it allowed miners to exploit, to take profits at public expense, to pay little in return for the privilege of mining on federal land, and to walk away from the scars and waste materials remaining when they abandoned their mines. The original nineteenth-century law allowed miners to stake a claim on potentially profitable public land and, should the claim prove to contain valuable minerals, to obtain a patent (legal title) for $2.50 an acre, and reap the profits. This law helped "win" the West by enticing thousands of prospectors to seek their fortunes in gold, silver, and other valuable commodities.

Since 1872 the law has allowed approximately $245 billion worth of the public's hardrock minerals to be wholesaled to private corporations with no consideration whatever to the environmental consequences of mining. The law

◆ FIGURE 13.17 The Ekati Diamond Mine™, Canada's first diamond mine, located in a remote Arctic tundra region known as the Barrenlands, is about 300 kilometers (180 mi) northeast of Yellowknife, Northwest Territories. The mine, with a potential life of about 25 years, is developed in kimberlite pipes similar to the famous diamond pipes of South Africa. Development of the property includes consultation with the leadership of the Aboriginal residents of the region and minimizing the impact on the environment. Remediation of disturbed lands is continually ongoing.

is staunchly defended today by some public officials and the mining industry who claim that it does a good job of providing the country with valuable minerals, employment, and economic development. The same law exists today, but it has been much modified over the years by more than 50 amendments, so that today's body of mining law fills six volumes. Two examples of the changes are (1) the Mineral Lands Leasing Act of 1920, by which the government retains title to all federal lands possessing energy resources (oil, gas, and coal) and assesses royalties on the profits from the lands leased by developers; and (2) beginning in 1992, a claim maintenance fee of $100 per claim per year that must be paid on claims staked under the 1872 law in order for the claims to remain valid. (This is the only return taxpayers receive for hardrock minerals taken from public lands, which amounts to $25 to $30 million per year, and the funds collected are used to enforce surface mining regulations.) In addition, environmental restrictions added in the 1970s and 1980s dictate that modern mining must be conducted within a different policy framework than the simply drafted law of 1872. Although interest in environmental protection and reclamation in coal mining districts began to develop in the 1930s, serious efforts at regulation and reclamation did not begin until the late 1960s due to the increasing amounts of unreclaimed land, and water pollution and problems arising from the lack of uniform standards among state mining programs. Consequently, the Surface Mining Control and Reclamation Act (**SMCRA**) was passed in 1977. This act established coordination of federal and state efforts to regulate the coal industry in order to prevent the abuses that had prevailed in the past. SMCRA regulations apply to all lands—private, state, and federal. SMCRA established a special fee that is charged on coal production and generates funds for reclaiming abandoned coal-mine lands. SMCRA also provides for state primacy once a state's coal mine reclamation laws and policies have been approved by the Secretary of the Interior. This gives state regulators primary SMCRA enforcement jurisdiction as long as the state programs meet federal standards.

Federal reclamation standards and other safeguards established by SMCRA for the coal-mining industry have no counterpart for noncoal underground and surface mineral mining on state or private lands. Although there are federal regulations for noncoal mining on federal lands, it is left to the individual state governments to codify safeguards and enforce noncoal mine reclamation requirements.

Mining on federal lands must comply with the regulations of the *National Environmental Policy Act* (NEPA) of 1969 (see Chapter 10). A mining company must file an operating plan and an *Environmental Impact Report* (EIR), and conduct public hearings before the administrating agency can approve an operating permit. Even though an applicant may spend several thousand dollars in complying with NEPA permit application requirements, there is no assurance that an application will be approved.

There is no specific fund for cleaning up abandoned noncoal mines on state and private lands, nor are there uniform regulatory requirements for these lands. Action is left to the states and the mining companies. However, once any state with coal production has implemented safeguards and standards for coal mining, it may use the SMCRA abandoned mine fund to address reclamation of abandoned mines that exploited other commodities. For example, Utah's abandoned mine reclamation program uses SMCRA funds to reclaim metals mines.

Environmental pollution abatement has become a major concern of domestic gold producers in some states, especially in Alaska, where there are many small placer mines. Reclamation of mined land has become an integral part of an increasing number of gold mine operating plans. Abandoned mine sites, as well as other contaminated industrial sites that are deemed to pose serious threats to health or safety, may be cleaned up under the Comprehensive Environmental Response, Compensation and Liability Act (**CERCLA**) of 1980, which is overseen by the *Environmental Protection Agency* (EPA). For example, the Forest Service, a branch of the United States Department of Agriculture, has CERCLA authority to clean up contaminated mines on National Forest System lands. It was CERCLA that established the so-called *Superfund* to clean up the worst toxic or hazardous waste sites that have been assigned to the National Priorities List (see Chapter 15).

Some states have enacted *State Environmental Policy Acts* (SEPAs) that are patterned after NEPA, while other states may have little regulation. The differences in the various states' permit procedures, their mining and reclamation standards, and their enforcement of underground and surface mining laws can be illustrated by comparing the laws of three states, each noted for over a century of mining activity: Colorado, Arizona, and Michigan. Colorado has strong regulatory provisions relating to mine operation, environmental protection, reclamation bonding, inspections, emergency response by the state, and even the authority over mine operation. In contrast, Arizona, the only state in the country without a mine-reclamation law until 1994, allows companies to self-bond their reclamation fund, to wait as long as 17 years after mining has stopped to begin reclamation, and to set their own reclamation standards. Furthermore, there is an escape clause that may allow companies a variance from any of the law's provisions. Michigan, which ranks sixth among all states in the production value of nonfuel minerals, has no comprehensive mining law. No financial bond is required to ensure that a mine site will be properly closed and reclaimed if the company abandons the project or declares bankruptcy. No permit is needed to open a mine except a local zoning permit and a federal water-discharge permit—permits that are required of *every* industrial operation. Reportedly, there is no enforcement of the state's voluntary reclamation law.

As we consider the environmental impacts of mining in the United States in this section, it is important to distinguish

◆ FIGURE 13.18 A large underground mining operation, the Barrick Goldstrike Meikle Mine on the Carlin Trend near Elko, Nevada. The mine, with 712,688 ounces of gold produced in 2001, had the highest production of any gold mine in the United States.

between (1) the legacy of careless exploitation that is obvious at long-abandoned mines with their residual scars and toxic wastes and (2) the current mining scene, in which, depending upon the particular state and whether the site is on state-owned land, land administered by the U.S. Forest Service or the Bureau of Land Management, more than 30 different permits may by needed in order to develop and operate a mine legally. Many mine owners consider the current permitting requirements to be excessive, and they question whether they will be able to continue to operate in the United States at all if the regulations become more restrictive.

Mining in the past was conducted largely by underground methods with a surface plant for milling and processing and for hoisting workers, ore, and equipment

(◆Figure 13.18). Surface mining methods, typified by vast open-pit excavations such as the Bingham Canyon mine (see Figure 13.7), have now largely replaced underground methods. The surface plant remains much the same, however, although it usually operates on a much larger scale than in the past. Both surface and underground mining create significant environmental impacts on the land and air and on biological and water resources. In addition, the needs for housing and services in mining areas have social impacts.

Impacts of Coal Mining

Much coal now mined in the United States is extracted by surface operations, which involve removing rock and soil that overlie the coal beds. Compared to underground mining (◆Figure 13.19), surface mining is generally less expensive, safer for miners, and facilitates more complete recovery of coal. Surface mining, however, causes more extensive disturbances to the land surface and has the potential for serious environmental consequences unless the land is carefully reclaimed. The three major methods of surface mining of coal are contour mining, area mining, and mountaintop mining. Each of these is explained here.

Contour mining is typical in the hilly areas of the eastern United States, where coal beds occur in outcrops along hillsides. Mining is accomplished by cutting into the hillside to expose the coal and then following the coal seam around the perimeter of the hill. Successive, roughly horizontal strips are cut, enlarging each strip around the hillside until the thickness of the overburden is so great that further exposure of the coal bed would not be cost-effective. At each level of mining is a *highwall*, a clifflike, excavated face of exposed overburden and coal that remains after mining is completed. Augers (giant drill bits) are used to extract coal from the areas beneath the highwall. The SMCRA requires that the sites must be reclaimed—that is, returned to the original

◆ FIGURE 13.19 The steps in conventional large-scale underground bituminous-coal mining.

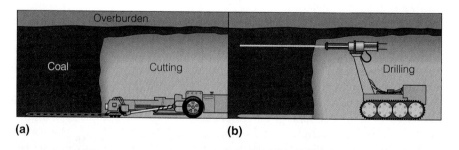

(a)   (b)

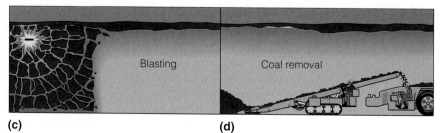

(c)   (d)

◆ FIGURE 13.20(a) A massive dragline (on the right) looks like a child's toy when compared to the huge scale of a mountaintop removal coal mining operation near Kayford Mountain, West Virginia.

contour, and highwalls must be covered and stabilized after contour mining is completed. This is accomplished by back-filling **spoil,** the broken fragments of waste rock removed in order to mine the coal, against the highwall, and spreading and compacting as necessary to stabilize the reclaimed hill-side. *Hydroseeding,* spraying seeds mixed with water, mulch, fertilizer, and lime onto regraded soil, is used on steep reclaimed slopes to aid in establishing vegetation that will help prevent soil erosion.

Area mining is commonly used to mine coal in flat and gently rolling terrain, principally in the Midwestern and Western states (◆ Figure 13.20). Because the pits of active area mines may be several kilometers long, enormous equipment is used to remove the overburden, mine the coal, and reclaim the land. Topsoil is stockpiled in special areas and put back in place when the mining is completed. After replacing it, it is tilled with traditional farming methods to reestablish it as

crop or pasture land. Often, the land is more productive after reclamation than before mining.

Mountaintop removal mining is used primarily in the eastern United States to recover coal that underlies the tops of mountains. After the coal is removed, the mined area is returned to its approximately original shape or left as flat ter-rain (◆ Figure 13.20a). Mountaintop removal mining involves removal and disposal of all rock and soil materials above a coal bed and allows nearly complete removal of the coal bed. A major problem with mountaintop removal min-ing is disposal of the waste rock, which includes carcinogenic chemicals used to wash the coal for market and toxic metals such as mercury and arsenic. Commonly, the waste is simply dumped in valleys, and in West Virginia and Kentucky many miles of streams have been polluted, ecosystems have been destroyed, whole valleys simply eliminated, and residents have been subjected to health hazards (◆ Figure 13.20b). In

◆ FIGURE 13.20(b) Waste from mountaintop removal coal mining forms an enormous valley fill in the Birchton Curve Valley, West Virginia. Such valley fills have buried many miles of streams, eliminating aquatic vegetation, fish and other wildlife habitat, as well as entire communi-ties and their residents who have lived for generations along now-buried valleys in West Virginia.

some instances, whole communities and the people that have lived for generations along the now-nonexistent stream banks are displaced with little or no financial compensation. Consequently, legal action charged that waste disposal practices in mountaintop removal coal mining violates the Clean Water Act and various mining regulations that prohibit unnecessary and undue degradation of the public's land and water. To counter this charge, a final clarification rule was issued by the federal government in 2002 that a company can apply to the Corps of Engineers to put virtually anything labeled as fill into U.S. waters (streams, rivers, and lakes) so long as it either turns a portion or all of a water body into dry land or changes the bottom elevation of U.S. waters. This final rule is astounding because it redefines fill and appears to violate both the stated goals of the Clean Water Act and mining regulations. Furthermore, the final rule applies broadly to all mining and industrial wastes including those that are hazardous to the environment and human health, and it may apply to many sites that would qualify as Superfund sites. A federal judge has reminded the administration that only Congress can rewrite federal law, and the issue is progressing through the courts.

Impacts of Underground Mining

The most far-reaching effects of underground mining are ground subsidence (Chapter 7), the collapse of the overburden into mined-out areas (◆Figure 13.21), and **acid mine drainage (AMD),** the drainage of acidic water from mine sites. AMD is generally considered to be the most serious environmental problem facing the mining industry today, as some acid drainage may continue for hundreds of years. AMD can result when high-sulfur coal or metallic-sulfide ore bodies are mined. In both cases, pyrite (FeS_2) and other metallic-sulfide minerals are prevalent in the walls of underground mines and open pits; in the **tailings,** finely ground,

sand-sized waste material from the milling process that remains after the desired minerals have been extracted; and in other mine waste. The reaction of pyrite or other sulfide minerals with oxygen-rich water produces sulfur dioxide (SO_2), which, facilitated by bacterial decomposition, reacts with water to form sulfuric acid (Figure 13.22 and ◌ Case Study 13.4 on page 377).

This not only acidifies surface and underground waters, but it also expedites the leaching, release, and dispersal of iron, zinc, copper, and other toxic metals into the environment. Such substances kill aquatic life and erode human-made structures such as concrete drains, bridge piers, sewer pipes, and well casings. (Zinc concentrations as low as 0.06 mg/L and copper concentrations as low as 0.0015 mg/L are lethal to some species of fish.) Estimates are that between 8,000 and 16,000 kilometers (5,000 and 10,000 mi) of U.S. streams have been ruined by acid drainage. The oxidized iron in AMD colors some polluted streams rust-red (◆Figure 13.23). Where polluted water is used for livestock or irrigation water, it may diminish the productive value of the affected land. Although expensive to implement, technologies do exist for preventing and controlling AMD (see ◌ Case Study 13.3 on page 375). The subject of surface- and ground-water protection is addressed later in the chapter.

> **CONSIDER THIS** *Pyrite (fool's gold) is common in deposits of gold, lead, zinc, silver, and copper. Why is a conspicuous rusty-colored outcropping often a clue to the existence of buried ore?*

Impacts of Surface Mining

A variety of surface-mining operations contribute to environmental problems. Hydraulic mining is largely an activity of the past, but **dredging**—scooping up earth material below a body of water from a barge or raft equipped to process or

◆ FIGURE 13.21 Store building subsiding into an abandoned underground coal mine south of Pittsburgh, Pennsylvania.

D. D. TRENT

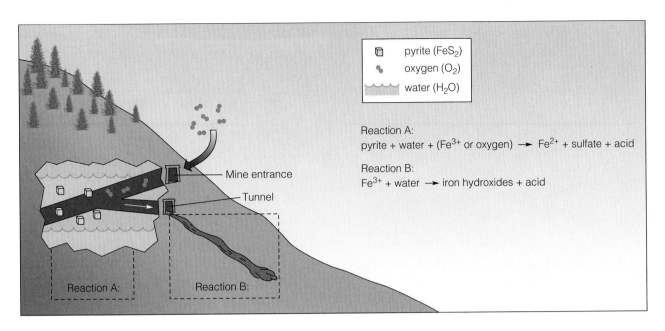

◆ **FIGURE 13.22** Pyrite reacts with air and water to produce sulfuric acid. Based on USGS data

transport materials—continues in the twenty-first century (◆Figure 13.24). Much of the current U.S. production of sand and gravel is accomplished by dredging rivers, and large dredges are used to mine placer tin deposits in Southeast Asia. Dredging causes significant disruption to the landscape; it washes away soil, leaves a trail of boulders, and severely damages biological systems. Scarification remains from former dredging in river bottoms in many Western states and elsewhere in the world (◆Figure 13.25). Few gold-dredge areas have been reclaimed, but an outstanding example of what can be accomplished appears in ◆Figure 13.26. Another example is a 73-square-kilometer (28-mi²) dredge-

scarified area near Folsom, California. The area was reclaimed by reshaping the waste piles, covering the surface with topsoil, and then building a subdivision of homes on it. Landscaped with trees and grasses, no evidence remains of the once scarified surface.

In **hydraulic mining,** or *hydraulicking,* a high-pressure jet of water is blasted through a nozzle, called a *monitor,* against hillsides of ancient alluvial deposits (Case Study 13.1). Hydraulicking requires the construction of ditches, reservoirs, a *penstock* (vertical pipe), and pipelines. Once constructed, it is possible to wash thousands of cubic yards of gold-bearing gravel per day. The gold is recovered in sluice boxes, where

◆ **FIGURE 13.23** Rust-red sludge of acid mine drainage from West Virginia coal mines that were abandoned in the 1960s. The acidic water has eaten away at the Portland cement–based concrete retaining wall, and the bridge supports will soon share that fate. Problems such as this are rarely caused by modern surface mines, because current mining and reclamation practices eliminate or minimize acid mine drainage.

◆ **FIGURE 13.24** Placer mining with a bucket-conveyer dredge near Platinum, Alaska, 1958. The dredge operated from the early 1930s until the late 1970s, during which time it was the major producer of platinum-group metals in North America.

◆ FIGURE 13.25 Scarified landscape from waste-rock remaining after dredging for gold along the Yuba River, California.

◆ FIGURE 13.26 Reclaimed dredge-mined site, a former placer gold mine; Fox Creek, Fairbanks District, Alaska. The State of Alaska awarded a certificate of commendation for the reclamation work.

mercury (the liquid metal that dissolves gold and silver to form an amalgam; it's explained in Case Study 13.1) usually is added to aid recovery at a low cost. Hydraulicking is efficient but highly destructive to the land (◆Figure 13.27). Because nineteenth-century hydraulicking was found to create river sediment that increased downstream flooding, clogged irrigation systems, and ruined farmlands, court injunctions stopped most hydraulic mining in the United States before 1900 (although its use continued in California's Klamath-Trinity Mountains until the 1950s). Hydraulicking is still being used in Russia's Baltic region for mining amber, and hydraulicking and mercury amalgamation is used to extract placer gold in remote areas of Brazil's Amazon basin. It is estimated that 100 tons of mercury is working its way into the ecosystem of the Amazon basin each year. The danger of using mercury is that when it escapes from mining activity it takes on different geochemical forms—including a suite of organic compounds, the most important being methylmercury, which is readily incorporated into biological tissues and is the most toxic to humans. In California's Mother Lode region, where hundreds of hydraulic placer-gold mines operated from the 1860s through the early 1900s, the mercury loss to the environment from these operations is estimated to have been 3 to 8 million pounds, and it still remains in the ecosystem.

Strip-mining is used most commonly when the resource lies parallel and close to the surface. The phosphate deposits of North Carolina and Florida are strip-mined by excavating the shallow, horizontal beds to a depth of about 8 meters (26 ft) (◆Figure 13.28). After the phosphate beds are removed, the excavated area is backfilled to return the surface to its original form.

Open-pit mining is the only practical way to extract many minerals when they occur in a very large low-grade-ore body near the surface. The process requires processing enormous amounts of material and is devastating to the landscape.

The epitome of open-pit mining is the Bingham Canyon copper mine in Utah, where about 3.3 billion tons of material— seven times the volume moved in constructing the Panama Canal—have been removed since 1906 (Figure 13.7). Now a half-mile deep, the pit is the largest human excavation in the world, and the waste-rock piles literally form mountains.

The environmental consequences of open-pit mining are several. The mine itself disrupts the landscape, and the increased surface area of the broken and crushed rocks from mining and milling sets the stage for erosion and the leaching of toxic metals to the environment. This is especially true of sulfide-ore bodies. They produce AMD, because the waste rocks and tailings are highly susceptible to chemical weathering.

Impacts of Mineral Processing

Except for some industrial minerals, excavating and removing raw ore are only the first steps in producing a marketable product. Once metallic ores are removed from the ground, they are processed at a mill to produce an enriched ore, referred to as a *concentrate*. The concentrate is then sent to a smelter for refining into a valuable commodity.

Concentration and smelting are complex processes, and a thorough discussion of them is well beyond the scope of this book. Briefly, the concentration process requires:

1. crushing the ore to a fine powder,
2. classifying the crushed materials by particle size by passing them through various mechanical devices and passing on those particles of a certain size to the next step, and
3. separating the desired mineral components from the noneconomic minerals by a flotation, gravity, or chemical method.

These means of separation are discussed individually here.

◆ FIGURE 13.27 A monument to the destructive force of hydraulic mining of bench placer deposits; Malakoff Diggins State Park, Nevada County, California. When the hydraulic mining operations here ended in 1884, Malakoff Diggins was the largest and richest hydraulic gold mine in the world with a total output of about $3.5 million. More than 41 million cubic yards of earth had been excavated to obtain gold. The site is now marked by colorful, eroded cliffs along the sides of an open pit that is some 7,000 feet long, 300 feet wide, and as much as 600 feet deep.

◆ FIGURE 13.28 A rich phosphate bed in Florida is strip-mined by a mammoth drag-line dredge. The bucket, swinging from a 100-meter boom (that's longer than a football field), can scoop up nearly 100 tons of phosphate at a time. The phosphate is then piped as a slurry to a plant for processing. Fine-grained gypsum, a by-product of processing, is pumped to a settling pond (*upper left*), where it becomes concentrated by evaporation.

Flotation

The **flotation** separation process is widely used, especially for recovering sulfide-ore minerals such as lead, zinc, and copper sulfides from host rock. The process is based on the principles of wettability of mineral particles and surface tension of fluids. After crushing and concentration, the wettability of the *undesired* mineral particles is increased by treating the crushed ore chemically—usually with liquid hydrocarbons—to ensure that the undesired minerals will sink. Air is then bubbled into the slurry of crushed ore and water, forming a froth that collects the *desired* mineral particles of low wettability. The froth, with the attached desirable mineral particles, is skimmed off the top of the flotation tank and dried; this is the concentrate. The undesired mineral particles, the tailings, sink to the bottom of the flotation tank. They are drawn off and piped to the tailings pond. Although they are usually environmentally undesirable, tailings are an unavoidable waste product of mining.

> **CONSIDER THIS** *As the geological sources of raw materials are exhausted, recycling of many metals will eventually be absolutely necessary. How would you educate the general public to the urgency of recycling?*

Gravity Separation

Gravity separation methods are used in recovering high-density ore minerals such as gold, platinum-group metals, tungsten, and tin. By this process, mineral particles mixed with water are caused to flow across a series of riffles placed in a trough. The riffles trap the desired high-density particles, and water carries away the undesired low-density minerals, the tailings. The rockers, sluice boxes, and dredges used in gold placer mining, explained in Case Study 13.1, are all examples of gravity separation methods.

Chemical Methods

For minerals whose physical properties make them unsuitable for separation by flotation or gravity methods, chemical processes are used, the major ones being leaching and cyanidation. *Leaching* is often used in treating copper-oxide ores. Sulfuric acid is added to crushed ore to dissolve the copper and produce a solution of copper sulfate. The dissolved copper is then recovered by placing scrap iron in the copper sulfate solution; the copper plates out onto the iron. The acidic waste materials are chemically neutralized by treating them with lime. Among the other potential toxic contaminants used in mineral processing are ammonia, benzene, bromine, chlorine, cyanide compounds, cyclohexane, ethybenzene, glycol, ethers, hydrazine, hydrochloric acid, naphthalene, nitric acid, phenol, propylene, sulfuric acid, thiourea, toluene, and xylene. It is beyond the scope of this book to elaborate on the use of these chemicals, but because of cyanide's widespread and controversial use in gold recovery, its use will be explained.

Cyanidation

Cyanidation, used to recover gold and silver since 1890, makes use of the special property of cyanide to dissolve gold and silver. An innovation of cyanide recovery, **cyanide heap-leaching,** began to be widely used in the United States in the 1980s. By the turn of the twenty-first century, about 100 heap-leach operations were active in the western United States and South Carolina. Although heap-leaching is efficient, it is controversial in environmental circles because the open-pit mining, waste-rock dumps, and tailings piles are destructive to the landscape. Furthermore, cyanide is perceived as a hazard to wildlife and a contaminant to ground and surface waters. In Montana alone, 51 cyanide releases were documented between 1982 and 1999. The use of cyanide is illustrated in ◆ Figure 13.29. Ore from an open-pit

◆ FIGURE 13.29 The major components of cyanide heap-leach gold recovery.

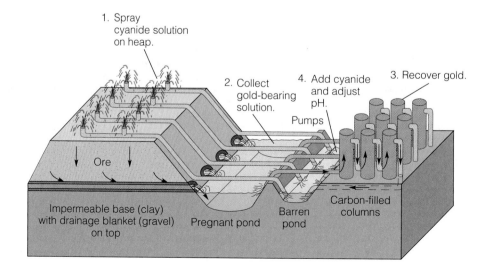

1. Spray cyanide solution on heap.

2. Collect gold-bearing solution.

4. Add cyanide and adjust pH.

3. Recover gold.

Pumps

Ore

Impermeable base (clay) with drainage blanket (gravel) on top

Pregnant pond

Barren pond

Carbon-filled columns

◆ FIGURE 13.31 A typical "pregnant pond" for collecting the cyanide solution containing dissolved gold from a leach heap. The face of the leach heap is at the upper right of the photograph. The pipes on the far side of the pond drain the "pregnant" solution from beneath the leach heap. The pond is lined with two layers of high-density polyethylene underlain by a layer of sand and a leakage detection system of perforated PVC pipe. In the rare chance that leakage should occur, it is trapped in the pregnant pond, and appropriate repairs are made. The pond is covered with fine-mesh netting to keep birds from the toxic solution.

◆ FIGURE 13.30 Cyanide drip lines on a leach heap. The cyanide concentration is 200 parts per million (one five-thousandth of one percent), and the solution is kept highly alkaline to prohibit cyanide gas from forming.

excavation is pulverized, spread out in piles over an impervious clay or plastic liner, and sprayed with a dilute cyanide solution, commonly about 200 parts per million (0.02 %) (◆Figure 13.30). The solution dissolves gold and silver (and several other metals) present in small amounts in the ore as it works its way through the heap to the pregnant pond (◆Figure 13.31). The gold and silver are recovered from the resulting "pregnant" solution by adsorption on activated charcoal, and the barren cyanide solution is recycled to the leach heap. The precious metals are removed from the charcoal by chemical and electrical techniques, melted in a furnace, and poured into molds to form ingots.

Monitoring wells are placed downslope from leach pads and ponds containing cyanide for detecting possible leakage. Maintaining the low concentrations of cyanide required by state regulations requires regular monitoring of the solution. The cyanide solution is kept highly alkaline by additions of sodium hydroxide, a strong base, in order to inhibit the formation of lethal cyanide gas. No measurable cyanide gas has been detected above leach heaps where instrumental testing

for escaping gas has been carried out. Unforeseen peculiarities of the ore chemistry and unusual weather have caused unexpected increases in cyanide values at some mines. This has required an occasional shutdown of operations until the chemistry of the solution could be corrected to the established standards. In a few cases, stiff fines have been assessed where cyanide levels exceeded the limits of the operator's permit. Upon abandonment of a leach-extraction operation, federal and state regulations require flushing and detoxification of any residual cyanide from the leach pile. Regular sampling and testing at ground-water-monitoring wells also may be required for several years following abandonment.

Securing the open cyanide ponds to keep wildlife from drinking the lethal poison is an environmental concern at several operations. After 900 birds died on the cyanide-tainted tailings pond at a mine in Nevada in 1989–1990, the company pleaded guilty to misdemeanor charges, paid a $250,000 fine, and contributed an additional $250,000 to the Nature Conservancy for preservation of a migratory bird habitat. Reported wildlife mortalities at cyanide-extraction gold mines in Arizona, California, and Nevada from 1980 to 1989 appear in ✳ Table 13.7. Of major concern are a number of endangered, threatened, or rare mammal species whose geographic ranges include cyanide-extraction mines. About 34 percent of the mammal mortalities reported in the table are bat species that are on the federal government's list of rare and endangered species.

✳TABLE 13.7 Mammal, Bird, Reptile, and Amphibian Mortalities Reported at Cyanide-Extraction Gold Mines in Arizona, California, and Nevada, 1980–1989

State	No. of Mines	Mammals	Birds	Reptiles	Amphibians	Total
Arizona	1	52 (12.7%)	357 (87.3%)	0	0	409
California	11	34 (5.2%)	606 (92.7%)	14 (2.1%)	0	654
Nevada	63	433 (6.6%)	6,034 (92.2%)	24 (0.4%)	55 (0.8%)	6,546
Totals	75	519 (6.8%)	6,997 (92.0%)	38 (0.5%)	55 (0.7%)	7,609

Source: D. R. Clark and R. L. Hothem, "Mammal Mortality at Arizona, California, and Nevada Gold Mines Using Cyanide Extraction," *California Fish and Game* 77 (1991), pp. 66–67.

Various techniques for discouraging wildlife from visiting cyanide-extraction operations have been tried: stringing lines of flags across leach ponds in an effort to frighten birds away, covering pregnant ponds with plastic sheeting, and blasting recorded heavy-metal rock music from loudspeakers. At one mine, the heavy-metal recordings were effective in keeping migratory waterfowl away, but resident birdlife adjusted to the din and remained in the area. One wonders what effect the blasting of recordings by Limp Bizkit or Nine Inch Nails might have. The current standard practice is to enclose the operations with chain-link fencing to keep out larger animals and to stretch netting completely across the ponds to protect birds (Figure 13.31). The fences and netting have been successful; California's 12 to 15 active heap-leach operations, for example, have reduced bird losses to only 50 to 60 a year in the cyanide ponds. This number is less than trivial when compared to the thousands of birds killed each year in the state when they fly into roof-mounted television antennas and the paths of automobiles.

Smelting

Historically, smelters have had bad reputations for causing extensive damage to the environment. Sulfurous fumes emitted as by-products of smelting processes have polluted the air, and toxic substances from smelting operations have contaminated soils and destroyed vegetation. Because water was necessary for operating early-day smelters, they were located near streams. The accepted practice at the time was to discharge mill wastes and tailings into streams or settling ponds, which commonly spilled over into streams. Improved smelting technologies are now eliminating these problems. For example, Kennecott's former smelter at the Bingham Canyon, Utah, copper mine released 2,136 kilograms (4,700 lbs) of sulfur dioxide (SO_2) per hour to the atmosphere. Kennecott's new smelter, designed to meet or exceed all existing and anticipated future federal and state emission standards, was put into service in 1995. SO_2 emissions were cut by 96 percent to only 91 kilograms (200 lbs) per hour, less than the rate for the world's cleanest smelters now operating in Japan.

Mine Land Reclamation

The volume of coal-mine spoil and hardrock-mining waste rock is usually greater than the volume occupied by the rock before mining. SMCRA requires that coal-mine spoil be disposed of in fills, usually in the upper reaches of valleys near the mine site. Because some settlement of this material can be expected over time, construction on these reclaimed sites may be risky. Waste rock from hardrock mining is usually deposited in great mesa-like piles near the mines. In recent years, many of these piles have been contoured to break up their unnatural, flat-topped appearance. In both cases the features are engineered and structured for stability with terraces and diversion ditches for controlling surface-water flow and preventing erosion, and then they are landscaped. Ground water within the fills is commonly channeled through subsurface drains.

Surface- and Ground-Water Protection

High-quality, reliable water is critical for domestic, industrial, and agricultural activities, and it is especially precious in arid and semiarid regions. SMCRA requires that a surface coal-mining operation be conducted in a way that will maintain hydrologic balance and ensure the availability of an adequate water supply for postmining use. In the early 1990s, nearly 1,500 abandoned U.S. coal mines had been identified as having water problems. Of these, about 500 have been reclaimed or have been funded to begin reclamation.

During surface coal mining all of the runoff water that collects in the pit is required to be collected and treated. Because surface mining destroys the original plant cover and exposes the soil to erosion, mitigation of erosion and sediment loss is needed during mining and reclamation. AMD problems have arisen from coal mining in Pennsylvania, West Virginia, Maryland, and Ohio and from the hardrock mining of sulfide ores in the Western states and in many areas of Canada. More than 100,000 abandoned hardrock mines in the West may pose AMD problems (see ○ Case Studies 13.2 and 13.3 on pages 375–377).

SMCRA requires coal mine operators to treat water from their mines before releasing it into streams and rivers. Control measures for AMD include:

- Holding mine drainage water in entrapment ponds and neutralizing it by adding alkaline materials from other industries such as kiln dust, slag, alkaline fly ash, or limestone before releasing it from the mine-permit area.

- Grading and covering acid-forming materials to promote surface-water runoff and inhibit infiltration.

- Backfilling underground mines with alkaline fly ash, which absorbs water and turns into weak concrete, in order to minimize the flow of water from the mine and reduce the amount of oxygen in the mine. Filling is done by pumping a slurry down several boreholes to reduce AMD and stabilize the ground surface of many communities built over mined-out areas.

- Using wetlands to treat AMD issuing from both operating and abandoned coal mines. Cattails and other wetlands plants have been found to be effective in removing some toxic metals and other substances from water. One such Pennsylvania wetland has saved an electric power company $50,000 in water treatment costs each year since it was built in 1985.

Wastewater from hardrock mines and from processed ore tailings is another source of contamination. These waters are usually treated by chemical or physical processes before disposal. When those processes proved unsuitable at the Homestake Gold Mine at Lead, South Dakota, the company developed a biological treatment. The bacterial bioxidation process breaks down residual cyanide from the mill's wastewater into harmless components that are environmentally compatible with the disposal creek's ecosystem. The creek now supports a viable trout fishery for the first time in over a century.

At some unprotected mine properties, siltation clogs streams and increases the threat of flooding. This problem may be mitigated by constructing settling ponds for trapping the sediment downstream from the source (◆Figure 13.32).

Of all mining operations, sand and gravel mining are the most widespread and the most obvious to urban dwellers—especially when the operations are along or within rivers. Such modification to rivers results in a variety of consequences: increased erosion to channels and river banks, degradation of water quality, disconnection of links between terrestrial and aquatic ecosystems, and creation of unacceptable habitats for spawning fish such as anadromous (migrating) salmonids (◆Figure 13.33). Fortunately, disrupted river channels can be restored, and more states are requiring restoration elements in mines' operating plans.

Mill and Smelter Waste Contamination

As ore is milled to produce concentrate, about 90 percent of the mined material is discarded as tailings. Smelters that process metallic sulfide concentrates release waste products that can contain thousands of times the natural levels of heavy metals such as lead, arsenic, zinc, cadmium, and beryllium. These contaminants remain in smelter slag (rendered inert by the smelting process so that it is not bioavailable) and in flue dust that commonly remains around former smelter sites long after smelting has ended. The primary concerns with such contamination are the environmental impact of stormwater runoff into nearby streams or lakes and the risks to human health that result when people unwittingly inhale or ingest the heavy metals. If the site is on the EPA's Superfund list, the cleanup will take one of two directions—either removal or in-place remediation such as

◆ FIGURE 13.32 A placer gold mining operation near Fairbanks, Alaska. A downstream settling pond traps placer tailings from the mine.

(a)   (b)

E. FRANK SCHNITZER, OREG. DEPT. OF GEOLOGY AND MINERAL INDUSTRIES

◆ FIGURE 13.33 (a) Site of a sand and gravel mining operation along the Molalla River in Oregon before restoration, and (b) after reclamation. Not only is the site more attractive, also the pit has been modified to decrease erosion, shallow ponds have been created to provide habitat suitable for migrating and spawning salmon, and the overall setting is improved for the general benefit of wildlife.

enclosing or encapsulating the heavy-metal sources—determined by which one is considered appropriate for preventing human contact. At one Superfund site the discovery of elevated levels of arsenic in the hair and blood of children in a small community built on contaminated soil resulted in the EPA's moving the entire community of about 30 families and destroying the houses. If a contaminated Superfund area is populated, the EPA may require an ongoing blood-testing program to monitor residents' blood levels of heavy metals. Contaminated stormwater runoff may require construction of a stormwater management system with engineered drainage channels, ponds, and artificial wetlands.

The EPA procedure for remediating contaminated smelter slag, flue dust, and tailings that pose a threat to surface or ground water is to remove the material and deposit it in specially constructed repository sites (◆Figure 13.34). Commonly these are pits underlain with heavy plastic liners

(◆Figure 13.35) into which the contaminated waste is placed and covered with crushed limestone, sealed with topsoil, and revegetated. To eliminate leaching of the material by rainwater or snowmelt, a subsoil drainage system collects excess water and disposes of it in such a way that it poses no threat to human health or the environment. (Disposal of toxic wastes is discussed thoroughly in Chapter 15.)

CONSIDER THIS *Why does mining that exposes limestone or marble as it removes ore not have AMD problems?*

Revegetation and Wildlife Restoration

SMCRA regulations for coal mining and most state reclamation regulations for hardrock mining require healthy vegeta-

◆ FIGURE 13.34 Removal of mill tailings at a Superfund site at Butte, Montana. Immense amounts of material must be removed from some Superfund sites.

ARCO

◆ FIGURE 13.35 Preparation of a repository site for contaminated smelter-flue dust near Anaconda, Montana. The pit will have a heavy polyethylene liner.

ARCO

tion to be reestablished once mining has ended. Permanent vegetation is the principal means of minimizing erosion and reducing stream siltation. The types of vegetation that are to be used in reclamation are stipulated in the original mine permit, based on premining vegetation and intended post-mining uses. Commonly, a straw mulch or chemical soil stabilizer is applied after seeding to inhibit erosion and retain moisture. Mine operators are responsible for maintaining the new plant cover until it is successfully reestablished; the SMCRA minimum is five years in the East and Midwest and ten years in the semiarid West.

Wildlife habitat is one of the most common uses of postmined land. Among the techniques the mining industry uses to meet the EPA, SMCRA, and state regulations for attracting and supporting desirable species of wildlife are contouring the land, introducing plant species that will support browsing and foraging, creating wetland habitats, and stocking ponds for sport fishing.

The Future of Mining

Concern about the future reserves of mineral resources has led to the obvious question, Are we running out? Numerous studies have been conducted over the years to determine reserves, and the answer to the question seems to be "Not yet." The question of availability is probably not the important one, however. Increasing needs for resources will require mining lower and lower grades of minerals. This will require improved technologies and larger, more powerful machines, which in turn will produce greater quantities of waste material. This will impose more stress on environmental systems. Because of these factors, the question that *should* be asked is: Can we afford the environmental and human costs required to satisfy our increasing need for minerals?

Today, we residents of industrialized countries enjoy living in comfortable homes, traveling by automobile, and having labor-saving appliances, all of which provide us with lifestyles that only 100 years ago would have been considered luxurious. These lifestyles are possible only because of the availability of inexpensive raw materials for manufacturing the products to which we have grown accustomed. Few of the earth's people realize that today's prices of mineral commodities typically reflect only the short-term, tangible costs of wages, equipment, fuel, financing, and transportation—just as they have since mining began in ancient times. Much of the *real cost* of exploiting mineral and fossil-fuel resources is intangible, and it has been externalized. Historically, the environment has borne a great deal of the cost of extracting raw materials; the consumer has paid only part of the real cost. The low prices of raw materials today do not cover the costs of polluted surface waters, dammed rivers, squalid mining towns, and devastated landscapes. Even in the United States, where regulations designed to protect the environ-

ment are imposed, mining and processing still cause substantial damage.

Worldwide, poor mining and mineral-processing practices contribute significantly to soil erosion, water contamination, air pollution, and deforestation, especially in developing nations, as illustrated by the examples in Table 13.3. Fortunately, the negative environmental impacts of mining and mineral processing are gaining more attention in the developing nations. For example, Mexico adopted new mining and environmental laws in 1992 before encouraging mining by privatizing that industry. And in 1994, Tanzania and Guinea began requiring environmental impact assessments to be submitted along with applications for mining licenses.

Mining generates twice as much solid waste as all other industries and cities combined, and most mining wastes in the United States are currently unregulated by federal law. In addition, the mineral industry is one of the greatest consumers of energy and a contributor to air pollution and global warming. It is estimated, for example, that the processing of bauxite into aluminum alone consumes about one percent of the world's total energy budget. Miners admit that mines leave holes in the ground, but they go on to point out that when they abandon their currently active mines in the United States, they will be restoring the site to as near a natural appearance as is reasonable, except for the open pits.

In evaluating mining and its impact on the environment, we must recognize that the only way not to disturb the landscape is *not to mine*. This would mean not having the raw materials to build automobiles, houses, farm machinery, airplanes, radios, television sets—virtually all of the objects we view as essential to a modern society. The question thus becomes, What must be done to minimize the impacts?

CONSIDER THIS *What is meant by the statement that the true cost of mining is externalized?*

Mining Legislation in the United States

In the 1970s some genuinely important concerns related to mining and processing mineral materials became obvious in the United States and many other countries, and laws began being written to address these concerns. Consequently, in the United States, the EPA now administers a number of acts regulating water quality and toxic wastes, and the Fish and Wildlife Service requires noninterference with rare and endangered species. Beginning in 2001, however, federal regulations dealing with some environmental laws and policies began changing, including regulations dealing with mining. Various constraints are imposed by numerous state and local laws requiring that mining operations meet certain air-quality standards. Many states will not issue a permit for an operation to begin without prior approval of an adequately funded reclamation plan specifying that the waste and tailings piles will be restored to resemble the surrounding topography to the extent that it is possible, that the disturbed ground and the tailings will be replanted with native species, that all buildings and equipment will be removed, and that the open pits will be fenced and posted with warning signs. Most states require that mining companies be bonded to ensure that funds will be available for reclamation in the event of bankruptcy or abandonment of their operation before remediation is completed (see ⊙ Case Study 13.5).

In 2003, California's State Mining and Geology Board approved permanent regulations requiring new open-pit metallic mines to backfill open pits and recontour their mine sites upon completion of mining. Subsequent to the new regulation, a Canadian mining company is filing suit against the U.S. government under the provisions of the North American Free Trade Act (NAFTA). NAFTA allows foreign investors to sue signatory governments for recovery of real or perceived financial losses caused by policies that foreign investors feel violate their rights. At issue is a group of mining claims in California's Mojave Desert on land that is held sacred to an Indian tribe and that the Canadian corporation would like to develop. However, because of the additional costs required by the new back-fill regulation, open pit mining of the property is now uneconomic.

It is claimed by some that enforcement of federal regulations pertaining to mining has been weak. Initially, the EPA did little to regulate mining wastes. Further, Congress specifically exempted hardrock-mining wastes and tailings from regulation as hazardous wastes in the Resource Conservation and Recovery Act (1976). Complicating the issues is the fact that instead of the federal government, it is the individual states that play the role of regulator on state and private lands.

Further reform of the Mining Law of 1872 may occur. Beginning in the 1980s public concern grew over the problems of protecting nonmineral values on public lands, the lack of meaningful federal reclamation standards, limited environmental protection, and the lack of royalty collection for exploited public land—issues of public-resource management that are not addressed in the current mining law. Bills revising the mining law have been introduced in nearly every session of Congress since the early 1990s. A point of contention in revising the mining law has been the imposition of royalties on mining income with the funds so generated earmarked for reclamation of some 100 thousand abandoned hardrock mines in the Western states, much as is already done by SMCRA with abandoned coal mines using the Abandoned Mine Fund. Some critics of mining law reform believe that charging royalties for hardrock operations on federal lands will close down U.S. mining and drive more mineral production to foreign countries. Nevertheless, efforts at reform are expected to continue.

The future may also bring some changes in the basic attitudes and assumptions that underlie our capitalistic society. The economic assumption that prosperity is synonymous with mineral production is now being questioned. Environmental deterioration from today's unprecedented rate of mineral production will, if continued, eventually overwhelm the benefits gained from increased mineral supplies. If, by 2050, everyone lives as the developed world does at present, mineral production would need to be 3 to 15 times current levels, energy production would need to be some five times the current level, and about 45 percent of the water cycle would be needed. It is possible that the use of marginal mineral resources could meet the necessary resource demands but the energy demands for extraction and processing would increase exponentially as the grade declines. It is imperative that by the middle of the twenty-first century when the world population has reached 10 to 15 billion (see Chapter 1), new technologies and economic strategies must be in place. The goals of protecting and managing the environment while at the same time exploiting and expanding the mineral resource base in an environmentally responsible manner are not necessarily mutually exclusive. But how can both goals be achieved? Society must recognize the need for sustainable development to replace the current materials-intensive, high-value, planned-obsolescent manufacturing processes that reduce our resource base. Sustainable development will occur only by using raw materials and fuels more efficiently, generating little or no waste, recycling what waste is generated, and developing inexpensive and widely available alternative energy sources. In the long run, sustainability is critical to maintain the viability of the biosphere. Economists often opine that there is no problem of shortages and that market forces will dictate needed changes, but it is earth scientists who perhaps have the best understanding of the natural resource base. However, they are not unified in their view of the future and resource depletion. Some focus on the identification, further exploration, exploitation, and use of resources, while others focus on the limits of resource availability, and overall future of society, and raise ethical questions about the inequality of lifestyles and resource-related wealth. Whether attitudes and technology can modernize fast enough to conserve natural resources at the same time that the world's wealth is increasing is the big question. This is the formidable challenge that must be faced jointly by industry, governments, and society.

CONSIDER THIS *Growth in consumption of natural (nonrenewable) resources increases exponentially, at a faster rate than the growth of earth's population (which doubled twice during the twentieth century). How long can this continue?*

Mining the California Mother Lode

Although rumors of gold in Spanish-held California were circulating as early as 1816, the first rush for California's gold did not begin until 1842, following ranchero Francisco Lopez's discovery of gold in what is now Placerita Canyon on Rancho San Fernando near Pueblo de Los Angeles. The lack of water at the site resulted in mining by only the crudest of methods until the arrival of experienced miners from Sonora, Mexico, who introduced the method of dry washing. The spurt of interest quickly waned, however, because the small amount of gold available was soon extracted, and the lack of water for washing the gravels was discouraging to all but the most experienced miners. In 1848, however, James Marshall's recognition of gold at Sutter's Mill on the American River electrified the world and triggered the greatest gold rush in history, ultimately altering the development of western North America and perhaps of the United States (◆ Figure 1).

The crude early mining methods by which knives and spoons were used to extract gold from placer deposits soon gave way to the gold pan and "rocker," an open-topped wooden box mounted on curved boards. Using a rocker was a two-person operation: one miner shoveled sand and gravel and poured water into the box, while the second rocked the box back and forth. The dirt would wash through the box and the gold would be caught behind "riffles," slats placed across the bottom of the box perpendicular to the flow of water. Rockers in turn led to the development of the elongated rocker, and the wooden sluice box, with similar transverse riffles (◆ Figure 2). The sluice was positioned in a stream so that water would flow down the length of the box. Miners shoveled in gold-bearing sand and gravel, and the gold was caught by the riffles. Cocoa matting was sometimes used to catch very small particles of gold, and oftentimes a copper plate coated with mercury was used to recover the very fine particles more efficiently by forming an *amalgam*, an alloy of mercury and gold—or silver; hence the reason that mercury is called *quicksilver*. When "cleaning up," the miners would heat the amalgam in a vessel to separate the dissolved gold from the amalgam by vaporizing and condensing the mercury, which could then be reused. The miners were careful to avoid breathing the fumes in recovering the vaporized mercury because they knew of its toxicity.

The easily obtainable gold was soon exhausted, and by 1853 the method of hydraulic mining—extracting gold by blasting a jet of water through a nozzle, called a *monitor*, against hillsides of ancient alluvial deposits—was developed (◆ Figure 3). Once the necessary ditches, reservoirs, penstocks (vertical pipes), and pipelines had been installed, it was possible to wash thousands of cubic yards of gold-bearing gravels without hand labor. An attempt that same year to mine alluvial gold by dredging failed; it was not until 1898 that the first real success was achieved with a bucket-conveyor dredge (Figure 13.24).

CONTINUED

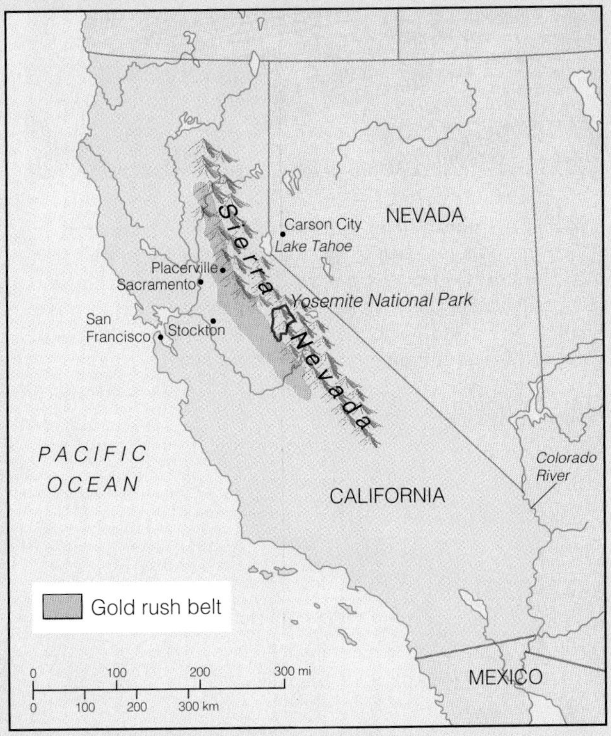

◆ FIGURE 1 The Mother Lode district, California's nineteenth-century gold-rush belt.

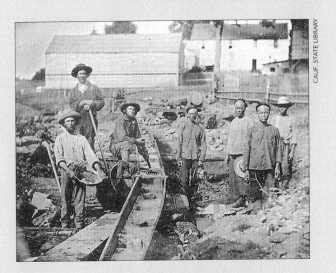

◆ FIGURE 2 American and Chinese gold miners working a sluice box at Auburn, California, in 1852.

Experienced miners immediately began searching for the *mother lode*—the bedrock source veins from which the placer deposits originated. The first gold-quartz lode vein was discovered in 1849 on Colonel John C. Fremont's grant near Mariposa. Discoveries of more gold-quartz veins of the lode system soon followed. In contrast to the simplicity of mining placer deposits, lode mining requires extensive underground workings (which require drilling, blasting, and tunneling) and a complex surface plant for hoisting men and ore from underground, for pumping water out of the mine, and for milling the raw ore. A few banks of old stamp mills (◆ Figure 4), batteries of heavy weights that were alternately lifted and dropped to pulverize the ore, remain today in the gold country. The gold was recovered by gravity separation from a slurry of powdered rock and water as it flowed over riffles and by amalgamation with quicksilver. In the 1890s, cyanide extraction was introduced into the milling process to make recovery even more efficient. It had been learned that gold dissolves readily in a dilute solution of sodium or potassium cyanide. Hence the new method mixed crushed ore with cyanide and removed the gold later by chemical methods.

The influx of thousands of forty-niners, as the fortune-seekers were known, to the Mother Lode district led to widespread prospecting and to the rapid and complete exploration of the previously little-known, sparsely populated region. Within a year of James Marshall's gold discovery at Sutter's Mill, California had attained statehood and become, almost overnight, the most important Western state. California's gold rush also served as the impetus for widespread prospecting rushes in the 1850s that led to important gold strikes in Nevada, Colorado, Montana, Australia, and Canada. A conservative estimate placed California's annual gold production at $750,000,000 by 1865. The major world impacts of the California gold rush were the widespread redistribution of population, with its resulting problems and challenges, and the general increase of money in circulation.

◆ FIGURE 3 Hydraulicking a bench placer deposit.

◆ FIGURE 4 Interior of a large stamp mill, like those used in the Mother Lode gold belt in the nineteenth century. Mills like this crushed ore 24 hours a day six days a week with a noise that must have been deafening. It is said that one could travel the length of the Mother Lode in the 1880s and never be out of earshot of the pounding stamp mills.

CASE STUDY 13.2

Acid Mine Drainage and Earth Systems

The origin of acid mine drainage (AMD) is an excellent example of the interaction of four of the spheres of the earth system. When sulfide ore bodies (components of the *lithosphere*) are mined, the surface area of the sulfide minerals in the underground workings, ore piles, waste heaps, and mill tailings is increased (due to their being broken). This increased exposure to chemical processes speeds up the minerals' reactions with both oxygen (an *atmosphere* component) and underground and surface water (components of the

hydrosphere). The chemical interaction of components of these three earth spheres produces sulfuric acid and liberates free metallic ions such as those of iron, lead, zinc, copper, and manganese (Figure 13.22). Additionally, certain bacteria (components of the *biosphere*) thrive in this acidic environment, deriving energy from the conversion of one type of iron (ferrous, Fe^{+2}) to another ferric, Fe^{+3}). In so doing, these bacteria cause a "cascading" effect in the conversion of Fe^{+2} to Fe^{+3}, increasing the rate by as much as a million times what

it would be if the bacteria were absent. Heat generated by the chemical and biochemical activity provides positive feedback to the entire process, further speeding up the reactions. The acidic water that is produced by the reaction interacts with other components of the biosphere—aquatic life in streams and rivers (see Case Study 13.3) and land plants and animals, including humans, that utilize the water—often with injurious effects.

CASE STUDY 13.3

Cleaning Up Historic Mining Sites in Colorado

Challenges of cleaning up abandoned mining regions are well-illustrated by ongoing activities in Colorado, where mine drainage is a very serious environmental problem. An estimated 1,300 miles of the state's streams are contaminated with metals due to contaminated drainage in mining areas. The water emitted from the mines, acid mine drainage (AMD), is commonly more acidic than lemon juice (citric acid) or vinegar (acetic acid)—often with pHs less than 4 (see Case Study 13.2)—and generally contains high concentrations of dissolved metals. The high metal concentrations are due to acid dissolution of the country rock and to oxidation of metallic sulfide minerals. These chemical reactions produce harmful concentrations of such elements as iron, manganese, zinc, lead, copper, arsenic, and cadmium in the waters (see Case Study 13.2). As water moves away from a mine site, additional chemical reactions may generate more

acid, causing red, orange, and yellow iron-rich precipitates to settle on aquatic plant life and stream bottoms.

The Argo Tunnel and Clear Creek
The Argo Tunnel was excavated between 1893 and 1910 to support the mining industry near Idaho Springs (◆ Figure 1). Its purposes were to drain waters from several mines and provide access for ore from those mines to the Argo processing mill adjacent to the tunnel opening. Ore last moved through the tunnel in the 1940s, but water laced with acid (sometimes with a pH as low as 2.5) and metals still drains from the tunnel continuously. Until the late 1990s, Argo water discharged directly into Clear Creek, the source of drinking water for more than 250,000 people north of Denver. Argo's drainage has sometimes carried as much as a ton of metals into Clear Creek per day. Argo's AMD legacy includes man-

ganese, iron, copper, zinc, arsenic, aluminum, and cadmium, with manganese and aluminum posing the greatest threats to aquatic life in Clear Creek. Until recently, drainage from the tunnel contaminated the water downstream for 35 miles. At that point, the drainage joined the waters of the South Platte river and became diluted.

In the 1980s the Argo Tunnel was identified as a Superfund site, and in 1998 a chemical water treatment plant began operating that will, continuously and forever, treat the metal-laden water issuing from the tunnel (◆ Figure 2). The treatment removes 99.89 percent of the metals (about a ton of metals per day from the Argo discharge), which are collected as a sludge and trucked to a landfill site. The cost of constructing the plant was $4.7 million, 90 percent of which was paid by the EPA, and the remainder by the

CONTINUED

◆ FIGURE 1 AMD draining from the Argo Tunnel flows into a drain that carries the water to the Argo Water Treatment Plant. With a pH of about 3, the water is almost as acidic as vinegar. Its yellow-orange color is due to precipitates of the metallic ions carried in the water.

State of Colorado. Annual operating costs are about $1 million. Similar Colorado AMD-water treatment plants operate at the Eagle Mine near Minturn, in California gulch near Leadville, and at Summitville in Del Norte County. A new, more efficient, water treatment plant is planned for the Summitville Mine Superfund site. The plant will cost as much as $8 million, is expected to be online by 2006, and will be capable of treating 1,200 to 1,400 gallons per minute regardless of the level of contaminant concentration. Complete cleanup of the Summitville site is anticipated to take 100 years, according to the EPA.

Remediation Studies for the Animas River Watershed

In the area of Silverton, in southwest Colorado, many millions of dollars' worth of lead, zinc, silver, and gold were produced between 1874 and 1991. By the late 1990s, most of the 2,000 mine sites within the Animas River watershed had been abandoned (◆ Figure 3), and metal-laden AMD issuing from the old mines had affected the river's aquatic life and water quality for more than 100 miles downstream.

One of the first objectives in cleaning up mining sites is *characterization*—that is,

determining the relative "metal-loading" contributions to the watershed from various sources and identifying the pathways of the toxic metal contaminants. Such information is needed for developing a hydrologic model for the area and determining the watershed's baseline conditions. For the Animas River watershed, these objectives were met by:

- studying surface-water quality;
- inventorying abandoned mine sites;
- collecting water samples from each subwatershed at low-flow and high-flow stages;
- developing an overview of point sources, transport pathways, and partitioning of toxic metals within the watershed; and
- determining specific sites of contamination by various means.

Collectively, these measurements provide toxic-loading data for specific sites. With this information, technical experts from the EPA, the Colorado Division of Minerals and Geology, the U.S. Forest Service, the Bureau of Land Management, the Colorado Department of Public Health and Environment, and the U.S. Geological Survey established priorities for site remediation. Prioritization was based on:

- contaminating sites' levels of environmental impact,
- amounts of dissolved metals in the drainage waters,
- existing and potential aquatic life,
- feasibility of treatment,
- cost-effectiveness, and
- preservation of the area's historic and cultural values.

Of the 300 or so abandoned mines in the watershed recognized as contributors of AMD, 25 were identified as "major" sources of AMD and toxic metals, and four as "critical."

The first stage of the remediation efforts began before the list of priority sites was finalized. It consisted of constructing simple barriers between water bodies and mine wastes in order to minimize or eliminate the production of AMD. The second stage is evaluating various remediation alternatives for the high-priority sites. From the least expensive to the most expensive, the options include cementing and sealing mine openings, in-place bioremediation using metal-reducing bacteria, and constructing a water-treatment facility such as the one at Idaho Springs. As of early 2000, no final solution had been reached for the cleanup of abandoned mines in the

Silverton area. Colorado's effort to eliminate toxic AMD from old mines, the costs incurred, and the number of government agencies involved illustrates well the challenges faced in reclamation of abandoned mines.

Sources: U.S.G.S. Circular 1097, 1994; Dunn, Russell, and Morrissey, Remediating Historic Mining Sites in Colorado, *Mining Engineering,* August 1999; and *Argo Water Treatment Plant,* Colorado Department of Public Health and Environment, August 1999.

◆ FIGURE 2 The Argo Water Treatment Plant at Idaho Springs, Colorado, continuously treats AMD from the Argo Tunnel before its discharge into Clear Creek. Operation of the plant began in April 1998, and the State of Colorado anticipates operating the plant for perpetuity. Idaho Springs and the Argo Mill Historical Mining Site are a short distance west of Denver on I-70.

◆ FIGURE 3 Longfellow Mine, at Red Mountain Pass, one of an estimated 2,000 abandoned mines in the Animas River watershed near Silverton, Colorado

CASE STUDY 13.4

The Mike Horse Mine—A Cleanup or Dilemma?

Representative of some 15,000 of the worst historic abandoned mining sites in the Western United States is the Mike Horse Mine located near Rogers Pass on the continental divide of Montana's Rocky Mountains. To environmentalists the Mike Horse is especially onerous because it is on a major tributary of the Blackfoot River, that notable trout stream made famous in Norman MacLean's novel, *A River Runs Through It.* The Mike Horse is the major source of toxic metal pollution—such contaminants as zinc, lead, copper, and arsenic—to the river that drains the valley in which the mine resides, and by 1990 the Mike Horse had become

a major concern of environmental activists (◆ Figure 1). Discovered in 1898 as a lead-silver deposit, the mine became a major source of lead and zinc during and following World War II, and in the 1960s it was being considered for development as a large open-pit copper mine. The mine's sulfide ore body includes galena, sphalerite, tetrahedrite, chalcopyrite, bornite, pyrite, marcasite, and arsenopyrite, from which the toxic contaminants are released in acid mine drainage (AMD). Lesser amounts of AMD and toxic metal leachate are emitted from other nearby abandoned mines but the most pernicious is the Mike Horse.

By the early 1990s, Montana's own version of the Superfund law, their Comprehensive Environmental Cleanup and Responsibility Act (CECRA), recognized two large mining corporations as the potentially liable persons (PLPs). The legal maneuvering, cleanup process, expense, and eventual results serve as a prime example of what state and federal agencies and mining corporations face in the cleanup of many thousands of abandoned and inactive mines in the American West. The remediation plan presented by the mining companies in 1993 consisted of isolation of the toxic metal sources and

CONTINUED

MONTANA DEPT. OF ENVIRONMENTAL QUALITY, REMEDIATION DIVISION

◆ FIGURE 1 Shed at portal of the level-300 adit, Mike Horse Mine, about 1990. Acidic seepage water, laden with dissolved lead, zinc, copper, and arsenic, drains into the headwaters of Mike Horse Creek.

other solid mine waste from uncontaminated surface and ground water and treatment of the contaminated water. This was accomplished by (1) recontouring and encapsulating waste rock with a layer of topsoil, (2) revegetating disturbed mining areas to increase surface runoff and decrease infiltration and the generation of leachate, (3) constructing adit bulkheads to control the flow of AMD, (4) removing acid-generating mine waste from sites where it affected surface water quality and relocating it to engineered repositories that would be shaped to natural contours, covered with topsoil, and reseeded, and (5) constructing treatment ponds.

Treatment of the AMD begins at the level-300 adit (horizontal tunnel) where the metal-laden AMD accumulates behind a plug of several hundred tons of crushed limestone that creates an oxygen-free acid-neutralizing environment and forces much of the iron and other metals to precipitate (◆ Figure 2). From here the water is piped to a vault, where a valve controls the discharge into a structure where the mine effluent is aerated and then flows into pretreatment ponds built on the old mine waste dump (◆ Figure 3). The ponds are lined with a supposedly impervious heavy-duty polyethylene (HDPE), and here the oxygenated and

neutralized water resides for four days during which iron and other metal hydroxides precipitate as sludge. About 10,000 cubic feet of sludge accumulates each year, and it is removed as necessary to a waste repository. The effluent then moves to an artificial wetland that incorporates biotechnology—namely, compost and sedges—to further remove any remaining toxic metals (◆ Figure 4). Finally, the treated water is released into the headwaters of the Blackfoot River.

When completed in 1996, the mining companies estimated optimistically that the metal loading would be reduced by about 90 percent, but by 2002 that was clearly not the case. One problem is artificial wetlands. The Montana Department of State Lands abandoned artificial wetlands in its mine reclamation efforts long ago when it discovered they were not effective at reducing metal loading. A further problem arises with artificial wetlands that utilize plants such as sedges and cattails that thrive in a metal-rich environment: when the plants die, how can they be disposed of without releasing the toxic metals back into the environment? A specific problem with the Mike Horse Mine is that acid mine waters accumulating behind the 300 adit plug have risen nearly to the level of the 200 adit. In 2002, metal-tainted springs were issuing from the waste rock dump below the 200 adit; these circumvent the pretreatment ponds and directly enter Mike Horse Creek, a tributary of the Blackfoot River. The springs are laced with aluminum, zinc, and copper compounds that precipitate in

◆ FIGURE 2 Generalized diagram of the remediation devices at the Mike Horse Mine.

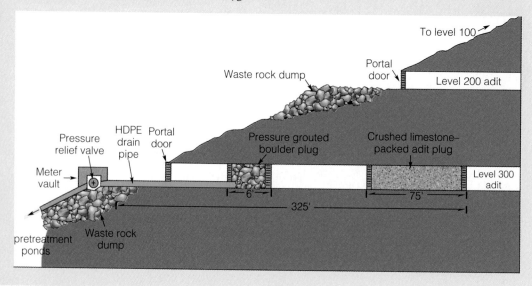

the creek bed. The table at the bottom of the page is a comparison between the quality of treated water from the Mike Horse Mine and Montana's state surface and groundwater protective standards.

The mining companies' wastewater discharge permit for their artificial wetlands committed the companies to meet protective standards by 2001. However, the 2002 data in the table clearly show that their remediation efforts had not met those standards; high metal loading of some contaminants is still a problem. Even low zinc concentrations are especially lethal to trout. Although court action may eventually require the companies to meet state water-quality standards, the cost of complete remediation of Mike Horse may preclude ever achieving those standards.

Sources: *Wounding the West—Montana, Mining and the Environment,* by David Stiller, 2000; David Bowers, Montana Department of Environmental Quality, personal communication; D. D. Trent, unpublished field notes.

◆ FIGURE 3 Pretreatment ponds constructed on the waste rock dump at level-300 adit, August 2002.

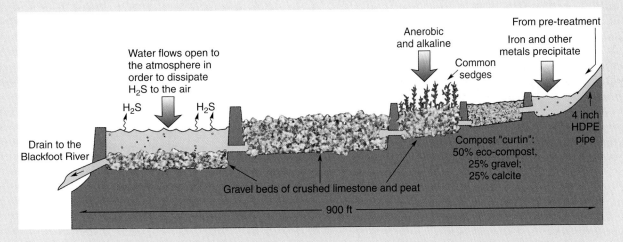

◆ FIGURE 4 Generalized diagram of the artificial wetlands at the Anaconda Mine. Pretreatment drainage from the Mike Horse Mine arrives at this site by a 1,500 meter (5,000 ft) pipeline. Wetlands incorporate biotechnology to remove metals and neutralize acid drainage from the mine wastewater.

Metal	Montana Groundwater Protective Standards	Minimum Montana Surface Water Standards that Protect Aquatic Life	Groundwater Concentrations Issuing from Mike Horse Waste Rock Dumps	Concentrations at Mike Horse Creek in the Spring
Copper	1.3 mg/l	5.2 at mg/l hardness	74 mg/l	4.2 mg/l
Zinc	2 mg/l	67 at mg/l hardness	286 mg/l	29 mg/l
Aluminum	No values	0.082 mg/l	70 mg/l	6.1 mg/l

Is a Company's Word as Good as Its Bond?

In recent years, mining companies have used corporate guarantees to promise they would reclaim mining sites after closure. But under new regulations initiated in 2000 by the Clinton administration, the Bureau of Land Management (BLM) no longer accepts corporate guarantees for mining operations on federal lands. The George W. Bush administration repealed many of the previous administration's environmental regulations, but not the ban on corporate guarantees. Mining companies must now post surety bonds for mine cleanup. With the exceptions of Alaska and Nevada, Western states no longer accept corporate guarantees for mining on nonfederal lands.

The need for surety bonding arose because several mining companies closed down mines and then declared bankruptcy, leaving the public to fund the cleanup. During 2000, Nevada had 10 companies go bankrupt that had been mining at 29 sites on BLM lands. Most of the mines were small, of minor consequence, but two were large. Estimates for the cleanup of one large mine, the Yerington, are as high as $250,000,000 and, because it is an old mine, it was essentially unbonded. The other large operation, the Paradise Peak, began operating in the 1980s with a total corporate guarantee of about $4,600,000 in secured and unsecured bonds. Thus the company, by declaring bankruptcy, took full advantage of corporate guarantees as the cost of cleanup is anticipated to well exceed the total corporate bond. Similar scenarios have been repeated elsewhere. For example, cleanup costs for the Zortman–Landusky open-pit, heap-leach gold mine in Montana that had shut down, when the mining company went bankrupt, are estimated to exceed $100,000,000.

Surety bonds cost only one or two percent of the total bond. Consequently, the surety companies require certainty that the company will not go bankrupt; otherwise the surety insurer is responsible for the cost of mine remediation. The economics of bonding are affected by many things beyond the control of the surety or the mining companies: the fluctuating price of a mineral commodity affects a company's economics and the eventual costs of cleanup may not be recognized until long after closure or abandonment. Consequently, beginning in 2002, mining companies, especially smaller ones, were having great difficulty obtaining surety bonds. Some observers argue that because mining is a high-risk and low-profitability industry—a survey in 1996 by Standard and Poor's revealed that the mining industry's average return on its investment was 4 percent, in contrast to publishing averaging 35 percent and the chemical industry averaging 29 percent profitability—and that the free market has failed the mining industry. As a result, the mining industry is now asking state and federal governments for assistance by returning to corporate guarantees rather than requiring surety bonds in order to remain in business. The record shows, however, that should this be done, in many cases the potential cleanup burden will be shifted to the public. Some people recognize such a change as externalizing the financial risk on the public in order that a corporation can make a private profit.

Another concern with mine cleanup is the "corporate veil" that mining companies utilize. Most mines, each supposedly owned by a mining company, actually are wholly owned subsidiaries. Thus the "corporate veil" legally insulates the parent company from any liability of its subsidiaries. As an example, the Dawn Mine, owned by one of the world's largest mining companies, is an inactive, open-pit uranium mine that is a Superfund site on the Spokane Indian Reservation in Washington State. The parent company disclaims any responsibility for the cleanup required by CERCLA.

The Legacy of Mines and Mining

The collage of photographs in this gallery illustrates some of the good and bad of past and present mining (Figures 1–5).

◆ FIGURE 1 Toxic legacies left for perpetuity by turn-of-the-century Colorado silver mining; waste-rock piles from abandoned mines above Silver Plume, Colorado. Although the mines have been shut down for decades, the barren piles still lack vegetation because of the accumulation of acid salts (indicated by the white coloration) due to evaporation of acidic seepage water that has percolated through the material. There are thousands of such abandoned mines in Colorado and perhaps hundreds of thousands in the Western states.

◆ FIGURE 2 Humans can be exposed to waters carrying high concentrations of toxic metals even in "undisturbed," natural settings. Several thousand people living in and near the Brooks Range of northern Alaska rely on subsistence fishing and hunting for food. Massive metallic sulfide deposits in the region pose a potential hazard to all life there if drainage from these deposits enters the rivers and streams. In this photo a scientist is collecting water samples downstream from the Drenchwater massive sulfide deposit in the northwestern part of the range. Water bodies' pH and concentrations of various metals are monitored in order to evaluate the health hazard of this lead-zinc-silver deposit.

◆ FIGURE 3 Even some national parks are not free of contamination from abandoned mining activity. An example is the Mariscal Mercury Mine in Big Bend National Park, Texas. Intermittently between 1900 and 1943 it produced mercury (quicksilver) from cinnabar ore. A detailed site investigation conducted in 1995 by the U.S. Bureau of Mines reports that minimal contamination remains in only a few isolated spots onsite, and there is minimal to no risk for short-term visitors and transient workers at the site.

◆ FIGURE 4 Creek near an open-pit copper mine at Morenci, Arizona. The water's distinct blue color reveals a high concentration of dissolved copper leached from mine wastes.

◆ FIGURE 5 The *entire* landscape visible in this photo is productive reclaimed land following extensive surface coal mining carried out here in the 1960s and 1970s; near Cologne, Germany.

Summary

Mineral Abundances, Definitions

Mineral Deposit

Locally rich concentration of minerals.

Ore

Metallic mineral resources that can be economically and legally extracted at the time.

Mineral Reserves

Known deposits of earth materials from which useful commodities are economically and legally recoverable with existing technology.

Factors That Change Reserves

Technology Changes

affect costs of extraction and processing.

Economic Changes

affect price of the commodity and costs of extraction.

Political Changes

cut off or open up sources of important mineral materials.

Aesthetic and Environmental Factors

may reduce reserves.

Distribution of Mineral Resources

Description

Mineral deposits are highly localized; they are neither uniformly nor randomly distributed.

Cause

Concentrations of valuable mineral deposits are due to special, sometimes unique, geochemical processes.

Origins of Mineral Deposits

Igneous Deposits

1. *Intrusive*—occur (a) as a pegmatite, an exceptionally coarse-grained igneous rock with interlocking crystals, usually found as irregular dikes, lenses, or veins, especially at the margins of batholiths; and (b) by crystal settling, the sinking of crystals in a magma due to their greater density, sometimes aided by magmatic convection.
2. *Disseminated*—a mineral deposit, especially of a metal, in which the minerals occur as scattered particles in the rock but in sufficient quantity to make the deposit a worthwhile ore.
3. *Hydrothermal*—a mineral deposit precipitated from a hot aqueous solution with or without evidence of igneous processes.
4. *Volcanogenic*—of volcanic origin, e.g., volcanogenic sediments.

Sedimentary Deposits

Mineral deposits resulting from the accumulation or precipitation of sediment. These include

1. Surficial marine and nonmarine precipitation.
2. Deep-ocean precipitation.
3. Placer deposits.

Weathering Deposits

1. *Lateritic weathering*—concentrations of minerals due to the gradual chemical and physical breakdown of rocks in response to exposure at or near the earth's surface.
2. *Secondary enrichment*—mineral development that occurred later than that of the enclosing rock, usually at the expense of earlier primary minerals by chemical weathering.

Metamorphic Deposits

Concentrations of minerals, such as gem or ore minerals, in metamorphic rocks or from metamorphic processes.

Categories of Mineral Reserves

Abundant Metals

Average continental crustal abundances are in excess of 0.1 percent.

Scarce Metals

Average continental crustal abundances are less than 0.1 percent.

Nonmetallic Minerals

Minerals that do not have metallic properties; include industrial minerals, agricultural minerals, and construction materials.

Environmental Impacts of Mining

1. Surface coal mining and land reclamation is regulated by federal law, the Surface Mining Control and Reclamation Act (SMCRA). No similar federal law regulates hardrock mining and land reclamation.
2. SMCRA requires surface coal mine pits to be backfilled with the mine waste (spoil) and then covered with topsoil and landscaped. Some states in which hardrock mining is conducted require similar handling of waste rock and tailings at currently operating mines.
3. Ground subsidence due to underground coal mining is less common than in the past because of the required backfilling. Surface collapse from hardrock (underground) mining is not common, and stabilization may be expensive.
4. Acid mine drainage (AMD) may be a problem at active and abandoned coal and sulfide-ore mines. Pit water in surface coal mines is required to be held and treated before it is released to streams or rivers. Sealing abandoned mines, covering waste rock with soil and landscaping, and installing drainage courses to direct surface drainage off tailings and

waste-rock piles are some methods of controlling AMD formation. Wells are installed below waste-rock and tailings dumps to monitor water quality.

5. Scarified ground, despoiled landscapes, and disrupted drainage that remain after dredging and hydraulic mining can be reclaimed by reshaping, adding topsoil, and landscaping.

6. Abandoned open pits with oversteepened side walls despoil the landscape and may pollute hydrologic systems. They are difficult or impossible to reclaim. The public is protected by surrounding the pits with chain-link fencing and posting warning signs. The pit floor may be partially backfilled and sealed with impermeable clay to protect ground water from pollution.

7. Cyanide heap-leach gold-extraction methods normally cause minimal environmental problems because they are strictly regulated by state and federal agencies. There are occasional leaks of cyanide into the environment, however. Wildlife losses have been mitigated by placing nets over the lethal ponds. Surface and ground waters are monitored by sampling and monitoring wells, respectively.

8. The main concerns associated with heavy-metal contaminated mill and smelter waste are the environmental impacts resulting from stormwater runoff into nearby surface waters and the human health problems caused by inhaling or ingesting the heavy metals in dust. Superfund cleanups require either removal or in-place remediation by encapsulating the heavy-metal sources. Wetlands with appropriate plantlife may be constructed, or water treatment plants constructed, to clean toxins from surface runoff.

The Future of Mining

Worldwide

Mining and processing of mineral commodities are contributing to soil erosion, water contamination, air pollution, and deforestation.

The United States

Mining and processing of mineral commodities must meet standards of air and water quality set by the EPA and other government agencies. The Mining Law of 1872, although much amended, is being criticized and evaluated in light of current economics, domestic needs, and conflicting land uses.

Key Terms

acid mine drainage (AMD)
alluvial placer
area mining
average crustal abundance
banded-iron formation
bauxite
bench placer
black smoker
CERCLA
concentration factor
contact-metamorphic deposit
contour mining

crystal settling
cyanide heap-leaching
disseminated deposit
dredging
evaporite
flotation
glacial outwash
gravity separation
high-grade deposit
hydraulic mining
hydrothermal deposit
lode

low-grade deposit
manganese nodule
massive sulfide deposit
mineral deposit
mineral reserves
mountaintop mining
open-pit mining
ore
ore mineral
pegmatite
placer deposit
porphyry copper

precipitation
regional metamorphism
secondary enrichment
SMCRA
smelter
spoil
strip-mining
tailings
vein deposit
volcanogenic deposit

Study Questions

1. Distinguish between mineral deposits, ores, and reserves.

2. Why do some metals, such as gold, require concentration factors in the thousands, while others, such as iron, require only single-digit concentration factors?

3. Of which metals does the United States have ample reserves? How do we obtain the scarce mineral resources we need but lack within our borders?

4. How do you expect domestic and foreign supplies of critical mineral resources to change in the next 25 years?

5. What would cause currently marginal or submarginal mineral deposits to become important mineral reserves?

6. Where are the major deposits of porphyry copper and porphyry copper-molybdenum deposits located? How do these deposits relate to plate tectonic theory?

7. What steps are involved in the origin of a placer gold deposit? Of a bench placer deposit?

8. What minerals may eventually be harvested from the deep-ocean floor? What might be the environmental consequences of such mining?

9. Explain why you wouldn't go to the Hawaiian Islands to prospect for gold.

10. What is hydrothermal activity? What metallic mineral deposits are commonly associated with hydrothermal action?

11. Contrast the gold mining and extraction techniques of the nineteenth century with those of today. What are the environmental legacies of the nineteenth-century mining methods? What environmental concerns accompany modern gold-extraction technology?

12. Contrast the environmental hazards and mitigation procedures of underground, open-pit, and surface mining.

13. Why are mercury and cyanide, very dangerous poisons if they are mishandled, used in gold extraction?

14. Briefly describe the cyanide heap-leach gold-extraction process.

15. It has been suggested that as society exhausts the earth's reserves of critical minerals, lower-grade mineral deposits will be mined and supply-and-demand economics will dictate prices and costs. This rationale could be extended to a scenario in which average rock would eventually be mined for critical materials. Is there a fallacy to such a rationale? Explain.

16. Discuss the ethics of changing the General Mining Law of 1872 to allow the federal government to collect royalties from current mine operations to be used for the cleanup of historic mines that were abandoned by companies that no longer exist.

For Further Information

Books and Periodicals

Alpers, Charles N., and Michael P. Hunerlach. 2000. *Mercury contamination from historic mining in California.* USGS fact sheet FS-061-00.

Baum, Dan, and Margaret L. Knox. 1992. In Butte, Montana, A is for arsenic, Z is for zinc. *Smithsonian* 23, no. 8:46–56.

Cutter, D. C. 1948. The discovery of gold in California. In Jenkins, O. P., ed. Geologic guidebook along Highway 49: Sierran gold belt, the Mother Lode country. *California Division of Mines bulletin* 141, (centennial ed.): 13–17.

Drew, Lawrence, and Brian K. Fowler. 1999. The megaquarry: A conversation on the state of the aggregate industry. *Geotimes* 44, no. 6:17–22.

Dunn, James, Carol Russell, and Art Morrissey. 1999. Remediating historic mine sites in Colorado. *Mining engineering* 51, no. 8:32–35.

Ernst, W. G., G. Heiken, Susan M. Landon, P. Patrick Leahy, and Eldrige Moores (co-convenors). 2003. The role of the earth sciences in fostering global equity and stability: *Report of Pardee Symposium K4,* 2002 GSA Annual Meeting, Denver, Colorado, October 28, 2002. *GSA Today,* March, pp. 27–28.

James, Patrick M. 1999. The miner and sustainable development. *Mining engineering* 51, no. 6:89–92.

King, Trude V. V., ed. 1995. *Environmental considerations of active and abandoned mine lands: Lessons from Summitville, Colorado.* U.S. Geological Survey bulletin 2220. Washington, D.C.: Government Printing Office.

National Research Council, 1999, *Hardrock mining on federal lands.* Washington, D.C.: National Academy Press.

Newmont Gold Company. 1995. *Newmont.* Newmont Gold Company, P. O. Box 669, Carlin, NV 89822-0669.

Office of Surface Mining Reclamation and Enforcement, Department of the Interior. 1992. *Surface coal mining reclamation: 15 years of progress, 1977–1992, Part 1.* Washington, D.C.: Government Printing Office.

Silva, Michael A. 1988. Cyanide heap leaching in California. *California geology* 41, no. 7:147–156.

Stewart, K. C., and R. C. Severson, eds. 1994. *Guidebook on the geology, history, and surface-water contamination and remediation in the area from Denver to Idaho Springs, Colorado.* U.S. Geological Survey circular 1097. Washington, D.C.: Government Printing Office.

Tilton, J. E. 1996. Exhaustible resources and sustainable development—two different paradigms: *Resources Policy,* vol. 22, no. 1/2, pp 91–97.

Wagner, Lorie A. 2002. *Materials in the economy—material flows, scarcity, and the environment.* US. Geological Survey circular 1221. Washington D.C.: Government Printing Office.

Wheeler, Gregory R. 1999. Teaching mineral resources as an aid to understanding international policy issues. *Journal of geoscience education* 47:464–468.

Wilshire, Howard, Jane Neilson, and Richard W. Hazlett. In press. *Tender land: A sourcebook for salvaging America's plundered west.* New York: Oxford University Press.

Videos

Hydraulic gold mining. California Department of Parks and Recreation, Publications Section, P.O. Box 942896, Sacramento, CA 94296-0001.

The magic of Homestake gold from the discovery of gold to mining today. Homestake Mining Tours, Lead, SD 57754

Mining: Discoveries for progress. American Mining Congress, Communications and Education Department, 1920 N Street, NW, Suite 300, Washington, DC 20036-1662.

Poison in the Rockies. 1992. NOVA. WGBH, P. O. Box 2284, South Burlington, VT 05407-2284; 800-255-9424.

Public lands, private profits. 1994. Overview of the controversy surrounding the Mining Law of 1872. Frontline, May 24. PBS Video, 1320 Braddock Place, Alexandria, VA 22314; 800-328-PBS1.

ENVIRONMENTAL
Geology ⇌ Now™

Assess your understanding of this chapter's topics with additional quizzing and comprehensive interactivities at http://earthscience.brookscole.com/pipkingeo4e

as well as current and up-to-date Web links, additional readings, and InfoTrac College Edition exercises.

Energy and the Environment

Flaming Ice, Energy, and the Greenhouse!

SETTING ICE BALLS ON FIRE IS U.S. GEOLOGICAL SURVEY OCEANOGRAPHER Bill Dillon's idea of fun, and it never fails to amaze his visitors. It is not ordinary ice that flames. Dillon uses hydrated methane (CH_4) recovered from sea-floor sediments deep below the ocean surface. This flame has become the focus of the energy-hungry nations of the world (◆Figure 1). What actually burns is methane gas released from a crystalline solid of methane molecules encased in a "cage" of ice (◆Figure 2). Methane is the gas that constitutes the bulk of the "natural gas" that is piped to your home. Methane hydrates are so extensive in deep-sea sediments that they have the potential to supply a share of the world's future fuel needs, *when and if* practical mining techniques can be developed. India and Japan, at present energy poor, are most interested in developing methane hydrates as an energy source. Experts agree that marine gas hydrates harbor twice the carbon contained in all known natural gas, petroleum, and coal deposits on earth—about 10,000 gigatons, or 10^{13} tons, of carbon. Extensive gas-hydrate deposits are currently known in hundreds of localities. Particularly

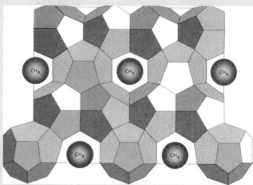

◆ FIGURE 2 *Crystalline structure of methane hydrate. Methane gas (CH_4) generated by bacterial digestion of organic matter in sea-floor sediments is sometimes trapped in "cages" of frozen H_2O (the large polygons) to form methane hydrate. Peter McCabe and others, 1993, USGS circular 115, p. 46*

◆ FIGURE 1 *A flame from ignited synthetic methane hydrate.*

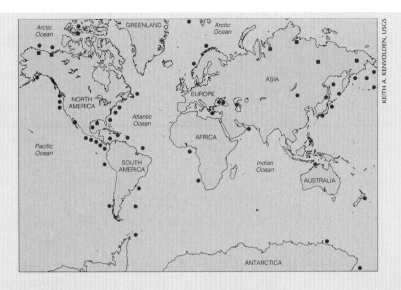

◆ FIGURE 4 *Methane hydrate nodule in marine sediment from the Blake Ridge off the Florida coast.*

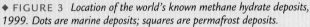

◆ FIGURE 3 *Location of the world's known methane hydrate deposits, 1999. Dots are marine deposits; squares are permafrost deposits.*

interesting for potential development are those of offshore Oregon, Costa Rica, New Jersey, Japan, and Florida (◆ Figure 3). Also intriguing are the significant deposits that have been encountered in oil drilling through Arctic permafrost areas in Canada, Alaska, and Siberia (see Chapter 6).

Gas hydrates in marine sediments have been known for several decades, but their full extent and potential were not realized until the 1990s (◆ Figure 4). Conditions necessary for their formation are high water pressure (water depth >530 m in low latitudes and >250 m in high latitudes) and low ambient temperatures (<7° C) at the water-sediment interface. These conditions are met on the sea floor's continental slope and rise. The exact origin of the gas is not known, but it is strongly suspected that bottom bacteria consume organic-rich detritus and generate methane as a waste product. Under particular conditions, the gas becomes trapped in ice in the sediment instead of being released to the overlying water column.

The bad news is that these hydrated molecular structures can release their caged methane into the atmosphere if there is a slight decrease in pressure (if sea level is lowered, for instance) or a slight warming of the surrounding sea water. Because methane is a powerful "greenhouse" gas linked to global warming (it is about 10 times as strong as carbon dioxide), melting of the hydrates and release of methane could have a drastic impact on the earth's surface temperature (see Chapter 11). One model suggests that as sea levels rose toward the end of the last ice age, the waters covered land that was underlain by permafrost containing the methane hydrates. As the warmer water temperature melted the permafrost ice, the caged greenhouse gas was released, increasing the global warming and the melting of glaciers.

Gas hydrates may also be a geologic hazard and may occupy much of the pore space in bottom sediments. When the hydrate-bearing sediment becomes deeply buried, the temperature may increase to the point at which gas hydrates are no longer stable. When a solid–gas mixture becomes a liquid–gas mixture, high pore pressure and unstable slopes result (see Chapter 7). This presents a danger to offshore drilling platforms, pipelines, communication cables, and other offshore structures. Current interest in methane hydrate is as an energy source and, as a factor in global climate change.

From the simplest algal scum to the most complex ecosystem, energy is essential to all life. Derived from the Greek word *energia* meaning "in work," energy is defined as the capacity to do work. The units of energy are the same as those for work, and the energy of a system is diminished only by the amount of work that it does.

Prosperity and quality of life in an industrialized society such as ours depend in large part on the society's energy resources and its ability to use them productively. We may illustrate this in a semiquantitative fashion with the equation:

$$L = \frac{R + E + I}{\text{population}}$$

where L represents quality of life, or "standard of living," R represents the raw materials that are consumed, E represents the energy that is consumed, and I represents an intangible we shall call *ingenuity*. As the equation expresses, when high levels of raw materials, energy, and ingenuity are shared by a small population, a high material quality of life results. If, on the other hand, a large population must share low levels of resources and energy, a low standard of living would be expected. Some highly ingenious societies with few natural resources and little energy can and do enjoy a high quality of life. Japan is a prime example. Some other countries that are self-sufficient in resources and energy–such as Argentina–are having difficult times. Thus we see that ingenuity, which is reflected in a country's political system, technologies, skills, and education, is heavily weighted in the equation, and it can cancel out a lack of resources and a large population.

To a physicist there is no energy shortage, because she or he knows that energy is neither created nor destroyed; it is simply converted from one form to another, such as from nuclear energy to heat energy. Fuels of all kinds are warehouses of energy, which can be tapped by some means and applied in some way to do work. Coal and oil, for example, are fossil fuels that have been storing solar energy in the lithosphere for millions of years (◆Figure 14.1).

Some forms of energy are **renewable;** that is, they are replenished at a rate equal to or greater than the rate at which they are used. Examples include solar, water, wood, wind, ocean and lake thermal gradients, geothermal, and tidal energy. The energy in all of these resources except for geothermal and tidal (gravitational) energy was originally derived from the sun. Renewable resources are dependable only if they are consumed at a rate less than or equal to their rate of renewal. If they are overexploited, some period of time will be required to replenish them. Peat, a fuel used extensively for space heating and cooking in Ireland and Russia, is estimated to accumulate at a remarkable 3 metric tons per hectare per year (1.3 tons/acre/year). Nonetheless, the conversion from plant litter to peat may take a hundred years. Wood energy may renew in a matter of a few decades, and water and wind are renewed continuously.

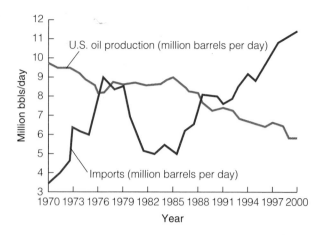

◆ FIGURE 14.1 Graph showing the relationship between oil production in the United States and the amount of oil imported by the United States. Adapted from Larry Fellows, Ariz. Geological Survey

Nonrenewable resources are not replenished as fast as they are utilized, and once consumed, they are gone forever. Crude oil, oil shales, tar sands, coal, and fissionable elements are nonrenewable energy resources. Their quantities are finite. Supplies of crude oil, for example, are within a few human generations of exhaustion. The problem is that, because of the use of gasoline-powered engines for transportation, demand far exceeds replenishment. Oil underground was discovered almost fifty years before the first automobile was operational. Prior to then, gasoline was a minor byproduct of refining oil for kerosene lamps. In 1885 Germans Gottlieb Daimler and Karl Benz independently developed gasoline engines, and in 1893 Massachusetts bicycle makers Charles and Frank Duryea built the first successful U.S. gasoline-powered automobile. By the turn of the century, production automobiles were hitting the roads, horses were being put out to pasture, and refineries were stepping up production to satisfy an increasing demand for gasoline. The transportation revolution to automobiles spawned the largest private enterprise on earth: the exploration and production of petroleum. This has led to major geopolitical and economic problems, because the nations that consume the most do not produce in comparable amounts.

Petroleum

Although the carbon content of the earth's crust is less than 0.1 percent by weight, carbon is one of the most important elements to the earth's people. It is indispensable to life, and it is the principal source of energy and the principal raw material of many manufactured products. Crude oil, or petroleum (Latin *petra*, "rock," and *oleum*, "oil"), is composed of many **hydrocarbon compounds,** simple and complex combinations of hydrogen and carbon (✳ Table 14.1). Petroleum

✱TABLE 14.1	Hydrocarbon Compounds Typically Found in Crude Oil		
Molecular Type			**Percentage of Weight In**
Name	**General Formula**	**Hydrocarbon Compound**	**Medium-Grade Crude Oil**
Paraffins	C_nH_{2n+2}	methane, CH_4 lighter	25
		ethane, C_2H_6	
		propane, C_3H_8	
		butane, C_4H_{10}*	
		pentane, C_5H_{12}**	
Aromatics	C_nH_{2n-6}***	benzene, C_6H_6	17
Naphthenes	C_nH_{2n}	asphalt	50
Asphaltenes	(solid hydrocarbons)	gilsonite heavier	8
			100

* Butane gives gasoline quick-starting capability.
** Pentane gives smooth engine warm-up.
*** Aromatics improve mileage and "knock" resistance.

occurs beneath the earth's surface in liquid and gaseous forms and at the surface as oil seeps, tar sands, solid bitumen (gilsonite), and oil shales. In addition to its use as a fuel, hydrocarbon compounds derived from petroleum are used in producing paints, plastics, fertilizers, insecticides, soaps, synthetic fibers (nylon and acrylics, for example), and synthetic rubber. Carbon combines chemically with itself and with hydrogen in an infinite variety of bonding schemes; some two million hydrocarbon molecules have been identified to date. The manufacturing process of separating crude oil into its various components is known as *refining*, or *cracking*.

Origin and Accumulation of Hydrocarbon Deposits

Carbon and hydrogen did not combine directly to form petroleum; they were chemical components of living organisms before their transformation to complex hydrocarbons in crude oil. Porphyrin compounds found in petroleum are derived either from chlorophyll, the green coloring in plants, or from hemin, the red coloring matter in blood, and their presence is solid evidence for an organic origin for crude oil. The fact that large quantities of oil are not found in igneous or metamorphic rocks also rules out an inorganic source for oil.

Four conditions are necessary for the formation and accumulation of an exploitable petroleum deposit in nature:

■ a source rock for oil,

■ a reservoir rock in which it can be stored,

■ a caprock to confine it, and

■ a geologic structure or favorable strata to "trap" the oil.

Even where these geologic conditions are met, the human elements of exploration, location, and discovery still remain. Oil companies utilize the skills of geologists to interpret surface and subsurface geology, to locate the potential oil-bearing structure or stratum, and to specify the optimal location for a discovery well. Until a well is drilled, the geologists' interpretation remains in doubt, much like a medical doctor's diagnosis of an ailment subject to surgery; the diagnosis is tentative until the patient is opened up.

A **source rock** is any volume of rock that is capable of generating and expelling commercial quantities of oil or gas. Source rocks are sedimentary rocks, mostly shales or limestones, usually of marine (ocean) origin and sometimes of *lacustrine* (lake) origin. The biological productivity (biomass) of surface waters must have been high enough to generate a settling "rain" of dead organisms, and the bottom waters must have been low enough in oxygen to prevent the deposited organic matter from being oxidized or consumed by scavengers. Almost all source rocks are dark-colored, indicative of high organic-matter content, and some of them carry a fetid or rotten-egg odor.

Favorable marine environments are rich in microscopic single-celled plants known as *diatoms* (phytoplankton), which form the largest biomass in the sea. Where there are diatoms, we also find zooplankton—tiny protozoans and larvae of large animals. These together with diatoms provide the molecules that make up crude oil. As the rocks are buried, they are heated. The conversion from organic matter to petroleum takes place mostly between 50° C and 200° C. Thus a proper thermal history, ideally between 100° C and 120° C, is necessary to form liquid petroleum—too cool, and oil does not form; too hot, and the hydrocarbons "boil" away.

After petroleum has formed in a source bed, it is squeezed out and migrates through a simple or complex plumbing system into a **reservoir rock.** This migration is a critical element in the formation of an economically exploitable accumulation of oil. Reservoir rocks are porous and permeable (see Chapter 8). Commonly they are sandstones, porous limestones, or in some cases, fractured shales. Reservoir porosities range from 20 to 50 percent, meaning that for each cubic foot of reservoir rock we will find 1 to 4 gallons of oil. The unit of oil volume is the barrel (equal to 42 gallons), and a so-called giant field, such as the north slope of Alaska, will yield a billion barrels of oil in its lifetime. Using standard recovery techniques, however, as much as 40 to 80 percent of the oil may be left in pore spaces and as films on mineral grains.

An impermeable **caprock** prevents oil from seeping upward to form tar pits at the surface or dissipate into other rocks. Such seals are analogous to aquicludes in groundwater systems and are most commonly clay shales or limestones of low permeability. Our discussion of the fourth requirement, a suitable geologic structure for trapping the oil, requires a separate subsection.

Geologic Traps—Oil and Gas Stop Here

Structural Traps

An **anticline** is an ideal structure for trapping gas and oil. It is a convex-upward fold in stratified rock (◆Figure 14.2a). Analogous to a teacup inverted in a pan of water that traps a layer of air inside it, an anticline holds a reservoir of gas and oil. This occurs because crude oil floats on water and natural gas rises to the top of the reservoir. The anticlinal theory of oil accumulation was not developed until 1900, forty-one years after oil was discovered. Strata dip away from the central axis of an anticline at the ground surface, and most anticlines with surface geologic expression have been drilled. Today the search for oil is more difficult, because less obvious and geologically more complex traps need to be discovered. Many times faults form impermeable barriers to hydrocarbon migration, and oil becomes trapped against them (see Figure 14.2a). Faulted and folded stratified rocks in a single oil field may thus contain many isolated oil reservoirs.

Salt domes are of much interest to geologists (◆Figure 14.2b). Not only do they create oil traps, but they are also valuable sources of salt and sulfur and may be potential underground storage sites for petroleum and hazardous waste. More than 500 salt domes have been located along the U.S. Gulf Coast, both offshore and on land (◆Figure 14.3). They rise as flowing fingers of salt, literally puncturing their way through the overlying rocks and buckling the overlying shallow strata into a dome (◆Figure 14.4). Some fingers rise as high as 13 kilometers (8 mi) above the "mother" salt bed, the Louann salt, and would be taller than Mount Everest if they were at the surface of the earth. Because the salt is considerably less dense than the overlying rock, buoyancy forces drive the salt upward—much as a blob of oil will rise through water. Brittle solids, such as salt or ice, will flow over long periods of time, and many salt domes are still rising measurably. The "mother" salt was deposited in Jurassic time by evaporation of seawater when the embryonic Gulf Coast was connected to the open ocean by a shallow opening. The opening allowed seawater to enter the basin but did not allow

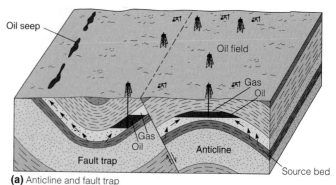

(a) Anticline and fault trap

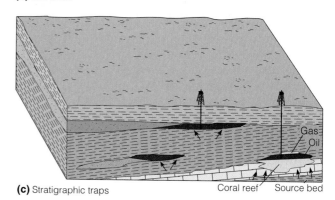

(b) Salt dome

(c) Stratigraphic traps

◆ FIGURE 14.2 (a and b) Common structural oil traps. (c) Stratigraphic oil traps.

the denser salt brines at the bottom to escape. Thus the brine became concentrated to the point of saturation, and the great thickness of salt that now underlies the entire Gulf Coast was deposited. Oil accumulates against the salt column and also in the dome overlying the salt. "Old Spindletop," a salt dome near Beaumont, Texas, that was discovered in 1901, is perhaps the most famous U.S. oil field. Soon after its discovery it was producing more oil than the rest of the world combined, and the price of oil plummeted to two cents a barrel.

Stratigraphic Traps

Any change in sedimentary rock lithology (its physical character) that causes oil to accumulate is known as a **stratigraphic trap** (in contrast to anticlines and salt domes, which are structural traps). Thus, if a stratum changes laterally

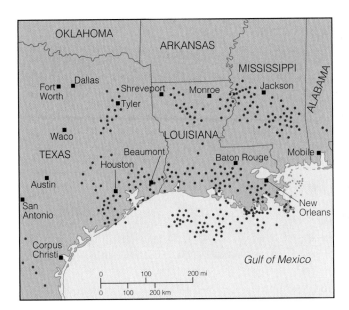

◆ FIGURE 14.3 Locations of salt domes of the U.S. Gulf Coast. More than 500 domes have been discovered on land and in the shallow parts of the Gulf of Mexico, and more are known to be in deep water offshore.

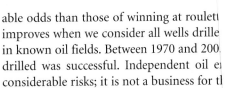

able odds than those of winning at roulett improves when we consider all wells drille in known oil fields. Between 1970 and 200 drilled was successful. Independent oil e considerable risks; it is not a business for tl

Most successful wells require pumping. .. gas and water pressure are sufficient, however, oil may simply flow to the ground surface. High reservoir pressures develop from water pressing upward (buoyancy) beneath the oil and gas pressure pushing downward on the oil (◆Figure 14.5). In some cases dissolved gases "drag" the oil along with them as they spew forth, as though from a bottle of champagne. If high pressures are not controlled, rocks, gas, oil, and even drill pipe may shoot into the air as a "gusher." High-pressure wells—749 in all—were set afire by Iraqi soldiers in Kuwait during Operation Desert Storm (◆Figure 14.6). Fewer than a dozen wells were set ablaze in the 2003 battle to free Iraq. Professional oil-fire-fighting firms such as the well-known Red Adair Company and Boots and Coots were called in. With the help of Kuwaiti *roughnecks* (the nickname applied to drilling-rig workers), the fires were extinguished in nine months, three months earlier than the most optimistic estimate.

It is possible to drill a well so that the drill hole slants from the vertical to penetrate reservoir rocks far from the drilling site. This is desirable where the oil structure is offshore and must be "slant-drilled" from land or from a drilling platform. **Slant drilling** is also employed for tapping reservoirs beneath developed land, such as in Beverly Hills, where a large number of oil wells are slant-drilled under fashionable commercial areas. Modern methods use "smart" drill bits that are remotely controlled. Producers can thus withdraw oil from below a large area more economically and at the same time minimize the visual blight of drilling towers.

from a permeable sandstone to an impermeable shale or mudstone, oil may be trapped in the stratum (Figure 14.2c).

Ancient coral reefs are ideal reservoirs, because they are porous and were biologically productive when they were living. Oil may be trapped in the porous, permeable debris on the flanks of the reef, and production from such fields can be measured in thousands of barrels per day (bbl/d). Many oil fields of the Middle East are of this type, and their production potential is tremendous. A comparison of reef production to that of sandstone reservoirs such as those of California or Texas, which typically yield only a few hundred barrels per day, explains why the Middle East can control oil production and therefore price.

Oil Production

The first successful oil well was drilled in Titusville, Pennsylvania, in 1859 (✪ Case Study 14.1 on page 409). In modern jargon, this well would be called a **wildcat well**, because it was the discovery well of a new field. There is 1 chance in 50 of a wildcat well being successful—less favor-

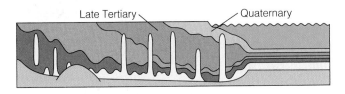

◆ FIGURE 14.4 Ancient salt deposits are buried deeply, and because salt is less dense than the overlying sediments, it rises buoyantly as pillars of salt, creating salt domes.

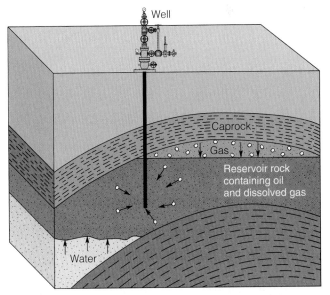

◆ FIGURE 14.5 Oil in an anticline is driven by gas pressure from above and by buoyant water pressure from below.

GURE 14.6 The Kuwaiti night skies
ere illuminated by 749 blazing oil wells
set afire by Iraqi troops during the 1991
Gulf War. The fires continued for nine
months.

The ultimate drilling technology of the 1990s was horizontal drilling, a technique by which the drill bit and pipe follow gently inclined reservoir sands and limestones. By this means a single borehole can provide access to reservoirs beneath a much larger area than provided by vertical drilling or slant drilling. This method is now superceded by multilateral drilling, which enables operators to reach multiple oil-bearing reservoirs by means of lateral extensions from one well (◆Figure 14.7). It is worth noting that some offshore drilling platforms are in water more than a mile deep and have strings of drill pipe dangling below them 8.5 kilometers (5 mi), about the height of Mount Everest. There are plans to drill exploratory holes in water depths greater than 3,100 meters (10,000 ft).

Secondary Recovery

Secondary recovery methods extract oil that remains in the reservoir rock after normal withdrawal methods have ceased to be productive. As much as 75 percent of the total oil may remain. Secondary recovery methods can be grouped into three categories: thermal, chemical, and fluid-mixing (miscible) methods (◆Figure 14.8). All of these methods require *injection* wells for injecting a fluid or gas and *extraction* wells for removing the remobilized oil.

Thermal methods include steam injection, which makes the adhering oil less viscous and thus more free to flow, and fire flooding—in which air is injected into the reservoir in order to set fire to the oil and thus produce gases and heat that will increase the flow of oil. Water injection is a chemical method that utilizes large-molecule compounds, which when added to water, thicken it and increase its ability to wash or sweep the adhering oil films and globules toward an extraction well. Fluids that mix with oil, that are **miscible,** are very effective in removing "stuck" oil from the reservoir. Miscible recovery methods use mixtures of water with

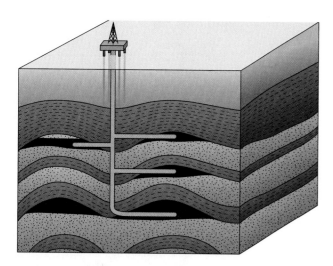

◆ FIGURE 14.7 Multilateral drilling from an offshore drilling platform allows many oil-producing zones to be tapped from one platform. In this illustration three zones are tapped by lateral pipes, and the bottom zone is tapped by a horizontal pipe.

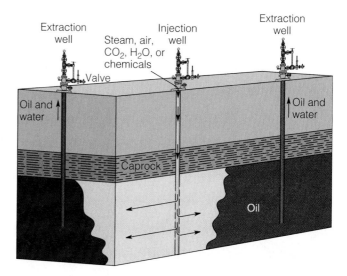

◆ FIGURE 14.8 Secondary recovery. Steam, air, carbon dioxide, or chemicals dissolved in water are injected into a sluggishly producing formation in order to stimulate the flow of oil to extraction wells.

*TABLE 14.2 Typical Composition of an A.P.I. Medium-Grade Crude Oil		
Common Name	Number of Carbon Atoms*	Percentage of Crude-Oil Weight
Gasoline	5–10	27
Kerosene	11–13	13
Diesel fuel	14–18	12
Heavy gas oil	19–25	10
Lubricating oil	26–40	20
Heavy fractions (tars, asphalt)	>40	18
		100

* Volatility decreases as the number of carbon atoms increases, C_3 to C_{40}.

propane or ethane extracted from natural gas or mixtures of CO_2 and water. Even after the use of these secondary recovery methods, as much as a fifth of the total oil may remain in the reservoir.

CONSIDER THIS *An acquaintance has offered you a share in a sure-shot wildcat oil-drilling venture in the Sierra Nevada. How should you respond to the offer?*

Quality and Price

The price of a barrel of crude (unrefined) oil varies with its grade—its quality—and with market demand at the time. Light oils bring higher prices, because they contain large proportions of paraffins and aromatic hydrocarbons, which are desirable for gasoline and bottled gases. Heavy crudes contain lower proportions of those components and greater proportions of the heavy, less valuable asphalts and tars (see Table 14.1). A scale of crude-oil quality has been established by the American Petroleum Institute based upon its weight.

Light crude oil is very fluid and yields a high percentage of gasoline and diesel fuel. Heavy crude, on the other hand, is about the consistency of molasses. The percentages of fuels and lubricating oils yielded by a barrel of medium-weight crude oil are shown in * Table 14.2.

The Future for Oil

Petroleum provides about 40 percent of the world's energy needs and as much as 90 percent of the energy needed for transportation. It is also critical to agriculture that provides energy to the six billion plus people on earth. Petroleum formed slowly millions of years ago and is finite as far as the human time scale goes. For these reasons it is not surprising that many people are involved in estimating the amount of oil left until the last drop is used.

Global oil reserves in the year 2001 were estimated to be 1,028 billion barrels of oil (◆Figure 14.9). (**Reserves** are the amount of an identified resource that can be extracted economically.) The number means little to us until we realize that approximately two-thirds of the proved reserves are located in just five countries, and about 72 percent of reserves are in predominantly Islamic nations. Of these Saudi Arabia has about one-fourth of the total, and Iran, Iraq, Kuwait, and the United Arab Emirates have about 100 billion barrels each of proved reserves (* Table 14.3).

The United States is the most energy-deficient of the largest oil-producing countries (* Table 14.4). We imported 56 percent of the 6.5 billion barrels of crude oil and natural gas liquids consumed in 2000. Natural gas liquids are condensates of natural gas that form when the gas depressurizes at the earth's surface. The liquids are processed to produce mostly butane and propane (natural gas liquids). Although domestic oil production has decreased about 30 percent over the past 20 years, natural gas liquids production has increased over 20 percent in the same period. At the current rate of production the U.S. reserves will last about ten years. Even though domestic production rates will continue to

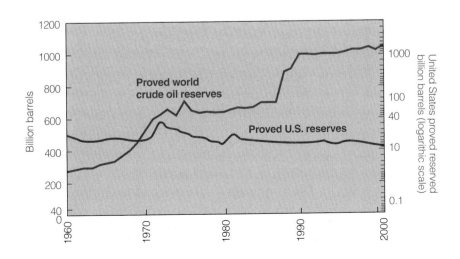

◆ FIGURE 14.9 World and U.S. proved petroleum reserves. Note the scale change for U.S. reserves.

✳TABLE 14.3 Global proved reserves, 2002

Country	Proved reserves (billion barrels)	Percentage of total world
Saudi Arabia*	262	25.5
Iraq*	112	10.9
United Arab Emirates*	98	9.5
Kuwait*	96	9.3
Iran*	90	8.8
Venezuela	77	7.5
Russia	49	4.8
Libya*	29	2.8
Mexico	28	2.7
China	24	2.3
Nigeria	22	2.1
United States	22	2.1
Qatar*	13	1.3
Norway	9	0.9
Algeria*	9	0.9
Brazil	8	0.8
Oman*	6	0.6
India	5	0.5
Indonesia*	5	0.5
United Kingdom	5	0.5
Kazakhstan*	5	0.5
Angola	5	0.5
Malaysia*	4	0.4
Yemen*	4	0.4
Canada	4	0.4
Rest of world	37	3.6
Total world	1,028	100.0
Total OPEC	814	79.2
Total for dominantly Islamic nations	742	72.2

*Dominantly Islamic nation
Source: Arizona Geological Survey, American Petroleum Institute (A.P.I.)

✳TABLE 14.4 2002 U.S. oil production*

State	Thousand barrels/day	Million barrels/yr	Percentage
Louisiana	1,513	552	25.7
Texas	1,400	511	23.8
Alaska	1,050	383	17.9
California	857	313	14.6
Oklahoma	193	70	3.3
New Mexico	176	64	3.0
Wyoming	167	61	2.8
North Dakota	90	33	1.5
Kansas	80	29	1.4
Other	355	130	6.0
Total	5,881	2,147	100.0
La., Tex., Alaska, Calif.	4,820	1,759	82.0
Gulf of Mexico OCS	1,420	518	24.1

*Not including natural gas liquids
Source: Arizona Geological Survey, American Petroleum Institute (A.P.I.)

decline (more imports), new reserves will be discovered so that there will not be an abrupt end to domestic oil production. (✿ Case Study 14.2 on page 410).

Arctic National Wildlife Refuge

The U.S. Geological Survey in 2002 estimated that there are between 5.7 and 16 billion barrels of oil in the Arctic National Wildlife Refuge (ANWR). They give a mean estimate of 10.4 billion barrels. The amount of oil that can be *economically* recovered depends upon the price of crude. At $13/barrel, the price of crude in early 2002, recovery would be minimal. At $35/barrel it is realistically estimated that about 70 to 90 percent of the recoverable oil could be extracted. There are a myriad of reasons, both political and environmental, why ANWR should not be opened for exploration and production of oil. However, the flip side is that with conservation measures and the added production from ANWR, the United States could reduce its dependence on imported oil 25 percent. It should be noted that even if the refuge were open tomorrow, it would be 7 years at the least before a single barrel of oil reaches the lower forty-eight states. The per capita energy use in the late 1990s United States should be kept in mind when evaluating the U.S. energy outlook. Some interesting facts are that each individual consumes:

- 4 tons of oil,
- 2.5 tons of coal,
- 2 tons (+) of natural gas, and
- the individual's weight in oil—every 7 days.

The United States has about 5 percent of the world's population but uses 33 percent of the world's energy supplies. However, even if ANWR were open to oil production today the relief would only be temporary. It is imperative to have a transition to a different kind of transportation and an energy source to drive it. Just how fast this transition takes place will determine how disruptive it will be.

The future is not so bleak when we factor U.S. ingenuity into the energy equation. Much oil remains to be discovered using current exploration techniques, although a crude-oil price of at least $25-$30/barrel is required to make the expensive searches feasible. In addition, there are nonconventional sources of petroleum such as oil shale, tar sands, and coal from which liquid fuel can be derived. Coal is also a source of fuel-grade natural gas at present.

Energy Gases and the Future

Some experts predict that the decline in the supply of *affordable* oil will begin as early as 2010, others say the crunch will not start until 2040 or 2050. It will happen, however, and the speculation about when the world will run completely out of oil is not relevant. What does matter is when production will fall off as demand inexorably continues to rise. In other words, price depends on the well-known supply and demand curves of the economists. ◆ Figure 14.10 reflects this. It shows the predicted decline in the use of oil in the twenty-first century and the increasing use of natural energy gases and alternative energy sources. Natural gas is currently plentiful and burns cleaner than coal, gasoline, or heating oil. It currently accounts for 75 percent of the total U.S. energy used in space heating and cooking. Natural gas is measured as cubic feet of gas at an atmospheric pressure of 14.7 pounds per square inch and a temperature of 60° F. A trillion cubic feet (Tcf) of gas would fill a cube 2 miles on a side (8 mi³), and the United States consumes about 17 Tcf per year. We have proven reserves of 177 Tcf, which suggests that we will "run out" in 10.4 years. However, it is estimated that about 2,400 Tcf of natural gas exists in the United States (see Fisher in the For

Further Information section). Many new discoveries will be made, and 40 percent of those will be at depths greater than 4,500 m (14,500 ft, almost 3 mi). About 300 Tcf will come from coal beds and 500 Tcf will come in the form of growth of existing fields. Fisher calls the approaching transition period from oil to gas the "methane economy."

In 2003 coal-bed methane was considered a world-class commodity by many states, and there is a 95 percent probability that 30 TCF are to be found in the coal regions of Wyoming, Colorado, Utah, and New Mexico. This form of coal energy is easy to extract by a drilled well as opposed to extracting the coal, which involves destructive surface mining (◆Figure 14.11). The opposing view is that gas production requires disruptive wells and roads and there is methane leakage, all of which degrade the environment. Speaking of methane, don't forget methane hydrates (see the chapter opener).

Hydrogen has a bad reputation because it was the gas that filled the German dirigible *Hindenburg* that exploded at Lakehurst, New Jersey, in 1937. Also, the testing of the first hydrogen bomb in 1952 is still remembered by many with fear and awe. Many experts believe it is the energy of the future, however, as it burns without pollutants, emitting just water vapor. Unfortunately, there are no large reservoirs of hydrogen, and at present it must be manufactured using fossil fuels. It offers promise as a fuel for airplanes and autos, and will be economically viable when it can be produced in large quantities using solar or hydroelectric energy. (Fuel cells that burn hydrogen gas are discussed later in the chapter.)

Coal

Coal is the carbonaceous residue of plant matter that has been preserved and altered by heat and pressure. Next to oil and oil shales, coal is the earth's most abundant reservoir of stored energy. Deposits are known from every geologic period since Devonian time and the appearance of widespread terrestrial plant life some 390 million years ago. Permian coal is found in Antarctica, Australia, and India—pre-continental drift Gondwanaland (Chapter 3). The large fields of North America, England, and Europe were deposited during the Carboniferous Period, so-named because of the extensive coal deposits in rocks of that age. In the United States the Carboniferous Period is divided into the Mississippian and Pennsylvanian Periods, named after the rocks found in those states. Tertiary coals are found in such diverse locations as Spitsbergen Island in the Arctic Ocean, the western United States, Japan, India, Germany, and Russia.

Coalification and Rank

The first stage in the process of *coalification* is the accumulation of large amounts of plant debris under conditions that

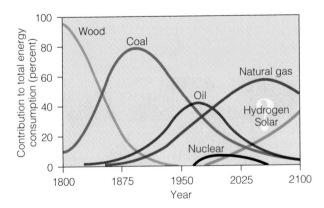

◆ FIGURE 14.10 Use of energy resources in the United States since 1800, with projections to 2100. The shifts from wood to coal and then from coal to oil and natural gas each took about 50 years. Many analysts believe a new shift, to increased use of solar energy and hydrogen gas, will occur over the next 50 years. Note the natural gas peak at about 2050.

◆ FIGURE 14.11 Schematic cross section showing conventional surface mining technology and relationship to coalbed methane production well. Colorado Geological Survey

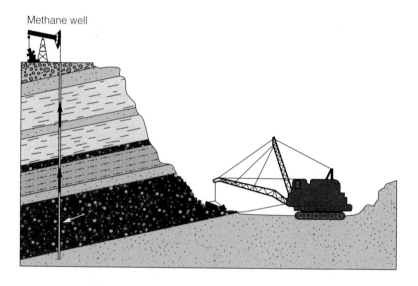

Methane well

will preserve it. This requires high plant production in a low-oxygen depositional environment, the conditions usually found in nonmarine brackish-water swamps. The accumulated plant matter must then be buried to a depth sufficient that heat and pressure expel water and volatile matter. The degree of metamorphism (conversion) of plant material to coal is denoted by its **rank.** From lowest to highest rank, the metamorphism of coal follows the sequence *peat,* to *lignite,* to *subbituminous,* to *bituminous,* to *anthracite.* The sequence is accompanied by increasing amounts of fixed carbon and heat (Btu) content, and a decreasing amount of quickly burned volatile material. Once ignited, it is the carbon that burns (oxidizes) and gives off heat, just as wood charcoal does in a barbecue pit.

Bituminous, or soft coal, usually occurs in flat-lying beds at shallow depths that are amenable to surface mining techniques. Anthracite, the highest rank of coal, is formed when coal-bearing rocks are subjected to intense heat and pressure—a situation that sometimes occurs in areas of plate convergence. In the Appalachian coal basin, high-volatile bituminous coals occur in the western part of the basin and increase in rank to low-volatile bituminous coals to the east. Anthracite formed close to the ancient plate-collision zone of North America and Africa (pre-Pangaea), where the coals were more intensely folded and subject to higher levels of heat. Traveling east from the Appalachian Plateau to the folded Ridge and Valley Province, one goes from flat-lying bituminous terrain to folded anthracite terrain. West Virginia, Kentucky, and Pennsylvania are the leading coal producers of the eastern states. The interior coal basins of the Midwest are characterized by high-sulfur bituminous coal, whereas the younger Western deposits are low-sulfur lignite and subbituminous coal. ✳ Table 14.5 projects the development of coal-mining activity in the United States. The expansion of mining activity will occur mainly in the Western coal basins, because the coal seams there are very thick and easily mined and the coals are low in sulfur. Wyoming, with its low-sulfur coal, was the leading coal producer in 1993 (190 million short tons), followed in order by West Virginia, Kentucky, Pennsylvania, and Illinois (◆Figure 14.12).

Reserves and Production

The future of coal-derived energy in the United States is far more optimistic than the outlook for petroleum. Coal constitutes 80 percent of U.S. energy stores but only 18 percent of present usage. In 1989, electric utilities accounted for 86 percent of the coal consumed, with residential and industrial use accounting for the remainder. It is estimated that there are 84 trillion (84×10^{12}) recoverable tons of coal in the world, the equivalent of 34 trillion barrels of oil. The United States has 283 billion tons of recoverable reserves, which could last 200–300 years at the current rates of production and use. These reserves could meet only 50 years of total U.S. energy demand, however.

It has been estimated that energy demand in the early twenty-first century will be between 100 and 150 quads (a quad is a quadrillion British thermal units, 10^{15} Btu), com-

✳TABLE 14.5 Active U.S. Coal Mines by Region, 1977 and 2000

Region	Surface Mines		Underground Mines		Total	
	1977	2000	1977	2000	1977	2000
Appalachia*	185	130–175	205	380–500	390	510–675
Midwest	91	95–135	54	120–180	145	215–315
West	141	700–1,005	13	80–110	154	780–1,115

*Encompasses parts of Pennsylvania, eastern Kentucky, West Virginia, Virginia, Maryland, Tennessee, and Ohio.
Source: U.S. Geological Survey

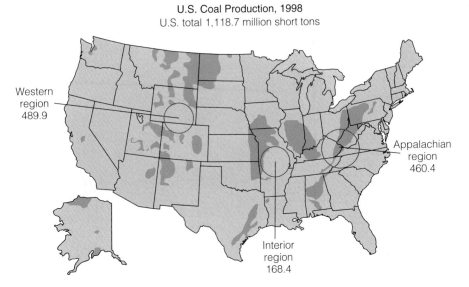

U.S. Coal Production, 1998
U.S. total 1,118.7 million short tons

Western
region
489.9

Appalachian
region
460.4

Interior
region
168.4

◆ FIGURE 14.12 Production in short tons for each coal-producing region and the total for the United States, 1998. Note the shift in production from the Appalachian region to the Western coal states. Energy Information Administration

pared to 73 quads in 1973 and about 80 quads in 1980 (✳ Table 14.6). Coal production will double to meet the demand, and this increase will be mostly in the Western coal basins, where the relatively inexpensive surface-mining techniques can be used. Underground mining costs more, requires a greater capital investment, and takes more time to get into production. Also, miners who work underground face greater risks than those who work at the surface.

The economic future of coal lies in its use as a substitute fuel for oil and gas. The technology is in place to make methane from coal—the product is called *synthesis gas*—and it has been used on a large scale to manufacture petrochemicals, including methanol (methyl alcohol). Methanol can be converted to gasoline using a mineral catalyst. Methanol is an interesting fuel: its combustion generates little nitric oxide (NO) and ozone, and it has a high octane number—about 110. Its high octane has made it attractive for years as a fuel for race cars and high-compression engines. Although it has a lower heat (Btu) content than gasoline, it burns more efficiently, which evens things out. On the negative side, methanol releases greater emissions of carbon dioxide. **Biomass** alcohol (ethanol) made from sugar cane has been used in Brazil for years as an auto fuel, and at one time almost 90 percent of the cars there burned it. As the use of lead in gasoline is phased out, methanol can be added as an octane enhancer.

Direct liquefaction of coal shows great promise. Using coal- or petroleum-derived solvents, coal is dissolved and hydrogenated, resulting in about 75 percent gasoline and the remainder propane and butane. About 5.5 barrels of liquid are derived from a ton of coal at the approximate cost of the most expensive oil during the war in Iraq, $35–$40/barrel. A major facet of our energy future lies in converting coal to liquid or gaseous **synthetic fuels,** often called simply *synfuels*.

CONSIDER THIS *Your Christmas stocking contains nothing but a lump of anthracite coal. Being an optimist, you rejoice in this gift. Why?*

Nonconventional Fossil Fuels

Tar Sands

Tar sands contain oil that is too thick and viscous to flow at normal temperatures. These deposits are found in Canada, Venezuela, Madagascar, and the United States and they are surface-mined or drilled, using steam to mobilize and recover the oil. Canada's Athabasca sands (see Figure 14.13a) constitute the largest oil field in the world; they are estimated to contain more than 1.7 trillion barrels of oil, more than all the reserves of Saudi Arabia. At Athabasca the sand is extracted by open-pit mining techniques, and the oil is separated from the sand using steam or other thermal methods. The amount of material moved in these excavations is

✳TABLE 14.6	Energy (Power) Units
Unit	**Explanation**
British thermal unit (Btu)	The amount of heat required to raise the temperature of 1 pound of water 1° F; ≈ energy released by a burning match
Quad (quadrillion)	10^{15} Btu = 172 million bbl/oil
Barrel of oil	5.8 million Btu = 42 gallons
Bituminous coal (average)	25 million Btu/ton
Natural gas	Variable Btu content, measured in cubic feet (cf)
Megawatt (MW)	1,000 kilowatts (kw), or a million watts
Gigawatt (GW)	1,000 megawatts, or a billion watts

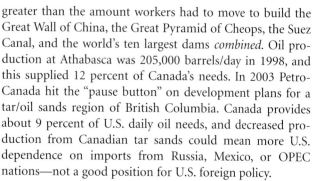

(a)

(b)

◆ FIGURE 14.13 (a) Locations of Alberta's tar sands: Athabasca, Cold Lake, and Peace River. (b) After extraction by huge machinery, the sands are washed with hot water to extract the oil adhering to the grains. Cleaned sand is then returned to the open pit, and the land is restored to its original contours.

greater than the amount workers had to move to build the Great Wall of China, the Great Pyramid of Cheops, the Suez Canal, and the world's ten largest dams *combined*. Oil production at Athabasca was 205,000 barrels/day in 1998, and this supplied 12 percent of Canada's needs. In 2003 Petro-Canada hit the "pause button" on development plans for a tar/oil sands region of British Columbia. Canada provides about 9 percent of U.S. daily oil needs, and decreased production from Canadian tar sands could mean more U.S. dependence on imports from Russia, Mexico, or OPEC nations—not a good position for U.S. foreign policy.

Tar sands that are too deep (500 m) for surface exploitation underlie Alberta's Cold Lake area. Steam at 300° C (575° F) is injected at high pressure and allowed to "soak" for a month. The oil released from the sand is then brought to the land surface by familiar "rocking-horse" pumps. More than 2,000 steam-injection and oil-production wells deliver 130,000 barrels of oil per day. The area is estimated to contain 220 billion barrels of recoverable oil.

Oil Shales

Oil shales are sedimentary rocks that yield petroleum when they are heated. Found on all continents, including Antarctica, they were originally deposited in lakes, marshes, or the ocean. The original rock oil used in kerosene lamps was produced from black oil shales. The most extensive U.S. deposits, the Green River shales, were formed during Eocene time in huge freshwater lakes in the present-day states of Wyoming, Colorado, and Utah. They are not oily like tar sands, but when heated to 500° C (900° F) they yield oil (◆Figure 14.14). The source of oil in the rock is **kerogen,** a

solid bituminous substance, and a ton of oil shale may yield 10–150 gallons of good-quality oil. Potential deposits of economic interest yield 25–50 gallons per ton, and the U.S. Geological Survey includes 3 trillion barrels of oil from shale in their estimates of reserves. Kerogen-rich lake deposits are also found in China, Yugoslavia, and Brazil.

Water is the limiting factor because processing the shale requires water volumes 3–4 times greater than that of the oil extracted. Furthermore, the extraction process destroys the natural landscape, and some of the richest shales are in scenic wilderness areas. At this time the oil shale industry cannot compete economically in the energy market.

Problems of Fossil Fuel Combustion

Air Pollution

Obtaining energy by burning fossil fuels creates environmental problems of global proportions. It produces oxides of carbon, sulfur, and nitrogen and fine-particulate ash. Carbon monoxide (CO), an oxide produced by combustion of all fossil and plant fuels, is converted to carbon dioxide (CO_2), which contributes to global warming. Burning coal also releases sulfur oxides (SO_x) to the atmosphere, where they form environmentally deleterious compounds. Nitrogen oxides (NO_x), mostly NO and NO_2, are products of combustion in auto engines and are the precursors of the photochemical oxidants, ozone and peroxyacetyl nitrate (PAN), which we associate with smog. Water and oxygen in the atmosphere combine with SO_2 and NO_2 to form sulfuric acid (H_2SO_4) and nitric acid (HNO_3), the main components of **acid rain**—rain with

◆ FIGURE 14.14 Oil shale from the Green River Formation in Wyoming and a beaker of the heavy oil that can be extracted from it.

increased acidity due to environmental factors such as atmospheric pollutants (⊙ Case Study 14.3 on page 411).

The Clean Air Act of 1963 as amended in 1970 and 1990 specifies standards for pollutant oxides and hydrocarbon emissions. By 1980 new cars were 90 percent cleaner than their 1970 counterparts. The Clean Air Act Amendments of 1990 (CAAA) set goals and timetables that are affecting how and what we drive. Beginning in 1992, fuel suppliers were required to sell only reformulated gasoline in 39 areas where winter air quality was a problem. Reformulations using methyl tertiary butyl ether (MTBE) add as much as 2.5 percent oxygen to the fuel, causing it to burn cleaner and create less ozone, at a cost of about 10 cents more per gallon. Unfortunately, reformulated gasoline (RFG) is stored in underground tanks that sometimes leak and pollute ground water. As a result a Blue Ribbon Panel appointed by EPA administrator Barbara Browner has recommended the following:

- Remove the Clean Air Act requirement for 2 percent oxygen in RFG.
- Enhance underground storage tanks, thus improving protection of the nation's drinking water.
- Reduce the use of MTBE nationwide, maintaining current air quality benefits.

So much for MTBE and reformulated fuels.

CONSIDER THIS *Historically, Pennsylvania crude oil (such brands as Quaker State and Pennzoil) has been known for its superior lubricating properties compared to, say, heavier California or Texas crude oils. Why do you suppose this is so? (Table 14.1 can help you deduce a reasonable answer.)*

Sulfur Emissions and Acid Rain

Sulfur occurs in coal as tiny particles of iron sulfide, most of which is the mineral pyrite, or "fool's gold" (◆ Figure 14.15). Coal also contains organic sulfur originally contained in the coal-forming vegetation. Upon combustion, mostly in coal-burning electrical generating plants, the sulfide is oxidized to sulfur dioxide (SO_2) and carried out through smokestacks into the environment. Emissions of SO_2 are known to have a negative impact on human health, particularly lung problems. Although most coal-burning in the United States occurs in the Midwest and the East, significant volumes of SO_x can travel 1,000 kilometers (600 mi) or more from their source. ◆ Figure 14.16 shows typical U.S. SO_2 emission densities, which reflect the pattern of coal usage very closely. Mobile sources using petroleum fuels emit less than 4 percent of the total.

Sulfur dioxide emitted from a smokestack may be deposited dry near the facility, where it will damage plant life and pose a health threat to animals. If the gas plume from the smokestack is transported away from the facility, its SO_2 will react with the atmosphere and eventually fall to the earth surface as acid rain. Acid rain can damage crops; acidify soils; corrode rocks, buildings, and monuments; and contaminate streams, lakes, and drinking water. Meanwhile, it is estimated that annual U.S. emissions of sulfur gases will increase almost 30 percent from 1990 levels by 2010.

Sulfur oxides can be removed from combustion gases in smokestacks by using a device called a *scrubber* that selectively reacts with SO_2 and absorbs or neutralizes it. Many coal-burning facilities are switching to a process known as fluidized-bed combustion. Coal and pulverized limestone are burned together, and the heat of combustion converts the limestone into calcium oxide (CaO) and CO_2. CO_2 is emitted, and the CaO combines with sulfur and oxygen to form calcium sulfate ($CaSO_4$, a gypsumlike substance) as a solid waste product. Sulfur emissions are reduced, but CO_2 emissions increase. It is thus a tradeoff between the health benefits of reduced atmospheric sulfur and the environmental effects of increased contributions to global warming.

Nitrogen Oxides and Smog

Nitrogen oxide emissions (NO_x) are significant from both stationary (coal-burning) and mobile (vehicular) sources. Mobile sources are the most difficult to control, and their contribution to total NO_x is increasing as the number of automobiles increases. NO_x forms when engine combustion causes nitrogen in air to oxidize to NO and NO_2, which then react with oxygen, free radicals (incomplete molecules that are highly reactive), and unburned hydrocarbons to produce **photochemical smog.** Photochemical smog requires sunlight for its formation. Although the word smog is a contraction of the words *smoke* and *fog*, the air pollution in Los Angeles, Mexico City, and Denver has little direct relationship to either smoke or fog.

Both NO_2 and ozone (O_3) are reactive oxidants and cause respiratory problems. A level of only 6 parts per million (ppm) of ozone can kill laboratory animals by pulmonary edema (water in the lungs) and hemorrhage within four hours. In humans, alcohol consumption, exercise, and high temperatures have been found to increase adverse reaction to ozone, whereas vitamin C and previous exposure seem to lessen the reaction. One investigator noted that the person most likely to succumb to ozone would be "an alcoholic with a 'snootful' arriving at LAX for the first time on a hot, smoggy day and jogging all the way to Beverly Hills." Such a scenario may not be all that unlikely in Southern California.

Energy-efficient cars and new fuels have gone a long way to decreasing NO_x emissions. Exhaust-system catalytic conversion of nitrogen oxides, carbon monoxide, and hydrocarbons into carbon dioxide, nitrogen, and water are mandatory automotive equipment on new cars in most states.

Domestic Coal Burning

China is the world's largest producer of coal, accounting for 25 percent of global output. It currently produces 10 percent of the global emissions of CO_2, and it is estimated that by 2025 it will emit more CO_2 than the United States, Canada, and Japan combined. Coal provides 76 percent of the huge country's commercial energy, and it is the only fuel millions of people use.

In a typical home coal is burned in unvented stoves for cooking, space heating, and heating water. In many areas, foods are brought indoors in the fall for drying with coal fires, as the climate is too cool and damp to dry them outdoors. At

◆ FIGURE 14.15 Small pyrite masses in a bituminous coal matrix (gray). The light-gray material is the cell walls of the original plant material, and the darker gray is the solid material that was contained within the cell cavities. Earth pressures compacted the plant material, squeezed out any water, and collapsed the cavities. That pyrite formed within the coal before completion of compaction is evidenced by plant materials bent around solid pyrite masses. The pyrite "blebs" are about 50 microns across.

least 3,000 people in Guizhou Province in southwest China suffer from severe arsenic poisoning caused by consuming chili peppers dried over high-arsenic coal fires (◆Figure 14.17). Although fresh chili peppers have less than 1 ppm

◆ FIGURE 14.16 Atmospheric concentration of sulfate for a 26-quad U.S. coal-use scenario in 1990. The point is that the heaviest concentrations are in the Northeast, where most coal is burned. After A. C. Stern, ed., *Air Pollution*, Vol. 1, San Diego, Academic Press.

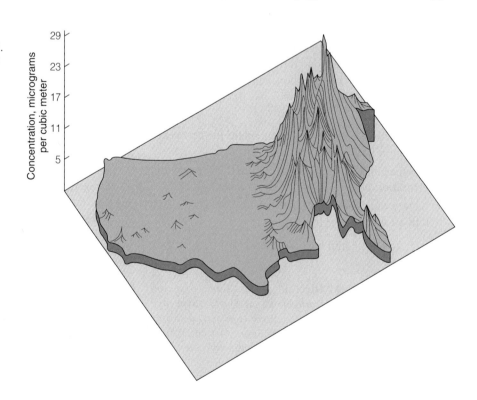

arsenic, chili peppers dried over high-arsenic coal have up to 500 ppm arsenic! More than ten million people in the same province suffer dental and skeletal fluorosis due to eating corn that has been cooked over coal containing large amounts of fluorine. About 3.5 billion people worldwide are exposed to toxic fumes indoors due to coal burning. Each year, an estimated 178,000 people die prematurely due to indoor air pollution in China alone. To combat these health hazards Chinese and U.S. Geological Survey geologists are working together to identify coal deposits whose high toxic-element concentrations make them unsuitable for domestic use.

Mine Collapse

Whenever coal is extracted from shallow underground seams, a collapse of the overburden into the mine opening is risked. Lignite and bituminous (soft) coals are most susceptible to collapse because they are exploited from near-horizontal beds at relatively shallow depths. The "room and pillar" method recovers about 50 percent of the coal, leaving the remainder as pillars for supporting the "roof" of the mine. As a mine is being abandoned, the pillars are sometimes removed or reduced in size, increasing the yield from the mine but also increasing the possibility of subsidence at some time in the future. Even when the pillars are left, failure is possible. The weight of the overlying strata can cause the pillars to rupture, and sometimes it physically drives the pillars into the underlying shale beds. Subsidence has been documented in Pittsburgh, Scranton, Wilkes-Barre, and many other areas in Pennsylvania. In 1982 an entire parking lot disappeared into a hole above a mined-out coal seam 88 meters (290 ft) below the surface. Wyoming, Montana, South Dakota, and other western "lignite" states are pockmarked with subsidence features, but human activities and human lives are less likely to be affected by subsidence, or collapse, in those areas.

Energy for the Future

Our oil-based economy is "mature," and certainly the energy scenario of the twenty-first century will be different and not carbon-based as it is now. The primary energy source will be, as always, the sun. The primary fuel may very well be hydrogen, the most common element in the universe and one that burns readily. This section examines the alternative energy sources that are available to us at this time.

Direct Solar Energy

Although the amount of solar energy that reaches the earth far exceeds all human energy needs, it is very diffuse. For example, the amount that strikes the atmosphere above the

◆ FIGURE 14.17 Drying chili peppers over a coal fire that emits arsenic fumes increases the arsenic content of the chilies up to 500 times, and eating them causes arsenic poisoning. Also, as families gather about coal fires at night, they are exposed to polycyclic aromatic hydrocarbons from incomplete combustion of the coal. Breathing these substances can lead to lung and esophageal cancer.

British Isles is 80 times Great Britain's energy needs; it averages about 1 kilowatt per square meter. Cloud cover reduces the amount by 80 percent, but there is still more than enough. So why aren't there solar-electrical plants in every city and hamlet? The reason is cost. The current technology requires a great deal of space, which makes installations expensive. There are essentially two ways to put the sun to work for us:

1. Solar-thermal methods utilize panels or "collectors" to warm a fluid, which is then used for heating or for generating electricity.
2. Photovoltaic cells use the second most common element on earth, silicon, to convert light directly into electricity.

In 1995, the electricity produced from photovoltaic cells cost about 16 cents per kilowatt hour (kwh), many times that produced by fossil fuels (✳ Table 14.7). However, it is estimated that photovoltaic electricity will soon cost less than 10 cents per kilowatt hour, and will compete with conventional electrical generation in this century.

Solar-thermal facilities called *power towers* illustrate a creative use of solar energy. The "Solar One" station in the Mojave Desert near Barstow, California, generated electricity between 1982 and 1988 for the Southern California Edison Company. It employed 1,818 suntracking mirrors focused on a water-filled receiver atop a tower where temperatures reached 900° C (◆Figure 14.18). The super-heated water was transferred to a plant where it flashed to steam and drove turbines that operated an electrical generator. Because of the success of Solar One, a 10-megawatt facility was built on the same site and began operating in mid 1996. Solar Two uses molten nitrate salt as the heat-transfer medium. Molten salt has excellent heat-storing properties; the heat can be used immediately to generate steam, or it can be stored and used later, during cloudy periods or after sundown. Solar Two generated enough power for 10,000 homes and operated from 1996 to 2000, when government funding ended. San Francisco has installed a $5.2 million solar energy project at the Moscone Convention Center. It was dedicated in late 2002, and it is estimated that it will cut energy costs as much as 38 percent and, after debt service, save the taxpayers $200,000.

Photovoltaic cells (PV cells) absorb pulses of light energy on semiconductor materials and turn that energy into an electric current. They are used widely as a power source for satellites, calculators, watches, and remote communication and instrumentation systems. In recent years the cost of PV cells has been reduced by 80 percent, and it is now possible to apply thin films of these materials to shingles, tiles, and window glass, enabling buildings to generate their own elec-

tricity. Costs need to fall another 50–75 percent to make them competitive with fossil-fuel generating plants. Several U.S. utility companies have installed PV systems to supply small users in out-of-the-way places without building costly power-line extensions.

Indirect Solar Energy

Wind Energy

Wind power is the fastest growing energy segment in the world and has been for the past 20 years. At the same time the cost of wind power has dropped approximately 90 percent, but present production pales in comparison to what electrical generation will be 20 years from now. Costs depend on the wind velocity, turbine design, and the diameter of the

✳TABLE 14.7	Cost of Electricity by Energy Source
Energy Source	**Cost, cents/kwh**
Sun	15.8*
Geothermal	13.3*
Biomass	13.0*
Wind	7.5**
Oil	4.4†
Gas	2.9†
Coal	1.2†
Nuclear	0.93†

* The rates that Southern California Edison Company was paying for purchased and manufactured energy in December 1995.
** 1998 value.
† The rates that Southern California Edison Company was paying for purchased and manufactured energy in December 1991 (more recent data not available).

RICHARD KEELER, SOUTHERN CALIF. EDISON CO.

◆ FIGURE 14.18 Solar One, a direct solar-energy power conversion facility in the Mojave Desert near Barstow, California. The heliostat (reflector) field directs sunlight to the central tower.

propeller. First, the energy that can be tapped from the wind is proportional to the cube of the wind velocity. For this reason a slight increase in the wind speed results in a large increase in the electricity produced. For instance, an increase of the wind speed from 14 to 16 miles/hour results in an increase in electrical generation of nearly 50 percent. Second, a larger blade size requiring a taller tower yields a good increase in production. Increasing the blade diameter from 10 meters (1980s technology) to 50 meters (2004 technology) gives a 55-fold increase in yearly electrical production. Part of this is because a taller tower is required, and wind speeds normally increase with distance from the ground. Finally, a bigger wind farm is less costly to operate than a small one and thus the electricity is cheaper. Nearly forty percent is the difference between a small 3-MW installation (0.059 ¢/kwh) and a larger 51-MW plant (0.036 ¢/kwh).

California leads the nation in wind power and produces enough electricity to light a city the size of San Francisco. Ninety-five percent of California wind turbines are in three locations: Tehachapi Pass southeast of Bakersfield, Altamont Pass east of San Jose, and Banning Pass near Palm Springs (◆Figure 14.19). These narrow passes through the mountains separate hot, desert-like valleys from areas with cooler maritime climates. As the heated air rises in the valleys, cooler air from the maritime regions is drawn through the narrow passes to replace it. Modern, "smart" windmills with larger props and programmed turbines produce electricity at a price competitive with other sources (✳ Table 14.7).

The wind potential of the lower 48 states is shown in ◆ Figure 14.20. One of the most attractive features of wind is that it can produce electricity during peak demand periods, on hot summer days and early in the evening in winter, thus avoiding construction of additional fossil fuel plants to meet the need. There are environmental problems that accompany wind generation, such as noise, land acquisition, TV interference, and most important, visual blight. Modern generators have largely solved the problem of noise and TV reception, but a couple of hundred windmills are hard to ignore. The feeling is that they destroy the visual aesthetics of a region. On the other hand, ranchers at Altamont love the windmills. They can still graze their cattle and at the same time receive land use royalties that have increased their pre-windmill land value.

Hydroelectric Energy

Falling water, our largest renewable resource next to wood, has been used as an energy source for thousands of years. First used to generate electricity about 100 years ago on the Fox River near Appleton, Wisconsin, today it provides a fourth of the world's electricity. In Norway, 99 percent of the country's electricity and 50 percent of its total energy is produced by falling water. The principle is relatively simple: impound water with a dam and then cause the water to fall through a system of turbines and generators to produce electricity. Because of this simplicity, electricity coming from

◆ FIGURE 14.19 Wind "farm" on the Mojave Desert side of Banning Pass, near Palm Springs, California. The San Bernardino Mountains are in the background.

existing facilities is also the cheapest source of this power. Aitupu Dam on the Parana River between Paraguay and Brazil, completed in 1982, is the largest-producing electrical complex in the world; its potential is 12,600 megawatts. It provides much more electricity than the present demand in this part of South America.

Because of land costs and environmental considerations, it is doubtful that any more large dams will be built in the United States (see Chapter 9, also Case Study 9.3). Therefore,

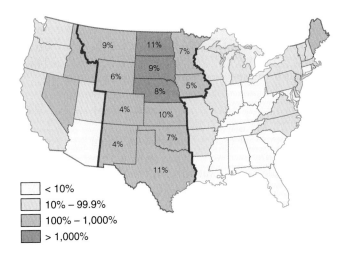

< 10%
10% – 99.9%
100% – 1,000%
> 1,000%

◆ FIGURE 14.20 Wind-electric potential of the 48 contiguous states as a percentage of each state's energy need. Twelve states in the central part of the country could provide about 90 percent of the total U.S. need. From "Energy from the Sun," by C. J. Weinberg and R. Williams, *Scientific American*, 1990.

dams that were built for other purposes have become attractive for retrofitting with generators. Hydroelectrical facilities are particularly useful for providing power during times of peak demand in areas where coal, oil, or nuclear generation provides the base load. Hydroelectricity can be turned on or off at will to provide peak power, and many utilities have built pumped-water-storage facilities for just that reason. During off-peak hours, when plenty of power is available, water is pumped from an aqueduct or other source to a reservoir at a higher level. Then during peak demand the water is allowed to fall to its original level, fulfilling the temporary need for added electricity.

Although hydroelectric energy is clean, dams and reservoirs change natural ecological systems into ones that require extensive management. One of the social consequences of dam building is displaced persons. This displacement is unacceptable in most societies today. Also unacceptable are recurring attempts to dam areas of great aesthetic value, such as the Grand Canyon. Finally, dam failures—most often geological failures of dams' foundations—result in floods that can take enormous human and economic tolls.

Geothermal Energy

The interior of the earth is an enormous reservoir of heat produced by the decay of small amounts of naturally radioactive elements that occur in all rocks. When the deep heat rises to shallower depths, this **geothermal energy** can be tapped for human use. Furthermore, this natural heat is cleaner than many other energy sources, and there are fewer negative environmental consequences of using it. Not surprisingly, the prospects for using geothermal energy are best at or near plate boundaries where active volcanoes and high heat flow are found. The Pacific Rim (Ring of Fire), Iceland on the Mid-Atlantic Ridge, and the Mediterranean belt offer the most promise. Today, the United States, Japan, New Zealand, Mexico, and countries of the former USSR are utilizing energy from earth heat. Iceland uses geothermal energy directly for space heating (one geothermal well there actually began spewing lava), and many other countries have geothermal potential. The versatility of earth heat ranges from using it to grow mushrooms to driving steam-turbine generators (◆Figure 14.21).

Geothermal energy fields may be found and exploited where magma exists at shallow depths and where there is sufficient underground water to form steam. Geothermal fields are classified as either *steam-dominated* systems or *hot-water-dominated* systems. Larderello, Italy, and The Geysers near Napa Valley in northern California (◆Figure 14.22) are examples of steam-dominated fields. Such fields are the rarest and most efficient energy producers. They occur where water temperatures are high and discharge is low so that steam forms. They are much like an oil reservoir in that porous rocks that hold steam or very hot water are overlain by an impermeable layer that prevents the upward escape of the

steam or water. ✳ Table 14.8 shows the electrical generating and direct-use capacities of selected U.S. geothermal fields in 1990. The Geysers is the largest producer of geothermal power in the world; its production is sufficient to supply electricity to a city of over one million people. More than 600 wells had been drilled there, some of them as deep as 3.2 kilometers (10,400 ft).

Geothermal energy is clean. The amount of CO_2 it emits is about one-tenth of the CO_2 emitted by coal-burning plants per megawatt of electricity produced, and about one-sixth of relatively clean-burning natural gas. Similar contrasts exist for emissions of sulfur and sulfur oxides.

The problems associated with geothermal energy production are related to water withdrawal and water quality. Some geothermal waters contain toxic elements such as arsenic and selenium and heavy metals such as silver, gold,

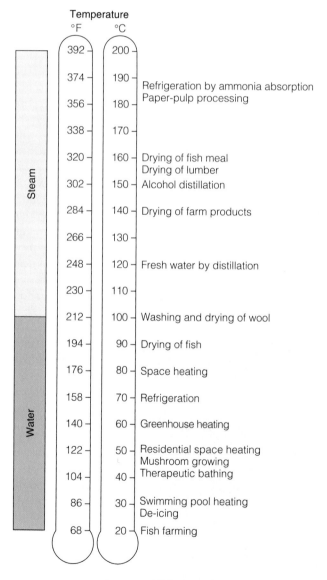

◆ FIGURE 14.21 Some established commercial uses of geothermal energy.

◆ FIGURE 14.22 The Geysers geothermal field in Sonoma County, California, is steam-dominated. It is the largest geothermal-energy producer in the world.

and copper. Technology is in place to transfer the heat of the toxic brine deep within the wells to a clean, working fluid that can be brought safely to the surface. In some geothermal areas removal of underground water can cause surface subsidence and perhaps even earthquakes. At The Geysers, subsidence of 13 centimeters (5 in) has been measured without noticeable impact. Reinjecting the cooled water back into the geothermal reservoir decreases the subsidence risk and helps to ensure a continued supply of steam to the electrical generators. All geothermal fields in California are being monitored for possible seismic activity related to steam production. Small quakes have been reported from The Geysers, but none of significant magnitude.

*TABLE 14.8	Electrical Capacities of Selected U.S. Geothermal Fields, 1990	
State	**Field**	**Electrical Capacity, MW**
California	The Geysers*	1,970[1]
	Coso	256[2]
	Salton Sea area*	214[2]
	East Mesa	10[3]
	Long Valley	43[3]
Hawaii	Puna	25[2]
Nevada	Dixie Valley	50[2]
	Steamboat Springs	18[2,3]
Utah	Cove Fort	13[2,3]

* 1993 data.
[1] Steam is piped directly to a power plant.
[2] Flash wells produce a mixture of steam and water that must be separated.
[3] Binary wells produce hot water that is piped to a power plant, where heat is transferred to another fluid.
Source: U.S. Geological Survey circular 1125, 1994.

The future for geothermal energy looks good. About 20 geothermal fields in the United States are generating electricity, and direct-use hydrothermal systems are being developed at more than 30 sites. Since electricity was first produced from geothermal energy in 1904, the production has grown to about 7,000 megawatts in 21 countries around the world. The United States alone produces 2,700 megawatts of electricity from geothermal energy, equivalent to burning 60 million barrels of oil each year to produce the same amount of electricity.

Nuclear Energy

Nuclear energy provides 16 percent of the world's electricity and 19 percent of the United States'. At one time it was considered *the* solution to the world's future energy needs—being clean, limitless, and "too cheap to meter." More than 400 nuclear power plants were built worldwide (◆Figure 14.23), 109 of them in the United States. Nevertheless, the Chornobyl accident in April 1986, following the 1979 Three Mile Island event in the United States, reinforced people's fears regarding nuclear power, resulting in the cancellation of all orders for new "nuke" plants in the United States.

Nuclear reactors and fossil-fuel plants generate electricity by similar processes. Both processes heat a fluid, which then directly or indirectly makes steam that spins turbine blades that drive an electrical generator. Uranium is the fuel of atomic reactors, because its nuclei are so packed with protons and neutrons that they are capable of sustained nuclear reactions. All uranium nuclei have 92 protons and between 142 and 146 neutrons. The common isotope uranium-238 (^{238}U) has 146 neutrons and is barely stable. The next most common isotope, ^{235}U, is so unstable that a stray so-called slow neutron penetrating its nucleus can cause it to split apart completely, a process known as **fission**—a term particle physicists borrowed from the biological sciences. Nuclei that are split easily, such as those of ^{235}U, ^{233}U, and plutonium-239 (^{239}Pu), are called **fissile isotopes.** Fission occurs when a fissile nucleus absorbs a neutron and splits into lighter elements, called *fission products,* while at the same time emitting several "fast" neutrons and energy, 90 percent of which is heat (◆Figure 14.24). The lighter fission products are elements that recoil from the split nucleus at high speeds. This energy of motion and subsequent collisions with other atoms and molecules creates heat, which raises the temperature of the surrounding medium.

Meanwhile, the stray fast neutrons collide with other nuclei, repeating the process in what is called a **chain reaction.** A controlled chain reaction occurs when one free neutron on average from each fission event goes on to split another nucleus. If more than one nucleus is split by each emitted neutron (on average), the rate of reaction increases rapidly, and the reaction eventually goes out of control. Because ^{235}U makes up only 0.7 percent of all naturally

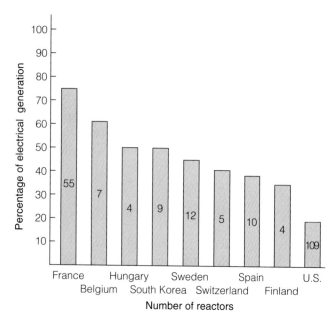

◆ FIGURE 14.23 Percentage of total energy need supplied by nuclear energy and the number of operating reactors for selected countries, 1994.

occurring uranium, the rest being ^{238}U, it must be "enriched" for use in commercial nuclear reactors.

The major risk associated with nuclear energy is the risk of contamination during processing, transportation, and disposal of ^{235}U and high-level nuclear waste products. The locations of these risks in the nuclear fuel cycle are apparent in ◆Figure 14.25.

Geological Considerations

One of the many criteria in the siting of nuclear reactors is geologic stability. The selected site must be free from landsliding, tsunami, volcanic activity, flooding, and the like. The Nuclear Regulatory Commission (NRC), the federal government body that licenses nuclear reactors for public utilities,

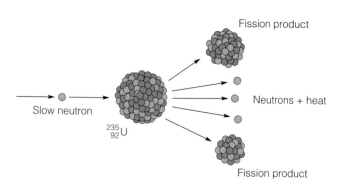

◆ FIGURE 14.24 Nuclear fission. A slow neutron penetrates the nucleus of a U-235 atom, creating fission products (also called *daughter isotopes*), 2 or 3 fast neutrons, and heat.

requires that all active and potentially active faults within a distance of 320 kilometers (200 mi) of a nuclear power plant be located and described. Reactors must be built a distance from an active fault determined by the fault's earthquake-generating potential. Further, the NRC defines an active fault as one that has moved once within the last 30,000 years or twice in 500,000 years. This is a very strict definition, and sites that meet these criteria are difficult to find along active continental margins. To provide additional safety and allow for geological uncertainty, reactors are programmed to shut down immediately at a seismic acceleration of 0.05 *g*. This is conservative indeed, and the shutdown would prevent overheating or meltdown should the earthquake be a damaging one.

Nuclear Waste Disposal

There is probably no more sensitive and emotional issue involved in geology today than the need to provide for safe disposal of the tons upon tons of radioactive materials that have accumulated over many decades. These wastes are the largest deterrent to further development of nuclear energy, as they present a legitimate hazard to humans, their offspring, and generations yet to come. Nuclear-waste products must be isolated, because they emit high-energy radiation that kills cells, causes cancer and genetic mutations, and causes death to individuals exposed to large doses. The importance of and problems encountered in locating a safe geological repository for high-level nuclear wastes will be treated in the next chapter.

Energy from the Sea
Ocean Thermal Energy Conversion

Ocean thermal energy conversion (OTEC) utilizes the temperature difference between warm surface water and colder deep waters, at least 22° C (72° F), to vaporize a low-boiling-point fluid and operate a generator. A potential OTEC site must be in a coastal area where water depth increases rapidly enough to obtain the needed temperature differential and close enough to shore for power transmission. The Island of Oahu in Hawaii has the population, need, and oceanographic conditions for this form of energy conversion. A mini-OTEC of about 50 kilowatts was operated in Hawaii in the 1980s. The electricity produced was not transmitted to the shore, but a continuously burning lightbulb on a float near the village of Mokapu testified to OTEC capability.

Wave Energy

Capturing wave energy and converting it to electricity is not a new idea, and its practicality has improved very little since 1900. Wave energy is so diffuse—spread over such a large area—that concentrating and utilizing it is difficult. In order

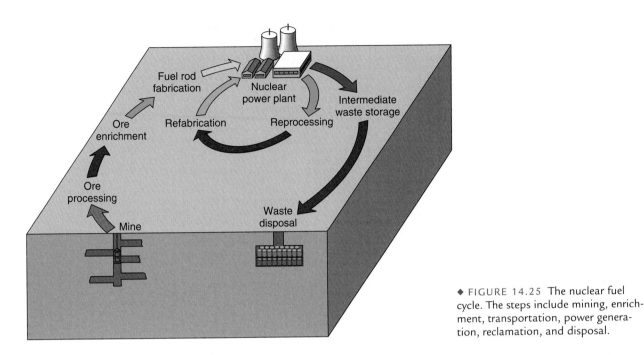

◆ FIGURE 14.25 The nuclear fuel cycle. The steps include mining, enrichment, transportation, power generation, reclamation, and disposal.

to convert a worthwhile amount of energy from wave motion, the energy must be focused by "horns," or funnels, extending seaward. In a 1971 experiment, a wave-powered generator hanging from a pier was used to power a string of lights of Pacifica, California. The Japanese currently market a wave-power machine for providing energy to lighthouses and buoys, and they have been working on wave-powered electrical-generating barges for offshore installation.

ENVIRONMENTAL
Geology⇌Now™

Click **Geology Interactive** to work through activities on Wave Properties, Wave Progression, and Wind–Wave Relationships through Waves, Tides, and Currents.

Tidal Currents

Tidal currents have been used by coastal dwellers to power mills for at least a thousand years. Restored, working tidal mills in New England and Europe are popular tourist attractions. Both flood (landward-moving) and ebb (seaward-moving) tidal currents are used. Modern designs utilize basins that fill with seawater at high tide and empty at low tide, generating electricity in the process. A potential site must have a large tidal range—the height difference between high and low tides—and adequate area for storing the elevated water. It is estimated that there are more than 100 suitable sites in the world that, together, are capable of generating 5 percent of the world's present generating capacity.

The first and best-known large-scale tidal power plant is on the Rance River near St. Malo on the English Channel in northwest France. The tidal range here varies from 9 to 14 meters (30–46 ft), and the power output is a reversible operation; that is, the plant generates power at both flood and ebb tides. Because the hydrostatic head (water drop) is only a 407few meters, 24 small 10-MW generators with a total capacity of 240 MW are used. On balance, the operation has been successful.

Annapolis Royal in Nova Scotia, Canada, is the only modern tidal generating plant in North America. It generates over 30 million kilowatts per year, enough electricity to support 4,500 homes. An island in the mouth of the Annapolis River was selected as the site for the powerhouse that opened in 1984 after four years of construction (◆Figure 14.26). It is an ideal location because the Bay of Fundy has the highest tides in the world and there was an existing causeway on the river with sluice gates that dammed a pond in the river. On the incoming tide the sluice gates are opened and the incoming seawater fills the pond. The sluice gates are then closed, which traps the seawater in the pond upstream of the generating turbine. As the tide recedes, the level of the water below the pond drops and a hydraulic head develops. When the difference is 1.6 meters or more the gates are opened and water flows through the turbine to generate electricity. Tidal power is clean and renewable and here in Canada, what started as a demonstration project has turned into a successful power-generating facility.

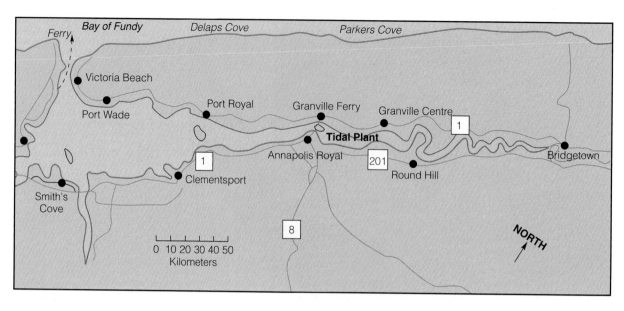

◆ FIGURE 14.26 Map of the site of the Annapolis Royal, Canada, tidal generating station. The Bay of Fundy has the highest tides in the world, which are put to work at the station to generate electricity for 4,500 homes. Nova Scotia Power, Inc.

CONSIDER THIS *What nonfossil-fuel energy sources are likely to be expanded in the United States in the coming century? In other words, if you wanted to make a long-term investment in one of them, which one would you choose?*

ENVIRONMENTAL
Geology⇌Now™

Click **Geology Interactive** to work through an activity on Tidal Forces and Tidal Shifts through Waves, Tides, and Currents.

Rock Oil and the Colonel

The history of the petroleum industry is full of interesting characters, and probably none was more charismatic than the self-styled Colonel of Titusville, Pennsylvania, Edwin L. Drake (♦ Figure 1). In the 1840s the demand for whale oil and lard had outstripped the supply and was being supplemented by "rock oil" from seeps and tar pits. The oil on northwest Pennsylvania's Oil Creek, so-named because oil bubbled to its surface, had long been skimmed by the Seneca Indians and other Native Americans and used to waterproof their homes and boats. Later, Europeans skimmed it for use as lamp oil, paraffin, liniments, and lubrication.

Drake was a railway conductor in New Haven, Connecticut, who had an interest in petroleum exploration but no experience. There are several versions of why Drake went to Titusville. The most credible one is that he had bought some shares in the Seneca Oil Company in New Haven and was sent by one of its founders, a friend of his, to stimulate the seepage at Oil Creek, which the company had leased from the Pennsylvania Rock Oil Company. Drake had a "Lincolnesque" appearance and he frequently wore a knee-length topcoat and stovepipe hat. The company bestowed upon him the honorary title of *Colonel* to enhance his credibility in Pennsylvania.

Colonel Drake tried digging trenches to stimulate the flow, and he installed a series of skimmers to collect the oil. The process was so inefficient that he persuaded the company to lease the land and attempt drilling. Drake enlisted the help of local "salt borers," who drilled for brine from which salt was extracted. Often these drillers had struck oil instead of water and they had been run out of town because of the terrible mess it created. The Colonel finally found an experienced driller named Uncle Billy Smith, who used percussion tools featuring a chisel on a cable to chip the rock and a bailer to remove the cuttings. Trouble with cave-ins ensued, and Drake ingeniously drove a large-diameter pipe into the caving zone and drilled the hole through it. On Friday, August 27, 1859, they struck oil, although they didn't know it until they checked the well on Sunday. A black goo had filled the 21-meter (69-ft)-deep hole, and the world had its first successful oil well.

Drake's well produced 10–35 barrels per day, almost doubling the previous world output of rock oil. The oil price jumped to $20/bbl, and a thousand oil pits sprang up around Titusville. Within a year the price had dropped to $0.20/bbl, and the Seneca Oil Company sold the Drake Well. Nearby towns such as Pithole City flourished for a few years and then became ghost towns. At Oil City a few miles south of Titusville, a Cleveland hay and grain dealer named John D. Rockefeller built the first commercial oil refinery in 1863. Seven years later he established the Standard (now Exxon/Mobil) Oil Company.

Because Edwin Drake had filed no patents and established no lease claims, he did not benefit from his vision and labor. He later lost his money speculating on oil stocks and became impoverished. He was forced to live on the generosity of friends. Learning of his dire need in 1873, Titusville citizens took up a collection for him. Three years later the Pennsylvania legislature voted to give him an annual income of $1,500. Drake was buried in Bethlehem, Pennsylvania, in 1880, but his body was reinterred in Titusville in 1901, ironically, the same year that a gusher in Texas, "Old Spindletop," drastically changed the oil supply of the world.

Today, Oil Creek between Titusville and Oil City is the site of the Oil Creek State Park. The ghost town of Pithole is just east of the park, and Drake Well Memorial Park is at the park's northern boundary.

DRAKE WELL MUSEUM, TITUSVILLE, PA.

♦ **FIGURE 1** Colonel Edwin L. Drake (right) in front of his well in 1861 with his friend, Titusville druggist Peter Wilson. The men in the background are probably drillers.

Tragedy in the Sound

Late in the evening of March 24, 1989, the supertanker *Exxon Valdez* left Valdez Fiord—in the area called the "Switzerland" of Alaska—and sailed into the designated outbound shipping lane on a southwest heading. The seas were calm and the winds light, and visibility was 10 miles. Everything was normal as the pilot left the ship outside the entrance to the fiord.

A short time later the officer in command radioed the Coast Guard, requesting permission to move over to the inbound lane so as to avoid a small iceberg that had calved off the nearby Columbia Glacier (see Chapter 11). The request was granted, and the ship altered its course 25° to port to change lanes. The ship continued on this more southerly heading for 30 minutes, sailing *past* the inbound lanes and into the shallow waters near Bligh Island (named by James Cook for Captain William Bligh of *Mutiny on the Bounty* fame). Realizing his error, the offi-

cer on the bridge gave urgent commands to turn starboard in order to return to the shipping lanes, but it was too late. At 4 minutes past midnight, the *Exxon Valdez* drove up onto Bligh Reef, tearing a gash in her hull and coming to a stop balanced on a pinnacle of rock (◆ Figure 1).

Within the next two days, 10.1 million gallons of oil spilled into a bay teeming with marine mammals, fish, waterfowl, and eagles. Exxon and Alyeska Pipeline Service Company, which jointly operate the terminal at Valdez, responded to the emergency, but their efforts to mobilize containment booms were slowed because the booms were buried under snow. By the time the booms could be put into place to keep the oil from spreading and skimmers could begin removing the oil from the sea surface, much damage had already been done to the shoreline and its inhabitants (◆ Figure 2).

It was estimated that 1,600 otters and 37,000 marine birds died from the oil. Some biologists believe this bird count represents only 10–30 percent of the actual number killed, that the losses were 100,000–350,000 birds. It would seem that restricted areas could easily be programmed into the navigation systems, and that audible warnings could be sounded when a vessel strays from the approved lane. The double hulls on newer supertankers offer assurance that at least some groundings will not result in oil spills. The final cost to Exxon of this "mistake" will probably be in excess of $4 billion ($1.025 billion of that in fines). As long as humans are at the controls, however, incidents such as this one can be expected.

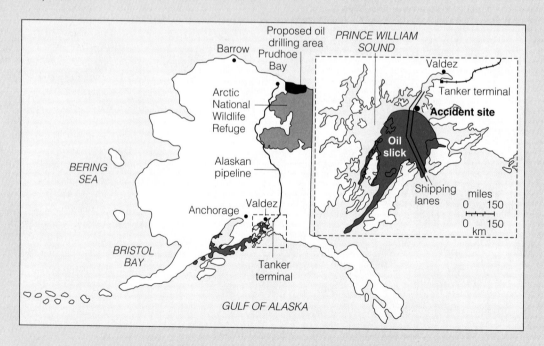

◆ FIGURE 1 Prince William Sound showing the tanker channel leaving Valdez. The *Exxon Valdez* went aground on Bligh Island and leaked millions of gallons of oil. From *Environmental Science*, 7th ed., by G. T. Miller, p. 329, Brooks/Cole.

BILL NATION/SYGMA

◆ FIGURE 2 An oil-soaked bird is rescued and prepared for clean-up after the spill. Thousands of other marine birds were not so lucky.

CASE STUDY 14.3

Baking Soda, Vinegar, and Acid Rain

The term *acid rain* refers to the atmospheric deposition of acidic substances including rain, snow, fog, dew, particles, and certain gases. Although volcanic activity (see Chapter 5) is the greatest source of materials that may form acids—sulfur, carbon dioxide, and chlorine—sulfur and nitrogen oxides introduced by human activities are approaching the amount of nature's contributions. Combustion of fossil fuels and the refining of sulfide ores are the major human sources of these contaminants. Acids form when these gases come in contact with water in the atmosphere or on the ground. Whereas carbon dioxide forms carbonic acid, a weak acid (see Chapter 6), chloride ion and the oxides of sulfur and nitrogen form the strong hydrochloric, sulfuric, and nitric acids, respectively.

The acidity or basicity (the chemical opposite of acidity) of an aqueous solution is referenced to the pH scale (◆Figure 1), a measure of the solutions' hydrogen-ion activity. A neutral solution will have a pH value of 7.0. The more hydrogen ions floating around in the solution, the more acidic

it is, and the lower its pH will be. Note that the scale is reversed, so to speak; a solution with a pH below 7.0 is acidic, and one with a pH above 7.0 is basic, or "alkaline." Acid-rain events with acidities below pH 2.8, the pH of vinegar, have been reported in large cities and heavily industrialized

areas. In contrast, lakes without outlets, such as Lake Natron in Africa, may become extremely alkaline and have pHs greater than 11.0—well in excess of the pH of a concentrated solution of baking soda and water.

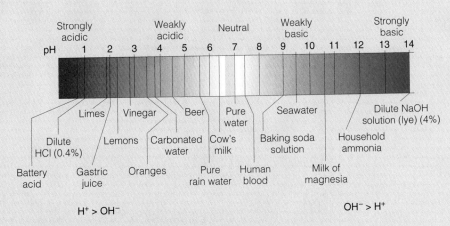

◆ FIGURE 1 The pH scale. A neutral solution has a pH of 7.0. A solution whose pH is less than 7.0 is acidic, and a solution whose pH is above 7.0 is basic, or alkaline.

Energy Is Where You Find It

Humans devote a great deal of mental and physical energy to harnessing the earth's energy. Energy, like fresh water, is invaluable, and when it is in short supply a family, a tribe, or a country will do almost anything to obtain it. The four photos in this gallery illustrate some of the earth's nonconventional energy sources and some relics of fossil energy. Keep in mind that all these energy resources derive from our sun.

◆ FIGURE 1 A community gathers around its solar-cell-powered television set; the Republic of Niger. More than likely they are watching a cricket or soccer match, not "Wheel of Fortune."

JOHN CHIASSON/GAMMA LIAISON

PETER BENTHAM, BP EXPLORATION

◆ FIGURE 2 This could be a scene in West Texas, Oklahoma, or Tampico, Mexico, in the 1920s and 1930s. Little regulation of the impact of oil production on the environment existed at the time. In reality, this is the Balakhany Oil Field in Northern Baku, Azerbaijan, in 2003.

B. PIPKIN

◆ FIGURE 3 A new windmill design, the Darrieus rotor, is utilized at Altamont Pass, California. The vertical-axis machine is nicknamed the "eggbeater" because of its appearance. The design is a major departure from horizontal-axis machines used to drive electrical generators. Darrieus rotors are more efficient and less expensive than propeller-driven types, and they do not need to face the wind, so they do not require pitch and pivoting mechanisms. Their principal drawback is their inability to self-start; a generator is needed to start the blades spinning.

D. D. TRENT

◆ FIGURE 4 The Blue Lagoon, a great hot-water pool 40 kilometers (24 mi) from Reykjavík, Iceland. It uses geothermal power-plant outlet water, which is salty and grows algae, giving it its blue color. Open every day, the lagoon attracts about 100,000 visitors annually.

Summary

Energy

Defined

Capacity or ability to do work. A society's quality of life depends in large part on the availability of energy.

Types

1. Renewable energy is replaced at least as fast as it is consumed; e.g., solar, wind, and hydrologic energy.

2. Nonrenewable energy is replenished much more slowly than it is utilized; e.g., energy derived from coal, oil, and natural gas (fossil fuels).

Petroleum

Defined

Volatile hydrocarbons composed mostly of hydrogen and carbon. Crude oil contains many different hydrocarbon compounds.

Origin and Accumulation

The necessary conditions for an oil field:

1. Source rock—origin as organic matter;
2. Reservoir rock—porous and permeable;
3. Caprock—impermeable overlying stratum; and
4. Geologic trap—such as an anticline, faulted anticline, reef, or stratigraphic trap.

Production

Drilled wells are either pumped or they flow if under gas and water pressure. Even with the best drilling and producing techniques, primary production may recover only half the oil.

Secondary Recovery

Oil remaining in pore spaces underground is stimulated to flow to a recovery well by injecting water, chemicals, steam, or CO_2. Much of the "stuck" oil can be stimulated to flow to the extraction well by these methods.

Value

Varies with weight. Light crude oil is the most valuable.

Reserves

At the 1988 rate of consumption, known world reserves of crude oil will last until 2050 and natural gas until 2110. The latest forecast is that world oil reserves will last until 2100.

Energy Gases

May be the fuel of the first half of the twenty-first century.

1. Methane hydrates, ice balls of natural gas
2. Natural gas from unconventional sources
3. Hydrogen gas

Coal

Defined

Carbonaceous residue of plants that has been preserved and altered by heat and pressure.

Rank

As coal matures, it increases in heat content and decreases in the amount of water and volatile matter it contains. The sequence is plant matter to peat, to lignite, to subbituminous, to bituminous, to anthracite.

Reserves

The United States has enough coal to last 200–300 years.

Synfuels

The manufacture of combustible gas, alcohol, and gasoline from coal has a promising future.

Other Fossil Fuels

Tar Sands

Sands containing oil that is too thick to flow and that can be surface-mined. Oil is then washed from the sand. One place the sands are mined is the Athabasca Field in Alberta, Canada.

Oil Shale

Eocene lake deposits containing light kerogen oil that can be removed by heating. Deposits are found in Wyoming, Utah, and Colorado.

Problems with Fossil-Fuel Combustion

Air Pollution

Products of combustion contribute to global warming, smog, and acid rain. Acid rain is a hazard in the U.S. East and Southeast and in eastern Canada, which receives some of the U.S. emissions. This is a major international air-pollution problem.

Mitigation

Sulfur content must be reduced below 1 percent. Methods include fluidized-bed combustion, neutralizing sulfur dioxide in smokestacks with limestone scrubbers, cleaning coals by gravity separation, and chemical leaching methods.

Subsidence and Collapse

Areas underlain by shallow lignite are subject to subsidence.

Alternative Energy Sources

Direct Solar

1. Solar heat collectors warm a fluid, which is then put to work.
2. Photovoltaic cells convert light directly into electricity.

Indirect Solar

Wind is produced by unequal heating of the earth's surface.

Geothermal

Facilities for generating electricity with geothermal energy currently are in place and may supply as much as 2 percent of U.S. electrical energy needs. Geothermal heat is also used worldwide for heating interior spaces, keeping sidewalks free of winter snow and ice, etc.

Hydroelectric

Hydroelectric dams harness the power of water distributed over the earth by the hydrologic cycle.

Oceans

The energy in waves, currents, and tides can be converted to do work for humans. Ocean thermal energy conversion (OTEC) shows some promise, but little has been done with waves and currents. Tidal power is being used successfully only at the Rance River, France, at present.

Fuel Cells

Fuel cells generate electricity fueled by hydrogen. Many think this is the fuel–motor combination of the future.

Nuclear Energy

Provides 16 percent of world and 19 percent of U.S. energy needs. Heat generated by fission of uranium isotopes and their daughter-isotopes such as plutonium is converted to electricity. There are 109 commercial nuclear power plants operating in the United States.

1. *Problems*—Mining, transportation, and disposal of high-level nuclear materials are the most dangerous aspects of the nuclear energy cycle. If the chain reaction runs out of control, the heat generated may lead to a central-core meltdown such as occurred at Chornobyl.

2. *Geological considerations*—Very careful geological site studies and investigations are performed to assess hazards from earthquakes, mass-wasting, subsidence, and coastal and riverine floods.

Key Terms

acid rain	geothermal energy	rank (coal)	source rock
anticline	hydrocarbon compound	renewable resource	stratigraphic trap
biomass fuel	kerogen	reserves	synthetic fuels (synfuels)
caprock	miscible	reservoir rock	tar sands
chain reaction	nonrenewable resource	salt dome	wildcat well
fissile isotopes	photochemical smog	secondary recovery	
fission	radon	slant drilling	

Study Questions

1. What four conditions are required for an oil deposit to form?
2. What oceanographic conditions favor the formation of a good source rock for oil?
3. Draw a geologic cross section that shows an anticline and a faulted anticline. Assume your cross section is looking north. What are the directions of the dips and strikes of the two limbs of the anticline? Dip and strike determination is explained in Appendix 3.
4. What nonrenewable energy resources exist in the United States, and which one is most abundant?
5. Describe atomic fission. How does it create heat for electrical generation?
6. What are methane hydrates and what is their potential as fuel for the twenty-first century?
7. What are the environmental consequences of burning fossil fuels? What will be the impact upon the global environment of continued dependence upon fossil fuels?
8. How can solar energy be collected and used for heating, generating electricity, and producing fuels?
9. What sector of the U.S. economy uses the most fossil fuel, and what international problem is connected with this use?
10. Explain the meaning of renewable energy. Which renewable energy sources have been developed, and which undeveloped renewable sources show the most promise?

For Further Information

Fact sheets may be obtained from the USGS free of charge. Write to U.S. Geological Survey (Information Services), P. O. Box 25286, Denver Federal Center, Denver, CO 80225.

Books and Periodicals

American Geological Institute. 2002. *The future of oil and gas* 47, no. 11, November.

Charlier, Roger. 1982. *Tidal energy.* New York: Van Nostrand Reinhold Co.

Churchill, Ron. 1997. Radon mapping. *California geology,* November–December:167–177.

Cooper, George A. 1994. Directional drilling. *Scientific American,* May:82–87.

Duffield, W. A., J. H. Sass, and M. Sorrey. 1994. *Tapping the earth's natural heat.* U.S. Geological Survey circular 1125.

Fellows, Larry D. 2001. Energy resources in Arizona. *Arizona geology* 31, no. 2.

Finch, Warren I. 1997. Uranium, its impact on the national and global energy mix. U.S. Geological Survey circular 1141.

Fisher, William L. 2002. The coming methane economy. *Geotimes* 47, no. 11, November.

Hafele, Wolf. 1990. Energy from nuclear power. *Scientific American* 263, no. 3.

Haq, Bilal U. 1998. Gas hydrates: Greenhouse nightmare? Energy panacea or pipe dream? *GSA today* (Geological Society of today (Geological Society of America) 8, no. 11 (November).

Hirsch, Robert L. 1987. Impending United States energy crisis. *Science* 235:1467–1473.

McCabe, Peter J., and others. 1993. *The future of energy gases.* U.S. Geological Survey circular 1115, Public issues in earth science series.

Mossop, Grant D. 1980. Geology of the Athabasca oil sands. *Science* 207, p. 145.

Reese, April. 1999. Bad air days. *E: The environmental magazine,* November–December: 28–35.

Ross, David. 1987. *Energy from waves,* 2d ed. New York: Pergamon Press.

Schmandt, Jurgen, J. Clarkson, and Hilliard Roderick, eds. 1988. *Acid rain and friendly neighbors: The policy dispute between Canada and the United States.* Durham, N.C.: Duke University Press.

Scientific American. 1990. Energy for planet earth. Scientific American special issue, September.

Sperling, Daniel. 1996. The case for electric vehicles. *Scientific American,* November: 54–59.

Spencer, Jon E., and Steven L. Rauzi. 2001. Crude oil supply and demand: long-term trends. *Arizona geology* 31, no. 4.

Suess, Erwin, Gerhard Bohrmann, Jens Greinert, and Erwin Lausch. 1999. Flammable ice. *Scientific American,* November: 76–83.

U.S. Geological Survey. 1998. Arctic National Wildlife Refuge (ANWR), 1002 Area, petroleum assessment 1998. Fact sheet FS-040-98, May.

U.S. Geological Survey. 2000. USGS world petroleum assessment 2000. Fact sheet FS-070-00, April.

U.S. Geological Survey. 2001. Coal-bed gas resources of the Rocky Mountain region. Fact sheet FS-110-01, November.

U.S. Geological Survey. 2002. Natural gas production in the United States. Fact sheet FS-113-01, January.

ENVIRONMENTAL
Geology ⇌ Now™

Assess your understanding of this chapter's topics with additional quizzing and comprehensive interactivities at http://earthscience.brookscole.com/pipkingeo4e

as well as current and up-to-date Web links, additional readings, and InfoTrac College Edition exercises.

Waste Management and Geology

Trash Talk

A WASTE-DISPOSAL SITE'S ABILITY TO isolate solid and liquid waste is determined almost entirely by its geologic and hydrologic conditions. The primary goal in site selection is to protect the local water supply. Surface and underground water become polluted when rain water percolates through solid wastes. Geologists assess a potential site's suitability through field observation, soil tests for percolation and strength, drill-hole sampling of underground rock materials, and evaluations of the region's underground water conditions based on tests of nearby wells. In addition, an environmental impact report must be submitted that addresses potential problems related to the local biology, air quality, traffic congestion, and visual blight.

Wrangell, Alaska, is a town of 2,000-some people on Wrangell Island in the state's beautiful southern panhandle. In 1999 the town's open trash dump was deemed unacceptable and was sealed (◆Figure 1). It was determined that the island has no geologically suitable landfill site and that the town's solid waste must be transferred by ocean barge to acceptable sites in the lower 48 states, hundreds of kilometers away. Much of the town's small budget now goes to pay for this new means of waste disposal.

In sharp contrast with the picture at Wrangell is an engineered sanitary landfill being manufactured in Southern California (◆Figure 2). Graded slopes are designed to provide stability, and a plastic liner has been placed beneath a compacted, impermeable clay layer. The clay-layer "seal" will prevent liquids from migrating out of the fill into the ground-water environment. Yes, geology is very important in the world of trash.

We aren't going to solve world peace if we can't figure out what to do with garbage.

Willie Brown, Mayor of San Francisco

◆ FIGURE 1 Wrangell's open trash dump before it was closed and sealed in 1999.

◆ FIGURE 2 An engineered solid-waste landfill; Southern California. Note the black plastic liner and tractor at the base of the slope.

Getting rid of trash is a major environmental problem in industrialized nations. Growing populations of consumers have created an explosion of solid, liquid, and hazardous wastes, much of which requires special handling. At risk due to careless waste disposal are ground- and surface-water purity, air quality, public health, and less threatening but still important, scenic beauty and land-surface integrity. Municipal solid waste generated in the United States in 2000 amounted to 2.1 kilograms (4.7 lbs) per person per day. This category of waste includes yard cuttings, garbage, construction materials, paper products, metal cans, plastics, and glass (◆Figure 15.1). It does not include wastes from agriculture, industry, utilities, or mining, which account for more than 96 percent of the solid waste generated in the United States (◆Figure 15.2). After incineration and recovery for recycling and composting, 57 percent went to a landfill. Just one day's total U.S. waste would cover 15 square kilometers (almost 6 mi^2) to a depth of 3 meters (10 ft). If it were loaded into ten-ton trucks lined up bumper-to-bumper, the trucks would stretch around the world 20 times. Those trucks containing only the day's *municipal solid waste* would circle the earth almost 3 times. The spectrum of solid-waste disposal problems ranges from the need to isolate highly dangerous nuclear waste to the challenge of dismantling and scrapping an astounding number of cars every year, estimated at 15 million in the United States alone. (One year 60,000 cars were abandoned on the streets of New York City.)

The amount of trash a country or governmental entity generates *per unit of land area* is just as important as the *total* amount it generates. If small countries with limited disposal sites generate large quantities of trash, major problems are

created. Although the United States generates the highest total amount of waste in the world, the tiny city-state of Singapore generates the most trash per unit area of land, almost 2,500 tons per square kilometer per year in 1985. This is in comparison to less than 2 tons per square kilometer for Canada and 20 tons per square kilometer for the United States. Poland generates the most *industrial* waste per unit area, followed by Japan, and Hungary produces the most *toxic* waste per unit area, followed by the United States.

The impetus to develop environmentally safe trash disposal sites, known as *landfills,* was the passage of two important laws, the National Environmental Policy Act of 1969 (**NEPA**) and the Resource Conservation and Recovery Act of 1976 (**RCRA**). NEPA requires an in-depth field study and issuance of an Environmental Impact Statement on the consequences of all projects on federal land. The RCRA mandates federal regulation of waste products and encourages solid-waste planning by the states. The RCRA addresses the problems of hazardous waste and its impact upon surface- and ground-water quality, and it created the framework for regulating hazardous wastes. These laws sent the message that the then-prevailing "out of sight, out of mind" approach to disposing of wastes of all kinds, hazardous and benign, was no longer acceptable. The laws mandated "cradle to the grave" systems for impounding wastes and monitoring them to ensure their low potential for migrating into fresh-water supplies. The Environmental Protection Agency (**EPA**) is charged with monitoring and policing the provisions of these and subsequent laws and amendments intended to prevent pollution of natural systems (◌ Case Study 15.1 on page 436).

Disposing of solid and liquid wastes, regardless of how they are generated, falls into two main categories:

- **isolation,** encapsulating, burying, or in some other way removing waste from the environment; and

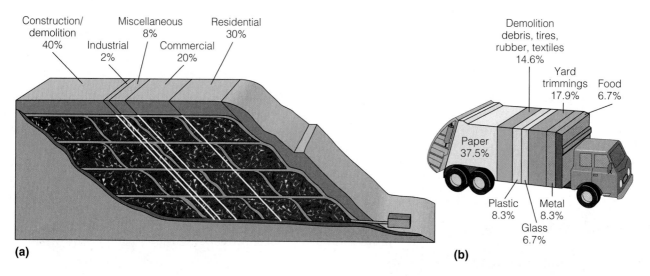

(a)   **(b)**

◆ FIGURE 15.1 Components of urban solid waste (a) by sector of the economy and (b) by material. Note that paper products—packaging and printed paper—comprise the largest volume of waste and that residential trash accounts for only about a third of all urban solid waste.

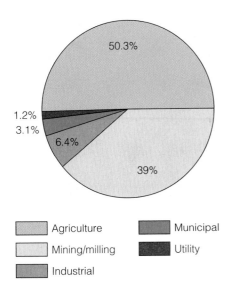

◆ FIGURE 15.2 Estimated relative contributions to total U.S. solid waste, 1992. The ratios had not changed appreciably by 2000. (EPA)

- **attenuation,** diluting, incinerating, or spreading trash or a pollutant so thinly that it has little impact.

Attenuation (dilution) is effective with sewage, some waste chemicals, and certain gaseous pollutants. About 10 percent of our waste, including sewage sludge, is incinerated in large, modern incinerators, usually found where land-disposal sites are not available. Historically, most of our solid waste has been isolated in dumps. Open dumps, the local "garbage dumps" of small towns such as the one pictured in the chapter opener, are no longer acceptable because of their attendant insect, vermin, odor, and air-pollution problems. In the mid 1990s, more than 70 percent of U.S. trash was being disposed of in 5,000 landfills, down from 7,900 landfills in 1988.

In 1993 alone almost 1,000 landfills were closed, most because they had reached their capacity, but many because they could not meet the stringent EPA standards for protecting ground water. Deep wells are used successfully for isolating hazardous liquid wastes, although the tremendous volume of used motor oil remains a problem. Today, good **waste management** consists not only of waste disposal, but also of reducing the amount of waste at its sources and promoting waste recovery and recycling programs.

Municipal Waste Disposal
Municipal Waste Disposal Methods

Sanitary Landfills

In 1912 the first solid-waste sanitary landfill was established in Great Britain. Known as "controlled tipping," this method of isolating wastes expanded rapidly and had been adopted by thousands of municipalities by the 1940s. The city of Fresno, California, claims to be the first city in the United States to employ a sanitary landfill for waste disposal. A **sanitary landfill** differs from an open dump in that each day's trash is covered with a layer of soil to isolate it from the rest of the environment. Although this sounds simple, the method involves spreading trash in thin layers, compacting it to the smallest practical volume with heavy machinery, and then covering the day's accumulation with at least 15 centimeters (6 in) of soil (◆Figure 15.3). When finished, a landfill is sealed by 50 centimeters (20 in) of compacted soil and is graded so that water will drain off the finished surface. This prevents water infiltration and the production of potentially toxic fluids within the fill (◆Figure 15.4). **Leachate** refers to the water that filters down through a landfill, acquiring (leaching out) dissolved chemical compounds

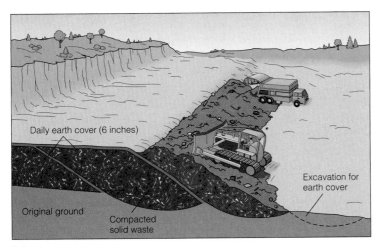

Daily earth cover (6 inches)

Original ground

Compacted solid waste

Excavation for earth cover

(a)

◆ FIGURE 15.3 (a) Method of dump and cover employed in a sanitary landfill. (b) An urban sanitary landfill in action; Palos Verdes, California.

(b)

◆ FIGURE 15.4 (a) Local-government landfill specifications for cells, final cover, and leachate-collection system. Note the barrier for reducing the visual blight of the landfill. (b) A full, graded landfill before planting; Puente Hills, Los Angeles County, California. Final grades and slope benches for intercepting runoff are visible in the photo.

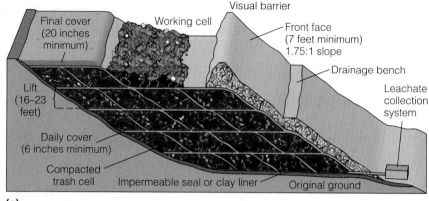

(a)

(b)

and/or fine-grained solid and microbial contaminants as it goes. Protecting the local environment from leachate is a major consideration in designing and operating waste-disposal sites. The dump-and-cover procedure carried out at sanitary landfills is known as the "cell" method, because it isolates waste from the environment in rhomb-shaped compartments, or cells. A well-operated sanitary landfill is relatively free of odors, blowing dust, and debris (⊙ Case Study 15.2 on page 436).

Landfills and individual disposal sites within a given landfill are classified on the basis of their underlying geology and the potential impact that leachate could have upon local ground or surface water. Each of the three classes is certified to accept specific wastes of differing degrees of reactivity (see * Table 15.1 and ◆Figure 15.5). Some landfill sites encompass all three classifications within their boundaries.

Incineration

Burning is the only proven way to significantly reduce the volume of garbage, and it can reduce the volume by as much as 90 percent. In 1993 about 10 percent of municipal solid waste was being burned at 162 modern municipal incinerators, up about a fifth from 136 in 1988. Use of the process is not expanding rapidly because of the high initial facility cost and the continuing costs of maintaining air-quality standards. In New York City, for example, burning costs are about $200 per ton, higher than the cost of landfilling waste in the mid-Atlantic states, the most expensive waste-disposal area in the United States. Heat energy is produced in the process, some of which is converted to electrical energy that can be sold to the local utility or used internally to run the plant. The use of incineration is highest in areas with high population density, little open land, or high water tables. Massachusetts, New Jersey, Connecticut, Maine, Delaware, and Maryland, for instance, burn more than 20 percent of their waste.

New York City is a leader in trash incineration, burning about 720,000 tons per year, or 11 percent of all its waste. Wastes must be burned at very high temperatures, and incinerator exhausts are fitted with sophisticated "scrubbers" that remove dioxins and other toxic air pollutants. Incinerator ash presents another problem, because it is in itself a haz-

*TABLE 15.1 Classifications of Disposal Sites and Waste Groups

Geology of Disposal Sites

Class I	No possibility of discharge of leachate to usable waters. Inundation and washout must not occur. The underlying lining material, whether soil or synthetic, must be essentially impermeable; that is, it must have a permeability less than 0.3 cm/year. All waste groups may be received (Figure 15.5, part a).
Class II	Site overlies or is adjacent to usable ground water. Artificial barriers may be used for both vertical and lateral leachate migration. Geologic formation or artificially constructed liners or barriers should have a permeability of less than 30 cm/year. Groups 2 and 3 waste may be accepted (Figure 15.5, part b).
Class III	Inadequate protection of underground- or surface-water quality. Includes filling of areas that contain water, such as marshy areas, pits, and quarries. Only inert Group 3 wastes can be accepted (Figure 15.5, part c).

Constituents of Waste Groups

Group 1	Consists of but not limited to toxic substances that could impair water quality. Examples are saline fluids, toxic chemicals, toilet wastes, brines from food processing, pesticides, chemical fertilizers, toxic compounds of arsenic, and chemical-warfare agents.
Group 2	Household and commercial garbage, tin cans, metals, paper products, glass, cloth, wood, yard clippings, small dead animals, and hair, hide, and bones.
Group 3	Non-water-soluble, nondecomposable inert solids such as concrete, asphalt, plasterboard, rubber products, steel-mill slag, clay products, glass, and asbestos shingles.

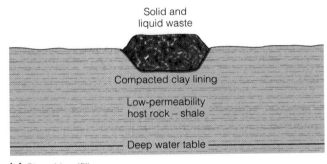

(a) Class I landfill

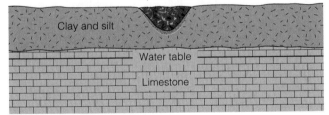

(b) Class II landfill

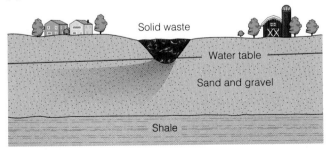

(c) Class III landfill

◆ FIGURE 15.5 Geology of the three classes of landfills. Note that a Class I fill offers maximum protection to underground water and that a Class III fill offers no protection.

ardous waste, containing high ratios of heavy metals that are chemically active (see Case Study 15.2).

The City of Philadelphia operates trash incinerators but lacks landfills for isolating the ash. In desperation, in the mid 1980s 14,500 metric tons (16,000 tons) of the ash from two of its incinerators was loaded onto the steamship *Khian Sea* and sent to sea (◆Figure 15.6). Like the sailor in Coleridge's "Rime of the Ancient Mariner," the *Khian Sea* sailed "alone, alone on a wide, wide sea" for two years, seeking a place to offload its unwanted cargo, with port after port refusing. In 1988 a miracle must have occurred, as the ship sailed into Singapore empty. No explanation has ever been given.

Ocean Dumping

Greek mythology relates how Herakles (Latin *Hercules*) was given the task of cleaning King Augeas's stables, which contained 30 years' accumulation of filth from 3,000 head of cattle. Inasmuch as one cow produces roughly 18 wet tons of manure per year, Herakles was up to his ears in 1.6 million tons of you-know-what, about the amount of sewage sludge that New York City generates in four months. Because Herakles was to perform the cleanup in a day, he ingeniously diverted the courses of two rivers to make them flow through the stables and wash the filth into an estuary. Had an environmental impact statement been required, it would have noted that a huge mass of solid sludge would cover a large area of the wetlands and bury bottom-dwelling organisms. In addition, phosphates and nitrates in the effluent would promote explosive algal blooms at the expense of other organisms, and dissolved oxygen in the water would decrease to the point of mass mortality of swimming and bottom-dwelling organisms.

Historically, all coastal countries have used the sea for waste disposal. This practice is based on the *assimilative capacity* approach to waste disposal; that is, the assumption that a body of seawater can hold a certain amount of material

◆ FIGURE 15.6 Greenpeace hung a protest banner on the loaded garbage barge in New York Harbor. In the past trash such as this was destined for sea burial. Disposal of solid waste is a problem in urban areas in all parts of the world.

without adverse biological impact. No U.S. municipality has dumped garbage in the sea since New York City stopped the practice in 1934.

Ocean pollution is governed by the Marine Protection Reserve and Sanctuary Act of 1972, better known as the *Ocean Dumping Act*. This act requires anyone dumping waste into the ocean to have an EPA permit and to provide proof that the dumped material will not degrade the marine environment or endanger human health. Hazardous substances entering the sea from the land (✳ Table 15.2) are contained in sewage sludge (the solid part of sewage), dredged materials, industrial effluents, and natural runoff. Sludge and other particulate matter have been found to create a biological imbalance in the ocean in three ways:

- they are consumed by marine life as a food source,
- they introduce heavy-metal and chlorinated organic compounds when they are attached to the particles, and
- they inhibit light in the water column.

These findings led to the Ocean Dumping Ban of 1988 that required total cessation of ocean dumping by 1991. Coastal cities that once ocean-dumped were forced to begin dewatering sludge and exporting it as a soil additive or depositing it in landfills. In the west Texas town of Sierra Blanca, instead of "Home on the Range," they now sing "sludge on the range." Two hundred and fifty tons a week of wet sludge is brought to Sierra Blanca to be sprayed on 78,500 acres of the contractor's (Merco) "sludge ranch." As an EPA employee and sludge critic said, "The fish in New York are being pro-

tected. The people in New York are being protected. The people of Texas are being poisoned." The feeling is that the odors coming off the application area are not just a nuisance and trespass, they are a health hazard. However, the EPA says the spreading is environmentally safe. Time will tell.

✳TABLE 15.2	Examples of Recognized Seawater Pollutants from Human and Natural Sources*
Organic Pollutants	
PCBs	polychlorinated biphenyls
PCDFs	polychlorinated dibenzofurans
PCDDs	polychlorinated dioxins
DDT	chlorinated hydrocarbon pesticide
Inorganic Pollutants	
Cd	cadmium
Pb	lead
Co	cobalt
Hg	mercury
Zn	zinc

* Pollutants originating in sludge, materials dredged from harbors or estuaries, and sewage-treatment and industrial wastewater that are known to pose health hazards.
Source: EPA (adapted).

Problems of Municipal Waste Isolation

Organic refuse in landfills ultimately decomposes. As it does, the ground surface settles; methane, carbon dioxide, and other gases are generated; and noxious or even toxic leachate from it may seep into the water table or bordering streams. Thus, a landfill's *pollution potential* depends upon the change in volume that accompanies decomposition, the waste's reactivity (which determines the chemistry of the leachate), the local geology, and the climate. Organic matter decomposes more slowly in cold, dry climates than in warm, wet ones.

Stabilization

Landfills continue to settle as decomposition goes on for many decades after burial. Typically, settlement is rapid at first and diminishes to a much slower rate after about ten years (◆Figure 15.7). The rate of subsidence depends upon many factors, including the climate, the kind of waste, and the amount of compaction during filling. Thirty-ton tractors left on a fill over a weekend have been known to settle as much as a meter into the trash, and total settlement of as much as 30 percent has been measured in landfills. Variation is great. A 6-meter-thick landfill in Seattle settled 23 percent in its first year, whereas a 23-meter-thick fill in Los Angeles settled only 1 percent per year in its first three years. Settlement is a problem, because it results in cracks, fissures, and infestation by vermin. It is monitored to assure that water does not pond and percolate downward, creating a high volume of polluting leachate.

As microorganisms break down organic material *(biodegradation),* they release heat, methane (natural gas), and other gases including hydrogen sulfide, carbon dioxide, and oxygen. Research at the University of Arizona and elsewhere has shown that biodegradation can be surprisingly slow, as illustrated by exhumed garbage containing virtually unblemished 10–20-year-old newspapers (◆Figure 15.8), a 14-year-old mound of recognizable guacamole, and positively ancient hot dogs (testimony to the effectiveness of sausage preservatives). The most efficient decomposers are the voracious oxygen-using aerobic bacteria, such as those that quickly turn grass cuttings into compost for our gardens. Most landfills are sealed off from the atmosphere, however, giving rise to anaerobic (no oxygen) microorganisms that manufacture methane and hydrogen sulfide in the slow process of decomposing cellulose. Pumping air into a fill, a procedure known as **composting,** stimulates aerobes, accelerates decomposition and settlement, and creates more space for trash.

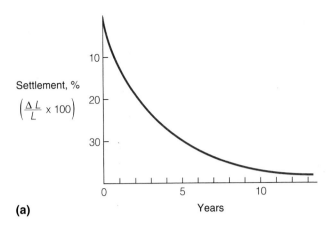

(a)

(b)

◆ FIGURE 15.7 (a) Graph of landfill settlement over time. The percentage of settlement is calculated by dividing the amount of settlement (ΔL) by the original thickness of the landfill (L) and multiplying by 100. The settlement values here are for illustration only and are not real values. (b) Landfill weigh station on the Palos Verdes Peninsula, California, built on trash-filled land. The station eventually collapsed, requiring that a new structure and scale be established at a different location.

Gas Generation

Methane (CH_4), a gaseous hydrocarbon, is the principal component of natural gas. Also known as *marsh gas,* it bubbles forth from stagnant ponds and swamps, the result of complex chemical fermentation of plant material by bacteria. It is the same gas that causes the explosions in coal mines that have taken innumerable lives. Methane is explosive when present in air at concentrations between 5 and 15 percent. Although enormous volumes of methane are generated by anaerobic microorganisms in a landfill, there is little danger of the fill exploding because no oxygen is present. Being lighter than air, methane migrates upward in a fill. Methane leaking through the cover of a landfill converted to a golf course near Bel Air, California, was known to "pop" from ignition when cigarettes were dropped on its greens. Methane can also migrate laterally, and it did so several hundred yards into the wall spaces of

◆ FIGURE 15.8 Underground newspaper—still readable after at least ten years' burial in a Pomona, California, landfill.

◆ FIGURE 15.9 Methane-extraction well and drill rig. The extracted gas is conveyed to a central location for conversion to electricity.

structures near a landfill south of Los Angeles. Fortunately, the potentially explosive situation was discovered, and wells were drilled to intercept and extract the gas (◆Figure 15.9). One strategy for handling the large volumes of methane generated at large landfills is to extract it through perforated plastic pipes, remove the impurities (mostly CO_2), and use it as fuel—the so-called refuse-derived fuel, *RDF*. This can be accomplished either by piping the gas away via pipelines or by using it on the site to generate electricity. Because the gas may be as much as 50 percent CO_2, it is more expedient to use the fuel directly to generate electricity than to clean out the impurities and put it into pipelines. The Los Angeles County (California) Sanitation District extracts landfill methane and generates sufficient electricity to supply up to 45,000 homes.

Leachate

The compositions of leachates ("garbage juice") in and leaving landfills are highly variable, but they are dangerous to the environment until established otherwise (◆Figure 15.10). A leachate may be such that a receiving body of water can assimilate it without any impairment in water quality. Some purification of a leachate occurs as it filters through clayey soils, and some purification also occurs when contained organic pollutants are oxidized. Ground-water-quality measurements taken in the vicinity of a landfill, at the fill, and downstream from the fill are shown in ＊ Table 15.3. Note that the leachate was causing deterioration of ground-water quality in the vicinity of the monitoring well. To correct this, the leachate must be intercepted by wells or a barrier must be installed between the leachate and the water table. Under existing regulations, a landfill that is in a geologic setting where leachate threatens local water bodies may accept only inert (Group 3) wastes.

The Completed Landfill

A completed state-of-the-art landfill is illustrated in ◆Figure 15.11. It is designed to collect leachate above the bottom liner

◆ FIGURE 15.10 The flow of leachate from a landfill can contaminate underground water, ponds, and streams.

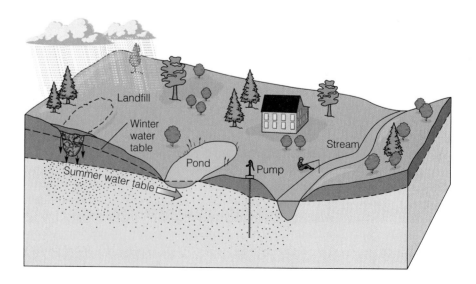

Water Characteristics	Measurement Site		
	Local Ground Water	Fill Leachate	Monitoring Well
Total dissolved solids, ppm	636	6,712	1,506
BOD, mg/L	20	1,863	71
Hardness, ppm	570	4,960	820
Sodium content, ppm	30	806	316
Chloride content, ppm	18	1,710	248

*TABLE 15.3 Ground-Water Quality Measured Near One Landfill

* Biological oxygen demand: the amount of oxygen per unit volume of water required for total aerobic decomposition of organic matter by microorganisms.
Source: D. R. Brunner and D. J. Keller, 1992, see For further Information.

accept waste for 10–40 years, and owners are required to monitor the fills for at least 30 years after closure. If they begin to leak after that period, the contamination problem belongs to whomever is living in the area at the time.

Recycling

Resource recovery—the removal of certain materials from the waste "stream" for the purpose of recycling or composting them—not only saves materials, it saves energy. Recycled aluminum cuts energy use 96 percent over processing aluminum ore. For recycled steel the saving is 74 percent, and for copper 87 percent. In 2002, 120 million tons of material were being recycled in the United States that saved energy equal to 4 months electrical demand. Remanufacturing, the process of disassembling a worn-out product, cleaning and replacing its own parts, then reassembling it, is estimated to save 85 percent of the energy required to make the product new from raw materials. Remanufacturing is estimated worldwide to be equal to the electricity generated by five nuclear power plants, and saves enough raw materials to fill 155,000 railroad cars forming a train 1,100 miles long. The

and to recover methane for use in power generation. Note the inclusion of leachate- and gas-monitoring wells in the design. Double-lined landfills have been required in the United States since 1996. These landfills are designed to

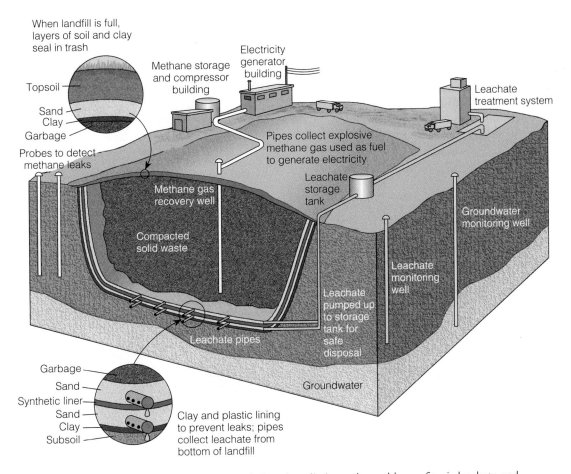

◆ FIGURE 15.11 State-of-the-art landfills are designed to eliminate the problems of toxic leachate and methane escaping into the environment. Note the double lining (clay and plastic layers) to prevent leakage. Unfortunately, 85 percent of U.S. sanitary landfills are unlined.

industrial nations are the "Saudi Arabias" of trash, and all have the same problems—dwindling resources and dwindling space for solid waste.

Until recently, **recycling** held little appeal; that is, it cost more to separate, sort, and recycle trash than to landfill it. Recycling is now mandated in many states, however, because existing landfills are filling, and acceptable sites for new ones are difficult to find. An effective way to extend the lifetimes of landfills and natural resources is to recycle municipal wastes. The consuming public is "thinking green" these days, and manufacturers are realizing favorable marketing results by using recycled materials in packaging their products. Because of a growing awareness that mineral resources are finite, recycling increased from 10 percent to 25 percent of municipal waste between 1985 and 1996. (◆Figure 15.12 summarizes resource recovery at landfills by category.) One indicator of the anticipated value of trash is that glass and high-grade plastics have joined corn and pork bellies as commodities bought and sold on the Chicago Board of Trade. Trash futures?

Curbside recycling programs grew enormously between 1988 and 1996, with more than 100 million people participating in 1993—almost 40 percent of the U.S. population. The "champion" city is Seattle, whose recycling rate is 45 percent. Many communities use a "pay as you throw" program, which bases collection charges on the amount of true waste that is added to a landfill; materials sorted out for recycling are collected free of charge. Many communities require separation of glass, metal, paper products, and plastics from other trash materials. Plastic recycling is difficult to handle and expensive, because there are seven kinds of plastic that must be separated by hand. Only 2 percent of waste plastic was recycled in 1995. Ingenuity pays, however. A small Massachusetts company grew into a $1.3 million business by recycling spent plastic ink-jet printer cartridges. It sells refurbished cartridges to consumers and the worn plastic and metal components to recyclers.

Not easily solvable is the problem of what to do with old automobile tires. There are at least 3 billion used tires stockpiled in the United States, a stockpile that is being added to at a rate of 250 million per year. When placed in landfills they tend to rise in the fill, a fact that caused 33 states to ban them there. In 2000 the nation used 32 million scrap tires (California alone contributes this many) in civil engineering projects, an increase of 23 percent of 1999. The incidental uses of old tires are construction of small dams, erosion control, houses (New Mexico is the leader), fencing, rifle ranges, playgrounds (◆Figure 15.13), and grain-storage structures. Used tires are being used experimentally as bedding for livestock corrals, as a replacement for stone in septic systems, and as reinforcement in flood-control structures. Because old tires are such an enormous environmental problem, many engineers and scientists are working on a technological "fix" that will rid us of this blight.

As more curbside separation of garbage for recycling is mandated, and as more states enact container-bill legislation encouraging the return of beverage containers, the flow of

◆ FIGURE 15.13 Three thousand tires were recycled as children's playground equipment and landscaping material at this park in Japan.

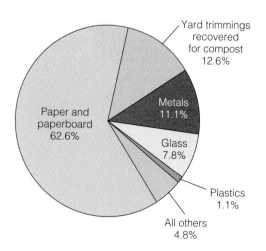

◆ FIGURE 15.12 Materials recovered at landfills as percentage of the total weight recovered, 1990. EPA data

Yard trimmings recovered for compost 12.6%

Metals 11.1%

Paper and paperboard 62.6%

Glass 7.8%

Plastics 1.1%

All others 4.8%

recyclables increases. The "champion" states (% recycled) in the United States are Minnesota (41%), New Jersey (39%), and Washington (35%). Taiwan is the world champion paper recycler; 98 percent of the paper used in that almost treeless country is recycled.

Composting

When decomposed by bacteria, biodegradable wastes, the so-called green garbage, can be used as a fertilizer and soil amendment. Yard cuttings, most kitchen wastes, and organic wastes from commercial operations are acceptable compost material. Because yard wastes are second only to paper in landfill tonnage (see Figure 15.12), many states and municipalities have mandated separate collection for them. As a result, the number of composting facilities in the United States grew dramatically from 651 facilities in 1988 to more than 3,000 in 1993.

> **CONSIDER THIS** *Composting is a "win–win" process, yet we have barely scratched the surface of its potential. What are the benefits of small-scale and large-scale composting to individuals, cities, and states?*

Multiple Land Use Strategies

Land, like trash, can be recycled. Where the local geology is amenable to it, abandoned rock quarries and sand and gravel pits have been converted to sanitary landfills. Population pressure and the need to conserve resources in the twenty-first century will require that we not only dispose of our waste efficiently, but also that we gain some benefit from the stored trash and the disposal site. The average new home requires about 65 cubic meters of concrete in its construction, and its inhabitants will generate about a ton of trash a year. Thus, it *makes sense* to place the household's trash in an opening created to build the house. After a landfill is completed, methane gas can be recovered from it and converted to energy, and the land may be reshaped into a golf course, athletic field, public garden, or the like. An ideal, beneficial cycle of multiple land uses is shown in ◆Figure 15.14.

As explained earlier, structures placed directly on a landfill are subject to settlement and to methane emanations. A restaurant and shops built upon a thin landfill (<8 meters thick) south of Los Angeles were placed on piles driven through the fill to keep them from settling. The walkways around the buildings were not, however, and after a few years they settled and had to be built up to the base of the structures (◆Figure 15.15). In the same commercial development, methane is extracted and used to illuminate exterior gas lanterns.

As cities expand and coalesce—such as in the New York-to-Boston corridor and the strip between Los Angeles and San Diego—waste-disposal-site selection must become very creative. Between 1940 and 1983 the Kaiser Corporation operated an open-pit iron mine at Eagle Mountain in the desert 115 kilometers (70 mi) east of Indio, California. The mining operation has now been abandoned, leaving an opening 3 kilometers (1.9 mi) long and more than 300 meters (about 1,000 ft) deep (◆Figure 15.16). It is the

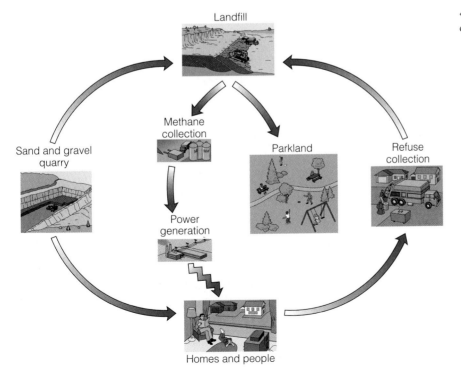

◆ FIGURE 15.14 Multiple land use, a cycle that can benefit everyone.

◆ FIGURE 15.15 Commercial development built on a landfill. Whereas the buildings are founded on piles through the fill to bedrock, the pavement was placed directly on the fill. The lanterns are fueled by methane from the fill.

proposed site for a landfill for all Southern California cities' household waste (no toxic or hazardous waste), which would come to it in closed railroad cars. The tracks that served the mine are still in place, and the site is known to be free from faults and potential for ground-water contamination. Far from any population center, the site could accommodate half of all the household trash generated in Southern California in the next 115 years! The open pit is in dense bedrock that would be lined with clay and high-density polyethylene plastic to protect ground water. The proposed site is acceptable geologically, and the only objections raised at public hearings regarding the proposal, from a technical point of view, seem to be entirely NIMBY, PIITBY, and NIMTOO (see Case Study 15.1).

CONSIDER THIS *During the past few decades in the United States, recycling waste has evolved from a concept to a major component of waste management for many communities. Why do some environmental groups object to waste-problem solutions that emphasize recycling?*

Hazardous Waste Disposal

Hazardous wastes are primarily such industrial products as sludges, solvents, acids, pesticides, PCBs (polychlorinated biphenyls; highly toxic organic fluids used in plastics and electrical insulation), and nuclear waste. EPA estimates are that U.S. industries were generating 260 million metric tons of hazardous waste per year in the mid 1990s. A typical rate at which liquid hazardous waste was generated in the early 1980s was 150 million metric tons per year, which can be thought of as 40 billion gallons. If that amount could be placed in 55-gallon drums and the drums placed end to end, the drums would encircle the earth 16 times (◆Figure 15.17). In the mid 1990s this would increase to almost 28 times. Examples of hazardous wastes individuals introduce to the environment include drain cleaners, paint products, used crankcase oil, and discarded fingernail polish (not trivial).

Hazardous-Waste Disposal Methods

Secure landfills are designed to totally isolate the received waste from the environment. Wastes are packaged and placed in an underground vault surrounded by a clay barrier, thick plastic liners, and a leachate removal system. Unfortunately, few of the nation's existing 75,000 industrial landfills have plastic liners, clay barriers, or leachate drains. Many older fills will either have to be retrofitted with modern environmental-

◆ FIGURE 15.16 Eagle Mountain Mine, site of the proposed Eagle Mountain landfill; Riverside County, California. As designed, the fill could accommodate up to 20,000 tons of trash per day, which would make it the largest-capacity landfill in the world. It would accept only nonhazardous waste and would have a recycling center.

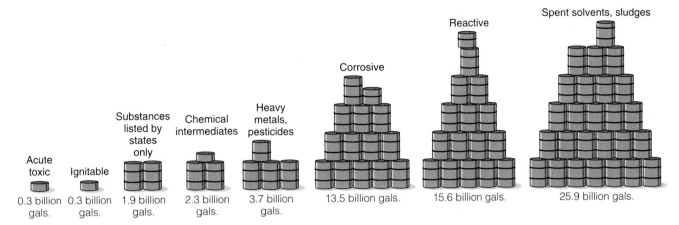

◆ FIGURE 15.17 Breakdown by category of the 40 billion gallons of toxic waste U.S. industries generated in 1981. The total of all categories is greater than 40 billion due to overlap; for example, a fluid may be both corrosive and reactive.

protection systems or shut down. Additional protection is provided by existing regulations that mandate ground-water monitoring at disposal sites. An ideal secure landfill for hazardous waste would be a site in a dry climate with a very low water table. In humid climates where water tables are high, alternative means of safe disposal must be found.

Deep-well injection into permeable rocks far below fresh-water aquifers is another method of isolating toxic and other hazardous liquid substances (◆Figure 15.18). Before such a well is approved, thorough studies of subsurface geology must be made in order to determine the location of faults, the state of rock stresses at depth, and the impact of fluid pressures on the receiving formations (⊙ Case Study 15.3 on page 437). The injection boreholes are lined with steel casing to prevent the hazardous fluids from leaking into fresh-water zones above the injection depth.

Disposal of hazardous wastes, both liquid and solid, is regulated by the Resource Conservation and Recovery Act of 1976 (RCRA). Exempt from the requirements of this act are businesses that generate less than a metric ton of waste per month. According to EPA estimates, about 700,000 of these small firms generate 90 percent of our hazardous waste. Consequently, most of these wastes end up in sanitary landfills.

Discharge into sealed pits is the least expensive way to dispose of large amounts of water containing relatively small amounts of hazardous substances. If the pit is well sealed and evaporation of water equals or exceeds the input of contaminated water, the pit may receive and hold hazardous waste almost indefinitely. Problems can arise from leaky seals and overflow of holding ponds during heavy storms or floods.

Superfund

What about the thousands of old and mostly unrecorded sites where hazardous waste has been dumped improperly in the past? Hazardous substances have been disposed of indiscriminately since the eighteenth century in the United States.

Superfund, the Comprehensive Environmental Response, Compensation, and Liability Act (CERCLA), was passed in 1980 to rectify past and present abuses from toxic-waste dumping. It authorizes the EPA to clean up a spill and send

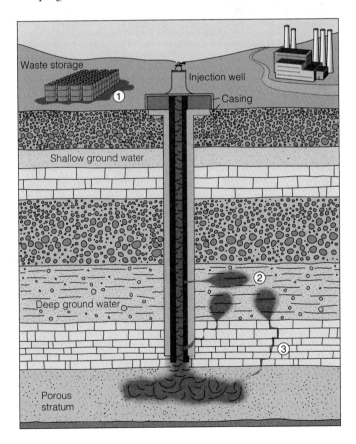

◆ FIGURE 15.18 Deep-well injection of hazardous waste. This method presumes that wastes injected into strata of porous rock deep within the earth are isolated from the environment forever. What could go wrong? (1) Toxic spills may occur at the ground surface. (2) Corrosion of the casing may allow injected waste to leak into an aquifer. (3) Waste may migrate upward through rock fractures into the ground water.

the bill to the responsible party. CERCLA also empowered the EPA to establish a National Priorities List (NPL) for cleaning up the worst of the 40,000 identified abandoned toxic and hazardous waste sites now known as *Superfund sites,* some of which date back many decades (⊙ Case Study 15.4 on page 438). By 1999 1,400 NPL sites had been designated or were pending. Of those, 427 sites had been cleaned up, and work was under way or had been completed on 474 others. New Jersey has the dubious distinction of harboring 6 of the 15 worst Superfund sites as ranked by impact on the environment—5 landfills and 1 industrial site. ✳ Table 15.4 lists the top 10 states with Superfund sites.

Although the costs of cleaning them up are staggering, toxic dump sites do not just "go away" on their own. They need to be cleaned up. Some problems with Superfund also need to be cleaned up. The Superfund concept was well intended, but the program has been poorly executed. As a result, 40 percent of its funding has gone to litigation expenses. Superfund has become a "lawyers' money pit," as one journalist put it.

| **CONSIDER THIS** | *Perhaps no other recent environmental legislation has received as much public* |

attention as the act with the catchy nickname "Superfund." What is Superfund? Who manages it, and what legislation is behind it? From an environmental standpoint, what is its main problem?

Nuclear Waste Disposal

There is probably no more sensitive and emotional issue in geology today than the need to provide for safe disposal of the tons upon tons of radioactive materials that have accumulated over many decades. These wastes are the largest deterrent to further development of nuclear energy, as they present a legitimate hazard to humans, their offspring, and generations yet to come. Nuclear-waste products must be isolated because they emit high-energy radiation that kills cells, causes cancer and genetic mutations, and causes death to individuals exposed to large doses. The need to develop the means for safe, permanent disposal led Congress to pass the Nuclear Waste Policy Act in 1982 (NWPA). This law gave the Department of Energy (DOE) responsibility for locating a fail-safe underground disposal facility, a **geological repository,** for high-level nuclear waste and designated nine locations in six states as potential sites. Based upon preliminary studies, President Reagan approved three of the sites for intensive scientific study (known as *site characterization*): Hanford, Washington; Deaf Smith County, Texas; and Yucca Mountain, Nevada. In 1987 Congress passed the Nuclear Waste Policy Amendments Act, which directed DOE to study only the Yucca Mountain site (Figure 15.19). Yucca Mountain in Nevada finally received approval to store over 70,000 metric tons of high-level nuclear waste, over the objection and veto by Nevada's Governor Kenny Guinn. In

✳TABLE 15.4	Top Ten States on the National Priority List (Superfund), 1990*	
State	**Sites**	**Percentage Distribution**
New Jersey	109	9.1
Pennsylvania	95	7.9
California	88	7.4
New York	83	6.9
Michigan	78	6.5
Florida	51	4.3
Washington	45	3.8
Minnesota	42	3.5
Wisconsin	39	3.3
Illinois	37	3.1

* Includes both proposed and final sites listed on the National Priorities List for the Superfund program as authorized by the CERCLA of 1980 and the Superfund Amendments and Reauthorization Act of 1986.
Source: *Statistical Abstract of United States, 1991,* p. 211.

July 2002, the U.S. Senate approved the site, which had received approval overwhelmingly by the House with President Bush's blessing. The DOE had spent over 20 years and $6.7 billion studying various ways and sites at which to store nuclear waste safely. Even if this site passes further legal objections by the State of Nevada, 2010 is the best that can be hoped for the first nuclear waste isolation. This site is discussed further later in the chapter.

Types of Nuclear Waste

High-Level Wastes

High-level wastes, by-products of nuclear power generation and military uses, are the most intensely radioactive and dangerous. About once a year, a third of a reactor's fuel rods are removed and replaced with fresh rods. Replacement is necessary because, as the fissionable uranium in a reactor is consumed, the fission products capture more neutrons than the remaining uranium produces. This causes the chain reaction to slow, and eventually it would stop. The radioactive rods, called *spent fuel,* are the major form of high-level nuclear waste. Currently they are stored in water-filled pools at the individual reactor sites, supposedly a temporary measure until a final grave is prepared for them. More than 63,000 MTHM of spent fuel is stored across the country, and at many of the reactors, pool storage capacity is nearly filled (◆Figure 15.20). Quantities of spent nuclear fuel are usually expressed in terms of metric tons of heavy metal (MTHM), mostly uranium. The number does not include other materials such as the tubes that contain the fuel and structural materials. One metric ton is 1,000 kilograms or 2,200

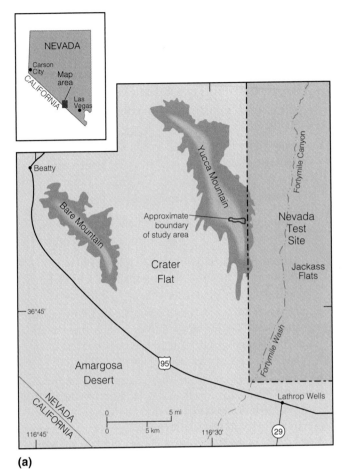

(b)

◆ FIGURE 15.19 (a) Location of the proposed Yucca Mountain nuclear repository. (b) Aerial view of the Yucca Mountain study area. The structures are at the site of the proposed nuclear waste repository. DOE data

pounds. Two major concerns about pool storage are that an unintended nuclear reaction might start in the pool and that the rods might deteriorate and release fuel pellets. It should be noted that although the fuel removed from a reactor every year weighs about 31,000 kilograms (65,000 lbs), because of its high density, it is only about the size of an automobile.

Alternatively, the rods are chopped up and *reprocessed* to recover unused uranium-235 and plutonium-239. However, the liquid waste remaining after reprocessing contains more than 50 radioactive isotopes, such as strontium (^{90}Sr) and iodine (^{131}I), that pose significant health hazards. In 1992, the Department of Energy (DOE) decided to phase out

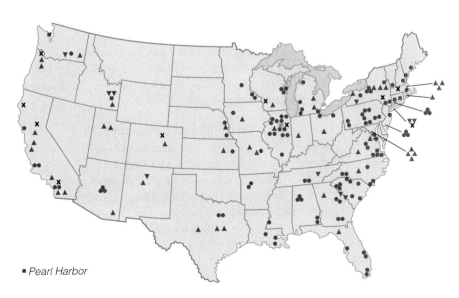

■ *Pearl Harbor*

◆ FIGURE 15.20 Location of temporarily stored spent nuclear fuel and other high-level nuclear waste awaiting permanent disposal, 2000. The symbols do not reflect precise locations. DOE data

reprocessing of spent fuel. Unfortunately, the volume of existing reprocessed waste in temporary storage would cover a football field to a depth of 60 meters (200 ft). These wastes are currently stored in metal tanks at four U.S. sites. The oldest site, the one at Hanford, Washington, originally stored wastes in single-walled steel tanks. In 1956 a tank leak was detected, and since then, 750,000 gallons of high-level waste have leaked underground from 60 of 149 storage tanks. It is said by some that "radioactive waste was invented here." Cleanup at Hanford may be the costliest environmental cleanup in world history; the cost has been estimated as at least $57 billion.

Low-Level Wastes

Low-level waste, by NCR definition, is anything that is not high-level waste or transuranic waste. Low-level wastes are created in research labs, hospitals, and industry as well as in reactors. Ocean dumping of these wastes in 55-gallon drums was a common practice between 1946 and 1970. Since then, shallow land burial has been the preferred disposal method, but many sites have been found to leak.

Mill Tailings

Uranium mill tailings, the finely ground residue that remains after uranium ore is processed, are typically found in mountainous piles outside the mill facility. Large volumes of tailings exist that still contain low concentrations of radioactive materials, including thorium (^{230}Th) and radium (^{226}Ra), both of which produce radioactive radon gas (^{222}Rn), the so-called hidden killer (⊙ Case Study 15.5 on page 439). Most of the abandoned tailings piles are in sparsely settled areas of the West. A former Atomic Energy Commission policy permitted the use of uranium-mill tailings in the manufacture of building materials such as concrete blocks and cement. For almost twenty years, 30,000 citizens of Grand Junction, Colorado, lived in homes whose radon levels were up to seven times the maximum allowed for uranium miners. Uranium tailings were used in building materials in Durango, Rifle, and Riverton, Colorado; Lowman, Idaho; Shiprock, New Mexico; and Salt Lake City, Utah.

The Uranium Mill Tailings Radiation Control Act (1978) makes the DOE responsible for 24 inactive sites in 10 states left from uranium operations. The act specifies that the federal government will pay 90 percent of the cleanup cost, with the respective state to pay the remainder. Uranium-mill tailings could become a serious problem, because they contain the largest volume of radioactive waste in the country.

Isolation of Nuclear Waste

Scientists agree that nuclear waste must be isolated from contact with all biological systems for at least 250,000 years in order for radioactive materials—particularly plutonium-239, whose half-life is 24,000 years—to decay to harmless levels. Should the ban on plutonium recovery be rescinded,

the conditions for high-level-waste disposal may change. Several methods of storing wastes have been proposed, but any method must meet these conditions:

- Safe isolation for at least 250,000 years,
- Safe from terrorists or accidental entry,
- Safe from natural disasters (hurricanes, landslides, floods, etc.),
- Noncontaminating to any nearby natural resources,
- Geologically stable site (no volcanic activity, earthquakes, etc.), and
- Fail-safe handling and transport mechanisms.

A number of disposal methods have been proposed. The more serious ones are discussed here.

1. *Continue the present tank storage.* This is not a permanent solution, because the tanks have limited lifetimes and leakage has already occurred.
2. *Dispose of waste in a convergent plate boundary (subduction zone).* The feasibility of this very imaginative proposal has not been demonstrated, because the rate of plate movement is very slow.
3. *Place containerized waste on the ice sheets of Antarctica or Greenland.* Radioactive heat would cause the containers to melt downward to the ice-bedrock contact surface, which could place the radioactive waste two miles beneath ice in either Greenland or Antarctica. This proposal assumes that long-term global warming will not occur in the 250,000-year period. This is unsound, because the last ice sheet of the Pleistocene melted back to Greenland in less than 10,000 years.
4. *Geologic repository: Salt mines or domes.* Salt is dry, flows readily, and is self-healing. Because it is a good conductor of heat, temperatures in the repository rocks would not become excessive. Salt is subject to underground solution, however, and some hydrated minerals in salt beds give up their water when heated. Salt beds and salt domes occur in geologically stable regions such as in Texas, Louisiana, and Mississippi.
5. *Geologic repository: Deep chambers in granite, volcanic rocks, or some other very competent, relatively dry rock formation.* This would require detailed site analysis to ensure that all geologic criteria are met.

Yucca Mountain Geologic Repository

The DOE's Civilian Radioactive Waste Management Program has been conducting a site characterization of Yucca Mountain, Nevada, as a high-level-nuclear-waste repository since the early 1980s. This is the site approved by the Congress and president in 2003. The 7-square-kilometer (2.7-mi^2) repository at Yucca Mountain is in pyroclastic rocks (tuffs, some welded, see Chapter 2), and would be placed more than 200 meters (650 ft) below the ground surface. The zone of aeration or vadose zone and the natural system operating there is shown in ◆Figure 15.21. All geologic studies indicate that the proposed repository is an acceptable site although there are detractors. The DOE plans to deliver its licensing application to the Nuclear Regulatory Commission in late 2004 and the agency is expected to take several years to review the request. The earliest that radioactive waste could find a permanent home at the repository is 2010. In the meantime, high-level nuclear waste will continue to grow and present a hazard to the public.

Health and Safety Standards for Nuclear Waste

There is general agreement that any nuclear-waste repository will leak radiation over the course of the next 10,000 years. Nuclear waste emits three types of radiation:

1. Alpha (α) particles, the most energetic but least penetrating type of radiation (they can be stopped by a piece of paper). If α emitters are inhaled or ingested, alpha radiation can attack lung and body tissue.
2. Beta (β^-) radiation, penetrating but most harmful when a β^- emitter is inhaled or ingested. For example, strontium-90 is a β^- emitter that behaves exactly like calcium (builds bone) in the food chain.
3. Gamma (γ) rays, nuclear X rays that can penetrate deeply into tissue and damage human organs (◆Figure 15.22).

Nuclear waste policy limits the radioactivity that may be emitted from a repository, as measured by cancer deaths directly attributable to leaking radiation. The mortality limit is 10 deaths per 100 years, and for this reason the maximum allowable radiation dose per year per individual in the proximity of a repository is set at 25 millirems. The *rem* is the unit used for measuring the amounts of ionizing radiation that affect human tissue, and a millirem (mrem) is 0.001 rem. Doses on the order of 10 rems can cause weakness, redden the skin, and reduce blood-cell counts. A dose of 500 rems would kill half of the people exposed. Rem units take into account the type of radiation—alpha, beta, or gamma—and the amount of radioactivity deposited in body tissue. ◆Figure 15.23 shows how the average annual dose of 208 mrems is partitioned for U.S. citizens and common sources of radiation. (Case Study 15.5 explains the hazards of high levels of radon gas in the home and methods for testing and mitigating them.) It is estimated that a person living within 5 kilometers (3 mi) of the Yucca Mountain repository would receive less than 1 mrem per year as a result of nuclear waste being stored there (currently no one lives that close to the site). The 25-mrem standard for maximum annual dosage is equal to the radiation a person would receive in 2 or 3 medical X rays.

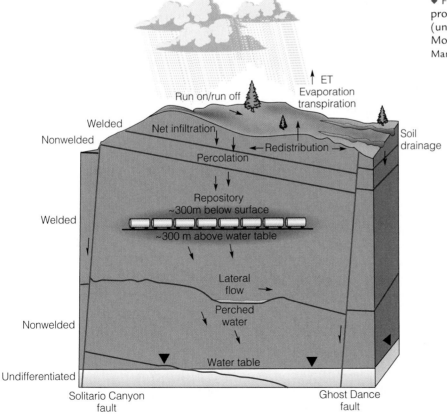

◆ FIGURE 15.21 Block diagram showing processes acting within proposed vadose (unsaturated) zone repository at Yucca Mountain. Office of Civilian Radioactive Waste Management, 1998

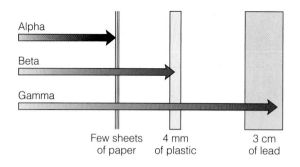

◆ FIGURE 15.22 Penetrating abilities of alpha, beta, and gamma radiation, the three types of radiant energy emitted from nuclear wastes.

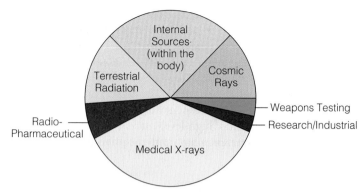

◆ FIGURE 15.23 The typical American receives 208 millirems of radiation per year from the various sources. Some medical, research, and industrial workers' exposures may exceed the 208-millirem average. National Academy of Sciences, NRC

Is There a Future for Nuclear Energy?

Nuclear reactors are very complex; a typical U.S. reactor may have 40,000 valves, whereas a coal-fired plant of the same size has only 4,000! This fact alone makes reactors relatively unforgiving of errors in operation, construction, or design. A newer generation of "standard" reactors are now under consideration that use natural forces such as gravity and convection instead of the network of pumps and valves currently required. For example, if the core of one of these new reactors should overheat, a deluge of water stored in tanks above the reactor would inundate the core. In addition, design would be standardized in the new reactors, which would greatly simplify construction, maintenance, and repair. Standardization would also facilitate better operator training and would eliminate the possibility of some human errors.

The ultimate, clean nuclear-energy source is released by **fusion,** whereby two nuclei merge to form a heavier nucleus and in the process give off tremendous amounts of heat. Deuterium ($_1^2$H) is a heavy isotope of hydrogen that at very high temperatures fuses to form helium ($_2^4$He). This is the very same nuclear process that energizes our sun. The fuel is inexhaustible, the reaction cannot run out of control, and it is environmentally nonpolluting.

Private Sewage Disposal

Modern household sewage-disposal systems utilize a septic tank, which separates solids from the effluent, and leach lines, which disseminate the liquids into the surrounding soil or rock (◆Figure 15.24). A septic tank is constructed in such a way that all solid material stays within the tank and only liquid passes on into the leach field. A leach field consists of rows of tile pipe surrounded by a gravel filter through which the effluent percolates prior to percolating into the soil or underlying rock. Sewage is broken down by *anaerobic* bacteria in the septic tank and the first few feet of the leach lines. As the effluent percolates into the soil or bedrock, it is filtered and oxidized, and aerobic conditions are established. Studies have shown that the movement of effluent through a few feet

of unsaturated fine-grained soil under aerobic conditions reduces bacterial counts to almost nil. Fine-grained soils have been found to be the best filters of harmful pathogens, whereas coarse-grained soils allow pathogenic organisms and viruses to disperse over greater distances.

Under the best of conditions, leach lines have a life expectancy of 10–12 years. Factors that reduce infiltration rates and thus limit the life of a septic system are dispersion of clay particles by transfer of sodium ions (Na^+) in sewage to clay particles in the soil, plugging by solids, physical breakdown of the soil by wetting and drying, and biological "plugging." This plugging occurs when the leach system becomes overloaded with sewage and anaerobic conditions invade the leach lines, resulting in a growth of organisms that plug the filter materials.

After passing through the leach field, the effluent must be kept underground until oxidation and bacteria can purify it. The geologic condition that most commonly causes a private sewage-disposal system to fail is a high water table that intersects or is immediately beneath the leach lines (◆Figure

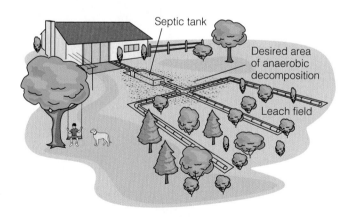

◆ FIGURE 15.24 A septic tank near the house drains into a leach field of drain tiles that are laid in trenches and covered with filter gravel and soil.

15.25a). This condition initiates the anaerobic growth that plugs filter materials and tile drains. Water tables may rise due to prolonged heavy precipitation or due to the addition of a number of septic tanks in a small area. One way to avoid exceeding the soil's ability to accept effluent is to design leach fields on the basis of percolation tests performed during wet periods. If necessary, a system's lifetime can be extended by constructing multiple leach fields whose use can be rotated regularly by switching a valve. Another problem is the improper placement of leach lines. This results in effluent seepage on hill slopes or stream valley walls (◆Figure 15.25b). Although typical plumbing codes specify a minimum of 5 meters (16 ft) horizontal distance from a hill face, this may not be sufficient if the vertical permeability is low and the hill slope is steep.

Nitrates in sewage effluent pose a eutrophication problem if households are built near a lake or pond (see Chapter 8). Cultural (human-caused) eutrophication accelerates the process and can convert an attractive, "real-estate" lake into a weed-choked bog in a short time. Nutrients from sewage systems and fertilized lawns must not be allowed to flow unchecked into bodies of water. Solving this problem requires the help of such experts as aquatic biologists and hydrogeologists.

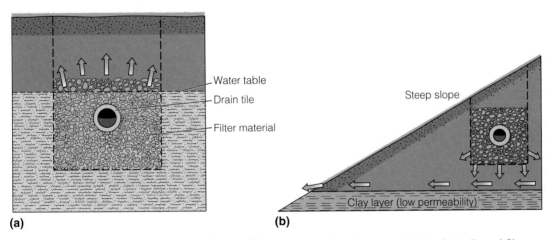

(a) **(b)**

◆ FIGURE 15.25 Conditions under which raw effluent can seep into the ground. (a) A drain tile and filter material intersect a high water table. When this occurs sewage rises, and the growth of anaerobic bacteria is promoted in the leach field. This leads to biological plugging of the filter material. (b) A leach line is underlain by low-permeability layers adjacent to a steep slope.

The Sociology of Waste Disposal

The EPA is charged with the Herculean tasks of monitoring existing land and ocean dump sites, approving new hazardous-waste disposal sites, and supervising the cleanup of historic toxic sites and accidental spills and leaks. In carrying out these responsibilities, EPA personnel have encountered the spectrum of political and public attitudes about waste disposal. A wall poster at EPA headquarters in Washington, D.C., provides the following acronyms for those attitudes:

NIMBY	Not In My Back Yard
NIMFYE	Not In My Front Yard Either
PIITBY	Put It In Their Back Yard
NIMEY	Not In My Election Year
NIMTOO	Not In My Term Of Office
LULU	Locally Unavailable Land Use
NOPE	Not On Planet Earth

The Highest Point between Maine and Florida

Fresh Kills landfill on Staten Island, New York, is the world's largest landfill, with 25 times the volume of the Great Pyramid of Giza, Egypt (◆Figure 1). Fresh Kills (Dutch *kil,* "stream") serves New York City, its suburbs, and parts of New Jersey. It accepts about 13,000 tons of trash daily. By 1998 it rose about 165 meters (500 ft) above sea level, the highest elevation on the eastern shore of the United States. The fill was established in 1948 on a salt marsh with no provisions for constraining leachate. For years, more than 4.2 million liters (1 million gal) of leachate leaked into the marsh and nearby waters each day. Recent cleanup has solved many of the pollution problems, but Fresh Kills remains a mountain of an eyesore to Staten Islanders.

What makes Fresh Kills unique, aside from being built on marshland, is that trash is taken to the fill by city-owned barges, each with a 600–700-ton capacity, that are loaded from garbage trucks at eight stations throughout the city and towed by tugboats. Specially built "skimmer" boats follow to pick up trash that falls from the barges. At the docks the refuse is unloaded by huge cranes into custom-made vehicles that carry it to a fill site. The refuse is dumped, sprayed with a deodorant, compacted by bulldozers, covered by clean earth, and eventually landscaped to "reclaim" the land.

New York City also has three active trash incinerators. In the past, ash residue from these was taken offshore and dumped

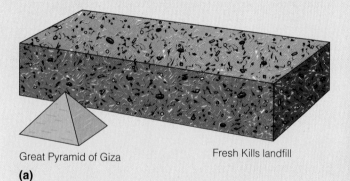

Great Pyramid of Giza   Fresh Kills landfill

(a)

(b)

◆ FIGURE 1 (a) Comparison of the volumes of the Great Pyramid of Giza and the Fresh Kills landfill. (b) The Great Pyramid (right) built by King Cheops at Giza on the west bank of the Nile River in the third century B.C. The length of each side at the base averages 756 feet, and its present height is 451 feet (original height, 481 feet). The smaller pyramid was built by Chephren, a later king. The automobiles give an idea of the immensity of these structures.

into the ocean. Now the ash is deposited at Fresh Kills along with raw garbage. Because the ash contains toxic metals, specially designed ash-only fill areas are planned, each with a double liner and a double leachate-collection system to keep toxic substances within the fill. A clay slurry "wall" has been excavated around the fill to further confine leachate if there are leaks. A waste-to-energy incinerator is also planned; its ash will be dumped in Fresh Kills (◆Figure 2).

Major efforts are under way to extend the landfill's life by recovering resources from the trash. Methane is extracted and used for heating and cooking in nearby homes, and construction debris is recycled. Yard waste from Staten Island and the Borough of Queens is composted and mixed with soil to promote healthy plant growth on the finished fill. Two more recycling facilities are planned, as is an intermediate processing center to sort and process recyclable glass, cans, paper, and plastics. At the present rate of delivery, Fresh Kills landfill will be filled to capacity early in this century.

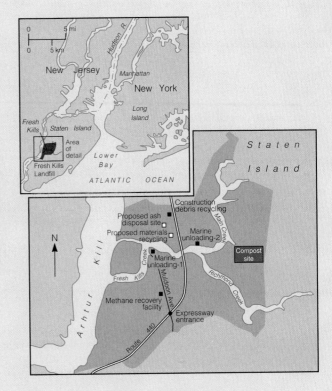

◆ FIGURE 2 Fresh Kills landfill on Staten Island.

Deep-Well Injection and Human-Caused Earthquakes

The U.S. Army's Rocky Mountain Arsenal in suburban Denver drilled a 3,764-meter (2½-mi) deep well for disposing of chemical-warfare wastes in 1961. The Tertiary and Cretaceous sedimentary rocks into which the well was drilled were cased off, and the lower 23 meters (75 ft) were open in Precambrian granite. About 40 kilometers (25 mi) to the west of the arsenal is the frontal fault system of the Rocky Mountains (◆Figure 1).

Injection began in 1962 at a rate of about 700 cubic meters per day (176,000 g/day). Soon after injection began, numerous earthquakes were felt in the Denver area. The earthquake epicenters lay within 8 kilometers (5 mi) of the arsenal and plotted along a line northwest of the well. A local geologist attributed the quakes to the fluid injection at the arsenal.

Although the army denied any cause-and-effect relationship, a graph of injection rates versus the number of earthquakes showed an almost direct correlation—especially during the quiet period when fluid injection was stopped (◆Figure 2).

The fractured granite into which the fluid was being injected was in a stressed

CONTINUED

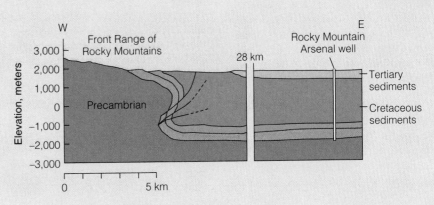

◆ FIGURE 1 Relationship between the Rocky Mountain Arsenal and the Rocky Mountain frontal fault system.

condition, and the high fluid pressures at the bottom of the hole decreased the frictional resistance on the fracture surfaces. Earthquakes were generated along a fracture propagating northwest of the well as the rocks adjusted to relieve internal stresses.

Thus earthquakes *can* be "triggered" by human activities, and the Denver earthquakes initiated research into the possibility of modifying fault behavior by fluid injection; that is, relieving fault stresses by producing a large number of small earthquakes. Seismic studies conducted during a water-injection (see Chapter 14) program at the nearby Rangely Oil Field confirmed that earthquakes *could* be turned on or off at will, so to speak, in areas of geologic stress. To date no group or government entity has assumed responsibility for triggering earthquakes by fluid injection, for obvious reasons.

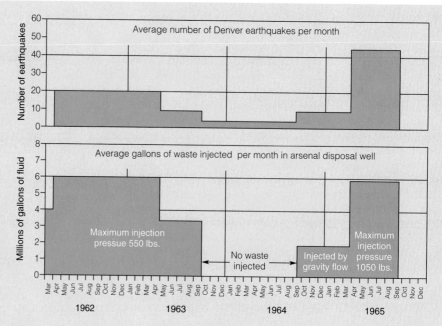

◆ FIGURE 2 Correlation of fluid waste injection and the number of earthquakes per month in the Denver area, 1962–1965.

CASE STUDY 15.4

Love Canal—A Precedent in Human-Caused Environmental Pollution

Love Canal, in the honeymoon city of Niagara Falls, New York, may sound very romantic, but it's not. Excavated by William T. Love in the 1890s but never finished, the canal is 1,000 meters long and 25 meters wide (3,300 ft × 80 ft). Unused and drained, Love Canal was bought by Hooker Chemical and Plastics Corporation in 1942 for use as a waste dump. In the next 11 years Hooker dumped 21,800 tons of toxic waste into the canal, which was believed to be impermeable and thus an ideal "grave" for hazardous substances.

After the canal was filled and covered with a clay cap, Hooker sold it to the Niagara Falls School Board for one dollar. The company inserted in the deed a disclaimer denying legal liability for any injury caused by the wastes. In 1957 Hooker warned the school board not to disturb the clay cap because of the possi-

ble danger from leakage of toxic chemicals. A school was built on the landfill, and hundreds of homes and all the infrastructure needed to support a suburban community were built nearby.

In the early 1970s, after several years of heavy rainfall, water leaked through the "impermeable" clay cap and into basements and yards in the Love Canal area, and also into the local sewer system. Soon the new community was experiencing high rates of miscarriages, birth defects, liver cancer, and seizure-inducing diseases among children. Residents complained that toxic chemicals in the water were causing these adverse health effects.

In 1978 the New York State Health Commissioner requested the EPA's assistance in investigating the chemistry of fluids that were leaking into a few houses around the canal. The study revealed the

presence of 82 toxic chemicals including benzene, chlorinated hydrocarbons, and dioxin. Five days later then President Carter declared it a federal disaster area, the first human-caused environmental problem to be so designated in the United States. The recommendation that pregnant women and children under the age of two be evacuated from the area followed, and New York State appropriated $22 million to buy homes and repair the leaks. In all, a thousand families were relocated (◆Figure 1).

Love Canal was a precedent-setting case, because it spurred Congress to enact CERCLA, which provides federal money for toxic cleanup without long appeals. It also authorizes the EPA to sue polluters for the costs of cleanup and relocation of victims. Although Love Canal has been essentially contained, some chemicals still infect a nearby stream and schoolyard, and maintenance costs amount to a half-million dollars annually. The EPA no longer ranks Love Canal high among the nation's most dangerous waste sites, and the New York State Health Commissioner has announced that recent federal-state studies found four of the seven polluted Love Canal areas to be habitable. In spite of the commissioner's statement, there has been no stampede to reinhabit the area.

After 16 years of litigation, the legal battles over Love Canal ended in 1995. Occidental Chemical Corporation, which purchased Hooker Chemical in 1968, agreed to pay Superfund $102 million, and the Federal Emergency Management Agency (FEMA) agreed to pay $27 million. The State of New York had made a $98 million settlement with EPA in 1994. The cleanup is estimated to have cost $275 million.

◆ FIGURE 1 Where the honeymoon ended and the nightmares began; Love Canal, Niagara Falls, New York.

MICHAEL J. PHILIPPOT/SYGMA

CASE STUDY 15.5

Radon and Indoor Air Pollution

Radon (^{222}Rn) is a radioactive gas formed by the natural disintegration of uranium as it transmutes step-by-radioactive-step to form stable lead. The immediate parent element of radon is radium (^{226}Ra), which emits an alpha particle (^{4}He) to form ^{222}Rn. Radon-222 has a half-life of only 3.8 days, but it in turn breaks down into other elements that emit dangerous radionuclides. *Emanation* is sometimes used to describe the behavior of this element, and concern has been growing about the health hazard posed by radon emanating from rocks and soils and seeping into homes. The EPA estimates that between 5,000 and 20,000 people die every year of lung cancer because they have inhaled radon and its decay products.

Radon is only a problem in areas underlain by rocks whose uranium concentrations are greater than 10 ppm—three times the average amount found in granite. For comparison, uranium ore contains more than 1,000 ppm of uranium. As radon disintegrates in the top few meters of rock and soil, it seeps upward into the atmosphere, or into homes through cracks in concrete slabs or around openings in pipes. The measure of radon activity is *picocuries per liter (pCi/L)*—the number of nuclear decays per minute per liter of air. One pCi per liter represents 2.2 potentially cell-damaging disintegrations per minute. Inasmuch as the average human inhales between 7,000 and 12,000 liters (2,000–3,000 gal) of air per day, high radon concentrations in household air can be a significant health hazard.

Cancer due to bombardment of lung tissue by breathing radon-222 or its decay products is a major concern of health scientists. The EPA has established 4 pCi/L as the maximum allowable indoor radon level. This is equivalent to smoking a half-pack of cigarettes per day or having 300 chest X rays in a year. ◆Figure 1 vividly illustrates the dangers of inhaling indoor radon (and of smoking). Such cancer is rarely apparent before 5–7 years of exposure, and its incidence is rare in individuals under the age of 40.

Widespread interest in radon pollution first appeared in the 1980s with the discovery of high radon levels in houses in Pennsylvania, New Jersey, and New York. The discovery was made accidentally when a nuclear-plant worker set off the radiation monitoring alarm when he arrived at work one day. Since his radioactivity was at a safe level, he did not worry about it. Returning to work after a weekend at home, he once again triggered the alarm. An investigation of his home revealed very high levels of radon. This was the first time radon was recognized as a public health hazard.

There are currently two types of radon "detectors." One is a charcoal canister that is placed in the household air for a week and then returned to the manufacturer for analysis. The other is a plastic-film alpha-track detector that is placed in

CONTINUED

pCi/L	Comparable Exposure Level	Comparable Risk *
200	1,000 times average outdoor level	More than 75 times nonsmoker risk of dying from lung cancer
100	100 times average indoor level	4-pack/day smoker / 10,000 chest X rays per year
40		30 times nonsmoker risk of dying from lung cancer
20	100 times average outdoor level	2-pack/day smoker
10	10 times average indoor level	1-pack/day smoker
4		3 times nonsmoker risk of dying from lung cancer
2	10 times average outdoor level	200 chest X rays per year
1	Average indoor level	Nonsmoker risk of dying from lung cancer
0.2	Average outdoor level	

* Based on lifetime exposure

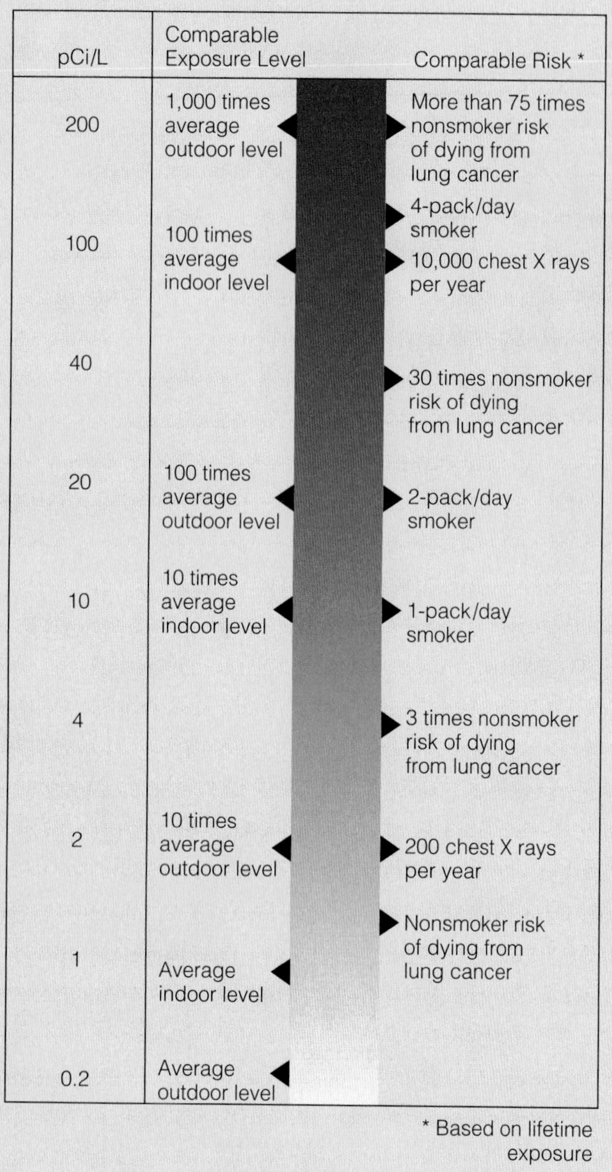

◆ FIGURE 1 Lung cancer risk due to radon exposure, smoking, and X-ray exposure. A few houses have been found with radon radiation levels greater than 100 pCi/L, and one in Pennsylvania had a level of 2,000 pCi/L.

the home for 2–4 weeks. It is recommended that the less-expensive canister type be used first, and that if a high radon level is indicated, a follow-up measurement with a plastic-film detector be made. Any reading above 4 pCi/L requires follow-up measurements. If the reading is between 20 and 200 pCi/L, one should consider taking measures to reduce radon levels in the home. Common methods are increasing the natural ventilation below the house in the basement or crawlspace, using forced air to circulate subfloor gases, and intercepting radon before it enters the house by installing gravel-packed plastic pipes below the floor slabs. For houses with marginal radon pollution, sealing all openings from beneath the house that go through the floor or the walls is usually effective.

Waste Disposal Then and Now

All humans generate solid waste, and disposing of it has always been a societal challenge. Safely disposing of containers of all kinds, old tires, large and small appliances, and just plain trash presents major problems today. As individuals we must recycle, demand recycled and recyclable containers, and actively assist in reducing the volume of waste (◆ Figure 1). Historians point out that Ancient Rome was a filthy city; its cobblestones were noisy, its air was polluted from thousands of wood fires, and municipal sewage disposal was nonexistent. Medieval castles, monasteries, and convents had primitive waste-disposal systems, most of which were unhealthy and emanated noxious odors (◆ Figure 2). Conditions are not much better in some parts of rural America today (◆ Figure 3). And it should be pointed out that sanitary landfills are not all work and no play (◆ Figure 4).

◆ FIGURE 1 Norris McDonald organized a major cleanup of the Anacostia River in Washington, D.C., and adjacent Maryland.

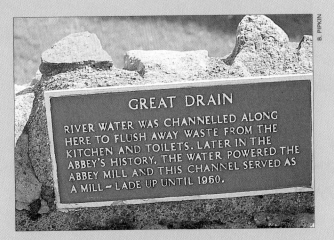

GREAT DRAIN

RIVER WATER WAS CHANNELLED ALONG HERE TO FLUSH AWAY WASTE FROM THE KITCHEN AND TOILETS. LATER IN THE ABBEY'S HISTORY, THE WATER POWERED THE ABBEY MILL AND THIS CHANNEL SERVED AS A MILL – LADE UP UNTIL 1960.

◆ FIGURE 2 An adjacent stream was diverted to this medieval abbey in Scotland to create what was called a "great drain." All of the abbey's wastes went into the drain, which carried them to the stream, thus polluting its water for downstream users. Little wonder that lifespans were so short in those days; diseases such as cholera, typhoid fever, and dysentery and pestilence caused by the lack of sanitation took their toll. The correlation between stream pollution and disease was not widely recognized until the 1800s.

◆ FIGURE 3 An all-too-common method of private sewage disposal in the rural United States.

◆ FIGURE 4 Caterpillar tractor operators play "king of the mountain" at a landfill. The 25-ton "Cats" are used to compact and flatten the debris to reduce its volume.

Summary

Waste Management

Management Components

Geology must be considered in all components:
1. Waste disposal
2. Source reduction
3. Waste recovery
4. Recycling

Rationale

Extending the lifetimes of landfills and natural resources.

Categories of Waste-Disposal Methods

Isolation and attenuation.

Municipal Waste Disposal

Sanitary Landfill

Trash is covered with clean soil each day.

Ocean Dumping

Favored in the past for disposal of sewage, sludge, dredgings, and canistered nuclear wastes. Now regulated by Marine Protection Reserve and Sanctuary Act (Ocean Dumping Act).

Landfill Problems

Leachate, settlement, gas generation, visual blight.

Resource Recovery

Amount of waste delivered to landfills can be reduced through recycling programs, composting, and implementing waste-reduction strategies.

Multiple Land Use Concept

A mine or quarry becomes a landfill, which when full may be converted to a park or other recreational area. Although some shallow landfills have been built upon, settlement is always a problem.

Hazardous-Waste Disposal

Secure Landfill

Lined with plastic and impermeable clay seal. Built with leachate drains.

Deep-Well Injection

Into dry permeable rock body sealed from aquifers above and below. Has triggered earthquakes in Denver area.

Superfund

Comprehensive Environmental Response, Compensation, and Liability Act (CERCLA) created in 1980 to correct and stop abuses from irresponsible toxic-waste dumping.

Private Sewage Disposal

Requires permeability and porosity and can lead to local pollution.

Radioactive Waste Disposal

High-Level Wastes

High-level nuclear waste from reactors (spent fuel rods), and 63,000 MTHM, is currently kept at individual reactors in pools of water. This is not a good situation. Yucca Mountain as the final geologic respository received approval in early 2003. But licensing by the Nuclear Regulatory Commission will take several years, and it will be at least 2010 before any waste is delivered to the site.

Low-Level Wastes

These are disposed of by shallow burial.

Mill Tailings

There are 24 inactive sites in 10 states, many of which are being cleaned up with the government paying 90 percent of the costs.

Health and Safety Standards

The standard for radiation that a person may receive is 25 mrem per year, about equal to the dosage that a person would receive from 2 or 3 medical X rays.

The Future

There is a future for nuclear energy and it lies in new technology and training.

Key Terms

attenuation (waste disposal)	eutrophication	low-level waste	sanitary landfill
biological oxygen demand (BOD)	fusion	NEPA	secure landfill
composting	geological repository	RCRA	Superfund
deep-well injection	high-level wastes	recycling	waste management
EPA	isolation (waste disposal)	resource recovery	
	leachate		

Study Questions

1. Explain the federal legislation known as NEPA and RCRA and describe their impacts on the environmental safety of landfills.

2. What is leachate, and what methods are used to prevent landfill leachates from entering bodies of water?

3. Explain the "cell" method of placing trash in a landfill.

4. How much solid waste does the average American generate each day in kilograms? In pounds?

5. What problems are associated with operating a sanitary landfill?

6. What resources are found in a landfill, and how can they be recovered? What can individuals do to assist in recovering and extending the lifetimes of certain limited natural resources?

7. Describe two common methods of isolating hazardous wastes from the environment. Explain the steps that are taken to keep these wastes from polluting underground water.

8. What constitutes a good geological repository, and do you think Yucca Mountain is a good choice? (See Figures 15.19 to 15.21 to help you answer this one.)

9. What are some of the health hazards connected with exposure to radiation, and how are dosages measured?

For Further Information

Books and Periodicals

Alexander, J. H. 1993. *In defense of garbage.* New York: Praeger.

Barber, D. A. 2002. *Building with tires.* University of Arizona, Report on Research. University Publications, Tucson.

Brown, Lester, and others. 1999. *State of the world.* New York: Worldwatch Institute and W. W. Norton.

Brunner, D. R., and D. K. Keller. 1972. *Sanitary landfill design and operations.* Washington, D.C.: U.S. Environmental Protection Agency.

Crowley, Kevin D., and John F. Ahearne. 2002. Managing the environmental legacy of U.S. nuclear weapons production. *American Scientist* 90, November–December.

Grove, Noel. 1994. Recycling. *National geographic* 186, no. 1 (July).

Lewin, David. 1991. Solid waste: Bury, burn, or reuse. *Mechanical engineering,* August:75.

McKinney, Michael, and Robert Schoch. 1996. *Environmental science.* St. Paul, Minn.: West Educational Publishing.

Miller, G. Tyler. 1999. *Environmental science.* Belmont, California: Wadsworth Publishing Company.

Rathje, William L. 1991. Once and future landfills. *National geographic,* May:114–134.

Rathje, William L., and Cullen Murphy. 1992. *Rubbish! The archeology of garbage.* New York: Harper Collins.

U.S. Department of Energy, Office of Civilian Radioactive Waste Management. 2002. Final Environmental Impact Statement for a Geologic Repository for the Disposal of Spent Nuclear Fuel and High-Level Radioactive Waste at Yucca Mountain, Nye County, Nevada. DOE/EIS-0250F, February.

U.S. Environmental Protection Agency. 1992. *Characterization of municipal solid waste in the United States: Final report.*

White, Peter T. 1983. The wonderful world of trash. *National geographic,* April.

ENVIRONMENTAL
Geology⇌Now™

Assess your understanding of this chapter's topics with additional quizzing and comprehensive interactivities at http://earthscience.brookscole.com/pipkingeo4e as well as current and up-to-date Web links, additional readings, and InfoTrac College Edition exercises.

Periodic Table of the Elements

Noble Gases

Period

Legend:
- 1 — Atomic Number
- H — Symbol of Element
- 1.008 — Atomic Mass (rounded to four significant figures)

Period	IA	IIA	IIIB	IVB	VB	VIB	VIIB	VIII			IB	IIB	IIIA	IVA	VA	VIA	VIIA	Noble Gases
1	1 **H** 1.008																	2 **He** 4.003
2	3 **Li** 6.941	4 **Be** 9.012											5 **B** 10.81	6 **C** 12.01	7 **N** 14.01	8 **O** 16.00	9 **F** 19.00	10 **Ne** 20.18
3	11 **Na** 22.99	12 **Mg** 24.31											13 **Al** 26.98	14 **Si** 28.09	15 **P** 30.97	16 **S** 32.06	17 **Cl** 35.45	18 **Ar** 39.95
4	19 **K** 39.10	20 **Ca** 40.08	21 **Sc** 44.96	22 **Ti** 47.90	23 **V** 50.94	24 **Cr** 52.00	25 **Mn** 54.94	26 **Fe** 55.85	27 **Co** 58.93	28 **Ni** 58.71	29 **Cu** 63.85	30 **Zn** 65.37	31 **Ga** 69.72	32 **Ge** 72.59	33 **As** 74.92	34 **Se** 78.96	35 **Br** 79.90	36 **Kr** 83.80
5	37 **Rb** 85.47	38 **Sr** 87.62	39 **Y** 88.91	40 **Zr** 91.22	41 **Nb** 92.91	42 **Mo** 95.94	43 **Tc**^{x} 98.91	44 **Ru** 101.1	45 **Rh** 102.9	46 **Pd** 106.4	47 **Ag** 107.9	48 **Cd** 112.4	49 **In** 114.8	50 **Sn** 118.7	51 **Sb** 121.8	52 **Te** 127.6	53 **I** 126.9	54 **Xe** 131.1
6	55 **Cs** 132.9	56 **Ba** 137.3	57 **La** 138.9 ●	72 **Hf** 178.5	73 **Ta** 180.9	74 **W** 183.9	75 **Re** 186.2	76 **Os** 190.2	77 **Ir** 192.2	78 **Pt** 195.1	79 **Au** 197.0	80 **Hg** 200.6	81 **Tl** 204.4	82 **Pb** 207.2	83 **Bi** 209.0	84 **Po**^{x} (210)	85 **At**^{x} (210)	86 **Rn**^{x} (222)
7	87 **Fr**^{x} (223)	88 **Ra**^{x} 226.0	89 **Ac**^{x} (227) ■	104 **Unq**^{x} (261)	105 **Unp**^{x} (262)	106 **Unh**^{x} (263)	107 **Uns**^{x} (262)	108 **Uno**^{x} (265)	109 **Une**^{x} (266)									

● Lanthanide series

58 **Ce** 140.1	59 **Pr** 140.9	60 **Nd** 144.2	61 **Pm**^{x} (147)	62 **Sm** 150.4	63 **Eu** 152.0	64 **Gd** 157.3	65 **Tb** 158.9	66 **Dy** 162.5	67 **Ho** 164.9	68 **Er** 167.3	69 **Tm** 168.9	70 **Yb** 173.0	71 **Lu** 175.0

■ Actinide series

90 **Th**^{x} 232.0	91 **Pa**^{x} 231.0	92 **U**^{x} 238.0	93 **Np**^{x} 237.0	94 **Pu**^{x} (244)	95 **Am**^{x} (243)	96 **Cm**^{x} (247)	97 **Bk**^{x} (247)	98 **Cf**^{x} (251)	99 **Es**^{x} (254)	100 **Fm**^{x} (257)	101 **Md**^{x} (258)	102 **No**^{x} (255)	103 **Lr**^{x} (256)

x: All isotopes are radioactive.

( ) indicates mass number of isotope with longest known half-life.

Some Important Minerals

Mineral	Chemical Composition	Color	Hardness	Other Characteristics	Primary Rock Occurrence
1. amphibole (e.g., hornblende) (hydrous)	Na, Ca, Mg, Fe, Al silicate	green, blue brown, black	5 to 6	often forms needlelike crystals; two good cleavages forming 60°–120° angle crystals hexagonal in cross section	igneous and metamorphic minor amounts in all types
2. apatite	$Ca_5(PO_4)_3$ (F, Cl, OH)	blue-green	5	crystals hexagonal in cross section	minor amounts, all types
3. biotite (a mica)	hydrous K, Mg, Fe silicate	brown to black	2½–3	excellent cleavage into thin sheets	all types
4. calcite	$CaCO_3$	variable; colorless if pure	3	effervesces in weak acid	sedimentary and metamorphic
5. dolomite	$CaMg(CO_3)_2$	white or pink	3½ to 4	powdered mineral effervesces in acid	sedimentary
6. galena	PbS	silver-gray	2½	metallic luster; cubic cleavage; heavy	sulfide veins
7. garnet	Ca, Mg, Fe, Al silicate	variable; often dark red	7	glassy luster	metamorphic
8. graphite	C	dark gray	1 to 2	streaks like pencil lead	metamorphic
9. gypsum	$CaSO_4 \cdot 2H2O$	colorless	2	2 directions of cleavage	sedimentary
10. halite	NaCl	colorless	2½	salty taste; cleaves into cubes	sedimentary
11. hematite	Fe_2O_3	red or dark gray	5½ to 6½	red-brown streak regardless of color	oxidation of iron-bearing minerals
12. limonite	$Fe_2O_3 \cdot 3H_2O$	yellow-brown	2 to 3	earthy luster; yellow-brown streak	hydration of iron oxides
13. magnetite	Fe_3O_4	black	6	strongly magnetic	minor amounts, all types
14. muscovite (mica)	K, Al, Si, silicate	colorless to light brown	2 to 2½	excellent basal cleavage	all types
15. olivine	Fe, Mg silicate	yellow-green	6½ to 7	glassy luster	igneous and metamorphic
16. plagioclase feldspar	Na, Ca, Al silicate	white to gray	6	2 cleavages at 90°, striations on cleavage surfaces	all types
17. orthoclase feldspar	K, Al silicate	white; or colored pink or aqua	6	two good cleavages forming 90° angle; no striations	all types
18. pyrite	FeS_2	brassy yellow	6 to 6½	metallic luster; black streak	all types
19. pyroxene (e.g., augite)	Na, Ca, Mg, Fe, Al silicate	usually green or black	5 to 7	two good cleavages forming 90° angle	igneous and metamorphic
20. quartz	SiO_2	variable; commonly colorless or white	7	glassy luster; conchoidal fracture; hexagonal crystals	all types

The Elements

Element	Symbol	Atomic Number	Element	Symbol	Atomic Number
Actinium	Ac	89	Mercury	Hg	80
Aluminum	Al	13	Molybdenum	Mo	42
Americium	Am	95	Neodymium	Nd	60
Antimony	Sb	51	Neon	Ne	10
Argon	Ar	18	Neptunium	Np	93
Arsenic	As	33	Nickel	Ni	28
Astatine	At	85	Niobium	Nb	41
Barium	Ba	56	Nitrogen	N	7
Berkelium	Bk	97	Nobelium	No	102
Beryllium	Be	4	Osmium	Os	76
Bismuth	Bi	83	Oxygen	O	8
Boron	B	5	Palladium	Pd	46
Bromine	Br	35	Phosphorus	P	15
Cadmium	Cd	48	Platinum	Pt	78
Calcium	Ca	20	Plutonium	Pu	94
Californium	Cf	98	Polonium	Po	84
Carbon	C	6	Potassium	K	19
Cerium	Ce	58	Praseodymium	Pr	59
Cesium	Cs	55	Promethium	Pm	61
Chlorine	Cl	17	Protactinium	Pa	91
Chromium	Cr	24	Radium	Ra	88
Cobalt	Co	27	Radon	Rn	86
Copper	Cu	29	Rhenium	Re	75
Curium	Cm	96	Rhodium	Rh	45
Dysprosium	Dy	66	Rubidium	Rb	37
Einsteinium	Es	99	Ruthenium	Ru	44
Erbium	Er	68	Samarium	Sm	62
Europium	Eu	63	Scandium	Sc	21
Fermium	Fm	100	Selenium	Se	34
Fluorine	F	9	Silicon	Si	14
Francium	Fr	87	Silver	Ag	47
Gadolinium	Gd	64	Sodium	Na	11
Gallium	Ga	31	Strontium	Sr	38
Germanium	Ge	32	Sulfur	S	16
Gold	Au	79	Tantalum	Ta	73
Hafnium	Hf	72	Technetium	Tc	43
Helium	He	2	Tellurium	Te	52
Holmium	Ho	67	Terbium	Tb	65
Hydrogen	H	1	Thallium	Tl	81
Indium	In	49	Thorium	Th	90
Iodine	I	53	Thulium	Tm	69
Iridium	Ir	77	Tin	Sn	50
Iron	Fe	26	Titanium	Ti	22
Krypton	Kr	36	Tungsten	W	74
Lanthanum	La	57	Uranium	U	92
Lawrencium	Lr	103	Vanadium	V	23
Lead	Pb	82	Xenon	Xe	54
Lithium	Li	3	Ytterbium	Yb	70
Lutetium	Lu	71	Yttrium	Y	39
Magnesium	Mg	12	Zinc	Zn	30
Manganese	Mn	25	Zirconium	Zr	40
Medelevium	Md	101			

Planes in Space: Dip and Strike

Geologists describe the orientation of a plane surface by the plane's "strike" and "dip." Geologists measure such tangible planes as those made by sedimentary strata, foliation, and faults, because it allows them to anticipate hazards during excavations for such things as tunnels, dam foundations, and roadcuts. Accurate surface measurements of planes also enable geologists to predict the underground locations of valuable resources such as oil reservoirs, ore deposits, and water-bearing strata.

Strike is the direction of the trace of an inclined plane on the horizontal plane, and the direction is referenced to north. In the illustrated example, the strike is 30° west of north. In geologic notation, this is written *N30° W*.

Dip is the vertical angle that the same inclined plane makes with the horizontal plane. Dip is assessed perpendicular to the strike, because any other angle will give an "apparent dip" that is less than the true dip. Because a plane that strikes northwest may be inclined to either the southwest or the northeast, the general direction of dip must be indicated. In our example, the dip is 60° to the northeast, which is notated *60° NE*.

Thus, a geologist's notes for the bedding planes in this example would describe their orientation as "strike N30° W, dip 60° NE." All other geologists would understand that the plane has a strike that is 30° west of north and a dip that is 60° to the northeast.

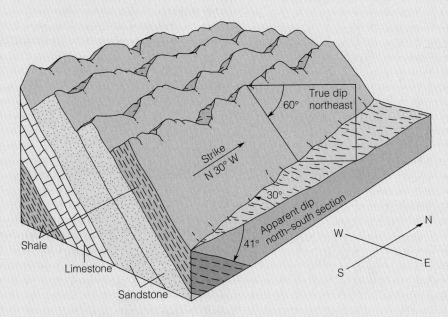

Block diagram showing sedimentary rock that is striking N30° W and dipping 60° northeast. The apparent dip is shown on the north–south cross section at the lower front of the diagram.

Modified Mercalli Scale

The Rossi–Forel intensity scale was the first scale developed for evaluating the effects of an earthquake on structures and humans. It was developed in Europe in 1878 and assigned Roman numerals from *I* (for *barely felt*) to *X* (for *total destruction*) for varying levels of intensity. This scale was first modified by Father Giuseppi Mercalli in Italy and later in 1931 by Frank Neumann and Harry O. Wood in the United States. The Modified Mercalli scale is the scale most widely used today for evaluating the effects of earthquakes in the field. An abbreviated version is given here.

I. Not felt except by a very few persons under especially favorable circumstances.

II. Felt by a few persons at rest, especially by persons on upper floors of multistory buildings, and by nervous or sensitive persons on lower floors.

III. Felt quite noticeably indoors, especially on upper floors, but many people do not recognize it as an earthquake; vibrations resemble those made by a passing truck.

IV. Felt indoors by many persons and outdoors by only a few. Dishes, windows, and doors are disturbed, and walls make creaking sounds as though a heavy truck had struck the building. Standing cars are rocked noticeably.

V. Felt by nearly everyone. Many sleeping people are awakened. Some dishes and windows are broken, and some plaster is cracked. Disturbances of trees, poles, and other tall objects may be noticed. Some persons run outdoors.

VI. Felt by all. Many people are frightened and run outdoors. Some heavy furniture is moved, and some plaster and chimneys fall. Damage is slight, but humans are disturbed.

VII. General fright and alarm. Everyone runs outdoors. Damage is negligible in buildings of good design and construction, considerable in those that are poorly built. Noticeable in moving cars.

VIII. General fright approaching panic. Damage is slight in specially designed structures; considerable in ordinary buildings, with partial collapse; great in older or poorly built structures. Chimneys, smokestacks, columns, and walls fall. Sand and mud are ejected from ground openings (liquefaction).

IX. General panic. Damage is considerable in specially designed structures, well-designed frame structures being deformed; great in substantial buildings, with partial collapse. Ground is cracked; underground pipes are broken.

X. General panic. Some well-built wooden structures and most masonry and frame structures are destroyed. Ground is badly cracked; railway rails are bent; underground pipes are torn apart. Considerable landsliding on riverbanks and steep slopes.

XI. General panic. Ground is greatly disturbed, with cracks and landslides common. Sea waves of significant height may be seen. Few if any structures remain standing; bridges are destroyed.

XII. Total panic. Total damage to human engineering works. Waves seen on the ground surface. Lines of sight and level are distorted. Objects are thrown upward into the air.

Glossary

aa A basaltic lava that has a rough, blocky surface.

ablation All processes by which snow or ice is lost from a glacier, floating ice, or snow cover. These processes include melting, evaporation, and calving.

ablation zone The part of a glacier or a snowfield in which ablation exceeds accumulation over a year's time; the region below the firn line.

abrasion The mechanical grinding, scraping, wearing, or rubbing away of rock surfaces by friction or impact. Abrasion agents include wind, running water, waves, and ice.

absolute age dating Assigning units of time, usually years, to fossils, rocks, and geologic features or events. The method utilizes mostly the decay of radioactive elements, but ages may also be obtained by counting tree rings or annual sedimentary layers.

accretion The mechanism by which the earth is believed to have formed from a small nucleus by additions of solid bodies, such as meteorites, asteroids, or planetesimals. Also used in plate tectonics to indicate the addition of terranes to a continent.

accumulation zone The part of a glacier or snowfield in which build-up of snow and ice exceeds ablation over a year's time; the region above the firn line. Also the B-horizon of a soil.

acid mine drainage (AMD) The acid runoff from a mine or mine waste.

acid rain Corrosive precipitation due to the solution of atmospheric sulfuric acid in rain, which makes it acidic. Although it is generally considered human-caused, it can also be natural, as from volcanic activity.

active layer The zone in permafrost that is subject to seasonal freezing and thawing.

active sea cliff A cliff whose formation is dominated by wave erosion.

adsorption The attraction of ions or molecules to the surface of particles, in contrast to *absorption,* the process by which a substance is taken into a clay-mineral structure.

aerosol A suspension of very fine, liquid and solid particles in the atmosphere.

alluvial fan A gently sloping, fan-shaped mass of alluvium deposited in arid or semiarid regions where a stream flows from a narrow canyon onto a plain or valley floor.

alluvial placer A placer deposit that is concentrated in a streambed.

alluvium Stream-deposited sediment.

alpha decay The transformation that occurs when the nucleus of an atom spontaneously emits a helium atom, reducing its atomic mass by 4 units and its atomic number by 2; for example, $^{238}_{92}U \rightarrow {}^{234}_{90}Th + {}^{4}_{2}a$.

amorphous Descriptive term applied to glasses and other substances that lack an orderly internal crystal structure.

andesite line The petrographic boundary between basalts and andesites. It runs roughly around the Pacific basin margin. It also marks the boundary between continental (explosive) and oceanic (Hawaiian-type) volcanoes in the Pacific Ocean.

angle of repose The maximum stable slope angle that granular, cohesionless material can assume. For dry sand this is about 34°.

anomaly A deviation from the average or expected value. In paleomagnetism, a "positive" anomaly is a stronger-than-average magnetic field at the earth's surface.

anthropogenic Relating to or resulting from the influence of humans on nature.

anticline A generally convex-upward fold that has older rocks in its core. (*Contrast with* syncline.)

aphanitic texture A fine-grained igneous-rock texture resulting from rapid cooling. Aphanites are fine-grained igneous rocks.

A.P.I. gravity (American Petroleum Institute) The weight of crude oil arbitrarily compared to that of water at 60° F, which is assigned a value of 10°. Most crude oils range from 5° to 30° A.P.I.

aquiclude A layer of low-permeability rock that hampers groundwater movement. Often applied to the confining bed in a confined aquifer.

aquifer A water-bearing body of rock or sediment that will yield water to a well or spring in usable quantities.

area mining The mining of coal in flat terrain (strip mining).

artesian pressure surface An imaginary surface defined by the level to which water will rise in an artesian well.

artificial levee An embankment constructed to contain a stream at flood stage. *See* levee.

ash Very fine, gritty volcanic dust: particles are less than 4 mm in diameter.

ash fall A "rain" of airborne volcanic ash.

asthenosphere The *plastic* layer of the earth, at whose upper boundary the brittle overlying lithosphere is detached from the underlying mantle.

atom The smallest particle of an element that can exist either alone or in combination with similar particles of the same or a different element.

atomic mass The sum of the number of protons and neutrons (uncharged particles) in the nucleus of an atom.

atomic number The number of protons (positively-charged particles) in an atomic nucleus. Atoms with the same atomic number belong to the same chemical element.

attenuation (waste disposal) To dilute or spread out the waste material.

average crustal abundance The percentage of a particular element in the composition of the earth's crust.

a-zone The upper soil horizon (topsoil) capped by a layer of organic matter. It is the zone of leaching of soluble soil salts.

back slope The gentler slope of a dune, scarp, or fault block.

ball mill A large rotating drum containing steel balls that grinds ore to a fine powder. Batteries of ball mills are used in modern milling operations.

banded-iron formation An iron deposit consisting essentially of iron oxides and chert occurring in prominent layers or bands of brown or red and black.

barchan dune An isolated, moving, crescent-shaped sand dune lying transverse to the direction of the prevailing wind. A gently sloping, convex side faces the wind, and the horns of the crescent point downwind.

barrier beach (barrier island) A narrow sand ridge rising slightly above hightide level that is oriented generally parallel with the coast and separated from the coast by a lagoon or a tidal marsh.

basalt A dark-colored, fine-grained volcanic rock that is commonly called *lava*. Basaltic lavas are quite fluid.

base isolation Building design option using a movable base under the building that allows the ground to move but minimizes building vibration and sway.

base level The level below which a stream cannot erode its bed. The ultimate base level is sea level.

base shear Due to a rapid horizontal ground acceleration during an earthquake. It can cause a building to part from its foundation, thus the term.

batholith A large, discordant intrusive body greater than 100 km^2 (40 mi^2) in surface area with no known bottom. Batholiths are granitic in composition.

bauxite An off-white, grayish, brown, or reddish brown rock composed of a mixture of various hydrous aluminum oxides and aluminum hydroxides.

beach cusps A series of regularly spaced points of beach sand or gravel separated by crescent-shaped troughs.

bedding *See* stratification.

bedding planes The divisions or surfaces that separate successive layers of sedimentary rocks from the layers above and below them.

bench placer Remnant of an ancient alluvial placer deposit that was concentrated beneath a former stream and that is now preserved as a bench on a hillside.

Benioff–Wadati zone A zone of earthquake epicenters beneath oceanic trenches that extends from near the surface to a depth of 700 km (430 mi). The zone is named after Hugo Benioff and is thought to indicate active subduction.

bentonite A soft, plastic, light-colored clay that swells to many times its volume when placed in water. Named for outcrops near Fort Benton, Wyoming, bentonite is formed by chemical alteration of volcanic ash.

beta decay The radioactive emission of a nuclear electron (beta ray, ß$^-$) that converts a neutron to a proton. The element is changed to the element of next-highest atomic number; for example, $^{14}_{6}C \rightarrow {}^{14}_{7}N + ß^-$.

biogenic sedimentary rock Rocks resulting directly from the activities of living organisms, such as a coral reef, shell limestone, or coal.

biological oxygen demand (B.O.D.) Indicator of organic pollutants (microbial) in an effluent measured as the amount of oxygen required to support them. The greater the B.O.D. (ppm or mg/liter) the greater the pollution and less oxygen available for higher aquatic organisms.

biomass fuels Fuels that are manufactured from plant materials; for example, grain alcohol (ethanol) is made from corn.

black smoker A submarine hydrothermal vent associated with a mid-oceanic spreading center that has emitted or is emitting a "smoke" of metallic sulfide particles.

block (volcanology) A mass of cold country rock greater than 32 mm (1¼ in) in size that is ripped from the vent and ejected from a volcano.

block glide (translational slide) A landslide in which movement occurs along a well-defined plane surface or surfaces, such as bedding planes, foliation planes, or faults.

body wave Earthquake waves that travel through the body of the earth.

bolide A large meteor that explodes as it passes through the earth's atmosphere.

bomb (volcanology) A molten mass greater than 32 mm (1¼ in) in diameter that is ejected from a volcano and cools sufficiently before striking the ground to retain a rounded, streamlined shape.

boreal The northern and mountainous regions of the Northern Hemisphere in which the average temperature of the hottest season does not exceed 18° C (64.4° F).

breakwater An offshore rock or concrete structure, attached at one end or parallel to the shoreline, intended to create quiet water for boat anchorage purposes.

breeder reactor A reactor that produces more fissile (splittable) nuclei than it consumes.

British thermal unit (Btu) The amount of heat, in calories, needed to raise the temperature of 1 pound of water 1° F.

caldera Large crater caused by explosive eruption and/or subsidence of the cone into the magma chamber, usually more than 1.5 km (1 mi) in diameter.

caliche In arid regions, a sediment or soil that is cemented by calcium carbonate ($CaCO_3$).

calving The breaking away of a block of ice, usually from the front of a tidewater glacier, that produces an iceberg.

caprock An impermeable stratum that caps an oil reservoir and prevents oil and gas from escaping to the ground surface.

capillary fringe The moist zone in an aquifer above the water table. The water is held in it by capillary action.

capillary water Water that is held in tiny openings in rock or soil by capillary forces.

carrying capacity The number of creatures a given tract of land can adequately support with food, water, and other necessities of life.

CERCLA The Comprehensive Environmental Response, Compensation and Liability Act of 1980. *See* Superfund.

chain reaction A self-sustaining nuclear reaction that occurs when atomic nuclei undergo fission. Free neutrons are released that split other nuclei, causing more fissions and the release of more neutrons, which split other nuclei, and so on.

chemical sedimentary rock Rock resulting from precipitation of chemical compounds from a water solution. *See* evaporite.

chlorofluorocarbons (CFCs) A group of compounds that, on escape to the atmosphere, break down, releasing chlorine atoms that destroy ozone molecules; for example, Freon, which is widely used in refrigeration, air-conditioning, and formerly was a propellant in aerosol-spray cans.

cinder cone A small, straight-sided volcanic cone consisting mostly of pyroclastic material (usually basaltic).

cinders Glassy and vesicular (containing gas holes) volcanic fragments, only a few centimeters across, that are thrown from a volcano.

clastic Describing rock or sediment composed primarily of detritus of preexisting rocks or minerals.

clastic sedimentary rock Rock composed of fragments of minerals or other rock materials.

clay minerals Hydrous aluminum silicates that have a layered atomic structure. They are very fine grained and become plastic (moldable) when wet. Most belong to one of three clay groups: kaolinite, illite, and smectite, the last of which are expansive when they absorb water between layers.

claypan A dense, relatively impervious subsoil layer that owes its character to a high clay-mineral content due to concentration by downward-percolating waters.

cleavage The breaking of minerals along certain crystallographic planes of weakness that reflects their internal structure.

climate controls The relatively constant factors that dictate a region's climate.

coastal desert A desert on the western edge of continents in tropical latitudes; for example, near the Tropic of Capricorn or Cancer. The daily and annual temperature fluctuations are much less than in inland tropical deserts.

columnar jointing Vertically oriented polygonal columns in a solid lava flow or shallow intrusion formed by contraction upon cooling. They are usually formed in basaltic lavas.

compaction The reduction in bulk volume of fine-grained sediments due to increased overburden weight or to tighter packing of grains. Clay is more compressible than sand and therefore compacts more.

complex landslide When both rotational and translational sliding are found within one landslide mass.

composting Bacterial recycling of organic material.

concentrate Enriched ore, often obtained by flotation, produced at a mill.

concentration factor The enrichment of a deposit of an element, expressed as the ratio of the element's abundance in the deposit to its average crustal abundance.

cone of depression The cone shape formed by the water table around a pumping well when pumping exceeds the flow of water to the well.

confined aquifer A water-bearing formation bounded above and below by impermeable beds or beds of distinctly lower permeability.

contact metamorphism A change in the mineral composition of rocks in contact with an invading magma from which fluid constituents are carried out to combine with some of the country-rock constituents to form a new suite of minerals.

continental drift The term applied by American geologists to Wegener's hypothesis of long-distance horizontal movement of continents he called "continental displacement."

contour mining The mining of coal by cutting into and following the coal seam around the perimeter of the hill.

convection The circulatory movement within a body of nonuniformly heated fluid. Warmer material rises in the center, and cooler material sinks at the outer boundaries. It has been proposed as the mechanism of plate motion in plate tectonic theory.

convergent boundary Tectonic boundary where two plates collide. Where a tectonic plate sinks beneath another plate and is destroyed, this boundary is known as a *subduction zone*. Subduction zones are marked by deep-focus seismic activity, strong earthquakes, and violent volcanic eruptions.

core Center of the earth, consisting of dense, metallic, iron-rich material.

Coriolis effect A condition peculiar to motion on a rotating body such as the Earth. It appears to deflect moving objects in the Northern Hemisphere to the right and objects in the Southern Hemisphere to the left, to an observer on the rotating body.

craton A part of the earth's continental crust that has attained stability—that has not been deformed for a long period of time. The term is restricted to continents and includes their most stable areas, the continental shields.

creep (soil) The imperceptibly slow downslope movement of rock and soil particles by gravity.

cross-bedding Arrangement of strata, greater than 1 cm thick, inclined at an angle to the main stratification.

crust A thin "skin" of aluminum and alkali-rich silicate rocks that surrounds the earth's mantle; it is still evolving.

crystal settling In a magma, the sinking of crystals because of their greater density, sometimes aided by magmatic convection. (*See* gravity separation).

cubic feet per second (cfs) Unit of water flow or stream discharge. One cubic foot per second is 7.48 gallons passing a given point in one second.

cutoff A new channel that is formed when a stream cuts through a very tight meander. (*See* oxbow lake).

cyanide heap leaching A process of using cyanide to dissolve and recover gold and silver from ore.

c-zone Soil zone of weathered parent material leading downward to fresh bedrock.

Darcy's Law A derived formula for the flow of fluids in a porous medium.

debris avalanche A sudden, rapid movement of a water–soil–rock mixture down a steep slope.

debris flow A moving mass of a water, soil, and rock mixture. More than half of the soil and rock particles are coarser than sand, and the mass has the consistency of wet concrete.

deflation The lifting and removal of loose, dry, fine-grained particles by the action of wind.

delta The nearly flat land formed where a stream empties into a body of standing water, such as a lake or the ocean.

desert An arid region of low rainfall, usually less than 25 cm (10 in) annually, and of high evaporation or extreme cold. Deserts are generally unsuited for human occupation under natural conditions.

desert pavement An interlocking cobble mosaic that remains on the earth surface in a dry region after winds have removed all fine-grained sediments.

desertification The process by which semiarid grasslands are converted to desert.

detrital *See* clastic.

dip The angle that an inclined geological planar surface (sedimentary bedding plane, fault, joint) makes with the horizontal plane.

discharge (Q) The volume of water, usually expressed in cfs, that passes a given point within a given period of time.

disseminated deposit A mineral deposit, especially of a metal, in which the minerals occur as scattered particles in the rock, but in sufficient quantity to make the deposit a commercially worthwhile ore.

divergent boundary Found at mid-ocean ridge or spreading center where new crust is created as plates move apart. For this reason they are sometimes referred to as *constructive* boundaries and are the sites of weak shallow-focus earthquakes and volcanic action.

doldrums The area of warm rising air, calm winds, and heavy precipitation between 5° north latitude and 5° south latitude. It is where the northeast and southeast tradewinds meet.

doubling time The number of years required for a population to double in size.

drainage basin The tract of land that contributes water to a particular stream, lake, reservoir, or other body of surface water.

drainage divide The boundary between two drainage basins.

drawdown The lowering of the water table immediately adjacent to a pumping well.

dredging The excavation of earth material from the bottom of a body of water by a floating barge or raft equipped to scoop up, discharge by conveyors, and process or transport materials.

drift A general term for all rock material that is transported and deposited by a glacier or by running water emanating from a glacier.

driving force (landslide) Gravity.

dune A mound, bank, ridge, or hill of loose, windblown granular material (generally sand), either bare or covered with vegetation. It is capable of movement, but maintains its characteristic shape.

Dust Veil Index (DVI) A measure of the impact of a volcanic eruption on incoming solar radiation and climate.

eccentricity The deviation of the Earth's path around the sun from a perfect circle.

effective stress The average normal force per unit area (stress) that is transmitted directly across grain-to-grain boundaries in a sediment or rock mass.

elastic Said of a body in which strains are totally recoverable, as in a rubber band. (*Contrast with* plastic).

elastic rebound theory The theory that movement along a fault and the resulting seismicity is the result of an abrupt release of stored elastic strain energy between two rock masses on either side of the fault.

element A substance that cannot be separated into different substances by usual chemical means.

end moraine (terminal moraine) A moraine that is produced at the front of an actively flowing glacier; a moraine that has been deposited at the lower end of a glacier.

engineering geology The application of the science of geology to human engineering works.

ENSO Abbreviation for the periodic meteorological event known as El Niño–Southern Oscillation.

environmental geology The relationship between humans and their geological environment.

eolian Of, produced by, or carried by wind.

epicenter The point at the surface of earth directly above the focus of an earthquake.

erosion The weathering and transportation of the materials of the earth's surface.

erratic A rock fragment that has been carried by glacial ice or floating ice and deposited when the ice melted at some distance from the outcrop from which it was derived.

estuary A semienclosed coastal body where outflowing river water meets tidal sea water.

eutrophication The increase in nitrogen, phosphorous, and other plant nutrients in the aging of an aquatic system. Blooms of algae develop, preventing light penetration and causing reduction of oxygen needed in a healthy system.

evaporite A nonclastic sedimentary rock composed primarily of minerals produced when a saline solution becomes concentrated by evaporation of the water; especially a deposit of salt that precipitated from a restricted or enclosed body of sea water or from the water of a salt lake.

evapotranspiration That portion of precipitation that is returned to the air through evaporation and transpiration, the latter being the escape of water from the leaves of plants.

exfoliation The process whereby slabs of rock bounded by sheet joints peel off the host rock, usually a granite or sandstone.

exfoliation dome A large rounded dome resulting from exfoliation; e.g., Half Dome in Yosemite National Park.

exotic terranes Fault-bounded bodies of rock that have been transported some distance from their place of origin and that are unrelated to adjacent rock bodies or terranes.

expansive soils Clayey soils that expand when they absorb water and shrink when they dry out.

extrusive rocks Igneous rocks that formed at or near the surface of the earth. Because they cooled rapidly, they generally have an *aphanitic* texture.

fault Fracture in the earth along which there has been displacement.

fault creep (tectonic) The gradual slip or motion along a fault without an earthquake.

felsic Descriptive of magma or rock with abundant light-colored minerals and a high silica content. The term is derived from *fel*dspar + *si*lica + *c*.

fetch The unobstructed stretch of sea over which the wind blows to create wind waves.

fiord (fjord) A narrow, steep-walled inlet of the sea between cliffs or steep slopes, formed by the passage of a glacier. A drowned glacial valley.

firn A transitional material between snow and glacial ice, being older and denser than snow, but not yet transformed into glacial ice. Snow becomes firn after surviving one summer melt season; firn becomes ice when its permeability to liquid water becomes zero.

fissile isotopes Isotopes with nuclei that are capable of being split into other elements (*see* fission).

fission The rupture of the nucleus of an element into lighter elements (fission products) and free neutrons spontaneously or by absorption of a neutron.

fjord *See* fiord.

flood frequency The average time interval between floods that are equal to or greater than a specified discharge.

floodplain The portion of a river valley adjacent to the channel that is built of sediments deposited during times when the river overflows its banks at flood stage.

floodwall A heavily reinforced wall designed to contain a stream at flood stage.

floodway Floodplain area under federal regulation for flood insurance purposes.

flotation The process of concentrating minerals with distinct nonwettable properties by floating them in liquids containing soapy frothing agents such as pine oil.

fluidization The process whereby granular solids (corn, wheat, volcanic ash and lapilli) under high gas pressures become fluid-like and flow downslope or can be pumped.

fluvial Pertaining to rivers.

focus (hypocenter) Point within the earth where an earthquake originates.

foliation The planar or wavy structure that results from the flattened growth of minerals in a metamorphic rock.

forecast Evaluation of the probability of an earthquake occurring during a given time period.

foreshore slope The zone of a beach that is regularly covered and uncovered by the tide.

fossil fuels Coal, oil, natural gas, and all other solid or liquid hydrocarbon fuels.

fracture Break in a rock caused by tensional, compressional, or shearing forces (*see also* joints).

frost wedging The opening of joints and cracks by the freezing and thawing of water.

fusion The combination of two nuclei to form a single heavier nucleus accompanied by a loss of some mass that is converted to heat.

Gaia The controversial hypothesis that proposes that life has a controlling influence on the oceans and atmosphere. It states that the earth is a giant self-regulating body with close connections between living and nonliving components.

gaining stream A stream that receives water from the zone of saturation.

geologic agents All geologic processes—for example, wind, running water, glaciers, waves, mass wasting—that erode, move, and deposit earth materials.

geological repository (nuclear waste) An underground vault in terrane area free from geologic hazards. Repositories in salt mines, granite, and welded tuff have been studied.

geothermal energy Energy derived from circulating hot water and steam; usually associated with cooling magma or hot rocks.

gigawatt (GW) A thousand million (a billion) watts; 1,000 megawatts.

glacial outwash Deposits of stratified sand, gravel, and silt that have been removed from a glacier by meltwater streams.

glacier A large mass of ice formed, at least in part, on land by the compaction and recrystallization of snow. It moves slowly downslope by creep or outward in all directions due to the stress of its weight, and it survives from year to year.

global warming The general term for the steady rise in the average global temperatures over the last 100 years.

Gondwana succession The fossil assemblage used by Alfred Wegener and others to discern the past existence of southern Pangaea.

graben An elongate flat-floored valley bounded by faults on each side.

graded stream A stream that is in equilibrium, showing a balance of erosion, transporting capacity, and material supplied to it. Graded streams have a smooth profile.

gradient (stream) The slope of a streambed usually expressed as the amount of drop per horizontal distance in meters per kilometer (m/km) or in feet per mile (ft/mi).

gravity separation See crystal settling.

greenhouse effect The warming of the atmosphere by an increase in the amount of certain gases in the atmosphere, primarily CO_2 and CH_4, that increase the retention of heat that has been radiated by the earth.

groin A structure of rock, wood, or concrete built roughly perpendicular to a beach to trap sand.

ground water That part of subsurface water that is in the zone of saturation (below the water table).

gullying The cutting of channels into the landscape by running water. When extreme, it renders farmland useless.

half-life The period of time during which one-half of a given number of atoms of a radioactive element or isotope will disintegrate.

hard rock (a) A term loosely used for an igneous or metamorphic rock, as distinguished from a sedimentary rock. (b) In mining, a rock that requires drilling and blasting for economical removal.

hardpan Impervious layer just below the land surface produced by calcium carbonate ($CaCO_3$) in the B horizon (*see* claypan).

high-grade deposit A mineral deposit with a high concentration of a desired element.

high-level nuclear waste Mostly radioactive elements in spent fuel with long half lives.

horse latitudes Oceanic regions in the *subtropical highs* about 30° north and south of the equator that are characterized by light winds or calms, heat, and dryness.

hot spot A point (or area) on the lithosphere over a plume of lava rising from the mantle.

hydration A process whereby anhydrous minerals combine with water.

hydraulic conductivity A ground-water unit expressed as the volume of water that will move in a unit of time through a unit of area measured perpendicular to the flow direction. Commonly called permeability, it is expressed as the particular aquifer's m³/day/m² (or m/day) or ft³/day/ft² (or ft/day).

hydraulic gradient The slope or vertical change (ft/ft or m/m) in water-pressure head with horizontal distance in an aquifer.

hydraulic mining, hydraulicking A mining technique by which high-pressure jets of water are used to dislodge unconsolidated rock or sediment so that it can be processed.

hydrocarbon One of the many chemical compounds solely of hydrogen and carbon atoms; may be solid, liquid, or gaseous.

hydrocompaction The compaction of dry, low-density soils due to the heavy application of water.

hydrogeologist A person who studies the geology and management of underground and related aspects of surface waters.

hydrograph A graph of the stage (height), discharge, velocity, or some other characteristic of a body of water over time.

hydrologic cycle (water cycle) The constant circulation of water from the sea to the atmosphere, to the land, and eventually back to the sea. The cycle is driven by solar energy.

hydrology The study of liquid and solid water on, under, and above the earth's surface, including economic and environmental aspects.

hydrolysis The chemical reaction between hydrogen ions in water with a mineral, commonly a silicate, usually forming clay minerals.

hydrothermal Of or pertaining to heated water, the action of heated water, or a product of the action of heated water—such as a mineral deposit that precipitated from a hot aqueous solution.

hypocenter See focus.

igneous rocks Rocks that crystallize from molten material at the surface of the earth (volcanic) or within the earth (plutonic).

inactive sea cliff A cliff whose erosion is dominated by subaerial processes rather than wave action.

indicated reserves The hydrocarbons that can only be recovered by secondary recovery techniques such as fluid injection. These reserves are in addition to the reserves directly recoverable from a known reservoir.

intensity scale An earthquake rating scale (I–XII) based upon subjective reports of human reactions to ground shaking and upon the damage caused by an earthquake.

intrusive rocks Igneous rocks that have "intruded" into the crust, hence they are slowly cooled and generally have phaneritic texture.

ion An atom with a positive or negative electrical charge because it has gained or lost one or more electrons.

isolation (waste disposal) Buried or otherwise sequestered waste.

isoseismals Lines on a map that enclose areas of equal earthquake shaking based upon an intensity scale.

isostatic rebound Uplift of the crust of the earth that results from unloading such as results from the melting of ice sheets.

isotope One of two or more forms of an element that have the same atomic number but different atomic masses.

J-curve The shape of a graphing of geometric (exponential) growth. Growth is slow at first, and the rate of growth increases with time.

jetties Structures built perpendicular to the shoreline to improve harbor inlets or river outlets.

joint Separation or parting in a rock that has not been displaced. Joints usually occur in groups ("sets"), the members of a set having a common orientation.

karst topography, karst terrane Area underlain by soluble limestone or dolomite and riddled with caves, caverns, sinkholes, lakes, and disappearing streams.

kerogen A carbonaceous residue in sediments that has survived bacterial metabolism. It consists of large molecules from which hydrocarbons and other compounds are released on heating.

kettle A bowl-shaped depression without surface drainage in glacial-drift deposits, often containing a lake, believed to have formed by the melting of a large detached block of stagnant ice left behind by a retreating glacier.

Kyoto Protocol 1997 agreement by which participating countries would reduce CO_2 emissions.

lahar A debris flow or mudflow consisting of volcanic material.

landslide The downslope movement of rock and/or soil as a semi-coherent mass on a discrete slide surface or plane (*see also* mass wasting).

lapilli (cinders) Fragments only a few centimeters in size that are ejected from a volcano.

lateral spreading The horizontal movement on nearly level slopes of soil and mineral particles due to liquefaction of quick clays.

laterite A highly weathered brick-red soil characteristic of tropical and subtropical rainy climates. Laterites are rich in oxides or iron and aluminum and have some clay minerals and silica. Bauxite is an Al-rich deposit of a similar origin.

lava Molten material at the surface of the earth.

lava dome (volcanic dome) Steep-sided protrusion of viscous, glassy lava, sometimes within the crater of a larger volcano (e.g., Mount St. Helens) and sometimes free-standing (e.g., Mono Craters).

lava plateau A broad elevated tableland, thousands of square kilometers in extent, underlain by a thick sequence of lava flows.

leachate The noxious fluids that percolate through sanitary landfills. If released to the surface or underground environment, leachate can pollute surface and underground water.

levees (natural) Embankments of sand or silt built by a stream along both banks of its channel. They are deposited during floods, when waters overflowing the stream banks are forced to deposit sediment.

liquefaction *See* spontaneous liquefaction.

lithification The conversion of sediment into solid rock through processes of compaction, cementation, and crystallization.

lithified Changed into stone, as in the transformation of loose sand to sandstone.

lithosphere The rigid outer layer from the earth's surface down to the asthenosphere. It includes the crust and part of the upper mantle and is of the order of 100 km (60 mi) thick. The "plates" in plate tectonic theory.

littoral drift Sediment (sand, gravel, silt) that is moved parallel to the shore by longshore currents.

lode A mineral deposit consisting of a zone of veins in consolidated rock, as opposed to a *placer deposit.*

loess A blanket deposit of buff-colored calcareous silt that is porous and shows little or no stratification. It covers wide areas in Europe, eastern China, and the Mississippi valley. It is generally considered to be windblown dust of Pleistocene age.

longitudinal wave (P-wave) A type of seismic wave involving particle motion that is alternating expansion and compression in the direction of wave propagation. Also known as *compressional waves,* they resemble sound waves in their motion.

longshore current (littoral current) The current adjacent and parallel to a shoreline that is generated by waves striking the shoreline at an angle.

losing stream A stream or reach of a stream that contributes water to the zone of saturation.

low-grade deposit A mineral deposit in which the mineral content is minimal but still exploitable at a profit.

low-level nuclear wastes Wastes generated in research labs, hospitals, and industry. Shallow burial in containers is sufficient to sequester this waste.

luster The manner in which a mineral reflects light using such terms as metallic, resinous, silky, and glassy.

M-discontinuity (Moho) The contact between the earth's crust and the mantle. There is a sharp increase in earthquake P-wave velocity across this boundary.

maar A low-relief, broad volcanic crater formed by multiple shallow explosive eruptions. Maars commonly contain a lake and are surrounded by a low rampart or ring of ejected material. They typically form from *phreatic eruptions.*

mafic Descriptive of a magma or rock rich in iron and magnesium. The mnemonic term is derived from *ma*gnesium + *f*erric + *ic.*

magma Molten material within the earth that is capable of intrusion or extrusion and from which igneous rocks form.

magnetic polarity A property indicating the position of Earth's magnetic poles relative to a rock during its formation, as expressed by the alignment of tiny iron-bearing grains contained within the rock.

magnitude (earthquake) A measure of the strength or the strain energy released by an earthquake at its source. (*See* moment magnitude, Richter magnitude.)

manganese nodule A small irregular black to brown, laminated concretionary mass consisting primarily of manganese minerals with some iron oxides and traces of other metallic minerals, abundant on the floors of the world's oceans.

mantle A thick shell of magnesium-silicate matter that surrounds the earth's core.

mantle lid Part of the earth's rigid outer shell that also includes the crust.

marine trenches Steep-walled abyssal valleys that fringe some ocean floors.

mass wasting A general term for all downslope movements of soil and rock material under the direct influence of gravity.

massive sulfide deposit Usually volcanogenic, often rich in zinc and sometimes in lead.

meander One of a series of sinuous curves in the course of a stream.

megawatt (MW) A million watts; 1,000 kilowatts.

meltdown *See* central-core meltdown.

metamorphic rocks Preexisting rocks that have been altered by heat, pressure, or chemically active fluids.

mid-latitude desert A desert area occurring within latitudes 30°–50° north and south of the equator in the deep interior of a continent, usually on the lee side of a mountain range that blocks the path of prevailing winds, and commonly characterized by a cold, dry climate (*see* rain-shadow desert).

mid-ocean ridges Belts of broad volcanic highlands that bisect the sea floor worldwide.

millirem A measure of radioactive emissions in the environment.

mineral Naturally occurring crystalline substance with well-defined physical properties and a definite range of chemical composition.

mineral deposit A localized concentration of naturally occurring mineral material (e.g., a metallic ore or a nonmetallic mineral), usually of economic value, without regard to its mode of origin.

mineral reserves Known mineral deposits that are recoverable under present conditions but as yet undeveloped. The term excludes *potential ore*.

mineral resources The valuable minerals of an area that are presently legally recoverable or that may be so in the future; include both the known ore bodies (*mineral reserves*) and the *potential ores* of a region.

miscible Soluble; capable of mixing.

modified Mercalli scale Earthquake intensity scale from I (not felt) to XII (total destruction) based upon damage and reports of human reactions.

Mohs hardness scale A scale that indicates the relative hardness of minerals on a scale of 1 (talc, very soft) to 10 (diamond, very hard).

moment magnitude (M_w or M) A scale of seismic energy released by an earthquake based on the product of the rock rigidity along the fault, the area of rupture on the fault plane, and the amount of slip.

moraine A mound, ridge, hill, or other distinct accumulation of unsorted, unstratified glacial drift, predominantly till, deposited chiefly by direct action of glacial ice in a variety of landforms.

mother lode (a) A main mineralized unit that may not be economically valuable in itself but to which workable deposits are related; e.g., the Mother Lode of California. (b) An ore deposit from which a placer is derived; the *mother rock* of a placer.

mountaintop mining The mining of coal that underlies the tops of mountains.

natural levee *See* levee.

negative anomaly Magnetization in ancient rocks caused by a magnetic field oriented opposite to the one existing at present; that is, the northern magnetic pole existed in the south, and the southern one existed in the north.

neutron An electronically neutral particle (zero charge) of an atomic nucleus with an atomic mass of approximately 1.

nonrenewable resource An energy or material resource that once used is not available for reuse in human time spans. Coal, oil, and metallic minerals are examples.

nucleus The positively charged central core of an atom containing protons and neutrons that provide its mass.

nuée ardenté (pyroclastic flow) French for "glowing cloud," it is a highly heated, almost incandescent cloud of volcanic gases and pyroclastic material that travels with great velocity down the slopes of a volcano. Produced by the explosive disintegration of viscous lava in a vent.

NWPA Nuclear Waste Policy Act of 1982 mandating selection of a repository for nuclear waste.

obliquity The tilt of the earth's axis.

open-pit mining Mining from open excavations, most commonly for low-grade copper and iron deposits, and coal.

ore deposit The same as a *mineral reserve* except that it refers only to a metal-bearing deposit.

ore mineral The part of an ore, usually metallic, that is economically desirable.

outgassing The release of gases and water vapor from molten rocks, leading to the formation of the earth's atmosphere and oceans.

overturn Process by which stagnant bottom waters in a lake come to the surface and are refreshed with oxygen.

oxbow lake A crescent-shaped lake formed along a stream course when a tight meander is cut off and abandoned.

pahoehoe Basaltic lava typified by smooth, billowy, or ropy surfaces.

paleoseismicity The rock record of past earthquake events in displaced beds and liquefaction features in trenches or natural outcrops.

Pangaea A supercontinent that existed from about 300–200 Ma, which included all the continents we know today.

passive margins Coastlines such as the Eastern Seaboard of the U.S. that are no longer active in terms of volcanism and mountain building.

pedalfer Soil of humid regions characterized by an organic-rich A horizon and clays and iron oxides in the B horizon.

pedocal Soil of arid or semiarid regions that is rich in calcium carbonate.

pedologist (Greek *pedo*, "soil," and *logos*, "knowledge") A person who studies soils.

pedology The study of soils.

pegmatite An exceptionally coarse-grained igneous rock with interlocking crystals, usually found as irregular dikes, lenses, or veins, especially at the margins of batholiths.

perched water table The upper surface of a body of ground water held up by a discontinuous impermeable layer above the static water table.

permafrost Permanently frozen ground, with or without water, occurring in arctic, subarctic, and alpine regions.

permafrost table The upper limit of permafrost.

permeability The degree of ease with which fluids flow through a porous medium (*see* hydraulic conductivity).

phaneritic texture A coarse-grained igneous-rock texture resulting from slow cooling. Phanerites are coarse-grained igneous rocks.

photochemical smog Atmospheric haze that forms when automobile-exhaust emissions are activated by ultraviolet radiation from the sun to produce highly reactive oxidants known to be health hazards.

phreatic eruption Volcanic eruption, mostly steam, caused by the interaction of hot magma with underground water, lakes, or seawater. Where significant amounts of new (magmatic) material are ejected in addition to steam, the eruptions are described as *phreatomagmatic*.

phytoremediation The use of plants to remove soil contaminants, particularly excess salt.

piping Subsurface erosion in sandy materials caused by the percolation of water under pressure. A somewhat similar effect is obtained by placing a garden-hose nozzle in a sandbox and jetting an opening through the sand.

placer deposit (pronounced "plass-er") A surficial mineral deposit formed by settling from streams of mineral particles from weathered debris. (*See also* lode.)

plastic Capable of being deformed continuously and permanently in any direction without rupture.

plate Broad region of the earth's crust that is structurally distinctive.

plate tectonics View of the structure and movement of the earth's outermost skin and landforms such as mountains and valleys.

Pleistocene The geologic Epoch of the Quaternary Period known as the Ice Age, falling by definition between 1.6 million and 12,000 years ago. Extensive glaciation is known as far back as the Pliocene Epoch 2.5 million years ago. The exact boundaries shift in time as new data are revealed.

Plinian column A steady, turbulent, nearly vertical column of ash and steam released from a vent at high velocity in an explosive eruption.

plutonic Pertaining to rock formed by any process at depth, usually with a phaneritic texture.

point bar A deposit of sand and gravel found on the inside curve of a *meander.*

polar desert The bitterly cold, arid region centered at the south pole that receives very little precipitation.

polar front The boundary in the atmosphere where cold polar surface air and mild-temperature surface air converge.

polar high The region of cold, high-density air that exists in the polar regions.

population dynamics The study of how populations grow.

population growth rate The number of live births less deaths per 1,000 people, usually expressed as a percentage; a growth rate of 40 would represent 4.0% growth.

pore pressure The stress exerted by the fluids that fill the voids between particles of rock or soil.

porosity A material's ability to contain fluid. The ratio of the volume of pore spaces in a rock or sediment to its total volume, usually expressed as a percentage.

porphyry copper A copper deposit, usually of low grade, in which the copper-bearing minerals occur in disseminated grains and/or in veinlets through a large volume of rock.

positive anomaly Magnetization in ancient rocks caused by a magnetic field oriented much as at present ("positive" pole in the north and "negative" pole in the south).

potential ore As yet undiscovered mineral deposits and also known mineral deposits for which recovery is not yet economically feasible.

potentiometric surface The hydrostatic head of ground water and the level to which the water will rise in a well. The water table is the potentiometric surface in an unconfined aquifer. Also called *piezometric surface.*

precession The slow gyration of the Earth's axis, analogous to that of a spinning top, which causes the slow change in the orientation of the Earth's axis relative to the sun that accounts for the reversal of the seasons of winter and summer in the Northern and Southern Hemispheres every 11,500 years.

precipitation The separation of a solid substance from a solution by a chemical reaction.

precursors (of earthquakes) Observable phenomena that occur before an earthquake and indicate that an event is soon to occur.

pregnant pond A catchment basin that holds a gold- and silver-bearing cyanide solution.

proglacial lake An ice-marginal lake formed just beyond the frontal moraine of a retreating glacier and generally in direct contact with the ice.

proton A particle in an atomic nucleus with a positive electrical charge and an atomic mass of approximately one.

protostar A star in the process of formation that has entered the slow gravitational contraction phase.

pumice Frothy-appearing rock composed of natural glass ejected from high-silica, gas-charged magmas. It is the only rock that floats.

P-wave *See* longitudinal wave.

pyroclastic ("fire broken") Descriptive of the fragmental material, ash, cinders, blocks, and bombs ejected from a volcano. (*See also* tephra.)

quad A quadrillion British thermal units (10^{15} Btu).

quick clay (sensitive clay) A clay possessing a "house-of-cards" sedimentary structure that collapses when it is disturbed by an earthquake or other shock.

radioactivity The property possessed by certain elements that spontaneously emit energy as the nucleus of the atom changes its proton-to-neutron ratio.

radiometric dating *See* absolute age dating.

radon A radioactive gas emitted in the breakdown of uranium.

rain-shadow desert A desert on the lee side of a mountain or a mountain range (*see* mid-latitude desert).

rank (coal) A coal's carbon content depending upon its degree of metamorphism.

recessional moraine One of a series of nested end moraines that record the stepwise retreat of a glacier at the end of an ice age.

recrystallization The formation in the solid state of new mineral grains in a metamorphic rock. The new grains are generally larger than the original mineral grains.

recurrence interval The return period of an event, such as a flood or earthquake, of a given magnitude. For flooding, it is the average interval of time within which a given flood will be equalled or exceeded by the annual maximum discharge.

regional metamorphism A general term for metamorphism that affects an extensive area.

regolith Unconsolidated rock and mineral fragments at the surface of the earth.

Regulatory Floodplain The part of a floodplain that is subject to federal government regulation for insurance purposes.

relative age dating The chronologic ordering of geologic strata, features, fossils, or events with respect to the geological time scale without reference to their absolute age.

renewable resource An energy or material resource that continually renews itself as it is being consumed. Timber, solar energy, wind, and water are examples.

replacement rate The birth rate at which just enough offsprings are born to replace their parents.

reserves Those portions of an identified resource that can be recovered economically.

reservoir rock A permeable, porous geologic formation that will yield oil or natural gas.

residence time The average length of time a substance (atom, ion, molecule) remains in a given reservoir.

residual soil Soil formed in place by decomposition of the rocks upon which it lies.

resisting force (landslide) Friction along the slide plane and normal forces across the plane.

resonance The tendency of a system (a structure) to vibrate with maximum amplitude when the frequency of the applied force (seismic waves) is the same as the vibrating body's natural frequency.

revetment A rock or concrete structure built landward of a beach to protect coastal property.

Richter magnitude scale (earthquakes) The logarithm of the maximum trace amplitude of a particular seismic wave on a seismogram, corrected for distance to epicenter and type of seismometer. A measure of the energy released by an earthquake.

rift zone (volcanology) A linear zone of weakness on the flank of a shield volcano that is the site of frequent flank eruptions.

rill erosion The carving of small channels, up to 25 cm (10 in) deep, in soil by running water.

riparian Pertaining to or situated on the bank of a body of water, especially of a river.

rock cycle The cycle or sequence of events involving the formation, alteration, destruction, and reformation of rocks as a result of processes such as erosion, transportation, lithification, magmatism, and metamorphism.

rocks Aggregates of minerals or rock fragments.

rotational landslide (slump) A slope failure in which sliding occurs on a well-defined, concave-upward, curved surface, producing a backward rotation of the slide mass.

runoff That part of precipitation falling on land that runs into surface streams.

runup The elevation to which a wind wave or tsunami advances onto the land.

Sahel Belt of populated semiarid lands along the southern edge of North Africa's Sahara Desert.

salt dome A column or plug of rock salt that rises from depth because of its low density and pierces overlying sediments.

sanitary landfill A method of burying waste where each day's dumping is covered with soil to protect the surrounding area. This is also monitored to ensure protection of local underground water.

sea-floor spreading A hypothesis that oceanic crust forms by convective upwelling of lava at mid-ocean ridges and moves laterally to trenches where it is destroyed.

seawall A rock or concrete structure built landward of a beach to protect the land from wave action. (*See also* revetment.)

secondary enrichment Processes of near-surface mineral deposition, in which oxidation produces acidic solutions that leach metals, carry them downward, and reprecipitate them, thus enriching sulfide minerals already present.

secondary recovery Production of oil or gas by artificially stimulating a depleted reservoir with water, steam, or other fluids in order to mobilize the remaining hydrocarbons.

sedimentary rock A layer of rock deposited from water, ice, or air and subsequently lithified to form a coherent rock.

seif (longitudinal) dune A very large, sharp-crested, tapering chain of sand dunes; commonly found in the Arabian Desert.

seismogram The recording from a seismograph.

seismograph An instrument used to measure and record seismic waves from earthquakes.

semiarid grasslands Extensive treeless grassland regions that are marginal to many mid-latitude deserts.

sensitive clay *See* quick clay.

settlement This occurs when the applied load of a structure's foundation is greater than the bearing strength of the foundation material (soil or rock).

sheet erosion The removal of thin layers of surface rock or soil from an area of gently sloping land by broad continuous sheets of running water (sheet flow), rather than by channelized streams.

sheet joints (*See* joint) Cracks more or less parallel to the ground surface that result from expansion due to deep erosion and the unloading of overburden pressure.

shield volcano (cone) The largest type of volcanoes, composed of piles of lava in a convex-upward slope. Found mainly in Hawaii and Iceland.

sinkhole Circular depression formed by the collapse of a shallow cavern in limestone.

slant-drilling (whipstocking) Purposely deflecting a drill hole from the vertical in order to tap a reservoir not directly below the drill site. The deflecting tool is a whipstock.

slide plane (slip plane) A curved or planar surface along which a landslide moves.

slip face The steeper, lee side of a dune, standing at or near the angle of repose of loose sand and advancing downwind by a succession of slides whenever that angle is exceeded.

slump *See* rotational landslide.

smelter An industrial plant that mechanically and chemically produces metals from their ores.

SMCRA The Surface Mining Control and Reclamation Act of 1977.

soil (a) Loose material at the surface of the earth that supports the growth of plants (pedological definition). (b) All loose surficial earth material resting on coherent bedrock (engineering definition).

soil horizon A layer of soil that is distinguishable from adjacent layers by properties such as color, texture, structure, or chemical composition.

soil profile A vertical section of soil that exposes all of its horizons or zones.

solution (weathering) The dissolving of rocks and minerals by natural acids or water.

source rock The geologic formation in which oil and/or gas originate.

specific yield The ratio of the volume of water drained from an aquifer by gravity to the total volume considered.

speleology The study of caves and caverns.

spheroidal weathering Weathering of rock surfaces that creates a rounded or spherical shape as the corners of the rock mass are weathered faster than its flat faces.

spontaneous liquefaction Process whereby water-saturated sands, clays, or artificial fill suddenly become fluid upon shaking, as in an earthquake.

stage The height of flood waters in feet or meters above an established datum plane.

steppe An extensive treeless grassland in the semiarid mid-latitudes of Asia and southeastern Europe. It is drier than the prairie that develops in the mid-latitudes of the United States.

storm surge The sudden rise of sea level on an open coast due to a storm. Storm surge is caused by strong onshore winds and results in water being piled up against the shore. It may be enhanced by high tides and low atmospheric pressure.

strain Deformation resulting from an applied stress (=force per unit area). It may be elastic (recoverable) or ductile (nonreversible).

stratification (bedding) The arrangement of sedimentary rocks in strata or layers.

stratigraphic trap An accumulation of oil that results from a change in the character (permeability) of the reservoir rock, rather than from structural deformation.

stratovolcano A large volcano that is shaped slightly concave upward and is composed of layers of ash and lava.

stress The force applied on an object per unit area, expressed in pounds per square foot (lb/ft^2) or kilograms per square meter (kg/m^2).

strike The direction west or east of north of the trace of the inclined plane (sedimentary bedding plane, fault, or joint surface) on the horizontal plane.

strip-mining Surficial mining, in which the resource is exposed by removing the overburden.

subduction The sinking of one lithospheric plate beneath another at a convergent plate margin. (*See* convergent boundary.)

subsidence Sinking or downward settling of the earth's surface due to solution, compaction, withdrawal of underground fluids, or to cooling of the hot lithosphere or to loading with sediment or ice.

subtropical desert A hot, dry desert occurring between latitudes 15°–30° north and south of the equator where subtropical air masses prevail. (*See* trade-wind desert.)

subtropical highs The climatic belts at latitudes of approximately 30°–35° north and south of the equator characterized by semipermanent high-pressure air masses, moderate seasonal rainfall, summer heat, dryness, and generally clear skies.

Superfund A spinoff of CERCLA insuring financial accountability for environmental polluters.

surcharge Weight added to the top of a natural or artificial slope by sedimentation or by artificial fill.

surface wave Seismic waves that travel the surface of the earth. They are generated by the unreflected energy of P- and S-waves striking the earth's surface. They are the most damaging of earthquake waves.

surging glacier A glacier that alternates periodically between brief periods (usually 1–4 years) of very rapid flow, called *surges,* and longer periods (usually 10–100 years) of near-stagnation.

suspension (a) A mode of sediment transport in which the upward currents in eddies of turbulent flow are capable of supporting the weight of sediment particles and keeping them held indefinitely in the surrounding fluid (as silt in water or dust in air). (b) The state of a substance in such a mode of transport.

sustainable society A population capable of maintaining itself while leaving sufficient resources for future generations.

sustained yield (hydrology) The amount of water an aquifer can yield on a daily basis over a long period of time.

S-wave *See* transverse wave.

syncline A concave-upward fold that contains younger rocks in its core. (*Contrast with* anticline.)

synthetic fuels (synfuels) A manufactured fuel; for example, a fuel derived by liquefaction or gasification of coal.

tailings The worthless rock material discarded from mining operations.

talus Coarse angular rock fragments lying at the base of a cliff or steep slope from which they were derived.

tar sand A sand body that is large enough to hold a commercial reserve of asphalt or other thick oil. It is usually surface excavated to remove the thick hydrocarbons.

tectonism (adj. tectonic) Deformation of the earth's crust by natural processes leading to the formation of ocean basins, continents, mountain systems, and other earth features.

tephra A collective term for all *pyroclastic* material ejected from a volcano.

terranes (*See* exotic terranes.)

texture The size, shape, and arrangement of the component particles or crystals in a rock.

till Unsorted and unstratified drift, generally unconsolidated, that is deposited directly by and beneath a glacier without subsequent reworking by water from the glacier.

tombolo A sandbar that connects an island with the mainland or another island.

total fertility rate Average number of children born to women of child-bearing age.

trace element An element that occurs in minute quantities in rocks or plant or animal tissue. Some are essential for human health.

trade winds Steady winds blowing from areas of high pressure at 30° north and south latitudes toward the area of lower pressure at the equator. This pressure differential produces the northeast and southeast trade winds in the Northern and Southern Hemispheres, respectively.

trade wind desert *See* subtropical desert.

transform boundaries Plate boundaries with mostly horizontal (lateral) movement that connect spreading centers to each other or to subduction zones.

translational slide *See* block glide.

transported soils Soil developed on *regolith* transported and deposited by a geological agent. Hence the term *glacial* soils, alluvial soils, *eolian* soils, *volcanic* soils, and so forth.

transverse dune A strongly asymmetrical sand dune that is elongated perpendicularly to the direction of the prevailing winds. It has a gentle windward slope and a steep leeward slope that stands at or near the angle of repose of sand.

transverse wave (S-wave) A seismic wave propagated by a shearing motion that involves oscillation perpendicular to the direction of travel. It travels only in solids.

troposphere That part of the atmosphere next to the earth's surface, in which temperature generally decreases rapidly with altitude and clouds and convection are active. Where the weather occurs.

tsunami Large sea wave produced by a submarine earthquake, a volcanic eruption, or a landslide. Incorrectly known as a tidal wave.

tuff A coherent rock formed from volcanic ash, cinders, or other *pyroclastic* material.

tundra (soil) A treeless plain characteristic of arctic regions with organic rich, poorly drained tundra soils and permanently frozen ground.

unconfined aquifer Underground body of water that has a free (static) water table; that is, water that is not confined under pressure beneath an aquiclude.

upwelling A process whereby cold, nutrient-rich water is brought to the surface. Coastal upwelling occurs mostly in trade-wind belts.

vapor extraction The use of vacuum pumps placed in shallow bore holes to remove vaporized organic contaminants in the soil, such as gasoline or TCE.

vein deposit A thin, sheetlike igneous intrusion into a crevice.

ventifact Any stone or pebble that has been shaped, worn, faceted, cut, or polished by the abrasive or sandblast action of wind-blown sand, generally under arid conditions.

Volcanic Explosivity Index (VEI) A scale for rating volcanic eruptions according to the volume of the ejecta, the height to which the ejecta rise, and the duration of the eruption.

volcanogenic deposit Of volcanic origin; for example, volcanogenic sediments or ore deposits.

water cycle *See* hydrologic cycle.

water table The contact between the zone of aeration and the zone of saturation.

wave base The depth at which water waves no longer move sediment, equal to approximately half the wavelength.

wavelength The distance between two equivalent points on two consecutive waves; for example, crest to crest or trough to trough.

wave of oscillation A water wave in which water particles follow a circular path with little forward motion.

wave of translation A water wave in which the individual particles of water move in the direction of wave travel, as in broken waves (surf).

wave period (*T*) The time (in seconds) between passage of equivalent points on two consecutive waveforms.

wave refraction The bending of wave crests as they move into shallow water.

weathering The physical and chemical breakdown of materials of the earth's crust by interaction with the atmosphere and biosphere.

welded tuff A glassy pyroclastic rock that has been made hard by the welding together of its particles under the action of retained heat.

wetted perimeter (stream) The length of stream bed and bank that are in contact with running water in the stream in a given cross section.

Wilson cycle An evolutionary sequence for oceans that identifies five stages in their birth and death; named for J. Tuzo Wilson.

zero population growth A population growth rate of 0; that is, the number of deaths in a population is equal to the number of live births on average.

zone of aeration (vadose zone) The zone of soil or rock through which water infiltrates by gravity to the water table. Pore spaces between rock and mineral grains in this zone are filled with air; hence the name.

zone of flowage The lower part of a glacier that deforms due to the glacier's own weight.

zone of saturation The ground-water zone where voids in rock or sediment are filled with water.

Credits

This page constitutes an extension of the copyright page. We have made every effort to trace the ownership of all copyrighted material and to secure permission from copyright holders. In the event of any question arising as to the use of any material, we will be pleased to make the necessary corrections in future printings. Thanks are due to the following authors, publishers, and agents for permission to use the material indicated.

Chapter 1 1, Steve Allen/Getty Images. 2, Fig. 1.1: data from Blue Planet Group; Fig. 1.2: from "Why the U.S. Is Becoming More Vulnerable to Natural Disasters" by G. van der Vink and others, 1998, *EOS* 70, no. 44. 3, Fig. 1.3: from "Why the U.S. Is Becoming More Vulnerable to Natural Disasters" by G. van der Vink and others, 1998, *EOS* 70, no. 44. 4, Fig. 1.5: D. D. Trent. 5, Fig. 1.6: USGS. 10, Fig. 1.9: data from Blue Planet Group. 14, Fig. 1: B. Pipkin; Fig. 2: Owen Frankyn/Sygma.

Chapter 2 17, NASA. 19, Fig. 2.1: D. D. Trent. 24, Fig. 2.7: (a) USGS, (b)–(d) Paul Silverman/Fundamental Photographs; Figure 2.8: top left, Michael Dalton/Fundamental Photographs; top right and bottom left, Sue Monroe; bottom right, Paul Silverman/Fundamental Photographs. 25, Figure 2.9: Sue Monroe. 27, Fig. 2.13: B. Pipkin. 28, Fig. 2.14: photos courtesy Sue Monroe. 29, Fig. 2.15: B. Pipkin. 30, Fig. 2.17: B. Pipkin. 31, Figs 2.18 and 2.19: D. D. Trent. 32, Fig. 2.21:B. Pipkin. 33, Fig. 2.22: Sue Monroe; Fig. 2.23: D. D. Trent. 35, Fig. 2.25: Adapted from A. R. Palmer, *Geology*, 1983, 504. 39, Figs 1 and 2: Janet Blabaum, Highland Geotechnical, AIPG. 41, Fig. 1: (a) B. Pipkin, (b) Martin Land, Science Photo Library/Photo Researchers, (c) Carl Frank/Photo Researchers; Fig. 2: © Gemological Institute of America; Fig. 3: (a) Smithsonian Institute, (b)–(e) Sue Monroe. 42, Fig. 1: © 1994 Margaret Miller/The National Audubon Society Collection/Photo Researchers. 43, Fig. 1: (a), (b) Reed Wicander. 44, Fig. 3: Theodore Roosevelt Collection, Harvard College Library; Fig. 4: B. Pipkin.

Chapter 3 47, Fig. 1: Earth Satellite Corp. 48, Fig. 4: Richard Hazlett. 49, Fig. 3.1: (a) Patricia Gensel, Univ. of N. Carolina. 50, Fig. 3.2: (c) Scott Katz. 53, Fig. 3.5: from B. Murphy and D. Nance, *Earth Science Today*, 2nd ed., Brooks/Cole, 1999. 58, Fig. 3.11: (a) from R. Wicander and J. Monroe, *Essentials of Geology*, 2nd ed., Brooks/Cole, 1999; (b) Photo Researchers; (c) USGS. 59, Fig. 3.12: Richard Hazlett. 63, Fig. 3.18: (b) from B. Murphy and D. Nance, *Earth Science Today*, 2nd ed., Brooks/Cole, 1999. 68, Fig. 1: Bildarchiv Preussischer Kulturbesitz; Fig. 2: Rodney Supple. 69, Fig. 3: (a)–(c) Woods Hole Oceanographic.

Chapter 4 73, Fig. 4.1: (b) Wendy Van Norden, Harvard Westlake School; (d) D. D. Trent; (f) USGS. 74, Fig. 4.2: (e) USGS; Fig. 4.3: from R. Wicander and J. Monroe, *Essentials of Geology*, 2nd ed., Brooks/Cole, 1999. 77, Fig. 4.5: B. Pipkin; Fig. 4.6: Trustees of the Science Museum, London. 78, Fig. 4.7: SCEC data. 80, Fig. 4.10: (a)

SCEC data; (b) USGS. 83, Fig. 4.12: (b) Dorothy Stout, Cypress College; (d) Dennis Fox. 84, Fig. 4.13: James C. Anderson, USC. 85, Fig. 4.15: B. Pipkin; Fig. 4.16: (a) NOAA; (b) NGDC 97. 86, Fig. 4.18: Dave Johnson. 87, Fig. 4.19: Metropolitan Museum of Art, H. O. Havemeyer Collection (JP 1847). 88, Fig. 4.20: (b) USGS. 89, Fig. 4.21: (b) Rod Combellick, Div. of Geological and Geophysical Surveys, State of Alaska. 90, Fig. 4.22: USGS. 91, Fig. 4.24: USGS data. 92, Fig. 4.25: Greg Davis, USC. 93, Fig. 4.26: Dennis Fox; Fig. 4.27: data from *EOS*–American Geophysical Union; Fig. 4.28: from M. L. Blair and W. W. Spangle, USGS prof. paper 941-B, 1979. 94, Fig. 4.29: (b) USGS fact sheet 168-95. 95, Fig. 4.30: USGS data. 96, Fig. 4.31: USGS; Fig. 4.32: Oreg. Dept. of Geology and Mineral Industries. 97, Fig. 4.33: USGS, modified and updated. 99, Fig. 4.36: Kerry E. Sieh, Seismological Laboratory, CIT. 100, Fig. 4.37: USGS; Fig. 4.38: USGS data; Fig. 4.39: D. D. Trent. 101, Fig. 4.40: D. D. Trent. 102, Fig. 4.41: FEMA. 104, Fig. 1: SCEC; Fig. 2: redrawn from F. Harp and R. Jibson, USGS. 105, Fig. 2: (b) B. Pipkin. 106, Fig. 1: B. Pipkin. 107, Figs 1 and 3: B. Pipkin; Fig. 2: Frederich Larson, *San Francisco Chronicle*. 108, Fig. 4: Yi-Ben Tsai, National Central Univ., Taiwan; Fig. 5: Ted Reeves, Chaffey High School, Upland, Calif.

Chapter 5 111, Fig. 1: Joyce M. Warren and USGS. 112, Fig. 2: (b) NASA; Fig. 3: (a) USGS, (b) David Schneider, Mich. Technological. 114, Fig. 5.1: after R. I. Tilling, C. Heliker, and T. L. Wright, *Eruptions of Hawaiian Volcanoes: Past, Present, and Future* (USGS, 1987); Fig. 2: after Hays, USGS prof. paper 1240-B, 1981. 116, Fig. 5.4: after T. Simpkin and L. Siebert, *Volcanoes of the World*, 2nd ed. (Tucson, Ariz.: Geoscience Press, 1994); Fig. 5.5: data from T. Simpkin and others, *Volcanoes of the World* (Dowden, Hutchinson, and Ross). 118, Fig. 5.7: B. Pipkin; Fig. 5.8: (a) Eros Data Center. 119, Fig. 5.10: (a) USGS; (b), (c) J. D. Griggs, USGS. 120, Fig. 5.11: J. B. Stokes, USGS; Fig. 5.12: (a) *Pele, Goddess of Hawaii's Volcanoes*, by artist Herb Kawainui Kane; (b) B. Pipkin. 121, Fig. 5.14: (a) Richard Hazlett; (b) Arturo Aburto Quesada. 122, Figs 5.15 and 5.16: B. Pipkin. 123, Fig. 5.17: (a) B. Pipkin; (b) Tao Rho Alpha; Fig. 5.18: USGS. 124, Fig. 5.19: B. Pipkin. 125, Fig. 5.20: USGS image, fact sheet 002-97, "What Are Volcano Hazards?" 126, Fig. 5.21: (a) B. Pipkin; (b) NOAA, National Geophysical Data Center, Boulder, Colo. 127, Fig. 5.22: Jim Mori, USGS; Fig. 5.23: USGS Photo Library. 128, Fig. 5.24: (a) Keith Ronnholm; (b) B. Pipkin. 129, Fig. 5.25: Richard Hazlett. 130, Fig. 5.26: B. Pipkin; Fig. 5.27: USGS data. 131, Fig. 5.28: B. Pipkin. 132, Fig. 5.29: modified from USGS fact sheet 169-97, 1997. 133, Fig. 5.30: USGS; Fig. 5.31 USGS data. 134, Fig. 5.32: USGS data; Fig. 5.33: (a) and (c) USGS; (b) and (d) Richard Hazlett. 135, Fig. 5.34: USGS data. 138, Fig. 1: (a) and (c) from Karen Williams, *Volcanoes of the Southwind: A Field Guide to the Volcanoes and Landscape of Tangariro National Park* (Tongariro National History Society); (b) David M. Bailey, Grant Co. Airport, Moses Lake, Wash. 139, Fig. 2: NOAA. 140, Fig. 1: Simon R. Young, Montserrat Volcano Observatory; Fig. 2: Montserrat Volcano Observatory Web site map, Jan. 1998, British Geological Survey.

141, Fig. 1: USGS data. 142, Fig. 2: John Rogie, ©1999; Fig. 3: B. Pipkin. 143, Fig. 1: (a) Mike Tuttle, USGS; (b) T. Orban/Sygma. 144, Fig. 1: John Mead, Science Photo Library/Photo Researchers; Fig. 2: Jack Green, Calif. State Univ., Long Beach. 145, Figs 3 and 6: J. D. Griggs, Hawaiian Volcano Observatory, USGS; Fig. 4: Jim Mori, USGS; Fig. 5: B. Pipkin.

Chapter 6 148, Fig. 1: UPI/Bettmann; Fig. 2: Bettmann Archive. 150, Fig. 6.3: D. D. Trent. 151, Fig. 6.4: (a), (b) John S. Shelton. 152, Fig. 6.6: (a), (b) B. Pipkin. 153, Fig. 6.7: B. Pipkin; Fig. 6.8: D. D. Trent. 154, Fig. 6.10: D. D. Trent. 155, Fig. 6.11: B. Pipkin. 157, Fig. 6.14: USDA data. 158, Fig. 6.16: (a) Walt Anderson, Visuals Unlimited; (b) Sue Monroe; Fig. 6.17, B. Pipkin. 160, Fig. 6.19: (a) Stew Monroe, (b) USDA. 162, Fig. 6.21: Howard D. Wilshire, USGS; Fig. 6.22 and 6.23: American Geological Institute. 163, Fig. 6.24: Doug Burbank, Univ. of California, Santa Barbara; Fig. 6.25: Edward B. Nuhfer and AIPG. 164, Fig. 6.28 and 6.29: Troy L. Péwé, USGS. 165, Fig. 6.30: Larry Hinzman, Univ. of Alaska, Fig. 6.31: (a), (b) B. Pipkin. 166, Fig. 6.32: (a), (b) ARCO. 167, Fig. 6.34: Douglas J. Hathwell, Cornell Univ. 168, Fig. 1: Visuals Unlimited. 169, Fig. 1: from J. M. Spencer, *Texas Journal of Science* 21 (1970). 170, Fig. 2: (a), (b) Verle Bohrman, Univ. of Nevada–Reno. 171, Fig. 1: (a), (b) Al Robb III. 172, Fig. 1: Alain Nogues/Sygma; Fig. 2: Robert Thayer/Projections.

Chapter 7 175, Fig. 1: B. Pipkin. 176, Fig. 7.1: from J. P. Krohn and J. E. Slosson, "Landslide Potential in the U.S.," *California Geology* 29, no. 10 (1976). 178, Fig. 7.3: B. Pipkin. 179, Fig. 7.4: Kevin Lamb; Fig. 7.5: (b) Randall W. Jibson, USGS. 180, Fig. 7.6: *California Geology* data from CDMG; Fig. 7.7: National Weather Service and NOAA. 181, Fig. 7.8: (a)–(c) B. Pipkin. 182, Fig. 7.10: (b) B. Pipkin, (c) D. D. Trent. 183, Fig. 7.11: B. Pipkin. 185, Fig. 7.13: USGS. 186, Fig. 7.15: NGDC. 187, Fig. 7.17: USGS. 189, Fig. 7.19: (a) Pam Irvine, CDMG; (b) Siang Tan, CDMG. 190, Fig. 7.20: (b) B. Pipkin; Fig. 7.21: (b) B. Pipkin. 191, Fig. 7.23: Bob Hollingsworth. 192, Fig. 7.24: data from Knox Williams, Avalanche Formation and Release, *Rock Talk*, Colorado Geological Survey 3, no. 4, Oct. 2000. 194, Fig. 7.26: T. L. Holzer, USGS. 195, Fig. 7.27: USGS Photo Library. 196, Fig. 7.28: (b) James L. Ruhle & Associates. 197, Fig. 1, 2: B. Pipkin. 198, Fig. 1: Robert Schuster, USGS. 199, Fig. 2: redrawn from Lynn Highland, *Earthquakes and Volcanoes* 24, no. 3 (1993); Fig. 3: redrawn from Lynn Highland, USGS. 200, Fig. 1: Ronny Jacques/Photo Researchers; Fig. 2: adapted from P. Gatto and L. Carbognin, "The Lagoon of Venice," *Hydrological Sciences Bulletin* 26, no. 4 (1981). 201, Fig. 1: City of Long Beach (California), Dept. of Oil Properties; Fig. 2: R. V. Dietrich; Fig. 3: USGS.

Chapter 8 205, Fig. 1: L. Lefkowitz/Taxi/Getty Images. 207, Fig. 8.2: from *Essentials of Geology*, 2nd ed., by R. Wicander and J. Monroe, Brooks/Cole, 1999. 208, Fig. 8.5: USGS. 211, Fig. 8.9: © Bettmann/CORBIS. 216, Fig. 8.19: D. D. Trent. 217, Fig. 8.20: USGS circular 1139. 219, Fig. 8.23: USGS circular 1186, p. 47. 220, Fig. 8.25: USGS. 221, Fig. 8.27: (a) Frank Kujawa, Univ. of Central Florida, Geophoto Pub. Co., (b) B. Pipkin. 226, Fig. 1: redrawn from Philip Micklin, Western Michigan Univ. 227, Fig. 2: Philip Micklin, Western Michigan Univ.; Fig. 3: redrawn from Philip Micklin, Western Michigan Univ. 229, Fig. 1: AIPG, Ground Water: Issues and Answers, 1985. 230, Fig. 2: Texas Geological Survey. 231, Fig. 1:

after B. L. Foxworthy, USGS prof. paper 950, 1978. 232, Fig. 2: after B. L. Foxworthy, USGS prof. paper 950, 1978; Fig. 3: Nassau County Dept. of Public Works. 233, Fig. 1: Reed Wicander; Fig. 2: Scott Fee, American Speleogical Society; Fig. 3: Visuals Unlimited.

Chapter 9 237, B. Pipkin. 238, (a), (b) B. Pipkin. 240, Fig. 9.4: Peter Kresan, Univ. of Arizona. 241, Fig. 9.6: NASA/JPL. 242, Fig. 9.7: (b) Victor Baker, Univ. of Arizona. 243, Fig. 9.10: Stew Monroe. 248, Fig. 9.19: (a), (b) Landsat imagery courtesy Earth Observation Satellite Co., Lanham, Maryland. 250, Fig. 9.20: (a), (b) and 9.21: B. Pipkin. 252, Fig. 9.25: USGS. 254, Fig. 1: EROS. 255, Fig. 2: B. Pipkin. 256, Fig. 1: Lloyd Olson. 257, Fig. 1: B. Pipkin. 258, Fig. 1: (a), (b) Security Pacific National Bank Historical Collection, Los Angeles County Library; Fig. 2, 3: B. Pipkin.

Chapter 10 261, Fig. 1: B. Pipkin. 262, Fig. 2: Karl and Jill Wallin, FPG International. 263, Fig. 10.3: Tom Servais, *Surfer* magazine. 264, Fig. 10.4: (b) B. Pipkin. 265, Fig. 10.6: B. Pipkin; Fig. 10.7: John S. Shelton. 266, Fig. 10.9: B. Pipkin. 268, Fig. 10.11: USGS data; Fig. 10.12: from *EOS* 79, no. 2, June 2, 1998. 269, Fig. 10.13: from *EOS* 79, no. 2, June 2, 1998. 270, Fig. 10.15: (a), (b) B. Pipkin. 271, Fig. 10.17: USAF. 272, Fig. 10.18: (a), (b) B. Pipkin. 273, Fig. 10.19: after R. Dolan and others, "Coastal Erosion and Acretion," USGS National Atlas, 1985. 274, Fig. 10.20: USGS data. 275, Fig. 10.21: from J. S. Williams and others, *Coasts in Crisis*, USGS circular 1075; Fig. 10.22, 10.23: B. Pipkin. 276, Fig. 10.24: (a) Jim Falls, California Dept. of Mines and Geology, (b) Geophoto Pub. Co.; Fig. 10.25: B. Pipkin. 277, Fig. 10.26: U.S. Army Corps of Engineers; Fig. 10.27: GEOPIC, Earth Satellite Corp. 279, Fig. 10.28: (b) NOAA. 280, Fig. 10.29: (b) NASA. 281, Fig. 10.30: (a), (b) NOAA. 282, Fig. 10.32: from *Meteorology Today* by C. D. Ahrens, Brooks/Cole, 2000; Fig. 10.33: NOAA. 283, Fig. 2: B. Pipkin; Fig. 3: USN. 284, Fig. 1: NASA; Fig. 2: Mike Booher. 285, Fig. 1: NOAA; Fig. 2: B. Pipkin; Fig. 3: Ruth Flanagan, ©1993, *Earth* magazine, Kalmbach Pub. Co.

Chapter 11 289, Fig. 1: (a) Offentliche Kunstammlung, Kupferstichkabinet, Basel, Switzerland; (b) D. D. Trent. 292, Fig. 11.3: Wilfried Haeberli. 293, Fig. 11.4: NPS data. 294, Fig. 11.5: John S. Shelton; Fig. 11.6: Richard P. Jacobs/JLM Visuals. 295, Fig. 11.8, 11.9: D. D. Trent; Fig. 11.10: B. Pipkin. 296, Fig. 11.11: C. Hillaire-Marcel; Fig. 11.12: after Dott and Batten, *Evolution on the Earth*, 3rd ed., p. 514, McGraw-Hill, 1981. 297, Fig. 11.13: data from Foundation for Glacier and Environmental Research. 298, Fig. 11.14: Rolf A. Larsson. 299, Fig. 11.15: after Weiss and Newman, "The Channeled Scablands of Eastern Washington, USGS pamphlet, 1971. 300, Fig. 11.16, 11.17: D. D. Trent. 301, Fig. 11.18: John S. Shelton. 302, Fig. 11.20: from B. Murphy and D. Nance, *Earth Science Today*, p. 331, Brooks/Cole, 1999. 305, Fig. 11.24: after Nanen and Lebedeff, "Global Trends of Surface-Air Temperatures," *Journal of Geophysical Research* 29, no. D11 (Nov. 20, 1987), Fig. 15, p. 13370. 307, Fig. 11.25: data from W. S. Broecher, "Chaotic Climate," *Scientific American* 273, no. 5 (Nov. 1995). 309, Fig. 11.27: data from J. R. Petit and others, 1999, "Climate and atmospheric history of the past 420,000 years from the Vostock ice core, Antarctica." *Nature* 399, pp. 429–36. 310, Fig. 11.28: courtesy of Jean Palutikof, Climatic Research Unit, Univ. of East Anglia. 313, Fig. 11.29: after R. Byerly, Jr., "The Policy Dynamics of Global Change," *EarthQuest*, Spring 1989. 317, Fig. 1: Jurg Alean. 319, Fig. 2: D. D.

Trent. 320, Fig. 1: data from GRIP, "Climate Instability during the Last Interglacial Period Recorded in the GRIP Ice Core," *Nature* 364, no. 6434 (July 15, 1993), pp. 203–207. 321, Fig. 1: Dan Guravich, National Audubon Society/Photo Researchers. 323, Fig. 1, 3, 4, 5: D. D. Trent; Fig. 2: Barislav Bosnjak.

Chapter 12 327, Fig. 1: British War Museum. 331, Fig. 12.5: Soil Conservation Service. 332, Fig. 12.6: Howard G. Wilshire, USGS; Fig. 12.7: D. D. Trent. 333, Fig. 12.8: Georg Gerster/Photo Researchers; Fig. 12.9, 12.10: D. D. Trent. 334, Fig. 12.11: SEAWIFS Project, NASA/Goddard Space Flight Center and Orbimage. 335, adapted from C. D. Ahrens, *Meteorology Today*, 6th ed., Brooks/Cole, 1999. 336, Fig. 2: Steve McCurry/Magnum; Fig. 3: Mike Hume, Climatic Research Unit, Univ. of East Anglia. 337, Fig. 4: (a), (b) Eden Foundation. 339, Fig. 2: George Steinmetz, National Geographic; Fig. 3: Frank Langfitt, Sunsource/*The Baltimore Sun*. 340, Fig. 1, 2, 4: D. D. Trent; Fig. 3: Joseph Allen.

Chapter 13 343, Fig. 1: D. D. Trent. 345, Fig. 13.1: data from Mineral Information Institute. 349, Fig. 13.3: B. Pipkin; Fig. 13.4: Wendell E. Wilson. 350, Fig. 13.5: Clifford Taylor, USGS bulletin 2156; Fig. 13.6: after "Alaska's Oil/Gas and Minerals," *Alaska Geographic* 9, no. 4, 1982. 351, Fig. 13.7 Kennecott Corp.; Fig. 13.8: Ralph White/CORBIS. 352, Fig. 13.9: Stew Monroe; Fig. 3.10: Bruce Heezen, Scripps Inst. of Oceanography, Univ. of Calif., San Diego. 353, Fig. 13.11: after "Alaska's Oil/Gas and Minerals," *Alaska Geographic* 9, no. 4, 1982; Fig. 13.12: T. K. Buntzen, Silverado Mines U. S., Inc. 354, Fig. 13.14: after "Alaska's Oil/Gas and Minerals," *Alaska Geographic* 9, no. 4, 1982. 355, Fig. 13.15: D. D. Trent. 358, Fig. 13.17: Jiri Hermann, BHP Billiton Diamonds, Inc. 360, Fig. 13.18: Barrick Gold Corp. 361, Fig. 13.20: (a), (b) Vivian Stockman. 362, Fig. 13.21: D. D. Trent. 363, Fig. 13.23: W. Virginia Geological and Economic Survey; Fig. 13.24: D. D. Trent. 364, Fig. 13.25: D. D. Trent; Fig. 13.26: T. K. Buntzen. 365, Fig. 13.27: Marc Jay Plumley; Fig. 13.28: IMC Fertilizer, Inc. 367, Fig. 13.30, 13.31: D. D. Trent. 369, Fig. 13.32: John E. Gray, USGS bulletin 2156. 370, Fig. 13.33: (a), (b) E. Frank Schnitzer, Oregon Dept. of Geology and Mineral Industries; Fig. 13.34, 13.35: ARCO. 373, Fig. 2: California State Library. 374, Fig. 3: Donald B. Sayner collection; Fig. 4: D. D. Trent collection. 376,

Fig. 1: D. D. Trent. 377, Fig. 2: Marion Galant, Colorado Dept. of Public Health and Environment; Fig. 3: D. D. Trent. 378, Fig. 1: Montana Dept. of Environmental Quality, Remediation Division. 379, Fig. 3: D. D. Trent. 381, Fig. 1, 2: D. D. Trent; Fig. 3: John Burghardt, National Park Service, U.S. Dept. of Interior. 382, Fig. 4, 5: D. D. Trent.

Chapter 14 386, Fig. 1, USGS; Fig. 2: from Peter McCabe and others, 1993, USGS circular 115, p. 46. 387, Fig. 3: Keith A. Kenvolden, USGS; Fig. 4: Dillon, USGS. 388, Fig. 14.1: adapted from Larry Fellows, Arizona Geological Survey. 392, Fig. 14.6: Steve Dutch, Univ. of Wisconsin–Green Bay. 396, Fig. 14.11: Colorado Geological Survey. 397, Fig. 14.12: Energy Information Administration. 398, Fig. 14.13: (b) Visuals Unlimited. 399, Fig. 14.14, U.S. Dept. of Energy. 400, Fig. 14.15: Hobart King, Mansfield Univ.; Fig. 14.16: after A. C. Stern, ed., *Air Pollution, Vol. 1*, Academic Press, San Diego. 401, Fig. 14.17: Robert Finkelman, USGS. 402, Fig. 14.18: Richard Keeler, Southern California Edison Co. 403, Fig. 14.19: D. D. Trent; Fig. 14.20, from "Energy from the Sun," by C. J. Weinberg and R. Williams, *Scientific American*, 1990. 405, Fig. 14.22: B. Pipkin. 408, Fig. 14.26: Nova Scotia Power, Inc. 409, Fig. 1: Drake Well Museum, Titusville, Pa. 410, Fig. 1: from *Environmental Science*, 7th ed., by G. T. Miller, p. 329, Brooks/Cole, 1999. 411, Fig. 2: Bill Nation/Sygma. 412, Fig. 1: John Chiasson/Gamma Liaison; Fig. 2: Peter Bentham, BP Exploration. 413, Fig. 3: B. Pipkin; Fig. 4: D. D. Trent.

Chapter 15 417, Fig 1, 2: B. Pipkin. 419, Fig. 15.3: (b) B. Pipkin. 420, Fig. 15.4: (b) B. Pipkin. 422, Fig. 15.6: Dennis Capolongo, Greenpeace. 423, Fig. 15.7: (b) B. Pipkin. 424, Fig. 15.8, 15.9: B. Pipkin. 426, Fig. 15.13: Mitsuhiro Wada/Gamma Liaison. 428, Fig. 15.15: B. Pipkin; Fig. 15.16: Richard Proctor. 431, Fig. 15.19: (a) Dept. of Energy data; (b) Dept. of Energy; Fig. 15.20: Dept. of Energy data. 433, Fig. 15.21: Office of Civilian Radioactive Waste Management, 1998. 434, Fig. 15.23: National Academy of Sciences, NRC. 436, Fig. 1: (b) B. Pipkin. 439, Fig. 1: Michel J. Philippot/Sygma. 441, Fig. 1: Robert Rathe; Fig. 2: B. Pipkin; Fig. 3: Joe Halbig, Univ. of Hawaii; Fig. 4: B. Pipkin.

Index

From the shrinking Arctic ice cap to the human condition . . .

This book will help you explore GEOLOGY

in ways that will affect you for years to come.

This fascinating and completely up-to-date new edition will take you on an exploration of the relationship between humans and the geologic hazards, processes, and resources that surround us. With an emphasis on prevention and remediation, the Fourth Edition will help you discover:

- *how geology can improve the human condition*
- *the vast implications of overpopulation*
- *what must be done to create a sustainable society for the next generation*
- *why "earthquakes don't kill people, buildings do"*
- *the evidence around you of global warming*
- *new information on tsunamis*
- ***and much more!***

http://earthscience.brookscole.com/pipkingeo4e

For up-to-the-minute online explorations and exceptional study tools

This site gives you a variety of additional learning resources designed to help you succeed in class and beyond. You'll find online study aids, Web-based critical thinking questions, maps, quizzes, daily feeds of the latest environmental news from NewsEdge, and the unique personalized tutorial system *Environmental GeologyNow.*

ALSO

Access to the ENVIRONMENTAL GeologyNow™

Web site is FREE with the purchase of this text! This outstanding resource will help you determine what you need to learn, how to learn it, and more—a true key to success.

InfoTrac® College Edition—also FREE with the text— is a world-class online library featuring thousands of top journals and publications. You will receive four months of unlimited access to this extensive Web resource when you purchase this book.

http://earthscience

THOMSON

BROOKS/COLE

Visit Brooks/Cole online at **www.brooks...**

For your learning solutions: **www.thomson...**